Perovskite Light Emitting Diodes

Perovskite Light Emitting Diodes

Materials and Devices

Hong Meng

WILEY VCH

Author

Prof. Hong Meng
Peking University
Building G 306
Lishui Road, Nanshan Disctrict
Shenzhen
China
518055

Cover Image: Courtesy of Prof. Hong Meng

Library of Congress Card No.: applied for

British Library Cataloguing-in-Publication Data
A catalogue record for this book is available from the British Library.

Bibliographic information published by the Deutsche Nationalbibliothek
The Deutsche Nationalbibliothek lists this publication in the Deutsche Nationalbibliografie; detailed bibliographic data are available on the Internet at <http://dnb.d-nb.de>.

Print ISBN: 978-3-527-35320-0
ePDF ISBN: 978-3-527-84493-7
ePub ISBN: 978-3-527-84494-4
oBook ISBN: 978-3-527-84495-1

Typesetting Straive, Chennai, India
Druck und Bindung CPI Group (UK) Ltd, Croydon, CR0 4YY

C9783527353200_090124

Contents

Preface

Perovskite materials, with their exceptional photophysical properties such as high photoluminescence quantum efficiency, broad absorption spectra, and adjustable narrow-band emission colors, have become the center of significant interest for researchers in the realm of optoelectronics. Their potential applications in diverse optoelectronic devices, ranging from solar cells and photodetectors to light-emitting diodes, have instigated profound research activity. Among these applications, perovskite light-emitting diodes (PeLEDs) are viewed as promising candidates for advancing display and lighting technologies.

Ever since the groundbreaking work on PeLEDs was reported by Tan et al. in 2014, there have been significant improvements in the external quantum efficiencies (EQEs) of the state-of-the-art infrared, red, green, and blue-emitting PeLEDs. They have seen EQE increases to 25.55%, 25.58%, 30.84%, and 18.65%, respectively. Depending on the emission wavelength, PeLEDs have vast potential for various applications, spanning from ultraviolet (UV) to visible and near-infrared (NIR) ranges. Additionally, they possess an unusually high color purity owing to their narrow emission spectra. The full-width at half maximum (FWHM) measures approximately 12, 20, and 40 nm for typical metal halide perovskite materials like $CsPbCl_3$, $CsPbBr_3$, and $CsPbI_3$ nanocrystals (NCs), respectively, indicating a promising future for these materials in display applications.

Despite these advancements and strong performance in areas like EQE, current/power efficiency, brightness, emission color, and spectral width, the widespread commercial deployment of PeLEDs still faces several obstacles. The most significant challenges include a drop in EQE at high current densities, limited lifetimes, and lower EQEs in blue PeLEDs compared to their green and red counterparts. The industry must address these hurdles before PeLEDs can truly revolutionize the future of light-emitting applications.

This book, *Perovskite Light-Emitting Materials and Devices*, offers a thorough overview of contemporary research concerning perovskite materials for light-emitting applications. Spanning 13 chapters, the content progresses from a basic understanding of the crystal structure and physical properties of perovskite materials to the latest developments in device performance enhancement and optimization strategies.

Chapter 1 acquaints readers with the crystal structure, exciton effects, size effect, and luminescence properties of perovskite materials, further discussing the elements that impact the efficiency of PeLEDs. Chapter 2 presents a detailed overview of perovskite material synthesis and preparation methods. Chapters 3–6 illuminate specific colors of light produced by perovskite LEDs, while Chapter 7 delves into the impact of metal ion doping on perovskite luminescent devices. Chapter 8 shifts focus to non-lead metal halide perovskite materials. The intricacies of white-light perovskite materials are discussed in Chapter 9. Chapter 10 emphasizes the crucial role of electron- and hole-transporting materials in the performance of perovskite LEDs. The current stability of perovskite LEDs and strategies to manage efficiency roll-off are outlined in Chapter 11. Finally, Chapter 12 explores the application of perovskite materials in laser technologies.

In essence, this book serves as an extensive resource for researchers and professionals in the field, detailing the fundamental concepts, fabrication, and optimization strategies of perovskite materials for light-emitting devices, with particular emphasis on LEDs and lasers.

It is my privilege to present this book, a culmination of years of in-depth research and dedicated writing. My aspiration is for this book to serve as a precious resource for students, academics, and anyone intrigued by this field. I am grateful to the students who contributed to the Chinese version of this book and the editors whose meticulous feedback has been instrumental in ensuring its high quality. I extend my heartfelt appreciation to my colleagues and family, whose unwavering support and insight have been a constant source of motivation. Lastly, to the readers choosing to explore this book, I hope the insights contained within these pages will guide you in your quest to become a more proficient learner and problem solver.

Please note that this book has been updated since the Chinese version was published in 2022, with the aim of providing even more comprehensive and current information on the subject matter.

November 2023

Hong Meng
Peking University

1

Structure and Physical Properties of Metal Halide Perovskites

1.1 Crystal Structure of Perovskite Materials

Since their discovery by Gustav Rose in 1839, perovskite materials have enjoyed a lengthy and fascinating history. Over the decades, a wide variety of perovskite materials have found uses in numerous applications, spanning from piezoelectric and optoelectronic devices to ferroelectric materials, superconductors, magnetoresistive sensors, catalysts, and ionic conductors, among others [1–7]. Halide perovskites, a specific category, have been the subject of extensive research for more than half a century [8]. As depicted in Figure 1.1, the general formula for perovskites is typically ABX_3, where A represents a monovalent cation (such as $CH_3NH_3^+$ (MA^+), $NH_2CHNH_2^+$ (FA^+), or Cs^+), B denotes a metal cation (generally Pb^{2+} or Sn^{2+}), and X represents a halogen anion (generally Cl^-, Br^-, or I^-).

Metal halide perovskite garnered significant attention from the scientific community following its application as a photosensitive layer in a dye-sensitized solar cell (DSSC) in 2009. Since then, comprehensive research has delved into its remarkable optical and electronic properties, marking it as an extremely promising material for diverse applications [10, 11].

Perovskite materials possess excellent physical properties, such as low defect density, high carrier mobility, long charge carrier lifetime, and long electron–hole diffusion length [12–15]. Therefore, perovskite-based solar cells have shown high defect tolerance, enabling them to achieve good photovoltaic performance even at relatively low purity levels of up to 90% [11, 16]. In addition, metal halide perovskites are not only used as a light-absorbing layer in solar cells but also widely utilized in the light-emitting layer of light-emitting diodes (LEDs) [11, 17–19]. Compared to conventional light-emitting materials, metal halide perovskites have unique optical properties such as tunable bandgap, high color purity, wide color gamut, and high fluorescence quantum efficiency [20–22]. Additionally, perovskite materials can be prepared as bulk single crystals, microcrystals, or nanocrystals (NCs) using a simple and cost-effective solution-based method [23].

Perovskite Light Emitting Diodes: Materials and Devices, First Edition. Hong Meng.

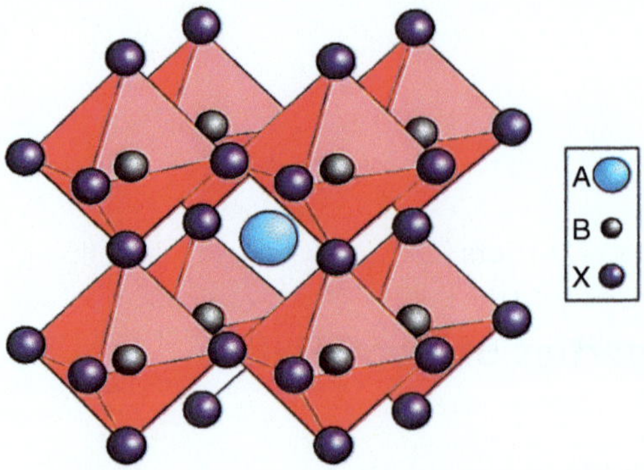

Figure 1.1 Crystal structure of perovskite. Source: Quan et al. [9]/ John Wiley & Sons.

1.2 Exciton Effects in Perovskite Materials

1.2.1 Definition of an Exciton

When a semiconductor absorbs a photon with energy greater than or equal to its bandgap energy, the electron is excited from the valence band to the conduction band, leaving behind a hole in the valence band. The electron and hole are attracted to each other due to the Coulomb force, forming an electron–hole pair, which is also known as an exciton. The distance between the electron and the hole is called the exciton Bohr radius, and its value depends on the material, typically ranging from 2 to 50 nm [24]. Excitons have different radii based on their size, which are divided into two types: free excitons (Wannier excitons) and bound excitons (Frenkel excitons) [25], as shown in Figure 1.2.

The prevalence of free excitons in most perovskite materials can be explained by the conventional hydrogen atom model, which is mainly dependent on the effective masses of electrons and holes in the semiconductor. The energy levels of excitons

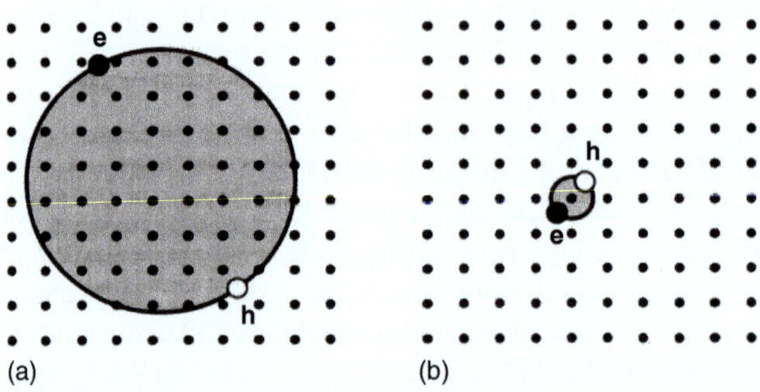

Figure 1.2 (a) Free exciton; (b) bound exciton. Source: Adapted from Dresselhaus [25].

can be expressed by Eq. (1.1).

$$E_n = E_g - \frac{R^*}{n^2} \tag{1.1}$$

where E_g is the forbidden band width and R^* is the effective Rydberg constant of the exciton.

In semiconductors, the effective mass of carriers is typically smaller than that of excitons, resulting in the screening of carriers by the lattice. This leads to the effective Rydberg constant being expressed as the normalized Rydberg constant (Eq. 1.2).

$$R^* = R_0\mu/m_0\varepsilon_r^2 \tag{1.2}$$

When the mass of the exciton is reduced to $1/\mu = 1/m_h + 1/m_e$ (m_h, m_e are the effective masses of holes and electrons, respectively, and ε_r is the relative permittivity of the crystal), it mainly originates from carriers that are screened by the lattice. As a result, the exciton binding energy R^* depends on the effective mass of the carriers and the dielectric constant of the crystal. Numerous studies have shown that the dielectric constants of perovskites can be very different in the static case and when photoinduced, in the case of halogenated perovskites, the dielectric constant is about $\varepsilon_s \approx 30$ at rest, whereas under high-frequency light irradiation, its dielectric constant $\varepsilon_\infty \approx 30$. Consequently, the calculation of the exciton binding energy of perovskites varies considerably between reports.

1.2.2 Self-Trapping Excitons in Perovskite Materials

Self-trapping excitons in perovskite materials are essential for the study of luminescent materials, particularly for white perovskite materials. In soft lattices, the lattice couples to carriers, causing electroelastic distortions and reducing the lattice's symmetry. These distortions create chiral carriers, also known as polaritons, which can be categorized into large and small polaritons based on their potential well depth and size [26]. Large polaritons exhibit a strong long-range Coulomb interaction and can be delocalized over several cells. Small polaritons, on the other hand, emerge from the short-range distortion of the lattice caused by localized carriers. Self-trapped excitons resemble small polaritons as their Bohr radius is small and closely linked to the lattice distortion (Figure 1.3).

The intrinsic self-trapped exciton is different from a permanent defect in the material [28], which can be imagined as a hard ball (electron/hole/exciton) falling onto a soft rubber sheet (distorted lattice); the ball falls into a potential well of its own formation, and the sheet distorts mainly from the appearance of the ball. After the ball leaves, the sheet returns to its original shape (Figure 1.4a). When there is already an indentation (permanent defect) in the sheet, it is also possible for the ball to fall into this indentation (Figure 1.4b). However, when the ball interacts with an indentation in its vicinity, a new defect is formed, and the ball will fall into a new depth, which is called an exotic excitation (Figure 1.4c).

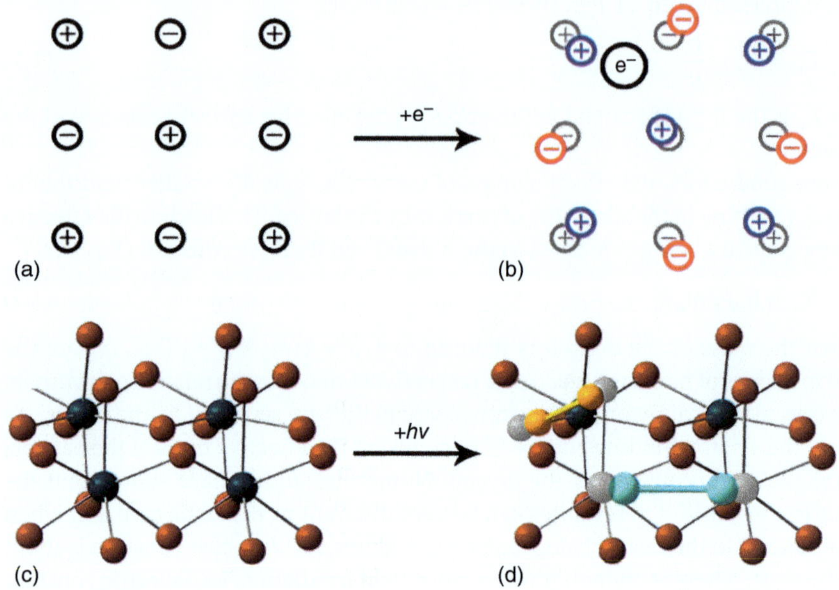

Figure 1.3 (a) and (b) Schematic diagram of the formation of a large polariton resulting from the addition of an electron to the ionic lattice, causing a long-range distortion of the lattice; (c) $PbBr_2$ crystal structure unit, with green and brown representing Pb and Br atoms, respectively; (d) self-trapped holes and self-trapped electrons formed by the orange Br_2^- and blue–green Pb_2^{3+} dimer, followed by the formation of photoexcitation. Source: Smith and Karunadasa [27]/American Chemical Society.

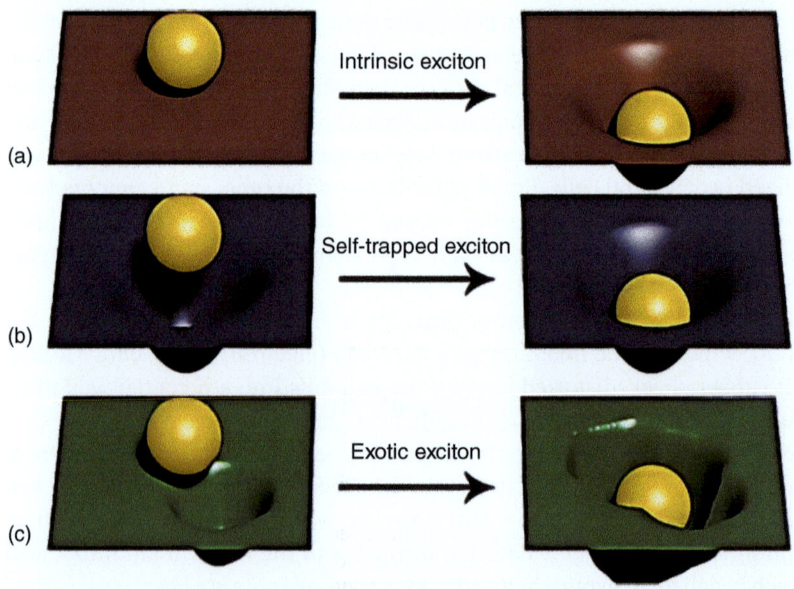

Figure 1.4 (a) Intrinsic exciton; (b) self-trapped exciton caught in a permanent trap; (c) exotic exciton (self-trapped exciton and permanent defect) interactions (in the figure, the ball represents the exciton and the rubber sheet represents the lattice). Source: Smith and Karunadasa [27]/American Chemical Society.

1.3 Size Effect of Perovskite Materials

For semiconductor nanocrystals that are extremely small or of zero size, the behavior of the exciton wave function – essentially, the behavior of electrons and holes – will be influenced by the limited space of the domain. As the size decreases, alterations occur in the density of the electronic states. Most notably, the forbidden band width increases, and the highest point of the valence band and the lowest point of the conduction band display discrete splitting energy levels. This is known as the quantum confinement effect [29]. As shown in Figure 1.5, as the size decreases, the bandgap of the semiconductor increases, resulting in a blueshift of the luminescence peak. Moreover, more discrete energy levels are generated, and the width of these discrete levels also increases.

The quantum confinement effect allows for tuning of the absorption and emission spectra by altering the size of particles while keeping the material components constant. Additionally, the scale of the quantum confinement domain varies in different directions due to variations in size and morphology, as shown in Figure 1.6 [31–33]. The size and shape of semiconductor nanocrystals greatly influence their optical and electrical properties, as changes in dimensionality can affect the bandgap and electrical conductivity of the material system. The confinement of

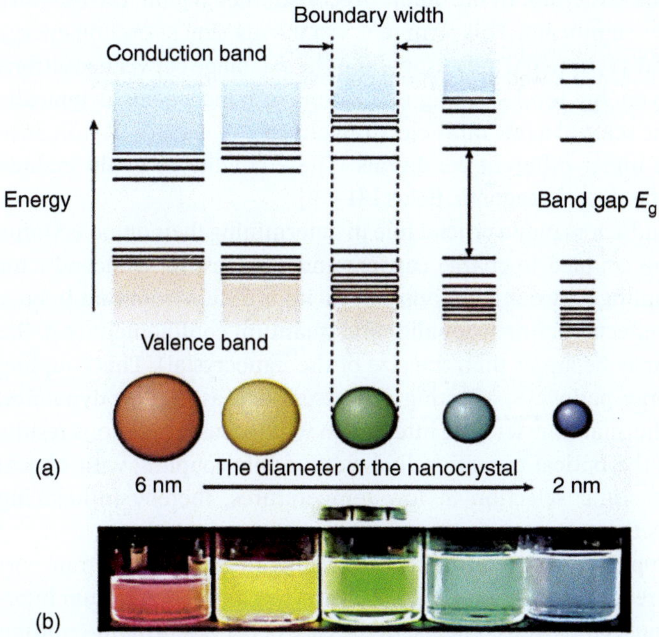

Figure 1.5 (a) Model of the quantum confinement effect: from this, it can be seen that as the bandgap of the semiconductor decreases in size, more discrete energy levels appear at the band edges, and the discrete width of the band edge energy levels increases further; (b) In the luminescence of five different sizes of CdSe quantum dot colloids under UV lamp irradiation, it can be seen that the luminescence changes from red to blue as the size changes from 6 nm to 2 nm. Source: Efros et al. [30], ©1996/Reproduced from American Physical Society.

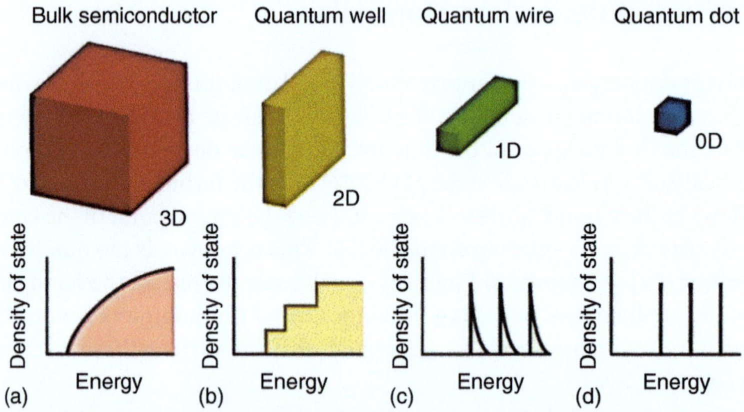

Figure 1.6 (a) Schematic diagram of the energy level structure of a semiconductor bulk; (b)–(d) semiconductor nanostructures of decreasing dimensionality (2D–1D–0D) and their electronic density of states. Source: Adapted from Efros et al. [30].

excitons in different directions results in the formation of zero-dimensional quantum dots, one-dimensional quantum lines, and two-dimensional quantum wells. As the quantum effect decreases, the growth of the material tends toward forming a three-dimensional bulk structure. In the quantum-constrained region, the exciton's fine structure becomes significant. This term refers to the division of exciton energy states due to the crystal's symmetry, anisotropy, and the exchange between electrons and holes. While the energy level splitting of excitons in nanocrystals is typically minor – usually on the scale of a few milli-electronvolts (meV), it can still influence the exciton's lifetime under different conditions. These conditions could include varying temperature fields and magnetic fields [31–39].

Phonons in semiconductors play a crucial role in determining their optoelectronic properties, as they are coupled to charge carriers and excitons. In semiconductor nanocrystals, the coupling between phonons and excitons differs from that in bulk materials due to the effects of dimensionality and quantum confinement (e.g. the phonon wavelength may be larger than the size of the nanocrystal). This coupling can lead to new energy pathways, affecting processes such as exciton dynamics, carrier cooling, and thermal transfer. The interaction with acoustic phonons results in the broadening of the optical transition linewidth, while coupling with optical phonons can induce chiral selection at low temperatures, thereby influencing phonon-assisted transitions.

In terms of device applications, metal halide perovskites meet the charge-transport and charge-injection requirements, with high diffusion lengths and optimum interface formation with most employed contacts. In the case of perovskite light-emitting diodes (PeLEDs), the emitter layer can consist of 3D, layered, or nanostructured perovskites, which are sandwiched between electron- and hole-transport layers and injection contacts [40] (Figure 1.7).

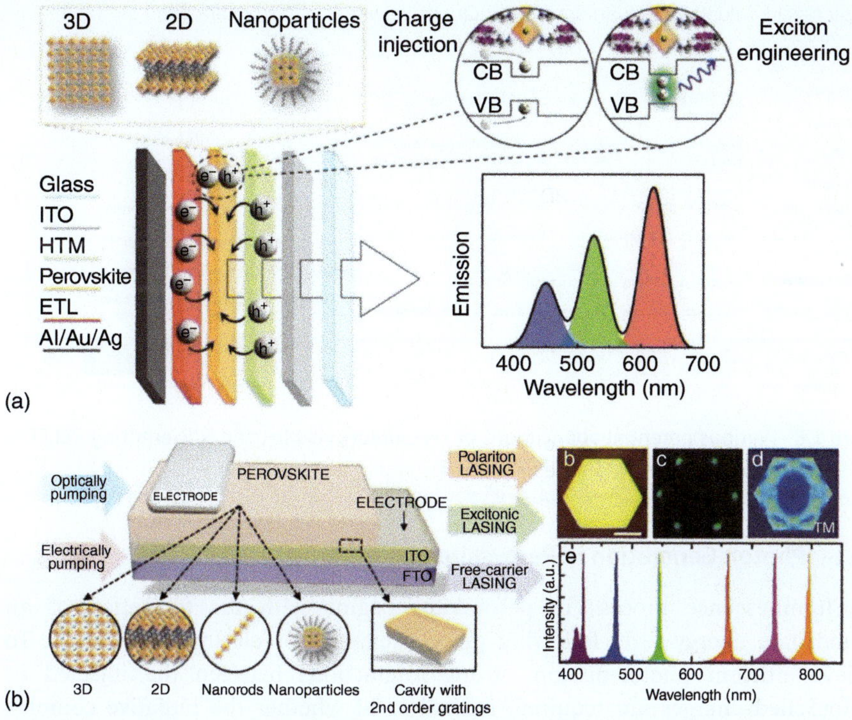

Figure 1.7 (a) Schematic representation of a PeLED device concept; (b) a perovskite laser device concept. Source: Veldhuis et al. [40]/Reproduced from John Wiley & Sons, Inc.

1.4 Luminescence Properties of Perovskite Materials

PeLEDs are experiencing rapid development and hold immense potential for various applications. As depicted in Figure 1.8, they are particularly notable in the visible region for applications such as display devices, high-brightness applications (e.g. projection displays and solid-state lighting), and augmented/virtual reality (AR/VR). With the shift in display technology from liquid crystal displays (LCDs) to active matrix organic light-emitting diodes (AMOLEDs), PeLEDs offer significant advantages due to their mild processing conditions and the possibility of direct integration onto thin-film transistor (TFT) backplanes. Furthermore, PeLEDs contribute to enhancing resolution, brightness, and color gamut in display technology [11].

In addition, PeLEDs are also promising for near-infrared applications [41], including night vision, telecommunications, and optical switching. In these applications, the emission wavelength within the NIR range (700–2500 nm) plays a crucial role, enabling specific uses such as *in vivo* bio-imaging, biomedical sensing, and optical wireless communications.

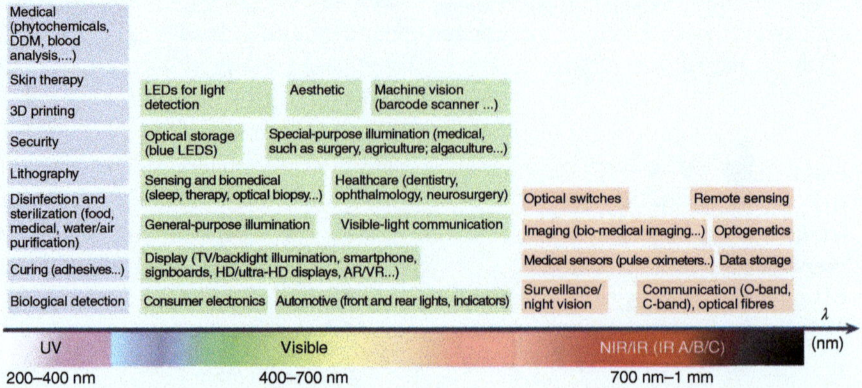

Figure 1.8 Various potential applications of UV-, violet-, visible-, and NIR-emitting PeLEDs. Source: Fakharuddin et al. [11]/Springer Nature.

1.4.1 Photon Generation in Perovskite Materials

The luminescence process in perovskites begins with the formation of an excited-state energy level following photo-excitation or electrical excitation. To achieve efficient photoemission, strong interactions between the injected or photoexcited charges are required, regardless of whether the radiative complex arises from free carriers or excitons [42]. In bulk materials, the large dielectric constants result in relatively low exciton binding energies, and radiative complexes are usually formed by free charge–carrier complexes [43, 44]. The luminescence process in bulk materials is characterized by the radiative recombination of free charge–carrier complexes, which is proportional to the quadratic carrier density. As a result of the large dielectric constants in bulk materials, the exciton binding energies are relatively low, making charge trapping during bimolecular radiative recombination more difficult. The internal bimolecular complexation constants depend on the cross-section and group velocity of the electron–hole exchange interaction. Radiation-free Shockley–Read–Hall (SRH) recombination arises from the trap density of the material, which is not an intrinsic property of the material. The efficiency of this recombination pathway is strongly dependent on the carrier density [42].

The abovementioned recombination pathways can be studied by analyzing the transient spectra of the material. This provides an insight into the rate constants of recombination linked with each pathway [42]. Transient spectra are typically measured at low excitation optical densities, where single-molecule processes are dominant in charge settling and decay, whereas bimolecular processes become more relevant at high excitation optical densities. These spectra can provide valuable information on the recombination constants associated with these different recombination pathways [45].

The interaction among various recombination mechanisms predicts that the highest internal quantum efficiency occurs at carrier densities of 10^{15}–10^{17} cm^{-3}, with a lower non-radiative recombination rate at higher carrier densities. To achieve high luminescence efficiency at carrier densities below $10^{13}\,cm^{-3}$, an extremely low non-radiative recombination rate with a value equivalent to the carrier lifetime at low densities, in the order of microseconds, is required. Alternatively, doping can enhance the emissivity of the perovskite luminescence layer [46]. Obtaining high luminous efficiency at low carrier densities is a challenge for perovskite diodes due to the thinness of the layer and the potential for carriers to cluster and form complexes at localized trapping centers [47]. However, the luminescence of perovskites with strong spatially confined excitons is not significantly affected by carrier density [48, 49].

The charge transport capability of bulk perovskite materials may be affected by perturbations in their crystal structure. To enhance the interaction between charges, the dielectric shielding within the lattice of bulk perovskites can be controlled by utilizing groups with different polarities. This approach enables the enhancement of luminescence efficiency without disrupting charge transport in the bulk crystal structure. However, the interactions between carrier trapping, exciton processes, and their mechanisms are not yet fully understood. The ionic behavior induced by light may alter the bounding potential of the local field and consequently influence the strength of carrier interactions. Further investigations are required to gain a more in-depth understanding of the radiative combination of carriers.

The strong spin–orbit coupling in perovskite materials may lead to Rashba-type energy band splitting, which could impede charge carrier complexation and hinder light transmission in circular photovoltaic effect measurements. Nevertheless, the complexation rates of bimolecules in perovskites remain consistent with those of absorption coefficients and comparable to those of second-order phases with complexation constants in groups III–V, without Rashba effects (a spin–orbit coupling phenomenon observed in solid-state systems, particularly in semiconductor and metal surfaces or interfaces, arising from the interaction between the spin of electrons and their motion in a crystal lattice under the influence of an electric field, leads to the splitting of electronic energy bands and the lifting of spin degeneracy.) Both are approximately on the order of 10^{-10} cm^3 s^{-1} [50]. To address these seemingly contradictory findings, it is crucial to take into account lattice dynamics or carrier localization effects. Additionally, more research is necessary to investigate the dependence of the bimolecular complex constant on carrier temperature and directly analyze the Rashba effect. Furthermore, it is important to examine the dynamical processes that occur after photoexcitation of halide perovskites. Over the past few years, several studies have explored the Rashba effect and its influence on carrier lifetimes in perovskites [51, 52].

As shown in Figure 1.9, the singlet energy level is lower than the triplet energy level due to the combined effect of spin and short-range electron–hole exchange. However, theoretical calculations indicate that significant Rashba coefficients can alter the fine structure of $CsPbX_3$ perovskite nanocrystals, leading to the generation of bright triplet excitons with energies lower than those of the dark monochromatic excitons. The ratio of bright triplet excitons to total excitons can reach as high as 3 : 4, indicating an η_r (radiative efficiency) of the $CsPbX_3$ nanocrystals as high as 0.75.

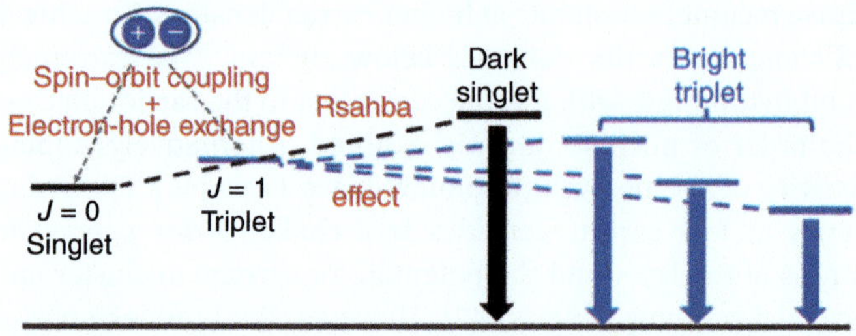

Figure 1.9 The fine structure of excitons in $CsPbX_3$ perovskite nanocrystals considering short-range electron–hole exchange and large Rashba effects [18]. Source: Copyright 2020, Chinese Chemical Society.

These nanocrystals exhibit rapid light emission with decay times on the order of picoseconds (ps) [18].

In recent years, the Rashba effect and its impact on carrier lifetimes in perovskites have been extensively studied. Current research is focused on understanding the underlying causes of the Rashba effect in perovskites, experimental verification of this effect, and its influence on carrier complexes under various environmental conditions [53, 54]. The researcher has also directed their attention toward investigating the Rashba effect in perovskites. However, there have been various viewpoints among researchers regarding the existence of the Rashba effect in perovskites [55]. Some researchers hold the view that the Rashba effect is not a significant issue in perovskites, which is based on arguments that suggest the Rashba effect either does not exist in perovskites or that the luminescence is not thermally activated. However, these seemingly contradictory findings can be reconciled by taking into account the lattice dynamics or carrier localization effects. To arrive at a conclusive understanding, it is essential to measure the dependence of the bimolecular complex constant on carrier temperature and conduct a more direct analysis of the Rashba effect. Additionally, it is necessary to investigate the dynamical processes that occur after photoexcitation of halide perovskites.

1.4.2 Photophysical Processes and Efficiency Calculations of Perovskite Luminescence

Perovskite luminescence involves a series of steps. Initially, under optical and electrical excitation, electrons and holes are generated in the conduction and valence bands. This is followed by a competition between radiative and non-radiative processes, which is influenced by the dimensionality and structure of the perovskite material (Figure 1.10). For instance, in the three-dimensional methylammonium lead iodide ($CH_3NH_3PbI_3$, $MAPbI_3$) perovskite, there is a dephasing time of approximately 220 fs after the initial excitation, during which the ground and excited states are in a superposition state. This dephasing time is three times longer than that observed in GaAs [56]. The dephasing time in $MAPbI_3$ perovskites is not influenced by the number of carriers, indicating that carrier scattering within the measurable

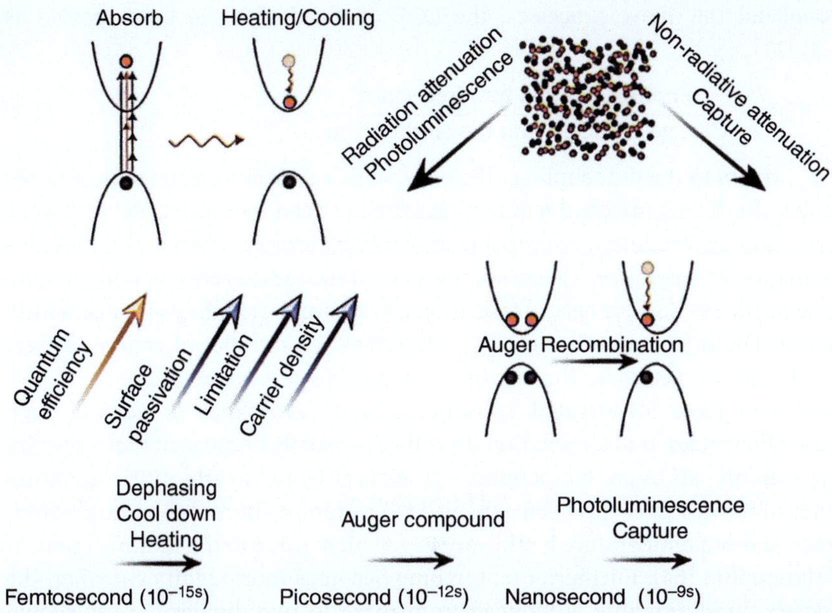

Figure 1.10 Physical processes in perovskite following photoexcitation, with intermittent combinations and non-radiative processes competing with the luminescence process. Source: Quan et al. [9]/John Wiley & Sons.

range is insignificant. However, for quasi-2D I-based perovskites, bulk scattering plays a more substantial role, and the dephasing time decreases with decreasing quantum confinement space [57].

Following dephasing, the excited state undergoes thermal exchange with other particles, such as carriers. In $MAPbI_3$, this process is primarily driven by carrier scattering, leading to thermal exchange times in the order of 10–85 fs [58]. After a duration of over 100 fs, carriers undergo a cooling process. However, under high-density light excitation, this time may be significantly prolonged due to the bottleneck effect of phonons [59–61]. For instance, cooling times of approximately 300 ps have been observed in amidine-based lead iodide ($NH_2CHNH_2PbI_3$, $FAPbI_3$) [55]. In $MAPbI_{3-x}Cl_x$, a decay time of approximately 45 ps was detected, and it stemmed from the thermal exchange process between the organic and inorganic sublattices [62].

At the conclusion of the ultrafast process described earlier, the competition between radiative and non-radiative decay pathways commences. In perovskite materials, non-radiative pathways result from charge trapping. In three-dimensional perovskites, the prevalence of charge trap recombination at low power can significantly limit quantum efficiency. At higher power, radiative recombination becomes more dominant, and close to 100% quantum efficiency can be achieved. This approach can be applied to LEDs. However, their external quantum efficiency (EQE) remains low due to the lack of charge-limited domains.

Through all the above processes, the EQE of the device can be expressed as Eq. (1.3) [11].

$$\mathrm{EQE} = \frac{\text{Photons emitted into free space}}{\text{Charges injected into the emitter film}} = \eta_{\mathrm{inj}} \times \eta_{\mathrm{rad}} \times \eta_{\mathrm{out}} \tag{1.3}$$

Here, η_{out} refers to the outcoupling efficiency, which depends on the optical losses resulting from the mismatched refractive indices between the various device layers, substrate, and air, leading to waveguide and substrate mode effects. The radiative recombination efficiency (η_{rad}) denotes the ratio of the radiatively recombined excitons, whose states allow for optical transitions, to the total excitons generated within the device. The injection efficiency (η_{inj}) describes the number of charge carriers passing the device electrodes that are injected into the perovskite EML.

Researchers have investigated two-dimensional perovskites to achieve high quantum efficiencies at low excitation densities, primarily because of their superior exciton stability at room temperature. However, while nearly 100% quantum efficiency of fluorescence has been achieved at low temperatures [63], the quantum efficiency at room temperature is still very low, which is due to the thermal quenching of the exciton [64]. Furthermore, trapping becomes more significant when the perovskite's dimensionality is reduced from three to two dimensions, as shown in research [65]. Mixed-phase quasi-two-dimensional perovskites demonstrate higher internal and external quantum efficiencies in the near-infrared and visible emission regions [66–68]. The improvement in luminescence efficiency is mainly due to the existence of an energy potential well, which localizes the charge to the lowest energy region, enabling the radiative recombination process to compete with charge trapping. However, for perovskite nanocrystals, surface traps can significantly reduce the quantum efficiency. Therefore, surface passivation has been identified as an effective approach to enhance the luminescence performance [69].

At high excitation power, the elevated carrier concentration and limited rate of radiative recombination can result in exciton annihilation, where excess carriers collide with each other during the diffusion process, leading to a non-radiative recombination process known as intermittent recombination. As the carrier concentration increases, radiative recombination competes with interstitial recombination until a threshold is reached, at which point the excess energy is rapidly lost due to cooling, resulting in a reduction in fluorescence quantum efficiency. The oscillatory complex increases as the exciton binding energy increases, and this effect is particularly pronounced when the dimensionality of the perovskite is reduced from three to two dimensions [70]. Similarly, an increase in the quantum-limited domain leads to a larger blueshift in the luminescence spectrum due to double exciton absorption [71].

As shown in Figure 1.11, metal halide perovskite films typically exhibit a distinct absorption edge in their absorption spectra [73–76], attributed to the phonon-coupling effects within the crystal. The Urbach energy (a parameter used to describe the absorption tail in the optical absorption spectrum of a material. It provides information about the disorder and imperfections in the material's crystal lattice or band structure) of perovskites is typically very low, at around 13 meV, which is comparable to other high-quality semiconductor materials such as GaAs

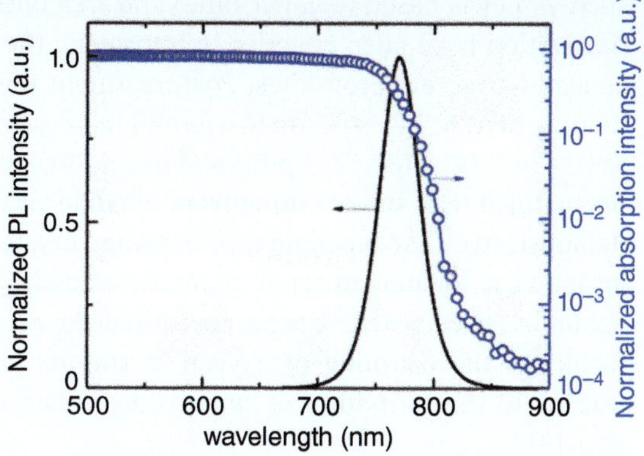

Figure 1.11 Photoluminescence spectrum (left axis) and absorption spectrum (right axis) of the $Cs_{0.06}MA_{0.15}FA_{0.79}Pb(I_{0.85}Br_{0.15})_3$ photothermal reaction deflection spectroscopy test. Source: Stranks et al. [72]/ John Wiley & Sons/CC BY 4.0.

(7.5 meV) and c-Si (9.6 meV) [77, 78]. These properties suggest that metal halide perovskites can maintain low-energy disordered states even at relatively high defect densities and are thus considered to be relatively clean semiconductor materials.

1.4.3 Non-radiative Combination Mechanisms at Surfaces and Interfaces

The EQE of typical perovskite bulk films, such as $MAPbI_3$ or mixed-cation perovskites $(Cs,FA,MA)Pb(I_{0.85}Br_{0.15})_3$, is typically in the range of 1–10% at 1 solar intensity, resulting in a quantum efficiency of less than 50% [45, 77, 78]. This value may increase with increasing carrier density, but it reaches a maximum plateau of less than 100%. This observation is consistent with the non-negligible sub-gap density of states, which in polycrystalline materials has a value of approximately 10^{13}–10^{15} cm^{-3} [63, 79], which dominates non-radiative and single-molecule complexes and determines the SRH complex constants. The sub-gap energy states become saturated when the carrier density is lower than the trap density, leading to dominance of radiative and bimolecular recombination processes. However, due to spatial inhomogeneity of trap density, the luminescence can exhibit significant variation between particles when observed on a microscopic scale [80, 81].

Although perovskites exhibit a large number of defects, resulting in many shallow defect states and explaining their defect tolerance, the presence of a large number of deep energy level traps also suggests potential non-radiative losses in such structures [82–85]. One possible origin of such trap states is halide vacancies, which are often found at grain boundaries and surfaces. However, the exact mechanism behind the formation of these trap states is not yet fully understood [77]. These hypotheses have been supported by subsequent studies, which have shown that surface passivation and defect modulation are effective in enhancing luminescence efficiency. For instance, adding KI to the precursors can increase the internal quantum efficiency of thin films up to 95% [77]. KI can serve as a source of iodine to fill vacancies in the halide and also immobilize excess halide with I.

Recent studies have shown that doping can be an effective way to achieve very high efficiencies in perovskite photovoltaic devices [86]. Passivation methods such as Lewis-based (e.g. pyridine) and trioctylphosphine oxide (TOPO) chemical passivation have been effective in improving the quantum efficiency and luminescence lifetime of perovskites. Post-treatment methods can also extend the carrier lifetime of $MAPbI_3$ films up to 8 μs and increase the internal quantum efficiency up to 90% [87, 88]. Here, it is proposed to combine an electron-donating group with an unoccupied lead site resulting from a halide vacancy [89, 90]. Several studies have demonstrated that exposing thin films and crystals to ambient conditions results in an increase in their internal quantum efficiency and luminescence lifetime, with values exceeding 90%. It is proposed that the presence of photogenerated electrons facilitates the coupling of oxygen in the air to surface vacancies in the halide, leading to the formation of superoxide, which effectively passivates these vacant sites [91].

Nanostructured $CsPbX_3$ exhibits high quantum efficiencies due to the combination of the charge-limited domain and surface passivation. The resulting EQE can exceed 90%. It is worth noting that in nanostructures with dimensions below the emission wavelength, the optical output coupling rate can approach 100%. Layered 2D and quasi-2D structures such as nanoplates can also achieve external quantum efficiencies comparable to those of quantum dots using this approach [92].

While the quantum efficiency and luminescence lifetime of thin films on glass and quartz substrates are important, the performance of the device is greatly impacted by the interfacial contact between the film and the electrode [78, 93]. Interfacial defects resulting from the contact between the layers can lead to non-radiative recombination and affect photon output coupling. In an ideal scenario, the interfacial contact should not reduce the quantum efficiency or carrier lifetime decay, and all combinations should result in radiative recombination, confined to the perovskite luminescent layer. While pure $(Cs,FA,MA)Pb(I_{0.85}Br_{0.15})_3$ films on ITO glass can achieve an EQE of over 20%, the efficiency decreases to 1% in a typical solar cell structure due to interfacial contact between n-type TiO_2 and p-type Spiro-OMeTAD. Therefore, optimizing interfacial contact is crucial for enhancing device performance [77]. Adding a passivation layer to the structure can effectively suppress efficiency losses in the n-type TiO_2 and p-type Spiro-OMeTAD contacts. This results in a photoelectric conversion efficiency of up to 15%. Furthermore, the loss of open circuit voltage due to radiation limitation is significantly reduced from 0.26 V to 0.11 V compared to structures without a passivation layer [77]. Passivation has been found to be a highly effective means of achieving a photoelectric conversion efficiency of 15% and extending the luminescent lifetime in full cells. However, in PeLEDs, TiO_2 and Spiro-OMeTAD have shown very low quantum efficiencies, despite no quenching of the substrate in clean films. This further highlights the presence of interfacial losses [47, 94].

Non-radiative losses at the interface can be reduced by insulating polymer films, such as polystyrene [95] and polyvinylpyrrolidone (PVP) [96]. The incorporation of PVP into the ZnO injection layer and $Cs_{0.87}MA_{0.13}PbBr_3$ film has been shown to enhance fluorescence intensity by a factor of 5 and increase fluorescence quantum efficiency to 55%, resulting in an EQE of 10.4% in device fabrication [96]. Furthermore, interfacial layers consisting of amines such as polyethyleneimine [97] and ethylenediamine [98] have been shown to reduce non-radiative losses and improve the performance of perovskite-based devices. This enhancement is mainly due to the combination of the N atoms in the amine with the bare lead–halogen framework, which effectively reduces the interfacial defect density. In addition, the use of TOPO can also improve the EQE of the device [89]. Other Lewis-based passivation layers, such as pyridine, have also been shown to have the same effect [99]. However, in these methods, a balance must be found between the increase in the passivation layer and the charge injection potential [19].

The intrinsic bimolecular recombination constant is determined by the cross-section and group velocity of the electron–hole interaction. Together with the non-radiative SRH composite (which depends on the trap density in the material and is an extrinsic property), these composite pathways contribute to the internal quantum efficiency, which is highly reliant on the carrier density in the emission layer (Figure 1.12a). The specific composite constant of a material can generally be determined through transient spectroscopy. Transient luminescence measurements are typically conducted at low excitation densities, where monomolecular processes

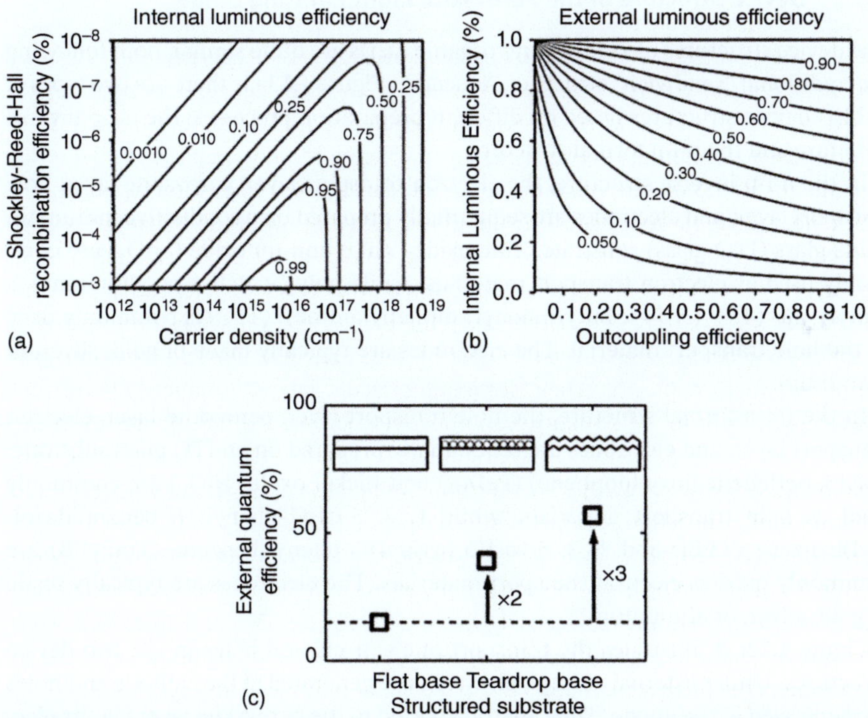

Figure 1.12 (a) Variation of internal quantum luminescence efficiency with carrier density and external Shockley–Read–Hall lifetime calculated using the reported bimolecular composite lifetime and Osher composite lifetime constants; (b) External luminescence quantum efficiency with output coupling efficiency and internal luminescence efficiency taking into account the phonon recycling process; (c) $MAPbBr_3$ film on a microcircuit substrate with external quantum efficiency enhancement. Source: Stranks et al. [72]/John Wiley & Sons/CC BY 4.0.

dominate the charge population decay dynamics. Bimolecular processes are more readily observed at higher excitation densities achievable with amplified laser systems. Moreover, the EQE is strongly influenced by the external coupling efficiency, as a significant portion of the light in the emissive layer can only be internally dissipated due to the distinct refractive indices of the device's various layers. The relationship between internal quantum efficiency, EQE, and external coupling efficiency is illustrated in Figure 1.12b. In general, for planar structures, the EQE limit of the device is approximately 20% (Figure 1.12c). The efficiency can be further improved through the use of mesoporous structures and periodic treatments of the substrate to enhance the external coupling efficiency, consequently enhancing the EQE.

1.5 Factors Influencing the Efficiency of Perovskite Light Emitting Diodes

1.5.1 Device Structure of the Perovskite Light Emitting Diode

The device structures of PeLEDs and organic LEDs are quite similar, both following the traditional "sandwich" structure. As seen in Figure 1.13a,b, there are two types of PeLED device structures based on different preparation processes: the n-i-p inverse structure and the p-i-n formal structure.

In the n-i-p inverse structure, the electron transport layer, perovskite layer, hole transport layer, and electrodes are sequentially prepared on a conductive indium tin oxide glass (ITO glass) substrate. Zinc oxide (ZnO) and tin oxide (SnO_2) are commonly used as electron transport materials, while poly[(9, 9-di-*n*-octylfluorenyl-2, 7-diyl) -alt- (4, 4′-(*N*-[4-*n*-butyl]phenyl)-diphenylamine)] (TFB) is commonly used as the hole transport material. The electrodes are typically made of gold, silver, or aluminum.

In the p-i-n formal structure, the hole transport layer, perovskite layer, electron transport layer, and electrodes are sequentially prepared on an ITO glass substrate. Poly(3,4-ethylenedioxythiophene) (PeDot) and nickel oxide (NiO_x) are commonly used as hole transport materials, while 1, 3, 5-tris(1-phenyl-1*H*-benzimidazol-2-yl)benzene (TPBi) and 1, 3, 5-tris[(3-pyridyl)-3-phenyl]benzene (TmPyPB) are commonly used as electron transport materials. The electrodes are typically made of gold, silver, or aluminum.

Figure 1.13c,d illustrates the transport of electrons and holes in the two device structures. Under external voltage, electrons are generated at the cathode and holes are generated at the anode. They are transported to the perovskite layer via the electron transport layer and hole transport layer, respectively, for radiation compound luminescence.

Figure 1.13e shows a typical architecture of a perovskite LED. The possible active layers constituting a PeLED are labeled on the left, and their contributions to various processes are listed on the right.

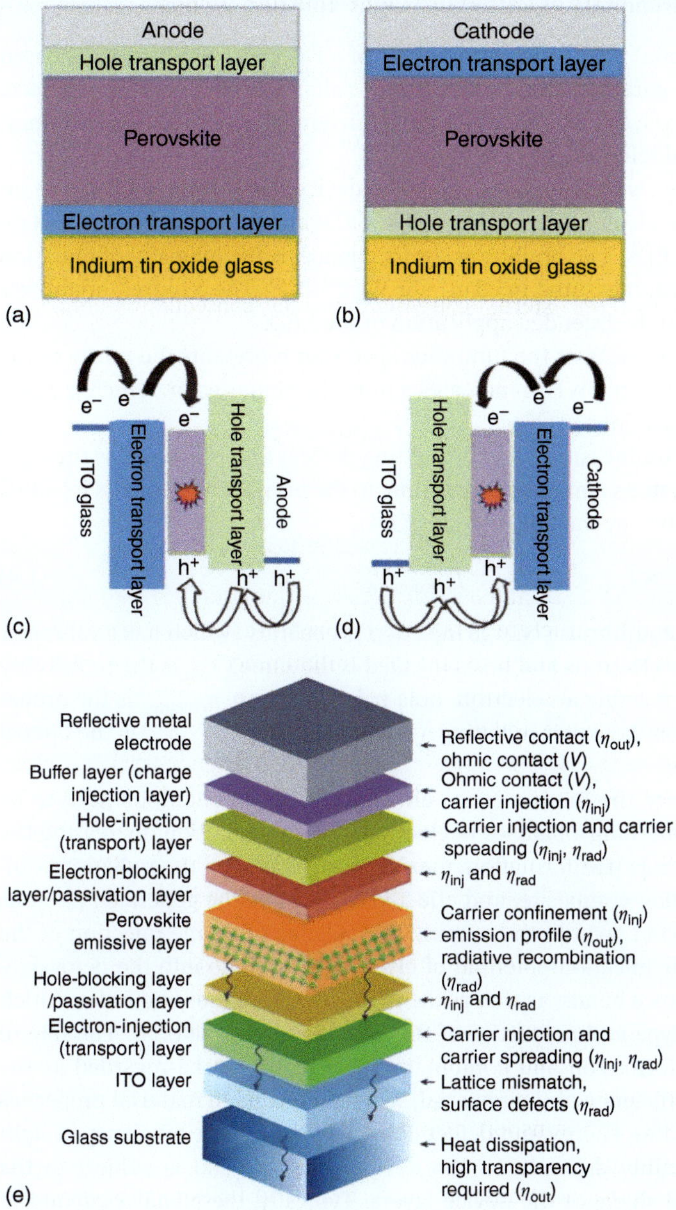

Figure 1.13 Schematic diagram of the structure of a perovskite light-emitting diode device: (a) n-i-p trans structure; (b) p-i-n formal structure; schematic diagram of charge transfer in a perovskite light-emitting diode; (c) trans structure; (d) formal structure; (e) A typical architecture of a perovskite LED. Source: Adapted from Fakharuddin et al. [11].

1.5.2 Physical Parameters of Perovskite Light-Emitting Diodes

Similar to conventional LEDs, the performance of a PeLED can be evaluated based on several physical parameters such as turn-on voltage, brightness or irradiance, peak luminous half-wave width (FWHM), EQE, current efficiency, lumen efficiency, energy conversion efficiency, and stability.

The turn-on voltage is the voltage at which the device starts to operate at a certain level of brightness, typically 1 cd m^{-2} for visible LEDs or when it starts to exhibit an EQE for UV or IR LEDs. The brightness, or irradiance, is the intensity of radiation emitted by the device, measured in cd m^{-2} or W sr^{-1} m^{-2}. The required brightness levels will depend on the intended application of the LED.

The peak-half-peak width of the luminous spectrum represents the purity of the emitted light. A narrower FWHM indicates a purer luminous color, which is desirable for display applications.

EQE is an important measure of light-emitting devices and is defined as the ratio of the number of photons emitted per unit time to the number of electrons injected. It can also be described by the following Eq. (1.4).

$$EQE = f_{\text{balance}} \times f_{\text{e-h}} \times \eta_{\text{radiative}} \times f_{\text{outcoupling}} \tag{1.4}$$

where f_{balance} is the equilibrium charge injection probability (which has a value of 1 when the number of electrons and holes injected is the same), $f_{\text{e-h}}$ is the probability of each carrier pair forming an electron–hole pair or exciton, $\eta_{\text{radiative}}$ is the probability of each electron–hole pair radiating a composite, and $f_{\text{outcoupling}}$ is the optical output coupling rate.

In order to enhance the luminescence efficiency, certain conditions need to be met, including maintaining a balance of electron and hole injection, increasing the chance of electron–hole pair formation, maximizing the ratio of the radiation composite to non-radiation composite, and effectively coupling the generated photons within the luminescent layer to the output. The balance of charge injection in the LED depends on the injection potential of electrons and holes into the perovskite conduction and valence bands, as well as the mobility of electrons and holes, which are limited by the type and properties of the transport layer material. The rate of electron–hole pair formation and combination efficiency can be regarded as the internal quantum efficiency of the material, which is related to material properties such as exciton binding energy and defect density of states. The efficiency of light coupling output mainly depends on the device structure and is subject to the refractive index and shape of the device layers. Typically, the planar structure of an LED limits the light-coupling efficiency to 25%, with most of the photon energy being dissipated in the form of a light waveguide within the device.

Current efficiency is the luminance of the light-emitting device and the ratio of current in cd A^{-1}; lumen efficiency is the ratio of the luminous flux emitted by the device to the electrical power input in lm W^{-1}; and energy conversion efficiency is the ratio of the photon energy emitted by the device to the total energy input.

Stability refers to the operating lifespan of an LED, which is typically measured by the time it takes for the device's brightness or EQE to decay to half of its initial value

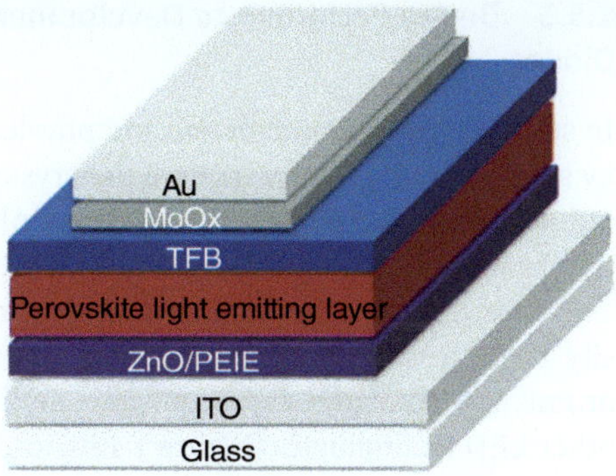

Figure 1.14 Structure diagram of a perovskite light-emitting diode device with ZnO/PEIE as the electron injection layer and TFB as the hole injection layer. Source: Stranks et al. [72]/John Wiley & Sons/CC BY 4.0.

at a constant voltage or current, also known as T50. There are also other measures, such as T90 and T80, which indicate the time taken for the brightness or EQE to decay to 90% and 80% of its initial value, respectively. Devices with better stability are more desirable for commercialization purposes.

Figure 1.14 illustrates the structure of a typical PeLED device. The perovskite light-emitting layer is sandwiched between the electron injection layer ZnO/PEIE and the hole injection layer TFB. When an external electric field is applied, electrons and holes are injected from the electron injection layer and hole injection layer, respectively, and they recombine in the perovskite layer, emitting light. To enhance the efficiency of electron injection, the electron affinity of the electron injection layer can be reduced. For example, doping Mg can form a ZnMgO structure in ZnO, leading to enhanced electron injection efficiency [100, 101]. Alternatively, an additional interface layer, such as polyethyleneimine, can be added between the injection layer and the light-emitting layer to improve electron injection. Similarly, the hole injection barrier can be reduced by doping the hole injection layer with perfluorinated lithium ion salt or by incorporating hole-transporting materials with high HOMO (highest occupied molecular orbital) energy levels, such as 4,4′-bis(9-carbazole) biphenyl, to lower the hole injection barrier [40, 102, 103].

The injection potentials of different layers can be determined by UV photoelectron spectroscopy. However, the energy level structure of perovskites is affected by the substrate work function, which means that UV photoelectron spectroscopy of perovskites requires the film to be deposited on a substrate that closely mimics the device structure. Alternatively, electron absorption spectroscopy can be used to measure the injection barrier in the device [104]. In electron absorption spectroscopy, the effect of the internal electric field on the transmitted light is mainly measured [105]. According to the single-electron Franz–Keldysh–Aspnes low electric field theory, the transmitted light varies with the square of the electric field, which can also be applied to $MAPbI_3$ and other bulk materials [106, 107]. The magnitude of the internal electric field can be measured by modulating the internal electric field with a DC bias. However, the effect of ion shielding has to be taken into account in this process [108].

1.5.3 Device Performance Development of Perovskite Light-Emitting Diodes

In summary, the luminescence and physical properties of perovskite are influenced by several factors. These include the crystal structure of the perovskite, size effects, radiative and non-radiative processes of photons, interface states, and charge injection balance.

In most PeLEDs, the EQE and light output tend to decrease as the current density (J) increases, primarily due to augmented Auger recombination, Joule heating, or imbalanced charge injection. This EQE roll-off phenomenon is also observed in other LED technologies (Figure 1.15b) and typically leads to a 50% decrease in EQE when J ranges from 100 to 1000 mA cm^{-2}. Only a limited number of devices have reported EQE values above 1 A cm^{-2}, and these devices are typically fabricated using pure-phase 2D or mixed-phase 2D/3D perovskite materials (Figure 1.15c). Significantly, PeLEDs with reduced EQE roll-off exhibit low turn-on voltages (V_{ON}), indicating barrier-free charge transport/injection. A low driving voltage can also help alleviate the quantum-confined Stark effect, which is another potential cause for EQE roll-off [11].

Changes in the dimensionality and crystal defects of the perovskite affect the state of the exciton and carrier transport, resulting in a significant change in device performance. In two-dimensional and one-dimensional perovskites, the exciton binding energy is stronger, resulting in a wider color gamut. In contrast, three-dimensional perovskites have a weaker exciton binding energy, leading to a narrower color spectrum. Crystal defects impact the carrier transport capacity, and higher defect densities lead to more exciton capture, resulting in more non-radiative output and reduced device efficiency.

Radiative and non-radiative processes of photons are closely linked to the crystal structure, size effects, exciton binding energy, and trap density, influencing the combination path of excitons. Excitation light density also affects radiation and decay processes. Surface and interface trap states significantly influence non-radiative processes, and surface state modulation can improve device luminous efficiency. Charge injection balance mainly depends on the energy level structure of the material layers and the ideal balance of injection potential at the interface, ensuring both stable device operation and improved EQE.

Although PeLED research started later than that of perovskite solar cells, it has experienced significant development in a short period. In just a few years, the EQE of PeLEDs has increased from less than 0.1% to over 20% [110–114]. To date, the maximum EQEs for near-infrared, red, green, blue, and white light-emitting perovskite diodes have reached 25.5%, 25.8%, 30.84%, 18.65%, and 12.2%, respectively [115–119] (Figure 1.16).

Figure 1.17a presents a comparison of the color gamut of LFMHP-based LEDs, demonstrating that the blue and red components have already fulfilled the requirements of the Rec. 2020 standard, while the green component still requires further improvement. Another critical factor limiting the commercialization of PeLEDs is device lifetime. Figure 1.17b illustrates the promising device stability

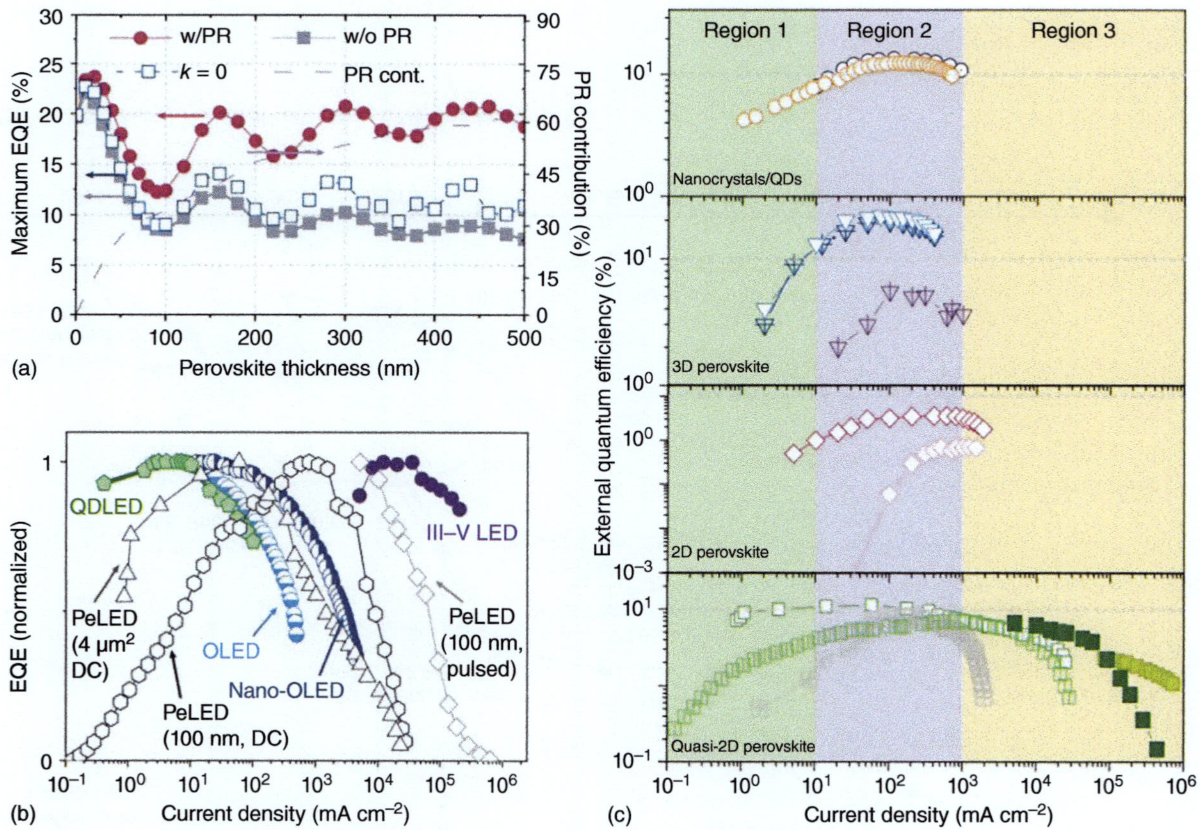

Figure 1.15 Photon-recycling and EQE roll-off in PeLEDs. (a) Calculated EQEs with and without photon recycling (PR; filled red circles and filled gray squares, respectively) and without reabsorption (open navy blue squares), as well as the relative photon recycling contribution (violet dashed lines) for an ideal PeLED (IQE = 100%) with perovskite EMLs of various thickness. *k* indicates the refractive index component connected to the absorption coefficient. Source: Cho et al. [109]/Springer Nature/CC BY 4.0. (b) A comparison of EQE roll-off in various thin-film LEDs (organic LEDs, OLEDs and nano-OLEDs, QDLEDs, and PeLEDs with state-of-the-art crystalline III–V LEDs). DC, direct current. LEDs are biased via steady-state voltage. Source: Fakharuddin et al. [11]/Springer Nature. (c) EQE roll-off trends of some notable PeLEDs reported in the literature employing NCs, 3D, 2D, and 2D/3D EMLs. The plot is divided into three regions based on injection current densities and the type of EML. Source: Fakharuddin et al. [11]/Springer Nature.

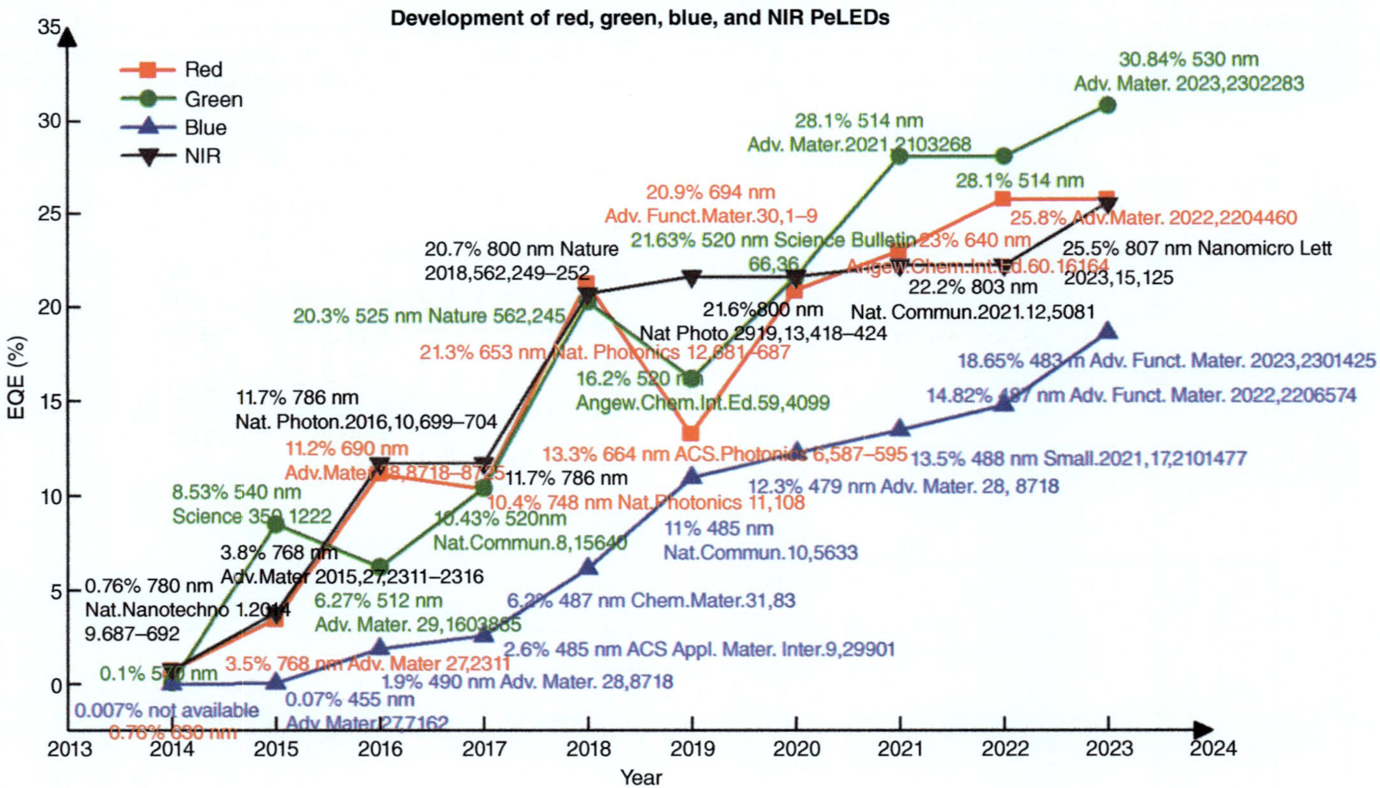

Figure 1.16 The development in EQEs of red, green, blue, and NIR PeLEDs.

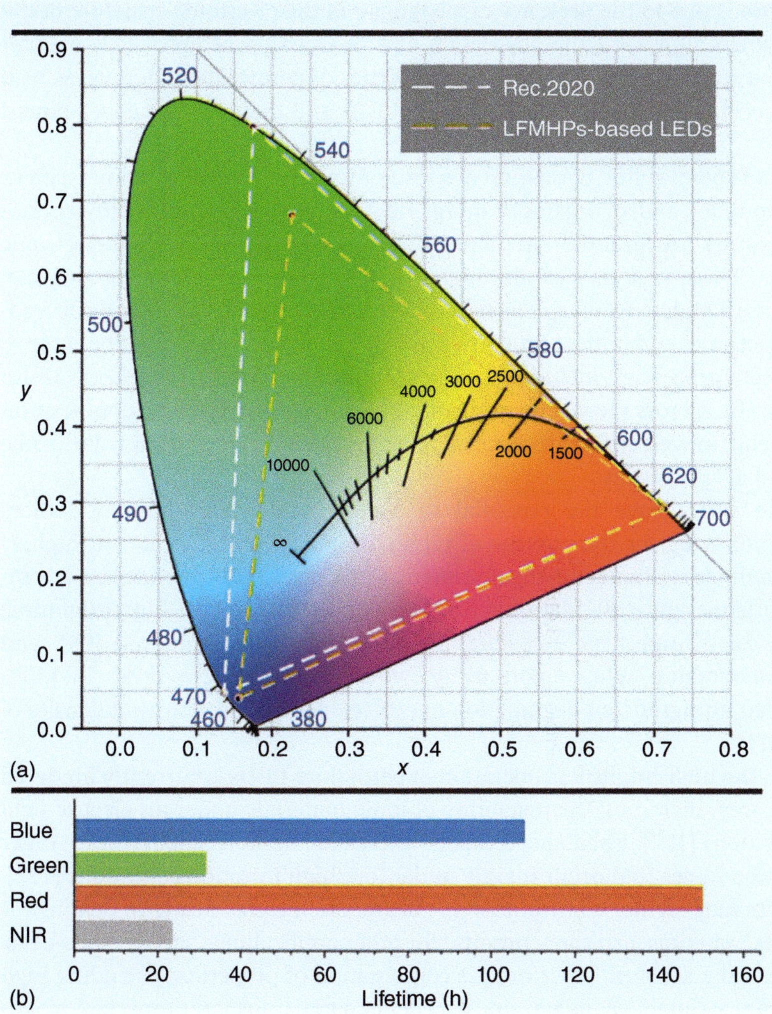

Figure 1.17 Current status of perovskite light-emitting diodes (PeLEDs) based on lead-free metal halide perovskite (LFMHP) emitters. (a) Color gamut coverage of Rec. 2020 and LFMHP-based PeLEDs. (b) Device lifetime of state-of-the-art NIR, red, green, and blue-emitting LFMHP-based PeLEDs. Source: Wang et al. [120]/Innovation Press Co., Limited.

of LFMHP-based LEDs, indicating their potential in overcoming the inherent chemical instability issue found in LHP-based LEDs [120].

1.6 Summary

Perovskite materials are highly sought after for their luminescent properties, including their ability to emit light when excited by an external energy source. This luminescence is due to the presence of a bandgap in the electronic structure of the material, which allows for the absorption and re-emission of light. This chapter discusses the structure and luminescent properties of perovskite materials, as well as some important physical parameters of PeLEDs and their current development status.

Metal halide perovskites, composed of a lead or tin cation, a halide anion (such as iodide or bromide), and an organic or inorganic A-site cation (such as methylammonium or cesium), are an important class of perovskite materials for optoelectronic applications. These materials exhibit high defect tolerance, which means they can accommodate a high density of defects and impurities without significantly degrading their electronic and optical properties. Metal halide perovskites exhibit a range of luminescent properties, including strong photoluminescence (PL) and electroluminescence (EL) across the visible spectrum. The crystal structure and composition of the material, as well as the presence of defects and impurities, strongly influence their PL and EL properties.

Perovskite LEDs have the potential to operate under ultra-high brightness conditions, allowing for the passage of high currents ($100\,cm^2\,V^{-1}S^{-1}$ or higher) through the diode without causing detrimental processes. This provides an advantage over organic and colloidal quantum dot semiconductors in high-brightness operations. Peak brightness exceeding $10^5\,cd\,m^{-2}$ has been achieved [96], and a recent encouraging aging report using two-dimensional perovskites in LEDs demonstrated no significant degradation in device operation at a current density of $A\,cm^{-2}$ [121].

However, the high-brightness operation of perovskite LEDs is currently hindered by other factors, including the instability of ionic materials under an electric field (ionic migration) [122], imbalanced carrier injection, and non-radiative and Auger recombination losses that occur mainly under low/high injection conditions [123]. These factors may further contribute to the degradation associated with heating.

Perovskite materials also show promise for other applications, such as solar cells, LEDs, lasers, and sensors. Their unique combination of properties, including high defect tolerance, tunable bandgap, and excellent charge transport properties, makes them an attractive candidate for a wide range of optoelectronic applications. Perovskite materials have also been explored for use in sensors, particularly for detecting gases and ions, due to their unique electronic properties, high surface area, and ability to selectively bind to certain molecules.

However, perovskite materials have some challenges that need to be addressed before they can be widely deployed in commercial applications. Their instability in the presence of moisture and oxygen can lead to degradation over time, which is a major obstacle for the development of perovskite-based solar cells. Researchers have made progress in addressing this issue through the development of encapsulation strategies and the use of stable electrode materials, but further work is needed to improve long-term stability. The toxicity of some of the elements used in perovskite materials, particularly lead, is another challenge. While there has been progress in developing lead-free perovskites, many of these materials have lower performance than their lead-based counterparts.

Furthermore, PeLEDs still face challenges in terms of the poor efficiency of blue devices, a decline in EQE under high-brightness conditions, and insufficient device lifespans. The low efficiency of blue PeLEDs is often attributed to halide segregation in mixed Cl–Br perovskite emitters. Some progress has been made through compositional adjustments and the incorporation of bulky organic cations in low-dimensional perovskites. Large organic cations can stiffen the perovskite crystal structure, reducing detrimental electron–phonon interactions and ion migration. However, these cations are typically insulating, resulting in inferior charge transport and reduced device performance. One potential solution is to replace these insulating ligands, commonly alkyl-chained, with conjugated semiconducting ligands, which could potentially alleviate this issue [11].

EQE roll-off and operational stability are significant challenges for PeLEDs, influenced by both intrinsic factors of the perovskite active layer and extrinsic parameters of the device. Ongoing strategies to address these challenges include compositional and dimensional engineering of the perovskite emitter, formation of defect-free emissive layers (EMLs) through passivation techniques, optimization of heterointerfaces to enable barrier-free charge transport and injection, achieving balanced charge injection, operating the devices at low voltages, and reducing the transport resistance in the emissive layer. Given the ionic nature of halide perovskites, the development of specific biasing schemes (e.g. pulsed biasing) may assist in minimizing ion migration and enhancing stability [11].

There is also a need for more research on the potential environmental impact of large-scale perovskite production and deployment, as well as the scalability and reproducibility of perovskite synthesis and device fabrication.

Despite these challenges, perovskite materials hold tremendous potential for a range of applications, and the field is likely to see continued rapid development and progress in the coming years. Advances in perovskite synthesis and device engineering have led to significant improvements in performance, and it is likely that perovskites will continue to be a focus of research and development. Further improvements in performance and stability, as well as the development of new applications and markets for perovskite materials, are expected as continued research and development in this area progresses.

References

1 You, Y.M., Liao, W.Q., Zhao, D. et al. (2017). An organic-inorganic perovskite ferroelectric with large piezoelectric response. *Science* 357: 306–309.

2 Møller, C.K. (1958). Crystal structure and photoconductivity of caesium plumbohalides. *Nature* 182: 1436–1436.

3 Bansal, V., Poddar, P., Ahmad, A. et al. (2006). Room-temperature biosynthesis of ferroelectric barium titanate nanoparticles. *Journal of the American Chemical Society* 128: 11958–11963.

4 Bednorz, J.G. and Müller, K.A. (1986). Possible high Tc superconductivity in the Ba-La-Cu-O system. *Zeitschrift für Physik B Condensed Matter* 64: 189–193.

5 Jin, S., Tiefel, T.H., McCormack, M. et al. (1994). Thousandfold change in resistivity in magnetoresistive La-Ca-Mn-O films. *Science* 264: 413–415.

6 Miao, X., Zhang, L., Wu, L. et al. (2019). Quadruple perovskite ruthenate as a highly efficient catalyst for acidic water oxidation. *Nature Communications* 10: 1–7.

7 Zhao, Y.C., Zhou, W.K., Zhou, X. et al. (2017). Quantification of light-enhanced ionic transport in lead iodide perovskite thin films and its solar cell applications. *Light: Science & Applications* 6: 16243–16243.

8 Mitzi, D.B. (2001). Thin-film deposition of organic-inorganic hybrid materials. *Chemistry of Materials* 13: 3283–3298.

9 Quan, L.N., Garcia de Arquer, F.P., Sabatini, R.P. et al. (2018). Perovskites for light emission. *Advanced Materials* 30: 1801996.

10 Kojima, A., Teshima, K., Shirai, Y. et al. (2009). Organometal halide perovskites as visible-light sensitizers for photovoltaic cells. *Journal of the American Chemical Society* 131: 6050–6051.

11 Fakharuddin, A., Gangishetty, M.K., Abdi-Jalebi, M. et al. (2022). Perovskite light-emitting diodes. *Nature Electronics* 5: 203–216.

12 Kang, J. and Wang, L.W. (2017). High defect tolerance in lead halide perovskite $CsPbBr_3$. *The Journal of Physical Chemistry Letters* 8: 489–493.

13 Yettapu, G.R., Talukdar, D., Sarkar, S. et al. (2016). Terahertz conductivity within colloidal $CsPbBr_3$ perovskite nanocrystals: remarkably high carrier mobilities and large diffusion lengths. *Nano Letters* 16: 4838–4848.

14 Shi, D., Adinolfi, V., Comin, R. et al. (2015). Low trap-state density and long carrier diffusion in organolead trihalide perovskite single crystals. *Science* 347: 519–522.

15 Brandt, R.E., Stevanović, V., Ginley, D.S. et al. (2015). Identifying defect-tolerant semiconductors with high minority-carrier lifetimes: beyond hybrid lead halide perovskites. *MRS Communications* 5: 265–275.

16 Mesquita, I., Andrade, L., and Mendes, A. (2018). Perovskite solar cells: materials, configurations and stability. *Renewable and Sustainable Energy Reviews* 82: 2471–2489.

17 Liu, A., Bi, C., Guo, R. et al. (2021). Electroluminescence principle and performance improvement of metal halide perovskite light-emitting diodes. *Advanced Optical Materials* 9.

18 Zhao, C., Zhang, D., and Qin, C. (2020). Perovskite light-emitting diodes. *CCS Chemistry* 2: 859–869.

19 Cheng, H., Feng, Y., Fu, Y. et al. (2022). Understanding and minimizing non-radiative recombination losses in perovskite light-emitting diodes. *Journal of Materials Chemistry C* 10: 13590–13610.

20 Shi, Z., Li, Y., Zhang, Y. et al. (2017). High-efficiency and air-stable perovskite quantum dots light-emitting diodes with an all-inorganic heterostructure. *Nano letters* 17: 313–321.

21 Han, B., Cai, B., Shan, Q. et al. (2018). Stable, efficient red perovskite light-emitting diodes by (α, δ)-$CsPbI_3$ phase engineering. *Advanced Functional Materials* 28: 1804285.

22 Ban, M., Zou, Y., Rivett, J.P.H. et al. (2018). Solution-processed perovskite light emitting diodes with efficiency exceeding 15% through additive-controlled nanostructure tailoring. *Nature Communications* 9: 1–10.

23 Protesescu, L., Yakunin, S., Kumar, S. et al. (2017). Dismantling the "red wall" of colloidal perovskites: highly luminescent formamidinium and formamidinium-cesium lead iodide nanocrystals. *ACS Nano* 11: 3119–3134.

24 Schweinberger, F.F., Berr, M.J., Doblinger, M. et al. (2013). Cluster size effects in the photocatalytic hydrogen evolution reaction. *Journal of the American Chemical Society* 135: 13262–13265.

25 Dresselhaus (2022). *Optical Properties of Solids*. Oxford University Press.

26 Zhu, X.Y. and Podzorov, V. (2015). Charge carriers in hybrid organic inorganic lead halide perovskites might be protected as large polarons. *The Journal of Physical Chemistry Letters* 6: 4758–4761.

27 Smith, M.D. and Karunadasa, H.I. (2018). White light emission from layered halide perovskites. *Accounts of Chemical Research* 51: 619–627.

28 Tokizaki, T., Makimura, T., Akiyama, H. et al. (1991). Femtosecond cascade-excitation spectroscopy for nonradiative deexcitation and lattice relaxation of the self-trapped exciton in NaCl. *Physical Review Letters* 67: 2701–2704.

29 Zhang, L., Sun, C., He, T. et al. (2021). High-performance quasi-2D perovskite light-emitting diodes: from materials to devices. *Light: Science & Applications* 10: 61.

30 Efros, A.L., Rosen, M., Kuno, M. et al. (1996). Band-edge exciton in quantum dots of semiconductors with a degenerate valence band: dark and bright exciton states. *Physical Review B* 54: 4843–4856.

31 Franceschetti, A., Fu, H., Wang, L.W. et al. (1999). Many-body pseudopotential theory of excitons in InP and CdSe quantum dots. *Physical Review B* 60: 1819–1829.

32 Leung, K., Pokrant, S., and Whaley, K.B. (1998). Exciton fine structure in CdSe nanoclusters. *Physical Review B* 57: 12291–12301.

33 Crooker, S.A., Barrick, T., Hollingsworth, J.A. et al. (2003). Multiple temperature regimes of radiative decay in CdSe nanocrystal quantum dots: intrinsic limits to the dark-exciton lifetime. *Applied Physics Letters* 82: 2793–2795.

34 Labeau, O., Tamarat, P., and Lounis, B. (2003). Temperature dependence of the luminescence lifetime of single CdSe/ZnS quantum dots. *Physical Review Letters* 90: 257404.

35 Donegá, C.M., Bode, M., and Meijerink, M. (2006). Size- and temperature-dependence of exciton lifetimes in CdSe quantum dots. *Physical Review B* 74: 085320.

36 Wang, H., Donegá, C.M., Meijerink, A. et al. (2006). Ultrafast exciton dynamics in CdSe quantum dots studied from bleaching recovery and fluorescence transients. *The Journal of Physical Chemistry B* 110: 733–737.

37 Zhao, Q., Peter, A., Wesley, B.J. et al. (2007). Shape dependence of band-edge exciton fine structure in CdSe nanocrystals. *Nano Letters* 7: 3274–3280.

38 Oron, D., Aharoni, A., Donega, C.M. et al. (2009). Universal role of discrete acoustic phonons in the low-temperature optical emission of colloidal quantum dots. *Physical Review Letters* 102: 177402.

39 Schaller, R.D., Crooker, S.A., Bussian, D.A. et al. (2010). Revealing the exciton fine structure of PbSe nanocrystal quantum dots using optical spectroscopy in high magnetic fields. *Physical Review Letters* 105: 067403.

40 Veldhuis, S.A., Boix, P.P., Yantara, N. et al. (2016). Perovskite materials for light-emitting diodes and lasers. *Advanced Materials* 28: 6804–6834.

41 Vasilopoulou, M., Fakharuddin, A., García de Arquer, F.P. et al. (2021). Advances in solution-processed near-infrared light-emitting diodes. *Nature Photonics* 15: 656–669.

42 Richter, J.M., Abdi-Jalebi, M., Sadhanala, A. et al. (2016). Enhancing photoluminescence yields in lead halide perovskites by photon recycling and light out-coupling. *Nature Communications* 7: 13941.

43 Lin, Q., Armin, A., Nagiri, R.C. et al. (2014). Electro-optics of perovskite solar cells. *Nature Photonics* 9: 106–112.

44 Saba, M., Cadelano, M., Marongiu, D. et al. (2014). Correlated electron-hole plasma in organometal perovskites. *Nature Communications* 5: 5049.

45 Deschler, F., Price, M., Pathak, S. et al. (2014). High photoluminescence efficiency and optically pumped lasing in solution-processed mixed halide perovskite semiconductors. *The Journal of Physical Chemistry Letters* 5: 1421–1426.

46 Wang, H., Zhang, X., Wu, Q. et al. (2019). Trifluoroacetate induced small-grained $CsPbBr_3$ perovskite films result in efficient and stable light-emitting devices. *Nature Communications* 10: 665.

47 Tan, Z.K., Moghaddam, R.S., Lai, M.L. et al. (2014). Bright light-emitting diodes based on organometal halide perovskite. *Nature Nanotechnology* 9: 687–692.

48 Congreve, D.N., Weidman, M.C., Seitz, M. et al. (2017). Tunable light-emitting diodes utilizing quantum-confined layered perovskite emitters. *ACS Photonics* 4: 476–481.

49 Li, G., Tan, Z.K., Di, D. et al. (2015). Efficient light-emitting diodes based on nanocrystalline perovskite in a dielectric polymer matrix. *Nano Letters* 15: 2640–2644.

50 Hutter, E.M., Gelvez-Rueda, M.C., Osherov, A. et al. (2017). Direct-indirect character of the bandgap in methylammonium lead iodide perovskite. *Nature Materials* 16: 115–120.

51 Etienne, T., Mosconi, E., and De Angelis, F. (2016). Dynamical origin of the Rashba effect in organohalide lead perovskites: a key to suppressed carrier recombination in perovskite solar cells? *The Journal of Physical Chemistry Letters* 7: 1638–1645.

52 Davies, C.L., Filip, M.R., Patel, J.B. et al. (2018). Bimolecular recombination in methylammonium lead triiodide perovskite is an inverse absorption process. *Nature Communications* 9: 293.

53 Niesner, D., Wilhelm, M., Levchuk, I. et al. (2016). Giant Rashba splitting in $CH_3NH_3PbBr_3$ organic-inorganic perovskite. *Physical Review Letters* 117: 126401.

54 Isarov, M., Tan, L.Z., Bodnarchuk, M.I. et al. (2017). Rashba effect in a single colloidal $CsPbBr_3$ perovskite nanocrystal detected by magneto-optical measurements. *Nano Letters* 17: 5020–5026.

55 Stroppa, A., Di Sante, D., Barone, P. et al. (2014). Tunable ferroelectric polarization and its interplay with spin-orbit coupling in tin iodide perovskites. *Nature Communications* 5.

56 March, S.A., Riley, D.B., Clegg, C. et al. (2017). Four-wave mixing in perovskite photovoltaic materials reveals long dephasing times and weaker many-body interactions than GaAs. *ACS Photonics* 4: 1515–1521.

57 Elkins, M.H., Pensack, R., Proppe, A.H. et al. (2017). Biexciton resonances reveal exciton localization in stacked perovskite quantum wells. *The Journal of Physical Chemistry Letters* 8: 3895–3901.

58 Richter, J.M., Branchi, F., de Almeida Camargo, F.V. et al. (2017). Ultrafast carrier thermalization in lead iodide perovskite probed with two-dimensional electronic spectroscopy. *Nature Communications* 8: 376.

59 Price, M.B., Butkus, J., Jellicoe, T.C. et al. (2015). Hot-carrier cooling and photoinduced refractive index changes in organic-inorganic lead halide perovskites. *Nature Communications* 6: 8420.

60 Yang, J., Wen, X., Xia, H. et al. (2017). Acoustic-optical phonon up-conversion and hot-phonon bottleneck in lead-halide perovskites. *Nature Communications* 8: 14120.

61 Yang, Y., Ostrowski, D.P., France, R.M. et al. (2015). Observation of a hot-phonon bottleneck in lead-iodide perovskites. *Nature Photonics* 10: 53–59.

62 Chang, A.Y., Cho, Y.J., Chen, K.C. et al. (2016). Slow organic-to-inorganic sub-lattice thermalization in methylammonium lead halide perovskites observed by ultrafast photoluminescence. *Advanced Energy Materials* 6: 1600422.

63 Stranks, S.D., Burlakov, V.M., Leijtens, T. et al. (2014). Recombination kinetics in organic-inorganic perovskites: excitons, free charge, and subgap states. *Physical Review Applied* 2: 034007.

64 Chondroudis, K. and Mitzi, D.B. (1999). Electroluminescence from an organic-inorganic perovskite incorporating a quaterthiophene dye within lead halide perovskite layers. *Chemistry of Materials* 11: 3028–3030.

65 Wu, X., Trinh, M.T., Niesner, D. et al. (2015). Trap states in lead iodide perovskites. *Journal of the American Chemical Society* 137: 2089–2096.

66 Xiao, Z., Kerner, R.A., Zhao, L. et al. (2017). Efficient perovskite light-emitting diodes featuring nanometre-sized crystallites. *Nature Photonics* 11: 108–115.

67 Yuan, M., Quan, L.N., Comin, R. et al. (2016). Perovskite energy funnels for efficient light-emitting diodes. *Nature Nanotechnology* 11: 872–877.

68 Quan, L.N., Zhao, Y., Garcia de Arquer, F.P. et al. (2017). Tailoring the energy landscape in quasi-2D halide perovskites for efficient light emission. *Nano Letters* 17: 3701–3709.

69 Pan, J., Quan, L.N., Zhao, Y. et al. (2016). Highly efficient perovskite-quantum-dot light-emitting diodes by surface engineering. *Advanced Materials* 28: 8718–8725.

70 Milot, R.L., Sutton, R.J., Eperon, G.E. et al. (2016). Charge-carrier dynamics in 2D hybrid metal-halide perovskites. *Nano Letters* 16: 7001–7007.

71 Yang, Y., Yang, M., Li, Z. et al. (2015). Comparison of recombination dynamics in $CH_3NH_3PbBr_3$ and $CH_3NH_3PbI_3$ perovskite films: influence of exciton binding energy. *The Journal of Physical Chemistry Letters* 6: 4688–4692.

72 Stranks, S.D., Hoye, R.L., Di, D. et al. (2019). The physics of light emission in halide perovskite devices. *Advanced Materials* 31: 1803336.

73 Stranks, S.D. and Snaith, H.J. (2015). Metal-halide perovskites for photovoltaic and light-emitting devices. *Nature Nanotechnology* 10: 391–402.

74 Green, M.A., Ho-Baillie, A., and Snaith, H.J. (2014). The emergence of perovskite solar cells. *Nature Photonics* 8: 506–514.

75 De Wolf, S., Holovsky, J., Moon, S.J. et al. (2014). Organometallic halide perovskites: sharp optical absorption edge and its relation to photovoltaic performance. *The Journal of Physical Chemistry Letters* 5: 1035–1039.

76 Sadhanala, A., Deschler, F., Thomas, T.H. et al. (2014). Preparation of single-phase films of $CH_3NH_3Pb(I_{1-x}Br_x)_3$ with sharp optical band edges. *The Journal of Physical Chemistry Letters* 5: 2501–2505.

77 Abdi-Jalebi, M., Andaji-Garmaroudi, Z., Cacovich, S. et al. (2018). Maximizing and stabilizing luminescence from halide perovskites with potassium passivation. *Nature* 555: 497–501.

78 Stranks, S.D. (2017). Nonradiative losses in metal halide perovskites. *ACS Energy Letters* 2: 1515–1525.

79 Wehrenfennig, C., Eperon, G.E., Johnston, M.B. et al. (2014). High charge carrier mobilities and lifetimes in organolead trihalide perovskites. *Advanced Materials* 26: 1584–1589.

80 Draguta, S., Thakur, S., Morozov, Y.V. et al. (2016). Spatially non-uniform trap state densities in solution-processed hybrid perovskite thin films. *The Journal of Physical Chemistry Letters* 7: 715–721.

81 Wu, C., Wu, T., Yang, Y. et al. (2019). Alternative type two-dimensional-three-dimensional lead halide perovskite with inorganic sodium ions as a spacer for high-performance light-emitting diodess. *ACS Nano* 13: 1645–1654.

82 Meng, F., Liu, X., Cai, X. et al. (2019). Incorporation of rubidium cations into blue perovskite quantum dot light-emitting diodes via FABr-modified multi-cation hot-injection method. *Nanoscale* 11: 1295–1303.

83 Ke, Y., Wang, N., Kong, D. et al. (2019). Defect passivation for red perovskite light-emitting diodes with improved brightness and stability. *The Journal of Physical Chemistry Letters* 10: 380–385.

84 Kovalenko, M.V., Protesescu, L., and Bodnarchuk, M.I. (2017). Properties and potential optoelectronic applications of lead halide perovskite nanocrystals. *Science* 358: 745–750.

85 Kirchartz, T., Markvart, T., Rau, U. et al. (2018). Impact of small phonon energies on the charge-carrier lifetimes in metal-halide perovskites. *The Journal of Physical Chemistry Letters* 9: 939–946.

86 Yang, W.S., Park, B.W., Jung, E.H. et al. (2017). Iodide management in formamidinium-lead-halide-based perovskite layers for efficient solar cells. *Science* 356: 1376–1379.

87 Braly, I.L., deQuilettes, D.W., Pazos-Outón, L.M. et al. (2018). Hybrid perovskite films approaching the radiative limit with over 90% photoluminescence quantum efficiency. *Nature Photonics* 12: 355–361.

88 Yang, X., Zhang, X., Deng, J. et al. (2018). Efficient green light-emitting diodes based on quasi-two-dimensional composition and phase engineered perovskite with surface passivation. *Nature Communications* 9: 570.

89 deQuilettes, D.W., Koch, S., Burke, S. et al. (2016). Photoluminescence lifetimes exceeding 8 μs and quantum yields exceeding 30% in hybrid perovskite thin films by ligand passivation. *ACS Energy Letters* 1: 438–444.

90 Sadaf, S.M., Ra, Y.H., Nguyen, H.P.T. et al. (2015). Alternating-current InGaN/GaN tunnel junction nanowire white-light emitting diodes. *Nano Letters* 15: 6696–6701.

91 Brenes, R., Eames, C., Bulovic, V. et al. (2018). The impact of atmosphere on the local luminescence properties of metal halide perovskite grains. *Advanced Materials* 30: 1706208.

92 Weidman, M.C., Seitz, M., Stranks, S.D. et al. (2016). Highly tunable colloidal perovskite nanoplatelets through variable cation, metal, and halide composition. *ACS Nano* 10: 7830–7839.

93 Tvingstedt, K., Malinkiewicz, O., Baumann, A. et al. (2014). Radiative efficiency of lead iodide based perovskite solar cells. *Scientific Reports* 4: 6071.

94 Chen, C., Gao, L., Gao, W. et al. (1927). Circularly polarized light detection using chiral hybrid perovskite. *Nature Communications* 2019: 10.

95 Wolff, C.M., Zu, F., Paulke, A. et al. (2017). Reduced interface-mediated recombination for high open-circuit voltages in $CH_3NH_3PbI_3$ solar cells. *Advanced Materials* 29: 1700159.

96 Zhang, L., Yang, X., Jiang, Q. et al. (2017). Ultra-bright and highly efficient inorganic based perovskite light-emitting diodes. *Nature Communications* 8: 15640.

97 Wang, J., Wang, N., Jin, Y. et al. (2015). Interfacial control toward efficient and low-voltage perovskite light-emitting diodes. *Advanced Materials* 27: 2311–2316.

98 Lee, S., Park, J.H., Lee, B.R. et al. (2017). Amine-based passivating materials for enhanced optical properties and performance of organic-inorganic perovskites in light-emitting diodes. *The Journal of Physical Chemistry Letters* 8: 1784–1792.

99 Wu, C.Y., Wang, Z., Liang, L. et al. (2019). Graphene-assisted growth of patterned perovskite films for sensitive light detector and optical image sensor application. *Small* 15: 1900730.

100 Hoye, R.L., Chua, M.R., Musselman, K.P. et al. (2015). Enhanced performance in fluorene-free organometal halide perovskite light-emitting diodes using tunable, low electron affinity oxide electron injectors. *Advanced Materials* 27: 1414–1419.

101 Hoye, R.L., Munoz-Rojas, D., Musselman, K.P. et al. (2015). Synthesis and modeling of uniform complex metal oxides by close-proximity atmospheric pressure chemical vapor deposition. *ACS Applied Materials & Interfaces* 7: 10684–10694.

102 Kim, Y.H., Cho, H., Heo, J.H. et al. (2015). Multicolored organic/inorganic hybrid perovskite light-emitting diodes. *Advanced Materials* 27: 1248–1254.

103 Kim, Y.H., Cho, H., and Lee, T.W. (2016). Metal halide perovskite light emitters. *PNAS* 113: 11694–11702.

104 Hoye, R.L.Z., Schulz, P., Schelhas, L.T. et al. (2017). Perovskite-inspired photovoltaic materials: toward best practices in materials characterization and calculations. *Chemistry of Materials* 29: 1964–1988.

105 Brown, T.M., Kim, J.S., Friend, R.H. et al. (1999). Built-in field electroabsorption spectroscopy of polymer light-emitting diodes incorporating a doped poly(3, 4-ethylene-dioxythiophene) hole injection layer. *Applied Physics Letters* 75: 1679–1681.

106 Ziffer, M.E., Mohammed, J.C., and Ginger, D.S. (2016). Electroabsorption spectroscopy measurements of the exciton binding energy, electron-hole reduced effective mass, and band gap in the perovskite $CH_3NH_3PbI_3$. *ACS Photonics* 3: 1060–1068.

107 Amerling, E., Baniya, S., Lafalce, E. et al. (2017). Electroabsorption spectroscopy studies of $(C_4H_9NH_3)_2PbI_4$ organic-inorganic hybrid perovskite multiple quantum wells. *The Journal of Physical Chemistry Letters* 8: 4557–4564.

108 Campbell, I.H., Hagler, T.W., Smith, D.L. et al. (1900). Direct measurement of conjugated polymer electronic excitation energies using metal/polymer/metal structures. *Physical Review Letters* 1996: 76.

109 Cho, C., Zhao, B., Tainter, G.D. et al. (2020). The role of photon recycling in perovskite light-emitting diodes. *Nature Communications* 11: 611.

110 Song, J., Li, J., Li, X. et al. (2015). Quantum dot light-emitting diodes based on inorganic perovskite cesium lead halides ($CsPbX_3$). *Advanced Materials* 27: 7162–7167.

111 Lu, M., Zhang, X., Zhang, Y. et al. (2018). Simultaneous strontium doping and chlorine surface passivation improve luminescence intensity and stability of $CsPbI_3$ nanocrystals enabling efficient light-emitting devices. *Advanced Materials* 30: 1804691.

112 Song, J., Fang, T., Li, J. et al. (2018). Organic-inorganic hybrid passivation enables perovskite QLEDs with an EQE of 16.48%. *Advanced Materials* 30: 1805409.

113 Chiba, T., Hayashi, Y., Ebe, H. et al. (2018). Anion-exchange red perovskite quantum dots with ammonium iodine salts for highly efficient light-emitting devices. *Nature Photonics* 12: 681–687.

114 Hou, S., Gangishetty, M.K., Quan, Q. et al. (2018). Efficient blue and white perovskite light-emitting diodes via manganese doping. *Joule* 2: 2421–2433.

115 Li, J., Duan, C., Zhang, Q. et al. (2023). Self-generated buried submicrocavities for high-performance near-infrared perovskite light-emitting diode. *Nano-Micro Letters* 15: 125.

116 Chen, Z., Li, Z., Chen, Z. et al. (2021). Utilization of trapped optical modes for white perovskite light-emitting diodes with efficiency over 12%. *Joule* 5: 456–466.

117 Bai, W., Xuan, T., Zhao, H. et al. (2023). Perovskite light-emitting diodes with an external quantum efficiency exceeding 30%. *Advanced Materials* 2302283.

118 Zhou, W., Shen, Y., Cao, L.X. et al. (2023). Manipulating ionic behavior with bifunctional additives for efficient sky-blue perovskite light-emitting diodes. *Advanced Functional Materials* 33: 2301425.

119 Jiang, J., Chu, Z., Yin, Z. et al. (2022). Red perovskite light-emitting diodes with efficiency exceeding 25% realized by co-spacer cations. *Advanced Materials* 34: 2204460.

120 Wang, Z., Zheng, S., Teng, Q. et al. (2023). Opportunity of lead-free metal halide perovskites for electroluminescence. *The Innovation Materials* 1: 100015.

121 Tsai, H., Nie, W., Blancon, J.C. et al. (2018). Stable light-emitting diodes using phase-pure Ruddlesden–Popper layered perovskites. *Advanced Materials* 30: 1704217.

122 Yuan, Y. and Huang, J. (2016). Ion migration in organometal trihalide perovskite and its impact on photovoltaic efficiency and stability. *Accounts of Chemical Research* 49: 286–293.

123 Zou, W., Li, R., Zhang, S. et al. (2018). Minimising efficiency roll-off in high-brightness perovskite light-emitting diodes. *Nature Communications* 9: 608.

2

Synthesis and Preparation of Perovskite Materials

2.1 Introduction

Metal halide perovskite materials were first discovered in 1893, but it was not until the 1990s that they caught the attention of the scientific and engineering communities [1]. The potential of perovskite materials in solar cells began to be extensively reported in 2012. At present, the photoelectric conversion efficiency of perovskite solar cells has exceeded 26.1% (2023.07.18 U. of Science and Technology of China, from website "https://www.nrel.gov/pv/interactive-cell-efficiency.html"). These materials boast exceptional optical and electrical properties, including high charge-carrier mobility, long free-carrier diffusion length, high photoluminescence quantum yield (PLQY), tunable bandgap, and solution-processability [2–9]. Demonstrating remarkable performance, perovskite materials have found a wide range of applications, from solar cells, photodetectors, and field-effect transistors to gas sensors, resistive-switching memory devices, lasers, and light-emitting diodes (LEDs) [10] (Figure 2.1). Notably, perovskites not only separate charges to generate electricity through photovoltaic means but also aggregate charges to emit electroluminescence (EL). With their relatively low non-radiative recombination rates, perovskite materials hold promise for greater energy efficiency and high color purity.

Regrettably, the low PLQY of three-dimensional perovskite structures is hindered by two significant factors: (1) non-radiative recombination resulting from bulk defects, and (2) a low exciton binding energy, which contributes to reduced electron–hole recombination rates. In response to these issues, researchers have adopted a synthetic strategy that transforms bulk perovskite materials into nanocrystals. This approach significantly reduces the density of electron defects in the perovskite materials and markedly enhances the fluorescence quantum efficiency of the perovskite films. However, this solution is not without its drawbacks. The introduction of organic cations or ligands has inadvertently caused a decrease in the films' electrical conductivity due to a reduction in dimensionality. This decrease in conductivity poses a new challenge to the fabrication of luminescent devices.

Perovskite Light Emitting Diodes: Materials and Devices, First Edition. Hong Meng.

(a) (b)

Figure 2.1 Applications of perovskite materials: (a) solar cells, Source: ABCDstock/Adobe Stock; (b) light-emitting diodes. Source: Igor Stepovik/Adobe Stock Photos.

2.2 Perovskite Materials Structures

2.2.1 3D Halide Perovskite Materials for Light-Emitting Diodes

The three-dimensional structure of halide perovskite materials is shown in Figure 2.2, consisting of ABX_3 cells, where the A is located at the corner position, the B is located at the center of the body, and the X is positioned at the center of the surface. The formation of a stable three-dimensional cubic phase depends on the Goldschmidt tolerance factor ($\boldsymbol{t}$) and the octahedral factor ($\boldsymbol{\mu}$). They are determined using Eqs. (2.1) and (2.2) as follows [10]:

$$\boldsymbol{t} = \frac{(R_a + R_x)}{(R_b + R_x)\sqrt{2}} \tag{2.1}$$

$$\boldsymbol{\mu} = \frac{R_b}{R_X} \tag{2.2}$$

where R_a, R_b, and R_x are the effective ionic radii of A, B, and X in the perovskite, respectively.

For oxide perovskite, $0.8 \le \boldsymbol{t} \le 1$, while for halide perovskite, $0.8 \le \boldsymbol{t} \le 1.11$; and for both oxide perovskite and halide perovskite, $\boldsymbol{\mu}$ must fall between 0.44 and 0.9 to form ABX_3 perovskite-like structures. When $0.9 \le \boldsymbol{t} \le 1$, a cubic crystal structure of perovskite is formed, and when $0.8 \le \boldsymbol{t} \le 0.89$, a distorted perovskite with a diamond, orthogonal, or tetragonal structure is formed. Figure 2.2b shows a plot of $\boldsymbol{t}$ and $\boldsymbol{\mu}$ of perovskite with stable perovskite phases.

2.2.2 Layered Halide Perovskite Materials

Derived from 3D perovskite, 2D perovskite has the basic formula $A'_2A_{n-1}MX_{3n+1}$. The formula of an ideal 2D perovskite can be expressed as $(R\text{-}NH_3)_2MX_4$, which is called a Ruddlesden–Popper perovskite (Figure 2.3). Generally, A is a long-chain organic molecule and plays a crucial role in separating the 3D structure into a 2D structure. The thinnest 2D perovskite was reported to be a single layer with

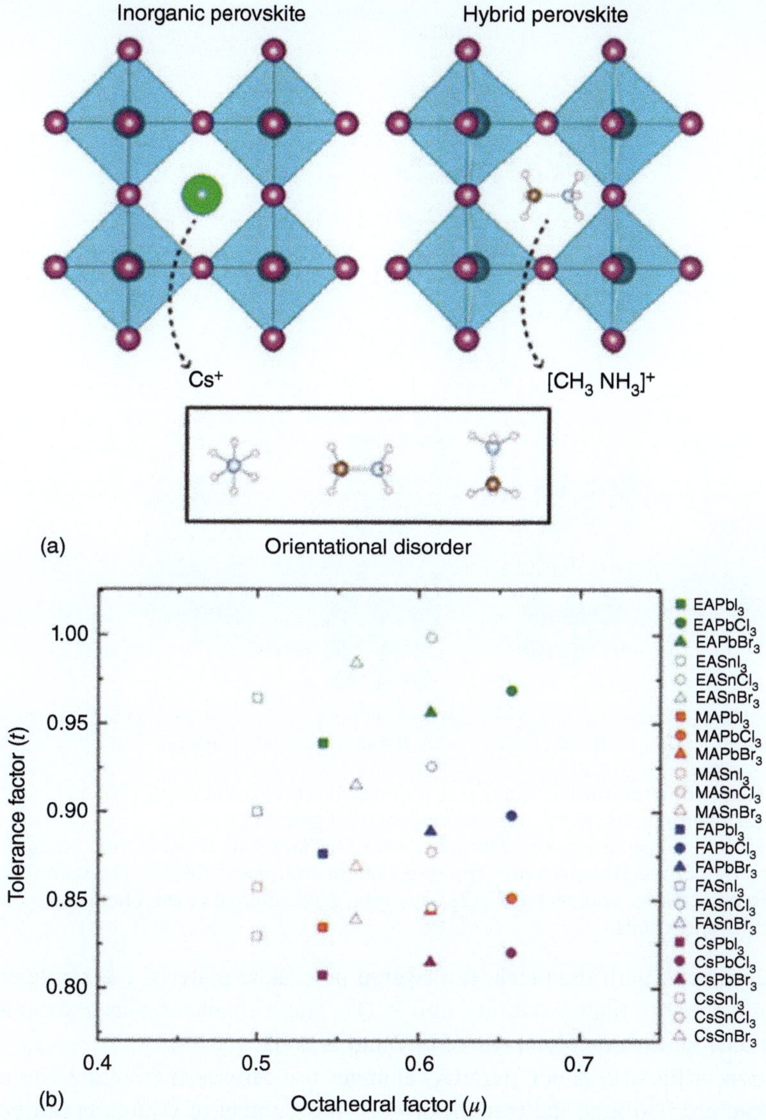

Figure 2.2 (a) Illustration of the perovskite structure based on corner-sharing BX_6 octahedra with either a monovalent metal (inorganic) or charged molecule (hybrid) at the center of the unit cell. For hybrids, there is an orientation dependence on the central cation. Source: Brivio et al. [2] Copyright, 2013, APL Materials. (b) a plot of the t and μ of perovskite with stable perovskite phases. Source: Chen et al. [11] Copyright, 2015, Nano Today.

a crystal structure oriented along the ⟨100⟩ direction, and the number of layers can be tuned by carefully choosing the organic cations. In 1994, Mitzi et al. reported the first layer-controllable perovskite, which had the chemical structure $(C_4H_9NH_2)_2(C_3NH_3)_{n-1}$ ($n = 1$–$5, \infty$) [13]. It is worth noting that layered perovskite exhibits different electrical properties, gradually transitioning from semiconductors

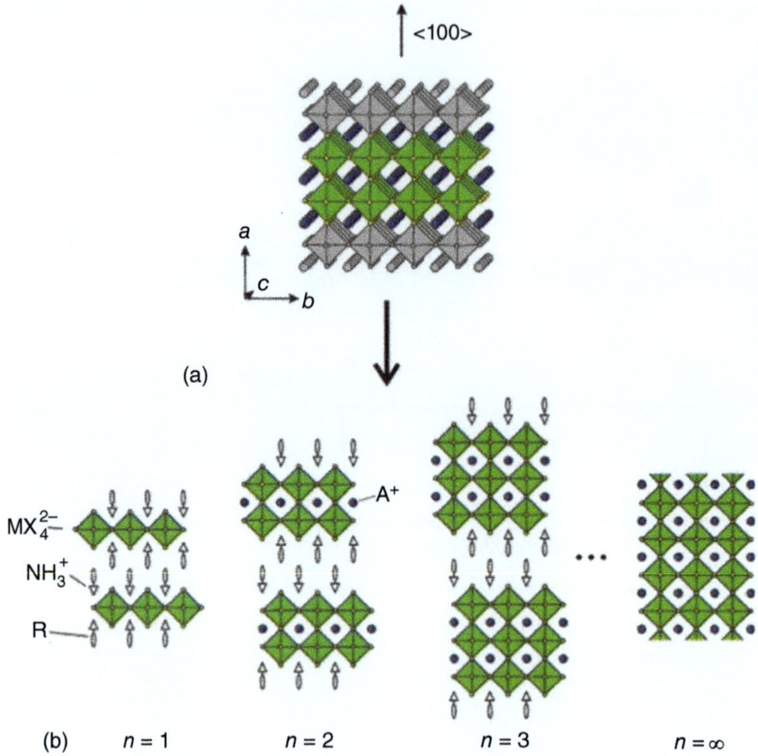

Figure 2.3 Family of intergrowth compounds formed from the parent AMX_3 (3D) and $(RNH_3)_2MX_4$ (2D) perovskites. The ⟨100⟩-oriented layered perovskite series, $(RNH_3)_2A_{n-1}MnX_{3n+1}$, can also be conceptually derived by: (a) taking n layers from along the ⟨100⟩ direction of the 3D perovskite structure, and (b) alternating these layers with bilayers of organic cations. Source: Mitzi [12] Copyright, 2001, Journal of the Chemical Society, Dalton Transactions.

to metals. Compared with the block, the layered perovskite material has stronger exciton binding energy, higher stability, and PLQY. These characteristics make it a strong candidate for the emission layer of LED lights [14].

The addition of PEA (PEA = $C_8H_9NH_3$) changes the perovskite from a 3D to a quasi-2D structure. Through the content of PEA, the number of layers (n) can be achieved from $n = 1$ to ∞.

Figure 2.4a shows that the conduction-band maxima (CBM) significantly decreased with a decreasing number of layers, while the valence-band maxima (VBM) remained unchanged. The optimized PLQY was up to 10% when $n = 3$–5, as shown in Figure 2.4c,b. The quasi-2D perovskite was not a homogeneous phase but rather contained multiple phases. For example, the film fabricated with an expectation of $n = 3$ is composed of $n = 2$, 3, 4, and 5, which was also found in the film fabricated with an expectation of $n = 5$ (Figure 2.4d). The multiple phases in the quasi-2D halide perovskite film may induce an energy funnel, facilitate charge transport, and increase the PLQY to improve the performance of the LEDs (Figure 2.4e).

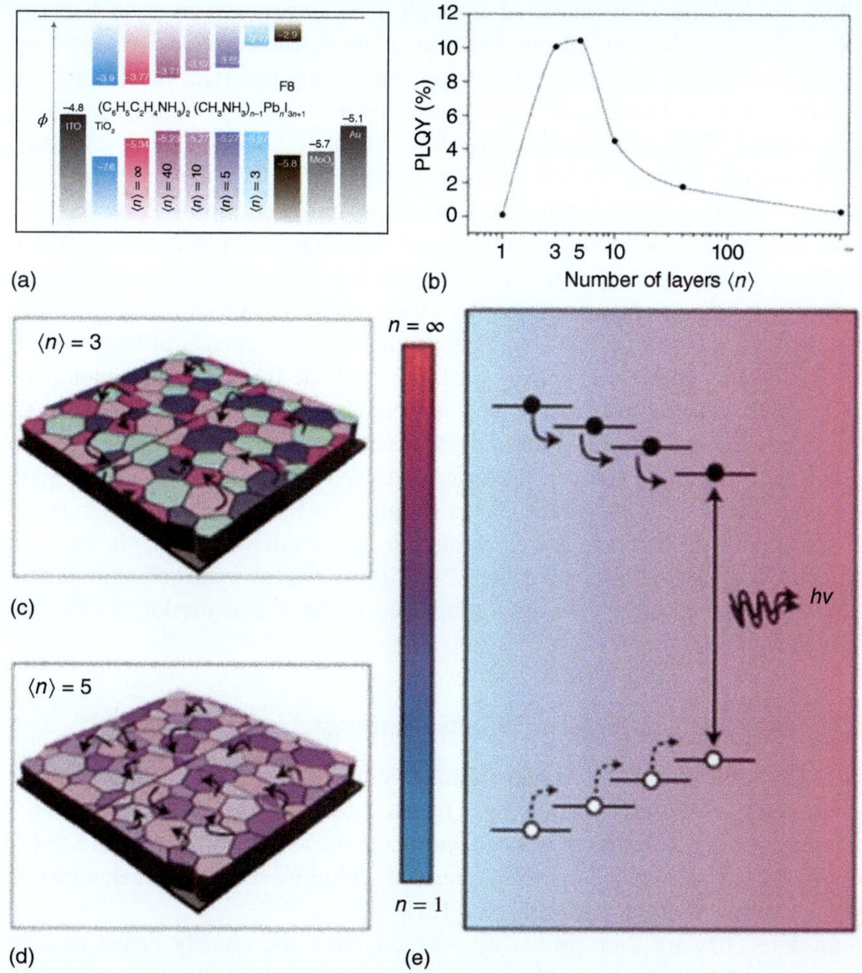

Figure 2.4 (a) Electronic band structure of perovskites with different n values, combined with the band structure of ITO, TiO_2, F8, MoO_3, and the Au electrode. Φ is the electric potential. (b) Summary of the PLQYs for perovskite films with different n values at a low excitation intensity (6 mW cm^{-2}). (c) The carrier-transfer process in the $n = 3$ perovskite. (d) Carrier-transfer process in the $n = 5$ perovskite. (e) Multiphase perovskite materials $PEA_2(CH_3NH_3)_{n-1}Pb_nI_{3n+1}$ channel energy across an inhomogeneous energy landscape, concentrating carriers to the smallest bandgap emitters. The arrows represent the carrier-transfer process. Source: Yuan et al. [15] Copyright, 2016, Nature Nanotechnology.

In addition to Ruddlesden–Popper perovskite, perovskite nanoplates and nanosheets (NSs) based on ABX_3 structure have also been manufactured and used as emitters for LEDs. Gas-phase growth using chemical vapor deposition and solution-phase growth induced by ligand-mediated size reduction are the two commonly used methods for obtaining perovskite nanoplates/NSs. Ha et al. reported for the first time the gas-phase growth of $MAPbX_3$ nanoplates [16]. First, PbX_2 nanoplates were deposited on Muscovite by chemical deposition. Then,

$MAPbX_3$ nanoplates were obtained through a heterogeneous reaction between MAX (gas) and PbX_2 nanoplates. Because of their uncontrollable morphology and poor uniformity, their use in LEDs has been rarely reported. In contrast, the application of two-dimensional perovskites grown in solution in LED has been widely studied. Tyagi et al. successfully obtained single-layer $MAPbBr_3$ NSs using octylammonium bromide as a long-chain organic ligand for the first time [17]. Generally, $PbBr_2$, methyl ammonium bromide, and octylammonium bromide (OABr) are added to the mixed solution of oleic acid (OA)and octadecene (ODE) at 80°C. The preparation method for $MAPbBr_3$ NSs is to quench the prepared mixture in acetone and then collect it by centrifugation. Figure 2.5a,b show the synthesis principle diagram of two-dimensional perovskite and TEM images of NSs at $n = 1$. By using these 2D perovskites, Ling et al. reported the first perovskite-nanoplate LEDs with a brightness of 10,590 cd m^{-2}, suggesting the potential of these materials in EL devices [18]. The structure and EL spectra of the LEDs are displayed in Figure 2.5c,d. Although the performance of 2D $FAPbBr_3$ is not particularly high, this study represents the "greenest" led (Figure 2.5f) with color coordinates of (0.168,0.773) of the International Commission on Illumination (CIE) in 1931. In addition, a large area of super-flexible LED with a bending radius of 2 mm is also displayed, which indicates the important role of layered perovskite in flexible display devices.

2.2.3 Halide Perovskite Quantum Dots/Nanocrystals

In 2012, Kojima et al. observed photoluminescence from perovskite quantum dots (perovskite QDs) for the first time [20]. In 2016, Dirin and co-workers reported a facile synthesis of perovskite QDs based on mesoporous SiO_2 templates with a PLQY of up to 50%. As shown in Figure 2.6, green photoluminescence was generated by spinning $CsPbBr_3$ onto mesoporous SiO_2 [21].

Perovskite QDs are then grown by solution methods, namely ligand-assisted precipitation (LARP) or thermal injection method at room temperature [22, 23]. The schematic diagram of synthesizing perovskite QDs through LARP is shown in Figure 2.7a [23]. Specifically, including AX and BX_2, perovskite precursors are dissolved in DMF along with long-chain organic ligands such as OA and n-octylamine. Then add the mixture dropwise to an anti-solvent solution, such as toluene. Dropping the perovskite precursor into toluene immediately observed the precipitation of strong photoluminescence particles, indicating the formation of QDs. This method can be applied to a wide range of high-pressure materials. However, due to the high reaction temperature, fully inorganic perovskites such as $CsPbX_3$ are usually prepared by thermal injection method [22]. Usually, CsX first reacts with OA in octadecene (ODE) solvent to form OA Cs at 150°C. The second step is to dissolve PbX_2 salt in ODE by reacting with OA and OAM at 120°C. Further, increase the temperature to 140–200°C, and then inject OA Cs to obtain perovskite QDs. The schematic diagram of the synthesis process is shown in Figure 2.7d [24]. Regardless of the synthesis method used, the obtained perovskites showed a high PLQY of up to 90%. The optical images and PL spectra of various

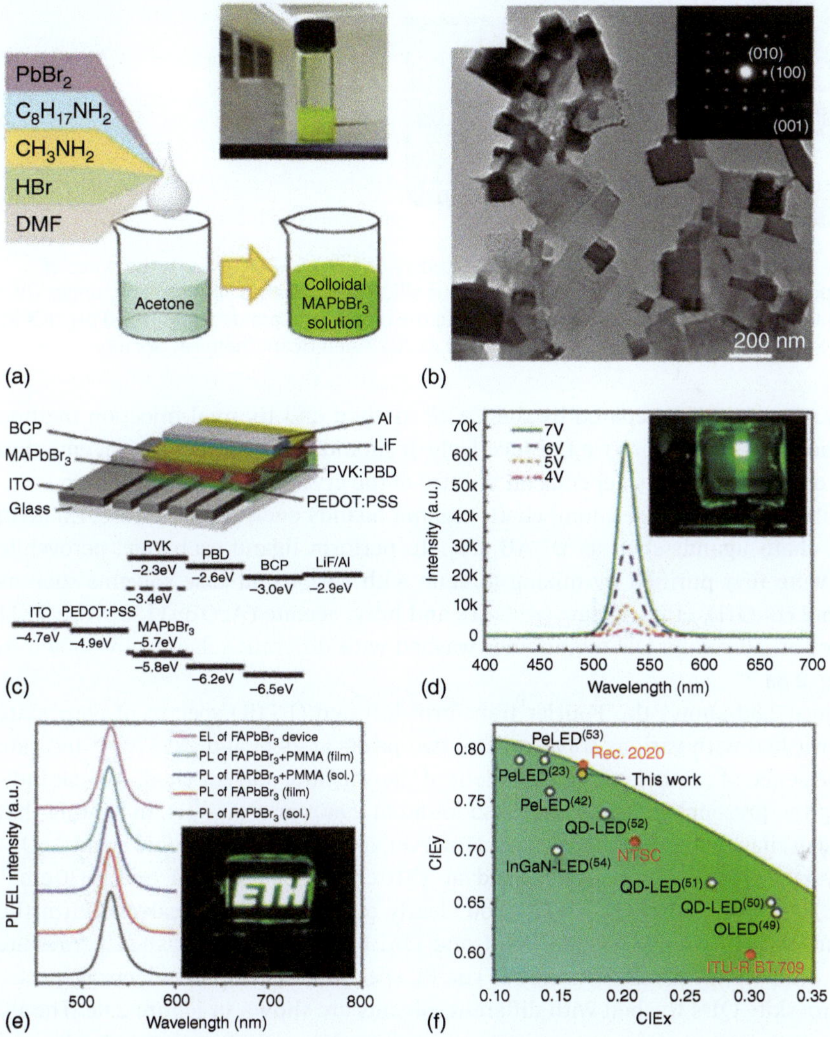

Figure 2.5 (a) Scheme of the one-pot synthesis of $MAPbBr_3$ nanoplatelets. Source: Photo from achieved colloidal solution. (b) TEM images and selected-area diffraction (SAED) patterns of colloidal MAPbBr3 nanoplatelets. (c) Device architecture for a perovskite with perovskite nanoplatelets as the emitter and the flat-band energy level diagram of the different layers of the materials, showing the conduction and valance band levels with respect to vacuum. (d) Electroluminescence spectrum from a standard device with a PVK:PBD weight ratio of 64/36 at an applied voltage of 4–7 V. Inset: photograph of a working device at 7 V. Source: (a)–(d) Ling et al. [18] Copyright, 2015, Advanced Materials. (e) PL spectra for the colloidal 2D $FAPbBr_3$ perovskites in toluene solution (black), thin film (red), a mixture with poly(methyl methacrylate) (PMMA) in toluene solution (blue), and a complex with PMMA in film (green). The EL spectrum of the LED device is also shown for comparison (pink). (f) The color coordinates of the LED device are based on the 2D $FAPbBr_3$ perovskites (this work) in the CIE 1931 color space. Source: (e, f) Kumar et al. [19] Copyright, 2017, Nano Letters.

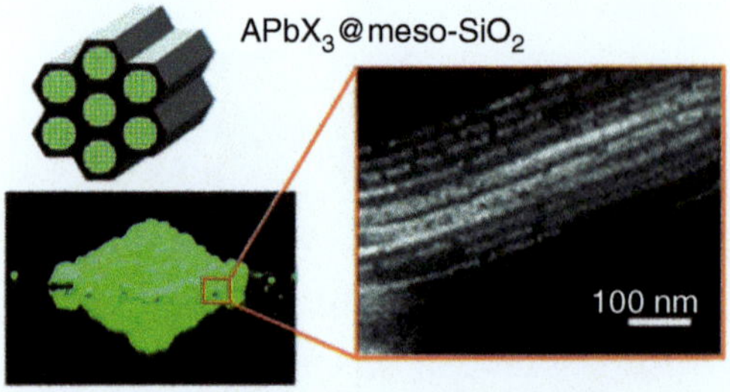

Figure 2.6 Schematic of the template-assisted synthesis of $APbX_3$ NCs in the pores of mesoporous silica. Photographs of mesoporous silica impregnated with $CsPbBr_3$ under UV illumination. HAADF-STEM images illustrating the effect of beam damage on $CsPbI_3$ NCs in 7 nm-SiO_2. Source: Dirin et al. [21]. Copyright © 2022 American Chemical Society.

types of perovskite prepared by the LARP method and thermal injection method are shown in Figure 2.7b,c,e,f, respectively. It should be noted that the PL emission peak of perovskite QDs depends on the size of the crystal and its composition.

Chiba et al. exchanged long-chain organic ligands such as OA and OAM with short chain ligands such as DDAB [25]. To perform ligand exchange, perovskite QDs were first purified by mixing toluene with various adverse solvents such as butanol (BuOH), ethyl acetate (AcOEt), and butyl acetate (AcOBu) [25]. The TEM images of perovskite QD membranes washed with different solvents are shown in Figure 2.8a–c.

Figure 2.8d shows the Fourier transform infrared (FTIR) spectra of perovskite QDs washed with various solvents. The two peaks at 2923 and 2855 cm^{-1} indicate the presence of methylene, independent of the washing solvent used. A peak indicating the presence of DDAB was also found at 2960 cm^{-1} (C—H), indicating that ligand exchange has been completed. However, in the spectra of BuOH – and AcOEt perovskite QDs, FTIR peaks located at 1710 cm^{-1} (C=O), 3004 and 1630 cm^{-1} (C=C), and 3310 cm^{-1} (N—H) can be clearly observed. These peaks confirm the presence of long alkyl ligands OAM and OA, but for AcOBu-washed perovskite QDs, the FTIR peaks are weakened. The PL spectra and UV photoelectron spectra of perovskite QDs washed with different solvents are shown in Figure 2.8e. The PL peaks of BuOH-, AcOEt-, and AcOBu-perovskite QDs were found at 517, 512, and 513 nm, respectively. The observed redshift is attributed to the different particle sizes of perovskite QDs affected by the washing solvent. Compared with BuOH- (20 nm) and AcOEt-perovskite QDs, AcOBu-perovskite QDs have the smallest PL spectrum and a FWHM of 19 nm. This is attributed to their greater monodispersity. These results indicate that AcOBu is the most effective washing solvent for removing OA and OAM. AcOEt-washed perovskite QD LEDs exhibit the best performance, with a current efficiency of 18.8 cd A^{-1} and an EQE of 8.73%.

Li et al. employed trimethylaluminum (TMA) as a cross-linking agent to achieve the formation of a perovskite QD film [26]. The TMA-treated QDs improve the surface coverage and improve the EQE of the LEDs by 5.7%. The morphologies of the perovskite QD films with and without the TMA treatment are shown in Figure 2.8g,h. Figure 2.8i provides a schematic diagram of the TMA cross-linking reaction.

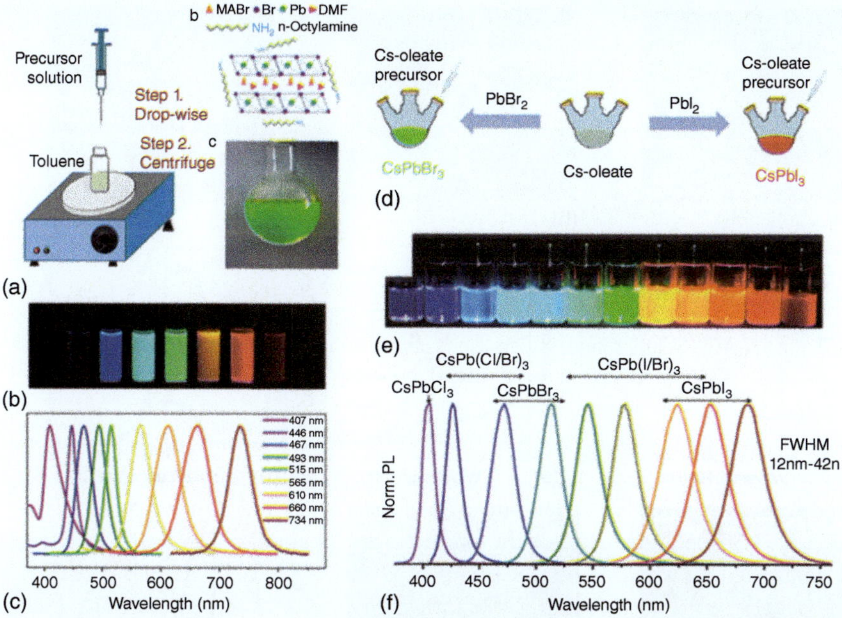

Figure 2.7 (a) Schematic illustration of the reaction system and process for the LARP technique. (b) Optical images of $CH_3NH_3PbX_3$ QDs under ambient light and a 365 nm UV lamp. (c) PL emission spectra of $CH_3NH_3PbX_3$ QDs. Source: (a–c) Zhang et al. [23] Copyright, 2015 ACS Nano. (d) Schematic of the synthesis of $CsPbX_3$ QDs via the hot-injection method. Source: Wang et al. [24] Copyright, 2016, Nanoscale. (e) Optical images of colloidal $CsPbX_3$ QD solutions in toluene synthesized by hot injection method under a UV lamp ($\lambda = 365$ nm). (f) Representative PL spectra ($\lambda_{exc} = 400$ nm for all samples except $CsPbCl_3$, for which $\lambda_{exc} = 350$ nm). Source: (e, f) Protesescu et al. [22] Copyright, 2015, Nano Letters.

2.2.4 Commercial Prospects of Perovskite Materials

Perovskite nanocrystals (PeNCs) have a unique band structure in which defective states are restricted to the valence or conduction bands and are essentially “inert,” resulting in PeNCs exhibiting remarkable efficiency in photoluminescence. As a result, PeNCs are often considered to be “fault tolerant.” In recent years, PeNCs have attained emission wavelengths covering the entire visible spectrum, and their fluorescence quantum yields have reached 100%. However, the fluorescence quantum efficiency of PeNCs varies based on the synthesis method used, and even minor modifications to a given synthesis route can significantly impact their optical properties. The reason for such substantial variations in their optical properties is that these properties are closely related to the stoichiometric ratio of ions in the perovskite structure as well as to the coating ligands. Nonetheless, research in these areas is still in its early stages.

The discovery of quantum size effects in the luminescence spectra of nano-sized semiconductors occurred in the early 1980s. A similar quantum confinement effect is present in metal halide perovskites. In 2011, a groundbreaking paper by Park was published, which explored the utilization of methylammonium lead iodide ($MAPbI_3$) in nanocrystalline form in solar cells [27]. Although not very

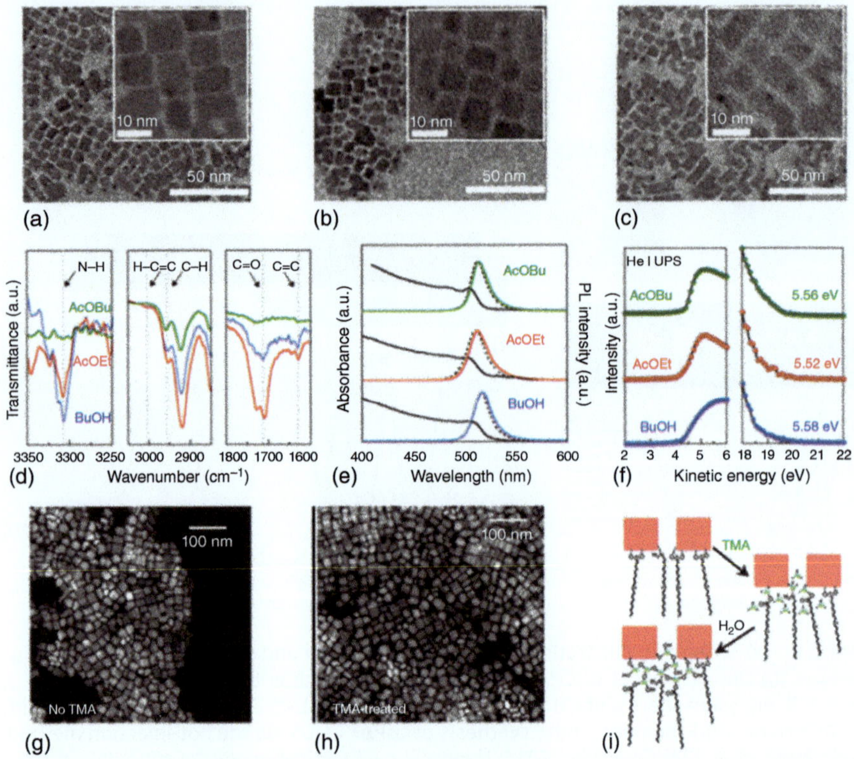

Figure 2.8 (a–c) TEM images of perovskite QD films washed with BuOH (a), AcOEt (b), and AcOBu (c). (d,e) FTIR spectra (d), and UV–vis absorption and PL spectra (e) of the perovskite QD films (dashed line: PL spectra of the solution), and UPS spectra of perovskite QDs (f). (g,h) High-angle annular dark-field scanning TEM (HAADF-STEM) images of untreated (g) and TMA-treated (h) $CsPbI_3$ nanocrystals, used for size-distribution analysis. (i) Reaction schematic of the TMA cross-linking process. Source: (a–f) Chiba et al. [25] Copyright, 2017, Applied Materials. (g-i) Li et al. [26] Copyright, 2016, Advanced Materials.

efficient initially, the potential of this material captured the interest of the materials chemistry community, leading to investigations into this direct bandgap semiconductor. A few years later, Schmidt et al. published the first colloidal solution-based method for preparing PeNCs. By the end of 2013, they had successfully synthesized $MAPbBr_3$ nanocrystals with a fluorescence quantum efficiency of 20% [22]. In 2015, Protesescu et al. demonstrated the synthesis of monodisperse $CsPbX_3$ nanocrystals using the standard thermal injection method commonly used for classical colloidal QDs like CdSe and PbSe. This discovery highlighted the excellent optical properties of PeNCs, including high fluorescence quantum efficiency (up to 90% without any post-processing), spectral half-peak (less than 100 meV), and the ability to adjust the wavelength of light throughout the visible spectrum simply by altering the type and proportion of halide ions in the structure.

In 2018, scientists developed a perovskite light-emitting diode (PeLED) with an EQE of approximately 20% through defect passivation of polycrystalline metal halide

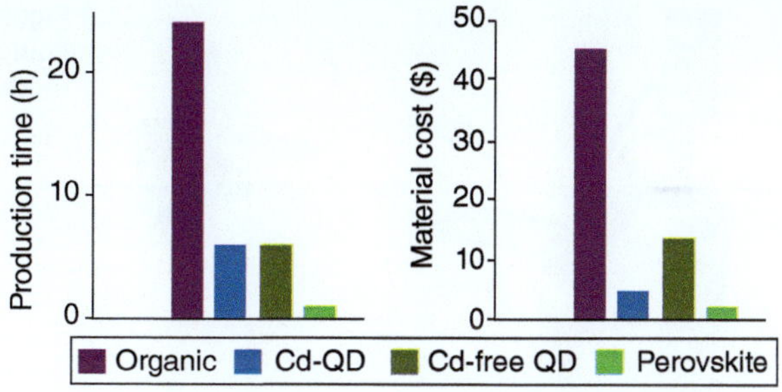

Figure 2.9 The comparison of production time and the precursor prices required for product preparation for perovskite, Cd QDs, Cd-free QDs, and $Ir(ppy)_3$ (organic luminescent materials). Material cost is in US dollars in 2020. Source: Han et al. [28] Copyright, 2022, Nature Reviews Materials.

perovskites (Mperovskites) films, synthesis of nanocrystals, ligand engineering strategies, as well as the use of mixed-dimensional perovskite films [28].

Currently, the majority of research on lead halide PeNCs has centered on three-dimensional $APbX_3$ crystal structures and compositions. However, to produce metal halide PeNCs with controlled size, morphology, and optical properties, various groups have been working on developing simple and dependable synthetic strategies in recent years [29, 30].

Various approaches can be used to synthesize perovskite nanocrystals (PeNCs). Ligand-assisted reprecipitation (LARP) and hot injection are currently common synthesis methods. Triple ligand synthesis, ultrasonic method, grinding method, and high-temperature sid-phase synthesis are other methods of synthesizing PeNCs at room temperature. LARP has the advantages of being fast and inexpensive, making it a good choice for commercial synthesis, as the reaction only takes a few minutes under ambient conditions at room temperature. However, the ligand-assisted precipitation method has drawbacks such as low PLQY and low production capacity. In red PeNCs, I^- strongly interacts with polar solvents; in blue PeNCs, Cl^- introduces deep trapped states and therefore has low solubility. In response to this, less coordinative polar solvents are used to improve the PLQY and stability of red PeNCs, and bringing soluble Cl-based precursors in nonpolar solvents to improve PLQY of blue PeNCs [28].

Using organic materials to produce emission layers will take several days. Due to slow reaction and shell-growing factors during the production process, the production of QDs will take about six hours. In contrast, the production time of perovskite materials is extremely short. Using LARP, each reaction takes approximately 10 minutes. It takes approximately one hour to use a hot injection. (including the precursor preparation and reaction time required for LARP and hot injection, respectively). The comparison of production time is shown in Figure 2.9.

We can also find advantages of perovskite materials based on the precursor prices required for product preparation. The average cost of precursor required to produce $CsPbBr_3$ perovskite is \$2/g, while the average cost of precursor required to produce Cd QDs is \$5/g. The cost of precursor required to produce Cd-free QDs is \$14/g, and the cost of the precursor required to produce $Ir(ppy)_3$ (organic luminescent materials) is \$45/g.

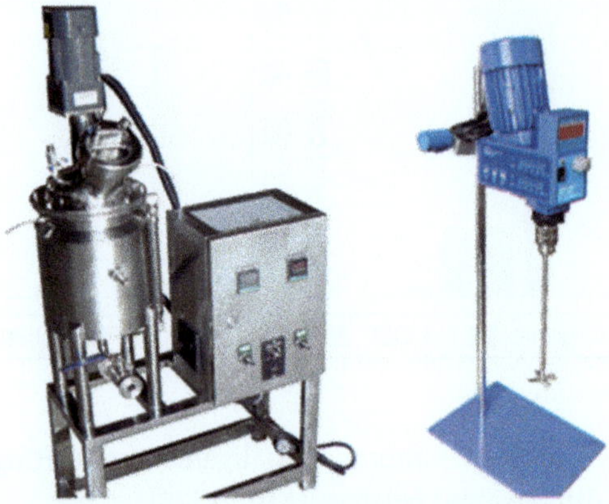

Figure 2.10 Commonly used equipment for mechanochemical methods.

This chapter will explore several popular methods used to synthesize metal halide PeNCs, including mechanochemical, spin-coating, ultrasonic, chemical exfoliation, thermal injection, and LARP.

2.3 Preparation of Perovskite Nanomaterials

2.3.1 Mechanochemical Method

In recent years, mechanochemistry has emerged as a method of synthesis that involves chemical reactions initiated by mechanical energy. This energy can come from several forces and modes of operation, such as impact, compression, or shear. In this chapter, we will focus on the mechanochemical synthesis of metal halide PeNCs, which can be achieved by grinding or milling their precursors. Examples include manual grinding with a mortar and pestle, as well as grinding with a shaker or planetary ball mill, or mechanical grinding by other methods such as twin-screw extruders (Figure 2.10). The first recorded example of a mechanochemical synthesis was in 1820, when Faraday reduced AgCl to Ag using solid zinc, copper, tin, or iron by grinding with a pestle and mortar without the use of a solvent.

Mechanochemistry is an emerging synthesis method that involves initiating chemical reactions through mechanical energy. This energy can come from various modes of operation, such as impact, compression, or shear. The mechanochemical synthesis of perovskites involves grinding or milling their precursors, using methods such as manual grinding with a mortar and pestle or mechanical grinding with a shaker or planetary ball mill. The exact mechanism of mechanochemistry is not yet fully understood, but it is believed that the grinding-induced particle size reduction exposes the active surface of the reactants, creating a suitable chemical reaction environment and promoting the reaction. Although mechanochemical methods are simple to operate, they have been neglected compared to other synthetic methods such as solvent synthesis, supercritical CO_2-assisted chemistry, or ionic liquids [32]. However, recent studies and reviews suggest that mechanochemical methods are gaining attention, and one of their main advantages is that they do not

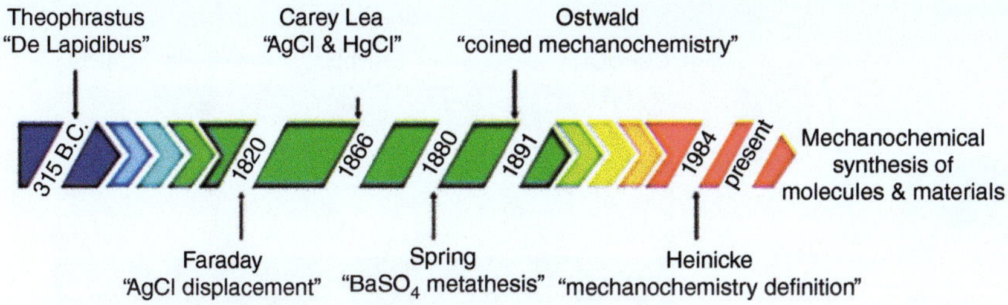

Figure 2.11 History of the development of mechanochemical methods. Source: Tan and Garcia [31] Copyright, 2019,Chemical Society Reviews.

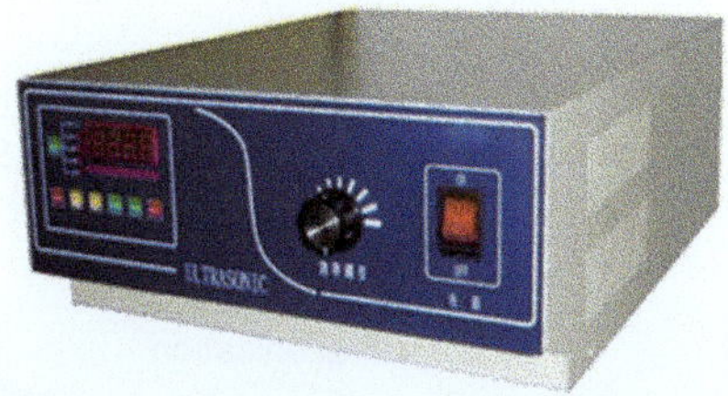

Figure 2.12 Common ultrasonic generators. Source: vladim_ka/Adobe stock.

require solvents, making them more environmentally friendly. Mechanochemical methods have been used to synthesize a variety of functional materials, including organic materials, exfoliated graphene, metal–organic skeletons, and metal alloys (Figure 2.11).

2.3.2 Ultrasonic Method

Ultrasound is a high-frequency sound wave with a frequency of 10^4–10^9 Hz. When applied to liquids, it generates strong tensile stresses that can break up and disperse particles, making it a useful tool for the preparation of small particles (Figure 2.12). Ultrasonic methods have been used since the early days of nanotechnology, for example, to prepare homogeneous iron nanoparticles using volatile organometallic iron compounds like $Fe(CO)_5$ as raw material and stabilizers like polyvinylpyrrolidone or OA. The resulting amorphous iron nanoparticles had a homogeneous particle size of approximately 8 nm.

In 2016, Feldmann's group demonstrated the synthesis of $CsPbX_3$ (X = Cl, Br, I) QDs using ultrasonication [33]. They prepared cesium oleate by reacting $Cs_2\ CO_3$ with OA under sonication and then adding PbX_2 in oleylamine solution to form perovskite. By replacing the halogen anion of perovskite, they achieved full-spectrum luminescence in the visible range of 400–700 nm, demonstrating the versatility of this synthesis method (Figure 2.13).

Huang's group employed a two-step process using oleic acid and oleylamine as a non-polar solvent system to synthesize perovskite nanocrystals with a fluorescence efficiency of 72% for green perovskite [34] (Figure 2.14).

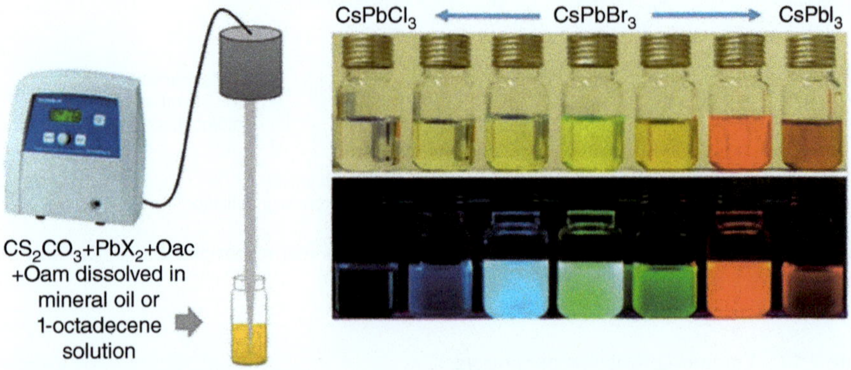

Figure 2.13 Ultrasound method for full-spectrum luminescence. Source: Tong et al. [33] Copyright, 2016, Angewandte Chemical Society Reviews.

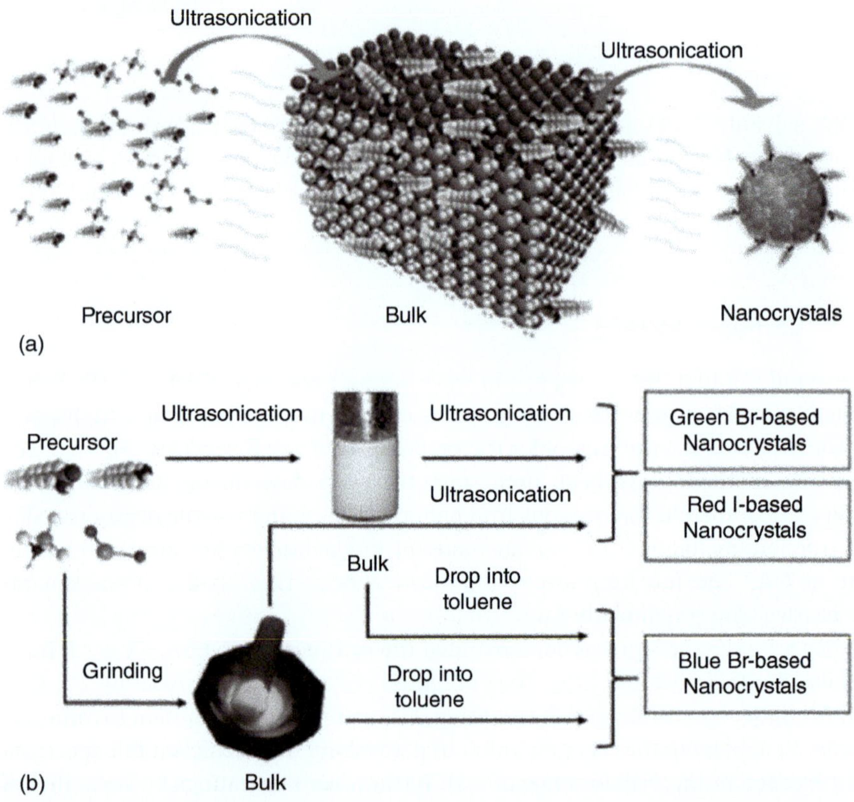

Figure 2.14 Ultrasonic two-step synthesis of perovskite nanocrystals: (a) schematic diagram; (b) flow chart. Source: Huang et al. [34] Copyright, 2017, Angewandte Chemical Society Reviews.

Figure 2.15 Common microwave reaction equipment.

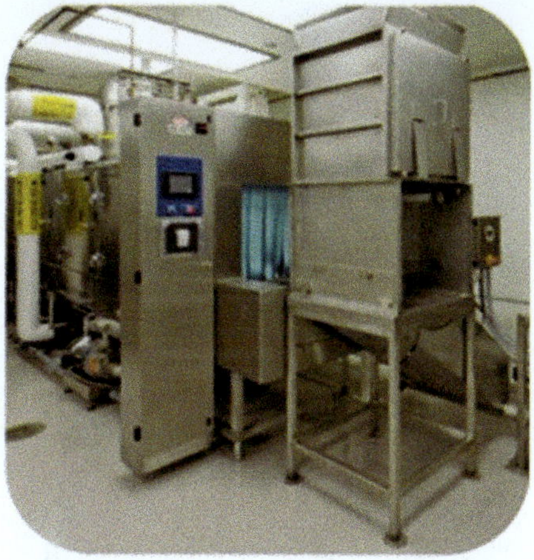

2.3.3 Microwave Method

Microwaves are electromagnetic waves with a frequency range of 300 kHz to 300 MHz, possessing properties such as penetration, reflection, and absorption. These waves can directly heat the contents of an object, without relying on the heat transfer of the surrounding material. Due to this, the heating is more uniform and rapid compared to other heating methods (Figure 2.15).

Microwave-assisted synthesis has been successfully used to prepare high-quality PeNCs with strong emission, such as CdSe nanocrystals, in both organic solvents and water. Recently, researchers have also reported the synthesis of metal halide perovskites using microwave radiation. In 2017, Rogach's group employed a microwave-assisted approach to synthesize $CsPbBr_3$ PeNCs and investigated the mechanism of nanocrystal formation. Microwave methods have been proven to be efficient and rapid and are increasingly being explored as a promising alternative to traditional synthetic methods [35]. Rogach's study investigated the effect of microwave power and reaction time on the synthesis of $CsPbBr_3$ PeNCs. They found that a 50-second reaction time at a certain power level led to optimal crystal formation, with shorter or longer reaction times affecting crystal morphology. Through SEM analysis, they concluded that microwave power affects the reaction rate and intermediate products in perovskite nanoparticle growth. The study also proposed a mechanism for the formation of $CsPbBr_3$, where lead and bromine ions first form a $PbBr_6$-positive octahedron, which is then filled with cesium ions (Figure 2.16).

Zhang's group observed that the morphology of perovskite crystals synthesized using microwave methods is temperature-dependent and has a significant impact on the morphology of perovskite crystals synthesized using the microwave method [36]. At a reaction temperature of 80°C, platelet structures were formed, while at elevated temperatures up to 160°C, rod-like structures were favored. They concluded that the accelerated dissolution of precursors at higher temperatures led to faster reaction kinetics, resulting in the formation of rod-like structures. They

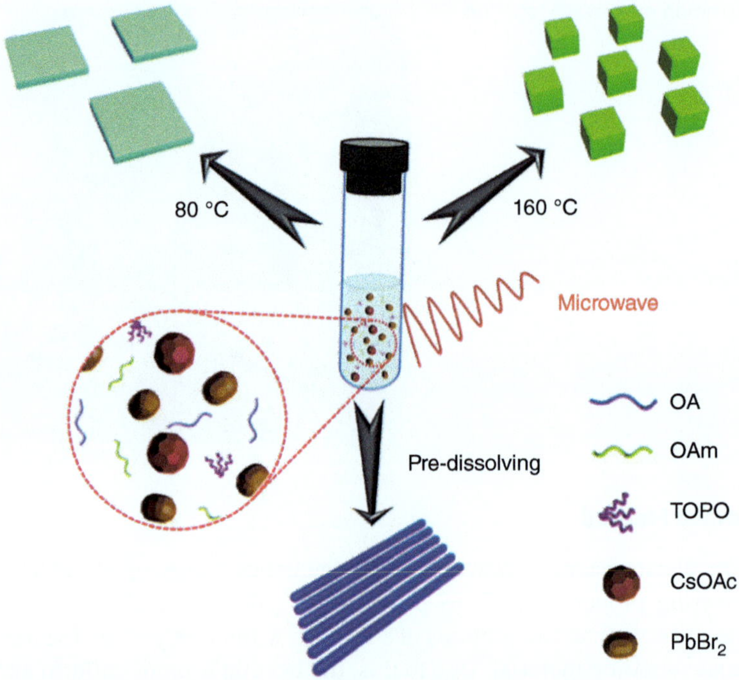

Figure 2.16 Effect of temperature on the preparation of $CsPbBr_3$ nanocrystals by microwave method. Source: Li et al. [35] Copyright, 2017, Journal of Materials Chemistry C.

suggested that pre-dissolving the precursors could further tune the reaction kinetics and promote the formation of rod-like structures (Figure 2.17).

2.3.4 Solvent Heat Method

Solvothermal is a synthesis method developed based on the hydrothermal method, where the original mixture is reacted in a closed system such as an autoclave using organic or nonaqueous solvents at a specific temperature and autogenous pressure. This method differs from the hydrothermal method in that organic solvents are used instead of water. While the hydrothermal method is suitable for the preparation and treatment of oxide functional materials or a few sulfur compounds that are not sensitive to water, the solvothermal method is ideal for preparing and treating water-sensitive compounds such as III–V semiconductors, carbides, fluorides, and new phosphate (arsenate) molecular sieve three-dimensional skeletal structure materials. Solvothermal synthesis has contributed to the creation and development of these materials (Figure 2.18).

The solvothermal method was first reported by Bindy in 1985 in Nature for the synthesis of zeolites using nonaqueous solvents in autoclaves, and it has subsequently gained considerable attention from researchers. In the past few decades, the solvothermal method has been widely recognized as a promising technique for the preparation of various nanocrystals due to its simplicity, precise morphological control, high crystallinity, and regenerative properties. In 2015,

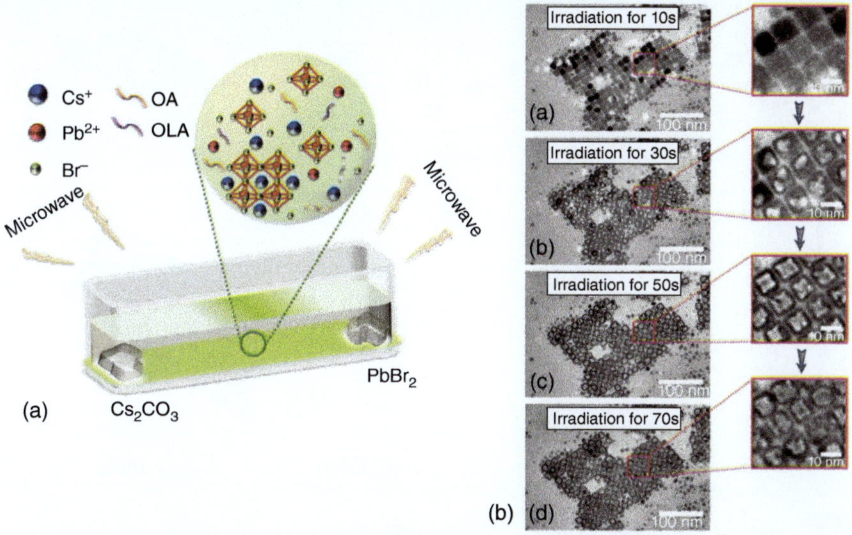

Figure 2.17 Effect of microwave radiation time on the size of $CsPbBr_3$ crystals. Source: Pan et al. [36] Copyright, 2018, Angewandte Chemical Society Reviews.

Figure 2.18 Common hydrothermal reaction equipment.

Yang's group successfully prepared high-quality $MAPbI_3$ perovskite crystals using the solvothermal method [37] (Figure 2.19).

Li's group successfully synthesized Mn-doped $CsPbCl_3$ nanocrystals with two-color light-emitting properties and air stability using the solvothermal method in 2018. These nanocrystals have potential applications in building pure white LEDs and temperature sensor media [38]. The solvent-thermal method has also been used to synthesize new nanocrystals by Tang Jiang's group at Huazhong University of Science and Technology. In particular, they have successfully produced efficient and stable warm white light-emitting devices using the solvothermal method with lead-free halogenated bicalcium $Cs_2Ag_xNa_{1-x}InCl_6$ (Figure 2.20).

2.3.5 Thermal Injection Method

The heat injection method for synthesizing cadmium sulfide nanocrystals was first reported as early as 1993 (the heat injection method was first used to synthesize

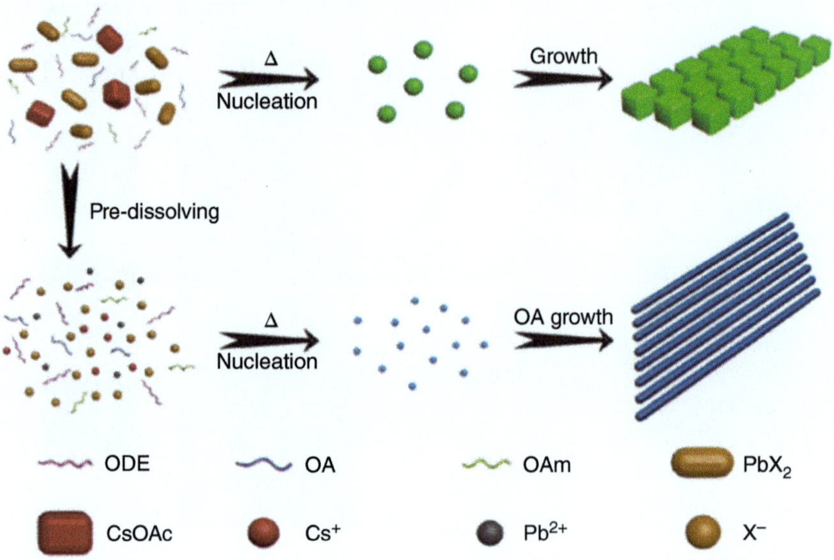

Figure 2.19 Growth process of perovskite materials in the solvothermal method. Source: Chen et al. [37] Copyright, 2017, Advanced Functional Materials.

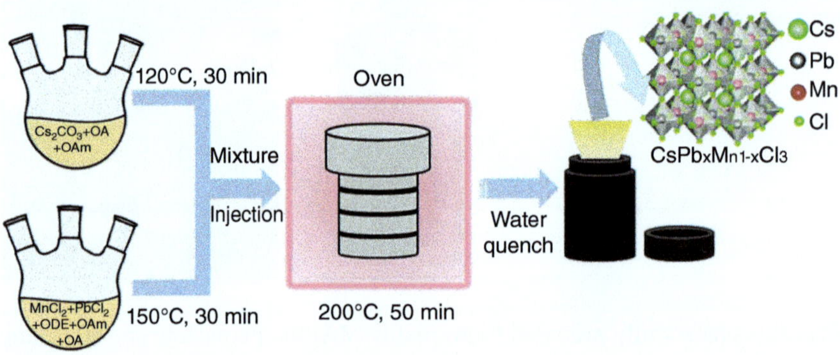

Figure 2.20 Effect of temperature on the synthesis of perovskite nanocrystals by the solvothermal method. Source: Chen et al. [38] Copyright, 2018, Journal of Materials Chemistry C.

cadmium sulfide nanocrystals) [39]. This method involves rapidly injecting a precursor into a high-temperature, high-boiling-point solvent containing the remaining precursors and ligands. The thermal injection method allows the nucleation and growth phases to be separated, resulting in small nanocrystals with a narrow size distribution. After the thermal injection, a rapid nucleation phase occurs, resulting in the formation of a large number of small nuclei. The nucleation phase ends when the monomers are rapidly consumed and the nuclei begin to grow. If the reaction is stopped when a large number of monomers are still present, nanocrystals with a narrow size distribution are produced. To control the size, distribution, and morphology of colloidal nanocrystals by thermal injection synthesis, it is crucial to control the ratio of surfactant to precursor, the injection temperature, the reaction time, and the precursor concentration.

In 2015, Protesescu and colleagues introduced the thermal injection method for the synthesis of $CsPbX_3$ QDs [22]. The main method involves injecting cesium oleate into an ODE solution containing lead halide, carboxylic acid, and primary amine at 140–200°C. Monodisperse nanocrystals are formed by adding a quantitative amount of carboxylic acid and a primary amine. The size of the nanocrystals can be adjusted by varying the reaction temperature. PeNCs with mixed halogen anions can be easily produced by adjusting the ratio of anions in the different halides. The size and ratio of anions can be adjusted to finely modulate the fluorescence emission of the resulting PeNCs throughout the visible region. Additionally, organic-inorganic hybridized $MAPbX_3$ PeNCs can be obtained by using a methylamine solution instead of cesium oleate and modulating the ratio of OA and oleylamine (as ligands) (Figure 2.21).

The thermal injection method has also been applied to the synthesis of perovskite-derived materials. For instance, Cs_4PbX_6 nanocrystals can be synthesized by introducing excess Cs^+ and oleylamine into the reaction environment. Similarly, $CsPb_2Br_5$ can be synthesized by introducing excess $PbBr_2$ and using OA and octylamine as ligands. To better understand the kinetics of the thermal injection method, Lignos et al. utilized a microfluidic platform to study the in-situ absorption and photoluminescence processes of $CsPbX_3$. They demonstrated that nucleation and growth of $CsPbX_3$ occurred within the first five seconds, highlighting the rapid reaction kinetics of these nanocrystals [40] (Figure 2.22).

A different approach was taken by Koolyk et al., who analyzed samples from different stages of the reaction using transmission electron microscopy (TEM) [41]. Koolyk et al. used TEM to analyze samples extracted from various reaction stages, and they observed a size-focused state of the nanocrystals during the nucleation phase that lasted approximately 20 seconds before crystals of different sizes started to form. Additionally, they found that the growth phase did not exhibit a size-focused state and expanded the particle size distribution from the beginning of growth, lasting up to 40 seconds of reaction time. Compared to classical group II–VI (CdTe, CdSe, CdS, HgS, etc.), group IV–VI (PbSe, etc.), and group III–V (InP, InAs, etc.) QDs, the nucleation and growth processes of metal halide perovskites are not well-defined. As ionic crystals that can form quickly in solution, PeNCs nucleate and grow rapidly, typically within one minute, making it challenging to control and homogenize the process. In recent years, the biggest challenge in studying perovskite has been controlling the nucleation and growth processes, which may be overcome by working under thermodynamic equilibrium rather than kinetic control, according to a recent study (Figure 2.23).

Liu et al. proposed a modified "three-precursor" thermal injection method to address some of the issues encountered with the traditional thermal injection method for the synthesis of $CsPbX_3$ nanocrystals. Instead of using PbX_2 as the lead precursor, they utilized PbO and NH_4X as the halogen precursors [42]. The "triple precursor" thermal injection method for the synthesis of $CsPbX_3$ nanocrystals was proposed by Liu et al. to address the issues with the thermal injection method. This method uses NH_4 X and PbO as halogen and lead precursors, respectively, instead of

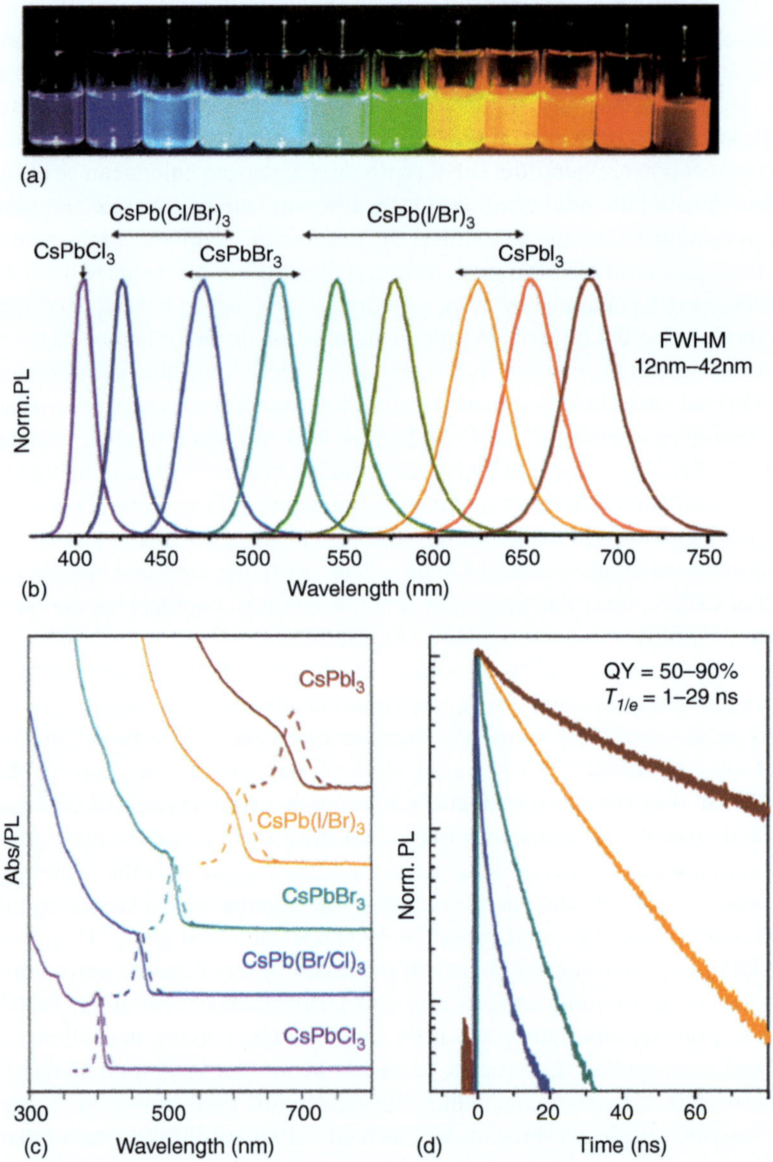

Figure 2.21 Colloidal perovskite $CsPbX_3$ NCs (X = Cl, Br, I) exhibit size- and composition-tunable bandgap energies covering the entire visible spectral region with narrow and bright emission: (a) colloidal solutions in toluene under UV lamp ($\lambda = 365$ nm); (b) representative PL spectra ($\lambda_{exc} = 400$ nm for all but 350 nm for $CsPbCl_3$ samples); (c) typical optical absorption and PL spectra; (d) time-resolved PL decays for all samples shown in (c) except CsPbCl3. Source: Protesescu et al. [22] Copyright, 2015, Nano Letters.

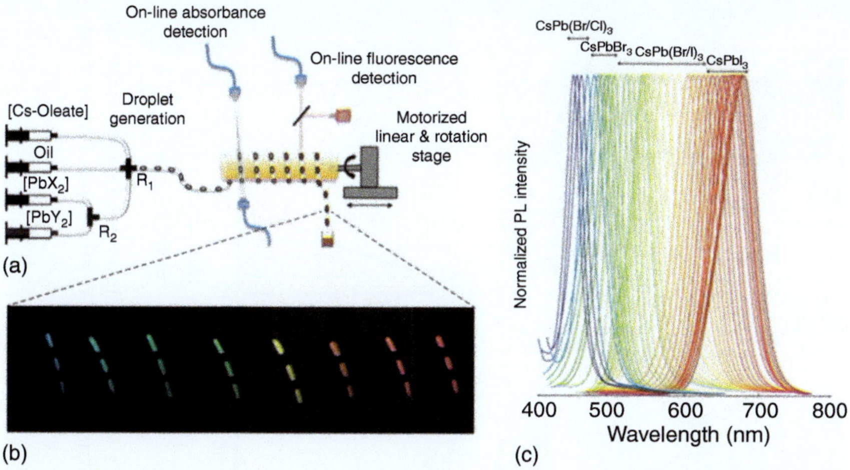

Figure 2.22 (a) Microfluidic platform (b) Fluorescence photograph of $CsPbX_3$ (c) Fluorescence luminescence spectrum of $CsPbX_3$. Source: Lignos et al. [40] Copyright, 2015, Nano Letters.

PbX_2 for the synthesis of perovskites. Excess NH_4 X is used to create a halogen-rich ionic environment, resulting in improved stability and optical properties of the nanocrystals as well as resistance to certain polar solvents during the purification process (Figure 2.24).

Yassitepe et al. extended the "three-precursor" thermal injection method by synthesizing OA-passivated $CsPbX_3$ QDs via the exclusion of organic amines from the precursor solution [43]. The Yassitepe et al. method for the synthesis of $CsPbX_3$ QDs involves the use of CsPbX as a precursor. Quaternary ammonium salt, such as tetraoctylammonium bromide, is reacted with cesium acetate and lead acetate, and these salts cannot be protonated to form amines, even in the presence of protons. By removing the oleylamine from the precursors, the reaction kinetics are significantly accelerated, allowing for the synthesis of $CsPbX_3$ nanocrystals at a lower temperature of 75°C. The resulting $CsPbBr_3$ QDs have a high PLQY of up to 70% and greater stability, but it was not possible to synthesize $CsPbI_3$ QDs with similar quality and stability (Figure 2.25). The Protesescu group subsequently used the three-precursor thermal injection method to synthesize colloidal $FAPbX_3$ nanocrystal solutions [44]. The Protesescu group used a two-precursor thermal injection method to synthesize colloidal $FAPbX_3$ nanocrystal solutions. Lead iodide, OA dissolved in ODE, and oleylamine were injected with a slight excess of formamidine. The resulting product was phase-pure $FAPbI_3$ QDs with better optical properties compared to the previous "three-precursor" method. However, the product still contained some impurities, likely due to the presence of residual lead iodide.

Although the "tri-precursor" method allows for unanchored metal ions from halogen ions, ensuring that the synthesis reaction proceeds according to the designed stoichiometry, some issues still remain. $CsPbI_3$, $CsPbCl_3$, and $MAPbX_3$ have not been synthesized by the three-precursor method, indicating that halogen

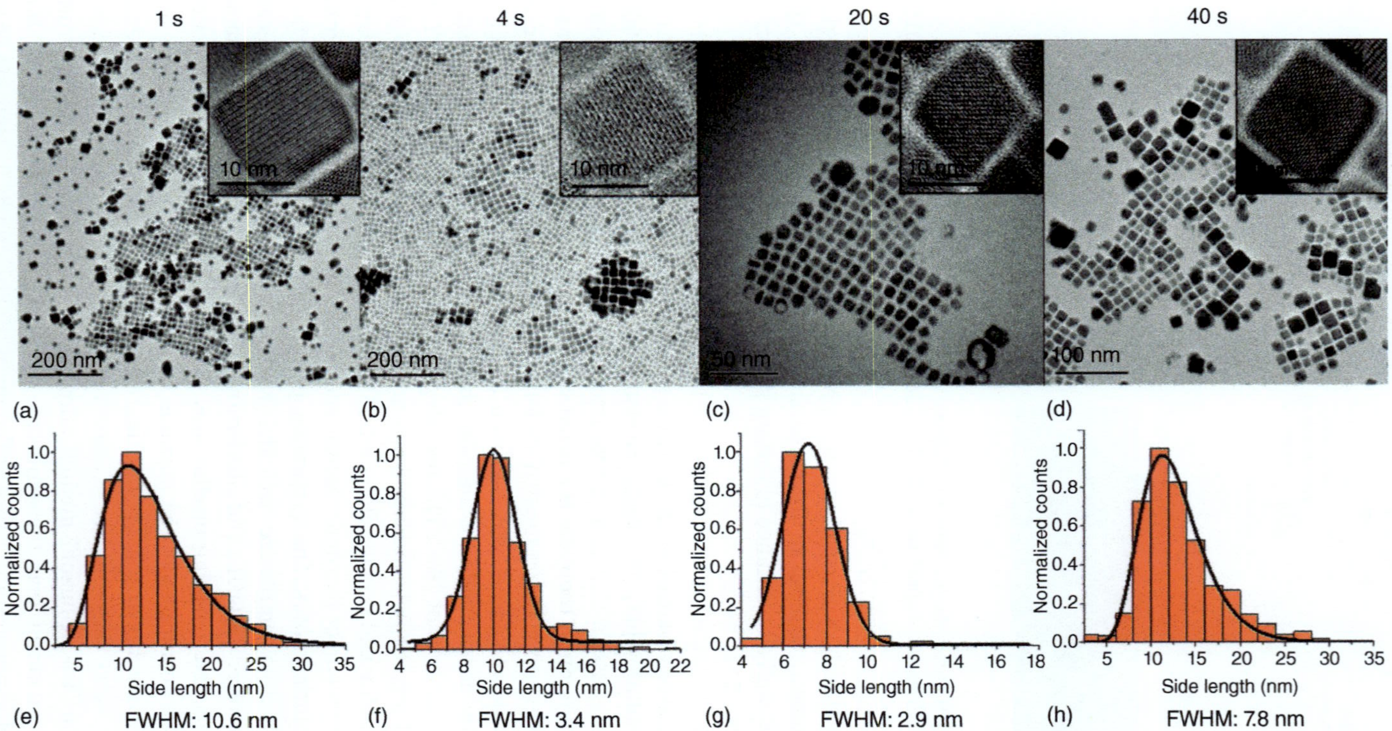

Figure 2.23 HR-TEM images of $CsPbBr_3$ NPs allowed to grow for (a) 1 second, (b) 4 seconds, (c) 20 seconds, (d) 40 seconds. Inset- HR-TEM of a single NP. Normalized size distribution histograms for NP populations of each growth time; (e) 1 second, (f) 4 seconds, (g) 20 seconds, (h) 40 seconds. FWHM of the size distributions is indicated. Source: Koolyk et al. [41] Copyright, 2016, Nanoscale.

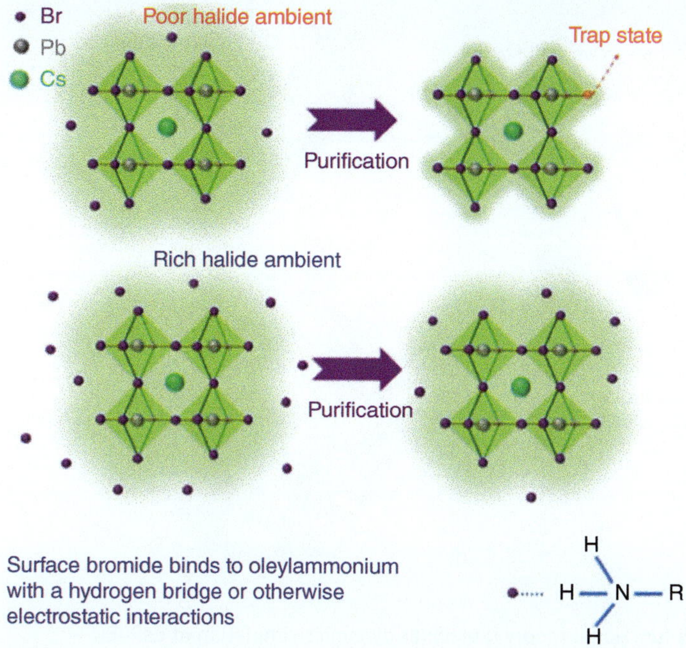

Figure 2.24 Preparation of perovskite nanomaterials in a halogen-rich environment. Source: Liu et al. [42] Copyright, 2017, Chemistry of Materials.

precursors are less reactive using this approach. Additionally, during the synthesis of $FAPbX_3$ nanocrystals, a significant number of unwanted secondary phases were generated, possibly from the decomposition of the alkylammonium halide precursors. To address these issues, Imran et al. and Creutz et al. proposed new methods using highly reactive halide ion precursors that can be easily injected into carboxylate solutions of metals, initiating nanocrystal nucleation and growth immediately after injection. Imran et al. utilized benzoyl halides to synthesize all-inorganic and organic/inorganic hybrid PeNCs with well-controlled size distribution and phase purity of the products. Creutz et al. utilized halogenated silanes to synthesize Cs_2AgBiX_6 lead-free PeNCs. These methods can be tailored to the desired stoichiometry ratio, and the synthesis of PeNCs can be carried out under halogen-rich conditions, addressing the issues of the tri-precursor method.

For the thermal injection method, the size and morphology of the synthesized nanocrystals can be adjusted by varying the type and ratio of ligands as well as controlling the reaction temperature. Typically, when using OA and oleylamine as ligands, at lower temperatures (90–130°C), the perovskites tend to grow nonuniformly, forming quasi-two-dimensional structures, often referred to as NSs. At higher temperatures (170–200°C), prolonged reactions result in the formation of nanowires (NWs). The impact of ligands on the morphology of $CsPbBr_3$ nanocrystals was examined by Pan et al. by systematically changing the length of the carbon chains of organic amines and carboxylic acids in the reaction [45]. Pan et al. investigated the effect of ligands on the morphology of $CsPbBr_3$ nanocrystals by varying the length

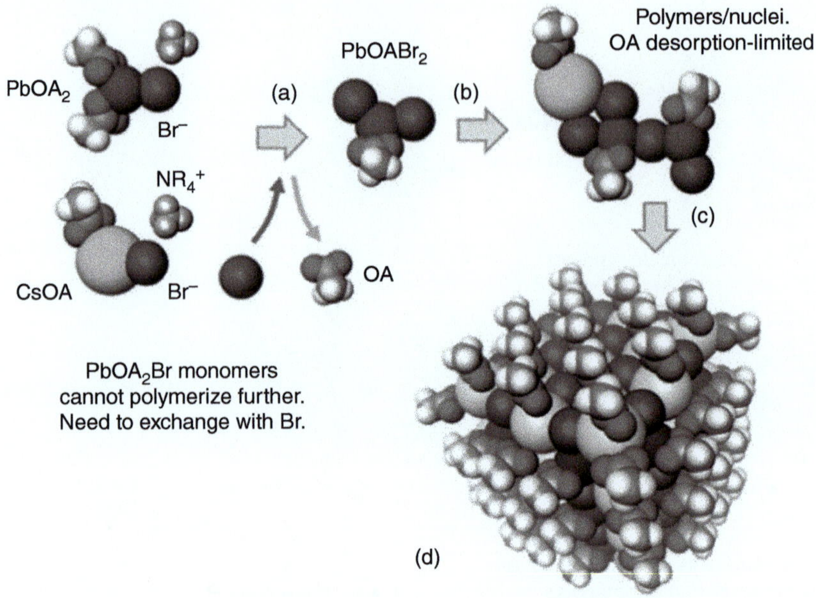

Figure 2.25 Density functional theory molecular dynamics simulation of $CsPbBr_3$ nucleation and growth steps (for simplicity, the synthesis is shown with acetate group instead of oleate). (a) Dissociated ammonium bromide ions form complexes with CsOA and $PbOA_2$. (b) Further reaction between these complexes requires one of the OA^- being replaced with Br^-. (c) Br^- enriched Cs and Pb complexes polymerize through multicoordinated Br. (d) CsPbBr3 nanocrystal passivated with OA ligands. Source: Yassitepe et al. [43] Copyright, 2016, Advanced Functional Materials.

of the carbon chains of organic amines and carboxylic acids in the reaction. They kept the amount of oleylamine constant and added different amounts of carboxylic acid to the reaction (at 170°C). Shortening the carbon chain of the carboxylic acid increased the side length of the $CsPbBr_3$ nanocubes obtained from 9.5 nm to 13 nm. Additionally, when the reaction was carried out with OA at a lowered temperature of 140°C, nanoflakes 20 nm wide and 2.5 nm thick were produced. In another series of experiments, the authors kept the OA constant and added different organic amines. NSs were produced in each experiment, particularly at a lower temperature (140°C) when oleylamine was used (Figure 2.26).

Although much research has been done to control the lateral dimensions of perovskite nanocubes, there have also been efforts to synthesize anisotropic nanostructures such as nanoflakes and NWs and to control their size and morphology. Song et al. successfully synthesized $CsPbBr_3$ atomic-scale flakes with a thickness of 3.3 nm and an edge length of 1 μm using a thermal injection method by using dodecylamine and OA as ligands and extending the reaction time to three hours [46]. Shamsi et al. utilized short-chain ligands, such as octylamine and octanoic acid, in addition to oleic and oleylamine ligands to prepare a $CsPbBr_3$ sheet with an adjustable lateral size ranging from 200 nm to a few millimeters while maintaining the thickness of several crystals. Meanwhile, Imran et al. demonstrated a thermal

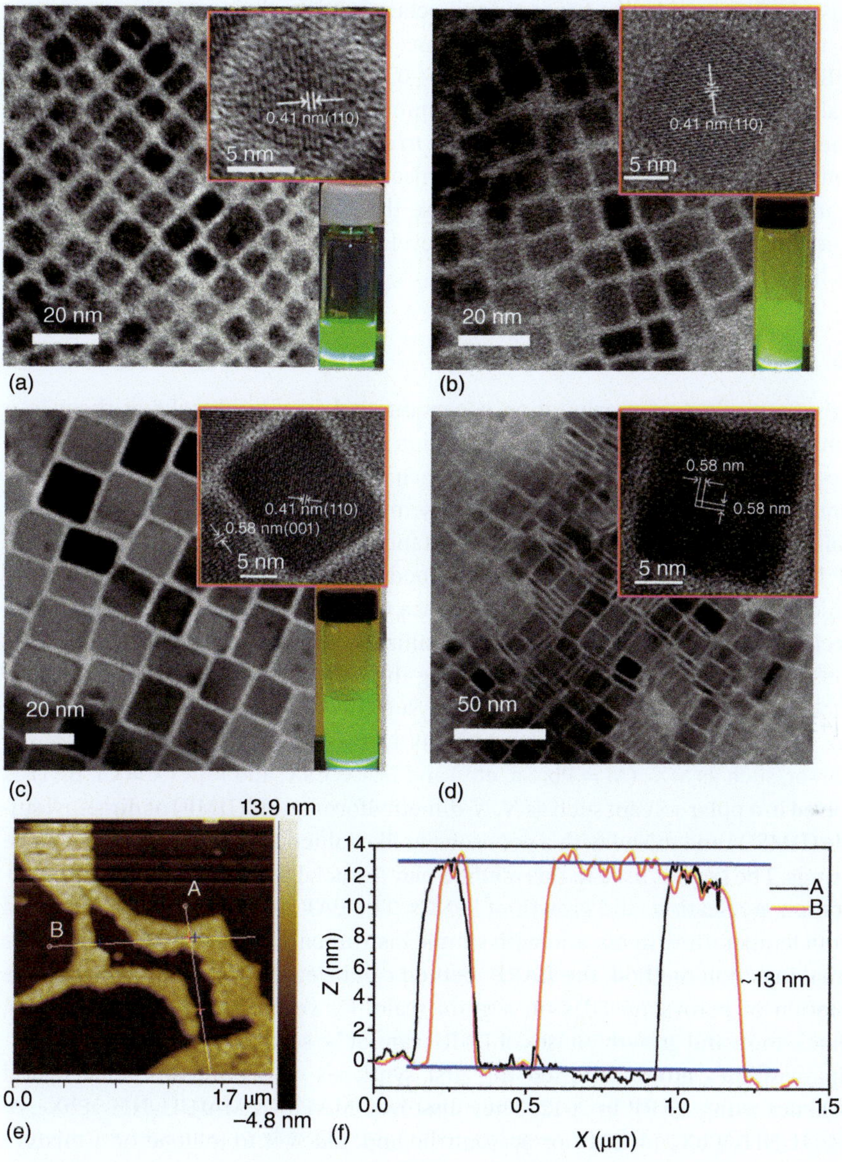

Figure 2.26 Low- and high-resolution TEM images of $CsPbBr_3$ nanocubes from 170 °C reactions using (a) C12A–C18B, (b) C8A–C18B, and (c) C6A–C18B, respectively. Insets in (a–c): Photographs of $CsPbBr_3$ nanocube colloidal solutions in hexane under UV light. (d) Low- and high-resolution TEM images of $CsPbBr_3$ nanoplatelets from the 140 °C reaction using C6A–C18B. (e) AFM and (f) line profiles of a monolayer film of nanocrystals from C12A–C18B. Source: Pan et al. [45] Copyright, 2016, ACS Nano.

injection method to synthesize $CsPbBr_3$ NWs with a tunable width down to a few crystals, using only octylamine and oleylamine as ligands. By introducing short chains of organic acids like hexanoic and octanoic acids, they were able to prepare NWs with radii ranging from 10 to 3.4 nm. Zhang et al. reported an ultra-thin $CsPbBr_3$ nanowire with a radius of only 2.2 ± 0.2 nm and a length of several microns by using oleylamine, OA, and dodecylamine as ligands for thermal injection synthesis, which was then purified stepwise to enhance fluorescence efficiency [47].

However, the drawbacks of thermal injection, such as the requirement for high temperatures and inert gases, increase the cost of synthesis and limit mass production. Therefore, there is a need to develop alternative methods to overcome these limitations.

2.3.6 Ligand-Assisted Reprecipitation

LARP is a method of precipitating supersaturated ions by dissolving them in a solvent to reach an equilibrium concentration and then rapidly inducing supersaturation. The supersaturated state can be achieved by changing the temperature, evaporating the solvent, or adding a poorly mixed-phase solvent with lower ionic solubility. In this state, the solute spontaneously precipitates and crystallizes until it returns to equilibrium. This method has been successful in producing organic or polymeric nanodot materials since the 1990s. The term "ligand-assisted reprecipitation" is used because the crystallization process is often regulated by various organic or inorganic ligands during the supersaturated precipitation. This method can also be used to prepare nanocrystals, QDs, and other similar materials.

The synthesis of perovskites using LARP is a straightforward process. The desired precursor, such as MX_2 (M = Pb, Sn, etc.) and MAX, CsX, and FAX (X = Cl, Br, I), is dissolved in a polar solvent such as *N*, *N*-dimethylformamide (DMF) or dimethyl sulfoxide (DMSO) and mixed with a poor solvent like toluene or hexane in the presence of ligands. The ligands used in this method play a crucial role in regulating the supersaturation, nucleation, and growth of PeNCs. The LARP method can be conducted at room temperature, in air, and with simple instrumentation. In comparison to the thermal injection method, the LARP method can be easily scaled up, allowing the generation of perovskite QDs on a gram scale. However, like thermal injection, the nucleation and growth phases of LARP cannot be separated in time.

Papavassiliou's group reported the first synthesis of organic-inorganic hybrid perovskites using LARP in 2012. They dissolved $MAPbX_3$, $MA(CH_3NH_3)PbX_7$, or $MA(C_6H_5NH_3)PbX_2$ in DMF or acetonitrile and added it to toluene or a mixture of toluene and polymethyl methacrylate. This process produced highly fluorescent nanocrystals with sizes ranging from 30 to 160 nm and fluorescence emission intensities that far exceeded those of the corresponding perovskite bulk materials [48]. The LARP method has since been extended to other ABX_3 nanocrystal systems, involving various combinations of A (such as MA, FA, Cs), B (such as Pb, Sn, Bi), and X (such as Cl, Br, I) elements (Figure 2.27).

In 2015, Zhang et al. systematically demonstrated the preparation of $MAPbBr_3$ QDs using the LARP method [23]. They dissolved MABr, PbBr, organic amines, and

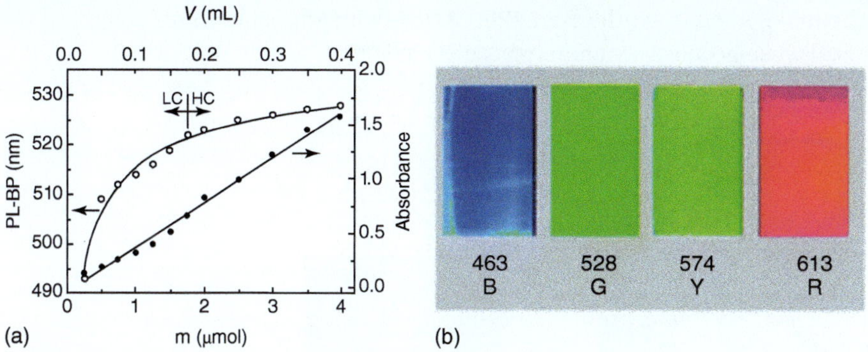

Figure 2.27 (a) PL-BP and the corresponding values of (optical) absorbance (A) of emitters, obtained from $(CH_3NH_3)(CH_3C_6H_4CH_2NH_3)_2Pb_2Br_7$, in toluene–PMMA as a function of the corresponding mass of the compound and the corresponding volume of its DMF-solution; solid lines are trend lines added to guide the eyes. (b) the photograph of the blue (463 nm), green (528 nm), yellow (574 nm), red (613 nm) emitters under UV-light (365 nm). Source: Papavassiliou et al. [48] Copyright, 2012, Journal of Materials Chemistry.

carboxylic acids in DMF to prepare precursors, and then dropped a certain amount of them into toluene under vigorous stirring to produce a colloidal solution of QDs at room temperature. To investigate the role of organic amines and carboxylic acids in this process, they used different organic amines (hexylamine, octylamine, dodecylamine, hexadecylamine) and carboxylic acids (butyric acid, octanoic acid, OA) for the experiments. The results showed that perovskites can be formed without organic amines, but size control of the nanocrystals cannot be achieved. Without carboxylic acids, PeNCs can be produced as usual, but they coalesce in a much shorter time. Based on these comparative experiments, the authors concluded that the role of organic amines is to regulate the crystallization kinetics for size control of the nanocrystals, while organic carboxylic acids act as stabilizers to prevent the coalescence of the nanocrystals (Figure 2.28).

In subsequent years, further research was conducted on similar experiments. Huang et al. discovered that the size and size distribution of the products could be further controlled by varying the reaction temperature [49] (Figure 2.29). Arunkumar et al. demonstrated that $MAPbX_3$ nanocrystals doped with Mn^{2+} can be synthesized by incorporating $MnCl_2$ into the precursor solution [50]. Other ligands have also been demonstrated to be effective in the synthesis of PeNCs using the LARP method. For example, Luo et al. used two branched ligands, 3-aminopropyltriethoxysilane(APTES) and [3-(2-aminoethyl)amino]propylhepta-isobutyl polystyrene sulfonate sodium-substituted caged polysesquioxane (NH_2-POSS), to prepare PeNCs with tunable sizes [51]. The authors observed that the use of APTES and NH_2-POSS as ligands could offer better control over the size of the PeNCs than straight-chain ligands. This is because these branched ligands are more effective in protecting the PeNCs from dissolution by DMF [52]. Luo et al. synthesized $MAPbBr_3$ nanocrystals using 12-aminolauric acid peptide as the exclusive ligand [53]. Moreover, the size of the nanocrystals can be controlled effectively using a peptide containing both amino and carboxyl groups as the

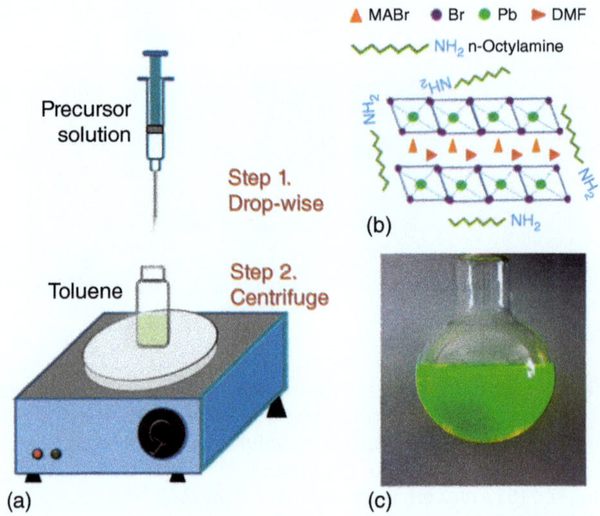

Figure 2.28 (a) Schematic illustration of the reaction system and process for LARP technique. (b) Schematic illustration of starting materials in the precursor solution. (c) Typical optical image of colloidal $CH_3NH_3PbBr_3$ solution. Source: Zhang et al. [23] Copyright, 2015, ACS Nano.

sole ligand, as demonstrated by Luo et al. Additionally, Shamsi et al. developed an optimized LARP method with minor modifications to the process [54]. One method that has been developed by Shamsi et al. involves dissolving PbX_2 in *N*-methylformamide (NMF) along with OA and oleylamine, and then heating the solution to 100°C for 10 minutes before slowly adding it drop by drop to a poor solvent like dichlorobenzene or chloroform. This method produces MA^+ in situ via a transamidation reaction, eliminating the need to synthesize MAX. This allows for the production of large $MAPbX_3$ crystals at room temperature without the use of undesirable solvents. Additionally, Dai et al. developed a special spray method to produce uniformly sized $MAPbBr_3$ nanocrystals. In this method, the precursor solution (consisting of MABr, $PbBr_2$, OA, and oleylamine dissolved in DMF) is sprayed directly into the undesirable solvent toluene [55]. The micron-sized droplets formed by the spray increase the contact area between the two solutions, allowing for more effective mixing.

Soon after the development of the LARP method for $MAPbBr_3$ QDs, researchers began working on optimizing their optical properties and controlling their size and morphology. In 2015, Sichert et al. introduced a modified LARP method that did not rely on the use of carboxylic acid ligands [14]. In the years following the introduction of LARP for the synthesis of $MAPbBr_3$ QDs, researchers focused on optimizing the optical properties of the materials and controlling their size and morphology. In 2015, Sichert et al. developed a new LARP method that did not require carboxylic acid ligands. Instead, they used only oleylamine and methylamine as ligands and found that the thinner the NSs prepared, the smaller the ratio of octylamine to methylamine (Figure 2.30). Moreover, when only octylamine was used, they

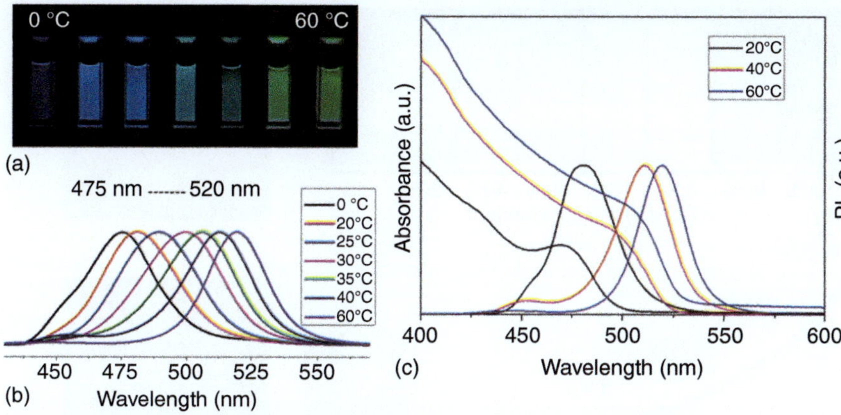

Figure 2.29 Effect of reaction temperature on luminescence color. Source: Huang et al. [49] Copyright, 2015, Advanced Science.

observed monolayer perovskite nanoflakes, similar to those found in layered perovskite crystals [56]. Cho et al. investigated the synthesis of $MAPbBr_3$ by LARP using various organic amines with different carbon chain lengths (such as butylamine, hexylamine, octylamine, dodecylamine, and oleylamine) in combination with OA [57]. In their investigation of the synthesis of $MAPbBr_3$ by LARP, Cho et al. observed that the carbon chain length and concentration of organic amines had a significant impact on the thickness of the resulting products. High concentrations of organic ammonium cations were found to effectively passivate the perovskite surface, preventing vertical growth and allowing for the production of perovskite nanoflakes with adjustable thickness. Longer-chain organoamines were shown to interact more strongly with the PeNC surface, resulting in stronger passivation. In 2017, Levchuk et al. employed OA and oleylamines as ligands and achieved good control over the thickness of the resulting $MAPbBr_3$ NSs by introducing chloroform as a poor solvent [58]. Ahmed et al. employed OA and oleylamine as ligands and further introduced pyridine to finely control the thickness of perovskite NSs [59]. Additionally, they conducted DFT calculations, which indicated that pyridine is a suitable ligand for calixarene NSs thickness control. The calculations showed that the N atom in pyridine could coordinate with the Pb^{2+} ion to restrict the longitudinal growth rate of the perovskite, leading to the formation of a two-dimensional nanosheet structure.

Along with controlling the longitudinal morphology of nanocrystals, efforts have also been made to control their lateral dimensions. Kirakosyan et al. found that the method of adding the precursor solution to a poor solvent had an impact on the size and shape of the resulting nanocrystals [60]. The size and shape of final nanocrystals were found to be influenced by the way precursor solution was added to a poor solvent. Kirakosyan et al. discovered that adjusting the amount of precursor solution added (1–8 drops, drop acceleration rate 45 drops per minute) could change the product size and that the lateral size of the prepared PeNCs could be modified from 3 to 8 nm while keeping the thickness constant (2.5–3 nm) (Figure 2.31). Meanwhile, Huang et al. investigated the synthesis mechanism and

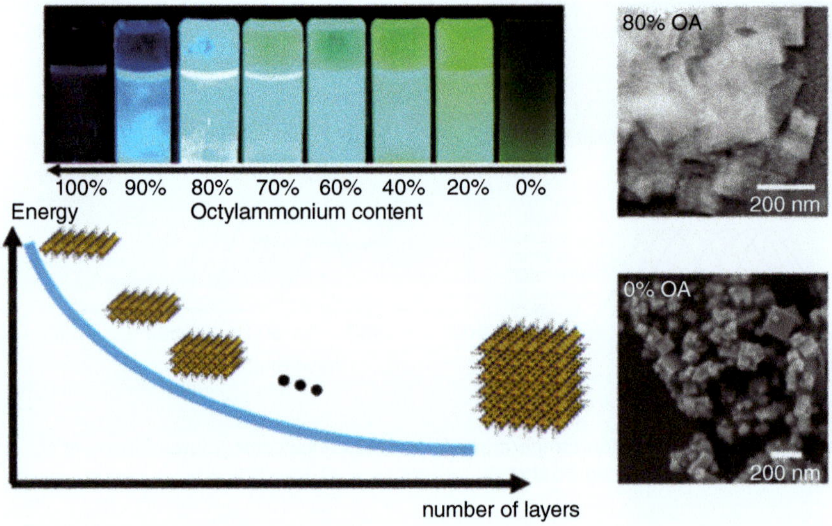

Figure 2.30 Preparation of monolayer nanoluminescent layers. Source: Sichert et al. [14] Copyright, 2015, Nano Letters.

systematically analyzed the impacts of total ligand amount, ligand-to-precursor ratio, reaction temperature, etc., on the $MAPbBr_3$ nanocrystals [61]. The size and thickness of $MAPbBr_3$ NSs can be finely controlled in the LARP method by adjusting the ratio of long-carbon-chain organic amine or pyridine ligands to precursors. This passivates the surface of the PeNCs, limiting their growth rate in the vertical direction while promoting growth in the lateral direction, resulting in perovskite nanomaterials with varying morphologies. Kirakosyan et al. showed that the size and shape of the final nanocrystals were also influenced by the way in which the precursor solution was added to a poor solvent. Huang et al. investigated the effects of various parameters such as the total amount of ligand, ratio of ligand to precursor, and reaction temperature on the $MAPbBr_3$ nanocrystals, concluding that optimal control of these parameters could result in the production of small, well-defined nanocrystals. At high ligand/precursor ratios, $MAPbBr_3$ nanocrystals were produced with controlled size and size distribution, while high precursor/ligand ratios resulted in polydisperse micron-sized crystals.

The LARP method was initially developed for $MAPbBr_3$ nanomaterials but has since been expanded to include the synthesis of all-inorganic PeNCs, such as $CsBX_3$ (where B = Pb^{2+}, Bi^{3+}, Sb^{3+}, etc.), Cs_4PbX_6, $CsPbX_{25}$, and $APbX_3$ mixed cation perovskite systems. In 2016, Weidman et al. demonstrated the use of LARP for the synthesis of all-inorganic or mixed cation ABX_3 (where A = Cs, MA, FA, and B = Pb, Sn) nanoflakes with thicknesses ranging from one to two monoatomic layers [62]. They dissolved precursors (AX and BX_2 salts) in DMF and added octylammonium halide and butylammonium halide. This precursor solution was added drop by drop to toluene and stirred vigorously to produce PeNCs immediately at room temperature. To achieve a uniform thickness and stabilize the colloid, they used an excess of ligand (ligand:BX_2:AX = 10:2:1). Perumal et al. also reported

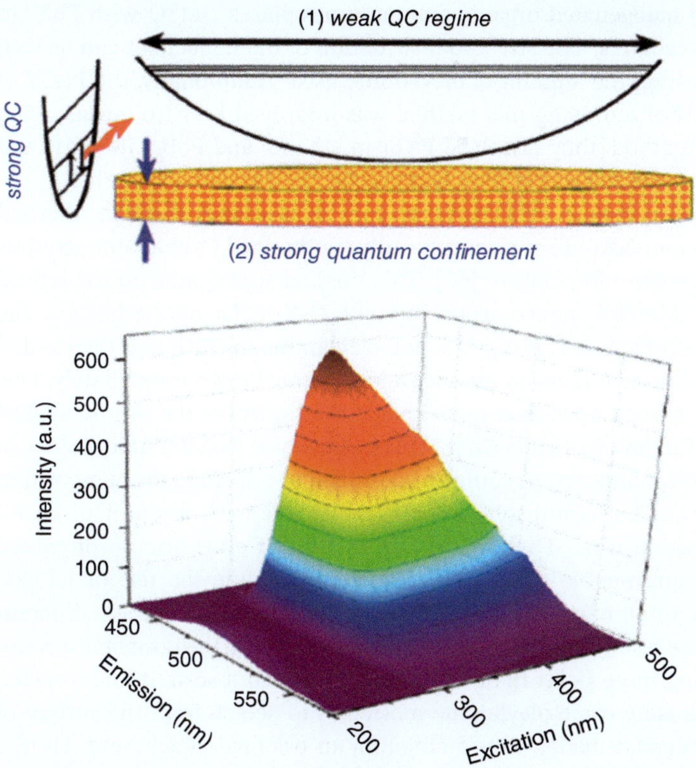

Figure 2.31 Control diagram for the growth direction of perovskite crystals grown laterally. Source: Kirakosyan et al. [60] Copyright, 2017, Crystal Growth and Design.

the synthesis of $FAPbBr_3$ nanocrystals using LARP [63]. They primarily focused on obtaining bright PL emission rather than controlling the morphology. In their method, FABr was dissolved with $PbBr_2$ in DMF and added dropwise to a solution of toluene, butanol, octylamine, and OA under stirring [64]. They developed a synthetic scheme for the preparation of $FAPbX_3$ (X = Cl, Br, I) nanocrystals, which was similar to their previously reported $MAPbX_3$ method with minor modifications. The method involved the rapid injection of a DMF solution containing PbX_2, FAX, OA, and oleylamine into chloroform at room temperature. The thickness of the nanocrystals could be controlled by adjusting the OA to oleylamine ratio, resulting in small nanocubes or nanoflakes with different morphologies. Interestingly, the presence of toluene prevented the formation of $FAPbI_3$ nanocrystals and caused rapid coalescence of $FAPbCl_3$ or $FAPbBr_3$ nanocrystals. It is worth noting that Levchuk et al. focused on obtaining bright PL emissions rather than controlling the morphology of the nanocrystals.

Minh et al. introduced a new approach for synthesizing $FAPbX_3$ nanocubes using PbX_2-DMSO complexes as precursors [65]. The complexes were dissolved with FAX and oleylamine in DMF and then injected into a mixture of toluene and OA. The synthesis mechanism is believed to involve an intramolecular exchange reaction in which halogenated organic ammonium replaces DMSO with PbX_2 to form perovskite structures. The size and distribution of the nanocubes can be well controlled by adjusting the amount of oleylamine used. Additionally, the PLQY of $FAPbBr_3$ NSs synthesized using this method was improved [19]. To deviate from the conventional method, they dissolved FABr in ethanol and $PbBr_2$ in DMF, and then added these solutions to a combination of toluene, OA, and octylamine.

Before 2015, all-inorganic PeNCs were only synthesized through thermal injection. However, in 2016, Li et al. successfully synthesized $CsPbX_3$ nanocrystals through LARP at room temperature [66]. This method was similar to the typical LARP method for $MAPbX_3$ nanocrystals, but with CsX replacing the MAX in the precursor. They dissolved CsX, PbX_2, OA, and oleylamine in DMF and then added the precursor solution to toluene to produce all-inorganic PeNCs immediately. This method was later refined by Seth et al. to enable the control of the morphological characteristics of $CsPbX_3$ nanocrystals [67]. They observed that the morphology of the resultant $CsPbX_3$ nanocrystals could be controlled by adjusting the composition ratio of the poor solvent (from toluene to ethyl acetate), the amount of ligand used, and the reaction time. They found that the use of ethyl acetate promoted the formation of quasi-cubic QDs, NSs, or nanorods, while the use of toluene tended to produce small nanocubes or NWs. The authors explained these different morphologies as the result of the interaction of the ligand with the nonpolar solvent. Ethyl acetate, being more polar than toluene, acts as both a solvent and a nucleophilic reagent, causing some oleylamine molecules to detach from the surface of the growing nanocrystal nuclei, thus facilitating an oriented attachment. On the other hand, in toluene, if the concentration of oleylamine is sufficient, the surface of the nanocrystal is protected in all directions, preventing surface incorporation or attachment. However, if the concentration of oleylamine is insufficient, the nanocrystal will exhibit anisotropic growth over a longer reaction time, resulting in the formation of NWs, which could be attributed to incomplete surface passivation.

In 2017, Kostopoulou et al. reported a method for synthesizing micron-length $CsPbBr_3$ NWs using a LARP approach [68]. Their approach focused on anhydrous and low-temperature conditions. The precursor solution, containing CsBr, $PbBr_2$, OA, and oleylamine in DMF, was injected into anhydrous toluene while the system was cooled using an ice water bath. The NWs grew gradually from short bullet-shaped nanorods to NWs after injection, with a width of 2.6 nm after 24 hours at room temperature. Thick NWs with a diameter of 6.1 nm were generated after one week of growth at room temperature.

In 2018, Zhang et al. incorporated water into the LARP method to investigate its effect on the growth rate and crystal morphology of $CsPbBr_3$ nanocrystals [69]. They hypothesized that H_3O^+ and OH^- ions can be considered as surfactants with higher activity than oleylamine and OA, resulting in the growth of PeNCs in diverse directions (Figure 2.32).

Although LARP allows for the direct synthesis of many different perovskite systems at room temperature, it has some drawbacks. Perovskites are very sensitive to polar solvents, and commonly used solvents in LARP, such as DMF and DMSO, can readily dissolve PeNCs, especially $CsPbI_3$, which is highly susceptible to

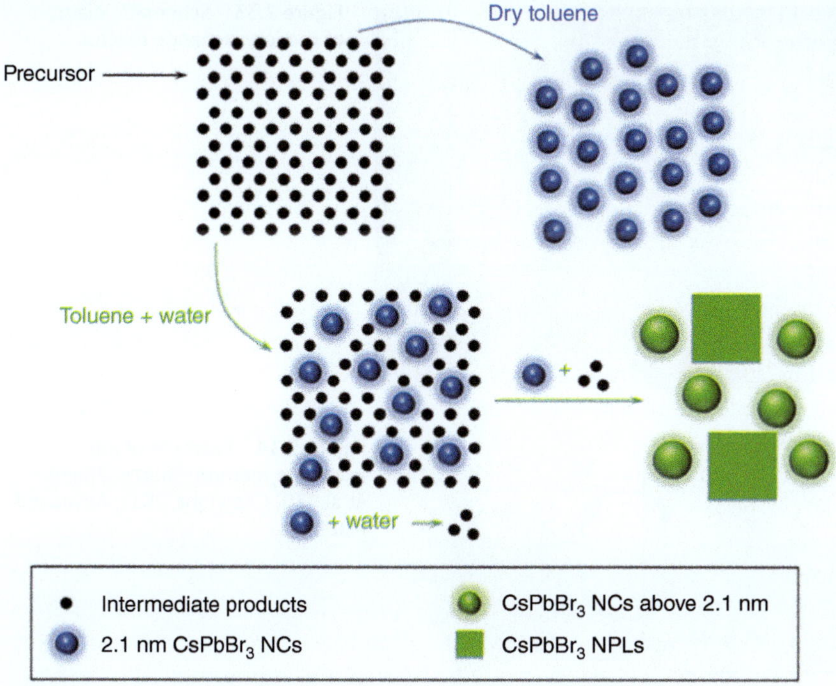

Figure 2.32 Effect of water introduction on the growth direction of $CsPbBr_3$. Source: Zhang et al. [69] Copyright, 2018, Angewandte Chemie International Edition.

degradation. In fact, precursor-polar solvent interactions play an important role in the generation of imperfect PeNCs. To elucidate this solvent effect, Zhang et al. investigated the effect of different polar solvents on the crystallization of $MAPbI_3$ nanocrystals. They found that PbI_2 generates stable intermediates in coordination solvents (e.g. DMF, DMSO, THF, etc.), while such complexes are not generated in non-coordination solvents (e.g. γ-butyrolactone and acetonitrile). Therefore, further research is highly recommended to achieve high-quality materials.

2.3.7 Ion Exchange Method

The anion exchange method is an indirect synthesis method that involves mixing pre-synthesized $CsPbX_3$ nanocrystals with solutions of different halogen compositions. During the reaction, halogen anions in the lattice migrate rapidly and replace each other, resulting in mixed halogen inorganic perovskites such as $CsPb(Br/I)_3$ or $CsPb(Br/Cl)_3$. While this synthesis method has a short reaction time and is simple to perform, it has limited controllability and reproducibility (Figure 2.33).

In 2017, Huang's group achieved a more stable hybrid halide passivation layer on the surface of $CsPbX_3$ QDs through ion exchange of CsPbX nanocrystals with surface chlorine-containing PbSe [70]. The improved hybrid halide passivation layer on the surface of $CsPbX_3$ QDs enhanced their fluorescence quantum yield and air stability. The maximum photovoltaic conversion efficiency of the solar cell was increased to

Figure 2.33 Schematic diagram of the ion exchange reactor.

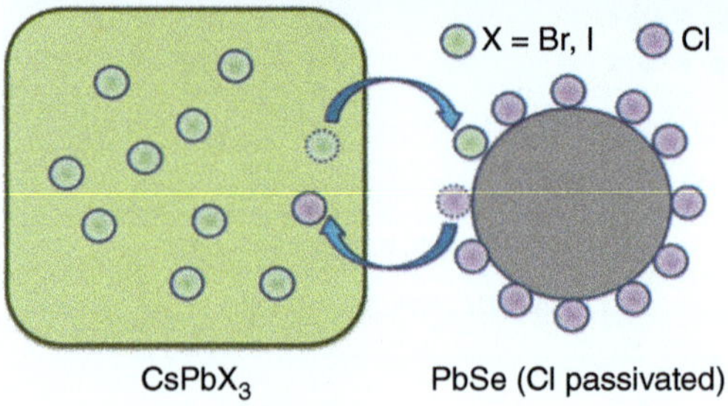

Figure 2.34 Example of ion exchange method. Source: Zhang et al. [70] Copyright, 2017, Advanced Materials.

8.2%, and the conversion efficiency remained at 95% of the initial value after 57 days of exposure to air (Figure 2.34).

In 2022, Kang Wang and Zih-Yu Lin demonstrate that extending the π-conjugation length and increasing the cross-sectional area of the ligand enables perovskite thin films with dramatically suppressed ion transport, narrowed phase distributions, reduced defect densities, and enhanced radiative recombination efficiencies. Consequently, we achieved efficient and stable deep-red LEDs with a peak external quantum efficiency of 26.3% (average 22.9% among 70 devices and cross-checked) and a half-life of ~220 and 2.8 hour under a constant current density of 0.1 and 12 mA cm^{-2}, respectively [71].

2.3.8 Laser Etching Method

Ultrashort pulse laser, that is, femtosecond laser, processing of transparent materials has been proven as an effective method to construct 3D photonic device structures for diverse applications in nonlinear optics, optical data storage and nanophotonics. Figure 2.34 shows the schematic of the femtosecond laser writing system for sample fabrication [72]. In 2021, Qiu Jianrong's group demonstrated a method to prepare PeNCs with tunable composition and band gap in doped metal oxide glasses using 3D direct lithography. This involved local melting and subsequent crystallization induced by ultrafast laser pulses [73]. They achieved this by inducing ultrafast laser pulses to locally melt and subsequently crystallize the glass. The composition of the

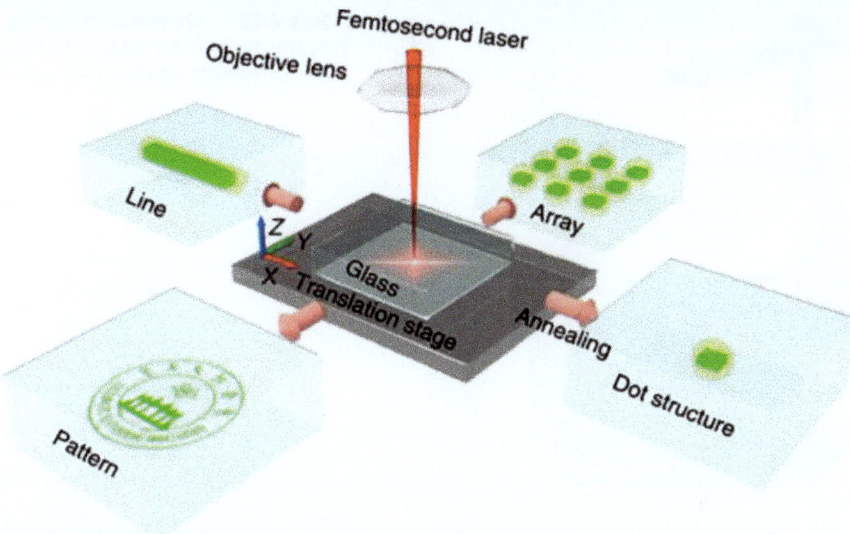

Figure 2.35 Schematic of the femtosecond laser writing system for sample fabrication. Source: Huang et al. [72] Copyright, 2019, Springer Nature Limited.

nanocrystals was transformed from $CsPb(Cl_{1-x}Br_x)_3$ to $CsPbI_3$ by the laser-induced liquid nanophase separation, resulting in a tunable photoluminescence bandgap ranging from 480 to 700 nm by controlling the distribution of halide ions at the nanoscale. The 3D-printed structures of this perovskite nanocrystalline glass material exhibit remarkable stability to UV irradiation, organic solvents, and high temperatures up to 250°C, making it a promising material for applications such as optical storage, miniature LEDs, and holographic displays (Figure 2.35).

2.4 Processing Technology for Large-Area Perovskite Films

Despite the many advantages of perovskite materials, there are still some challenges in achieving commercial display applications of PeLEDs (perovskite light-emitting diodes). The high nonuniformity of the large-area active layer is the main obstacle encountered in the production of perovskite thin films on a large scale. Compared to small Mperovskite films, large-area Mperovskite films exhibit increased roughness, decreased crystal uniformity, and decreased optical performance, which limits the large-scale production of PLEDs and hinders their many applications [74].

Below, we will introduce some commonly used processing techniques for large-area perovskite thin films.

2.4.1 Spin Coating Method

Spin coating is a coating process that utilizes centrifugal force and gravity generated by the rotation of the workpiece to evenly spread the coating material over the

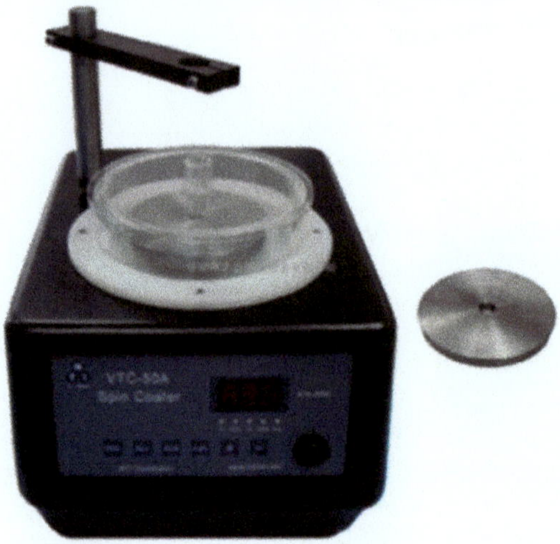

Figure 2.36 Common spin coating equipment. Source: henyang Kejing Auto-instrument Co., Ltd.

surface (Figure 2.36). The process involves various physicochemical processes such as fluid flow, wetting, volatilization, sticking, and dispersion. Solution spin coating is commonly used for the fabrication of metal halide perovskite polycrystalline films, including calcium titanate. Two types of solution spin coating methods are used: one-step and two-step. The one-step method was first proposed by Miyasaka's team in Japan, where they spin-coated perovskite and lead halide precursors on a titanium dioxide layer to form $MAPbI_3$ [75]. The preparation methods for perovskite films can be generally divided into three steps: (1) dissolving the precursors in a certain ratio using solvents such as DMF or DMSO, (2) spin-coating the precursor solution onto a clean substrate to form a film, and (3) annealing the wet film to obtain a well-crystallized polycrystalline film. Although some precursors may have low solubility and uneven spin coating or poor film formation (e.g. pinholes) can occur, these issues can be mitigated using solvent engineering techniques, annealing, or additive treatments. Solution spin coating is a simple and inexpensive method, and it is currently the most common method used to prepare optoelectronic devices. The efficiency of solar cell devices based on solution spin-coated absorber layers has exceeded 22%, and the efficiency of LED devices exceeds 20%. For instance, CsBr and $PbBr_2$ were dissolved in DMSO solvent to form a 0.3 M precursor solution, and various amounts of polyepoxyhexane (PEO) were added to improve film formation.

2.4.2 Vacuum Thermal Vapor Deposition Method

Vacuum evaporation is a method commonly used to prepare metal halide perovskite films. It involves heating the evaporation material under vacuum to vaporize it and deposit it onto the surface of the substrate, forming a thin film. In this process, the substrate is placed in a vacuum chamber, and the film material is heated by resistance, electron beam, laser, or other means, causing the material to evaporate or sublimate into gaseous particles with a certain energy (0.1–0.3 eV). The gaseous particles travel to the substrate in a linear motion with few collisions, and some of them are reflected while others are adsorbed and undergo surface diffusion.

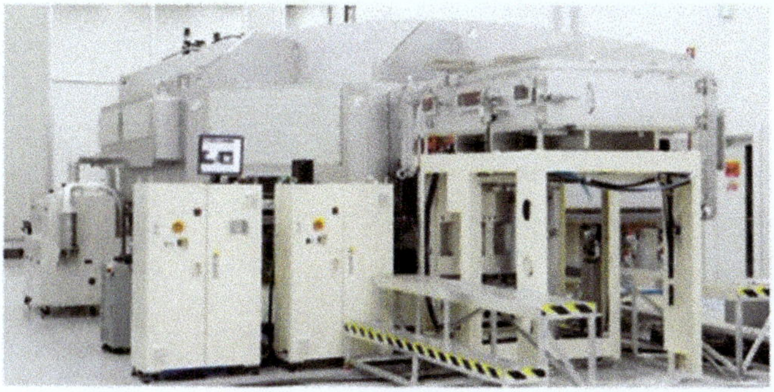

Figure 2.37 Equipment diagram for the vacuum thermal evaporation method.

These adsorbed particles create two-dimensional collisions, forming clusters of atoms, molecules, or clusters of atoms and molecules, which may remain on the surface for a short time before evaporating again. These clusters of particles continuously collide with the diffusing particles, either by adsorbing single particles or by releasing them. When the number of aggregated particles exceeds a certain threshold, they become stable nuclei and continue to adsorb diffused particles and grow gradually, eventually forming a continuous film through the contact and merging of neighboring stable nuclei. Although vacuum evaporation has the advantage of producing high-quality, uniform films, it requires high vacuum conditions, which can limit the size and type of substrate and can also limit the scalability and reproducibility of the method (Figure 2.37).

2.4.3 Printing Method

Printing is a versatile method for preparing large-area metal halide perovskite films. In this technique, a perovskite precursor solution and an organic polymer are typically mixed and printed onto a substrate. For instance, Yu et al. from Florida State University utilized printing to fabricate $MAPbBr_3$ luminescent layers and construct LEDs with excellent performance. In their study, $PbBr_2$ and MABr were dissolved in anhydrous DMF at a molar ratio of 1 : 1.5 to prepare a perovskite precursor solution with a concentration of 500 mg ml^{-1}. Poly(ethylene oxide) was also dissolved in DMF to obtain a concentration of 10 mg ml^{-1}. The two solutions were then mixed in the desired proportions, stirred at 70°C for 30 minutes, and then printed onto an ITO/glass substrate. The resulting composite film was spread evenly with a spatula and dried on a hot plate at 80°C for 5 minutes. The LEDs constructed using the composite film as the light-emitting layer exhibited an open circuit voltage of only 2.6 V and a maximum brightness of 21,014 cd m^{-2}, with a maximum external quantum efficiency of 1.1% [76] (Figure 2.38).

2.4.4 Vapor -Phase Deposition Method

Vapor deposition is a chemical process that utilizes one or more gas-phase compounds or monomers containing thin film elements to deposit thin films onto the surface of a substrate (Figure 2.39).

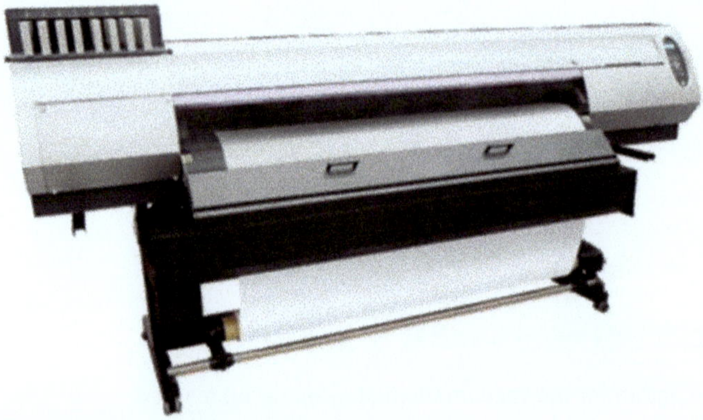

Figure 2.38 Diagram of the printing method equipment. Source: Dmitry Vereshchagin/Adobe Stock.

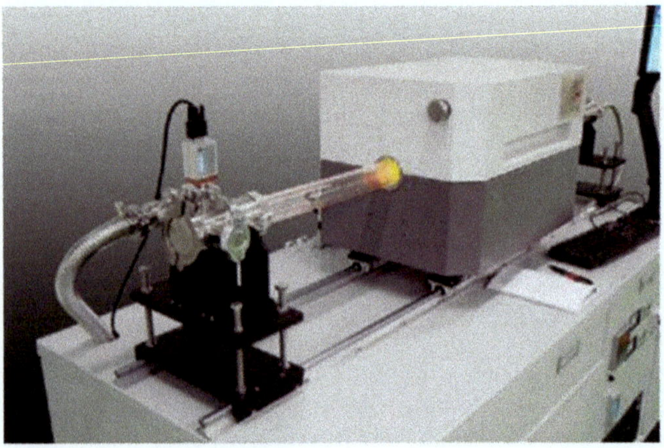

Figure 2.39 Chemical deposition method equipment. Source: polyparadigm/Wikimedia Commons.

In 2014, Xiong's group at Nanyang Technological University in Singapore reported the first successful preparation of methane-lead halide perovskite microsheets using vapor phase deposition [16]. They demonstrated the preparation of methane-lead halide perovskite microsheets through vapor-phase deposition, which showed promising optical properties for potential use in photovoltaic devices. The preparation process involved a two-step method: first, lead halide NSs were synthesized, and second, they were transformed into methylamine-lead halide micron sheets through a gas-solid heterogeneous reaction with ammonium methyl halide molecules.

To synthesize lead halide microsheets, PbI_2, $PbBr_2$, or $PBCl_2$ powders were used as raw materials and placed in the upstream area of a quartz tube, while a freshly cleaned white mica substrate was placed downstream. The quartz tube was first evacuated at a base pressure of 2 mTorr, and then 5% high-purity hydrogen–argon gas was passed through at a flow rate of 30 sccm. The temperature and pressure in

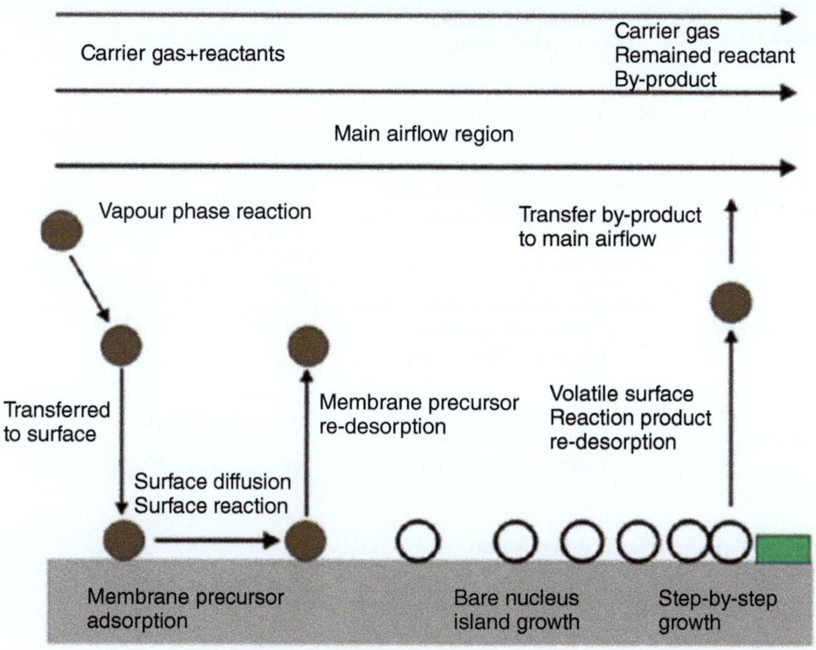

Figure 2.40 Chemical deposition process.

the tube were set to the desired values based on the type of lead halide used, and the synthesis time was controlled within 20 minutes. The furnace was then allowed to cool naturally to ambient temperature.

To synthesize methylamine lead halide micron flakes, the methylamine halide was prepared using the solution method and placed in the central area of the quartz tube. The prepared lead halide micron flakes were placed downstream from the central area, and the quartz tube was again evacuated to a base pressure of 2 mTorr. 5% high-purity hydrogen–argon gas was then passed through the tube at a flow rate of 30 sccm, and the pressure in the tube was stabilized at 50 Torr. The tube was then heated to 120 °C and held for one hour. The furnace was allowed to cool naturally to room temperature at the end of the reaction.

The resulting $MAPbI_3$ micrometer flakes showed excellent optical properties, with an electron diffusion length of 200 nm, approximately twice that of solution-treated films. This promising result suggests that metal halide perovskites hold great potential for use in photovoltaic devices (Figures 2.40 and 2.41).

Jin et al. at the University of Wisconsin, United States of America, demonstrated the production of large-area continuous $CsPbBr_3$ films with controlled film thickness in the micron range using vapor deposition on titanate substrates [77]. The thickness of the films can be controlled in the micron range, and the resulting films display exceptional crystal quality and photophysical and electrical properties, including a slow charge carrier complexation rate, low surface complexation rate, and low bulk defect density (Figure 2.42).

Device type	Heating method (Temperature range, °C)	Schematic diagram
Tubular furnace type	Resisitive heating mode(~1000)	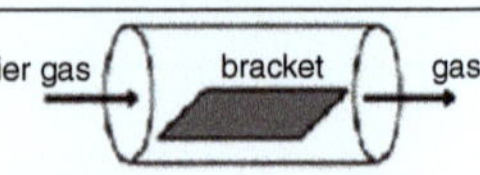
Vetical type	Plate heating mode(~500) Induction heating mode(~1200)	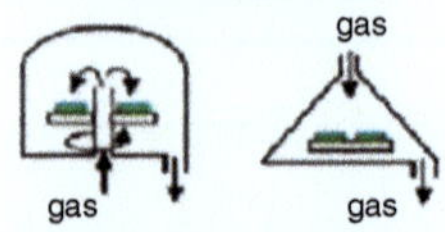
Cylindrical type	Induction heating mode(~1200) Infraread radiation heating mode(~1200)	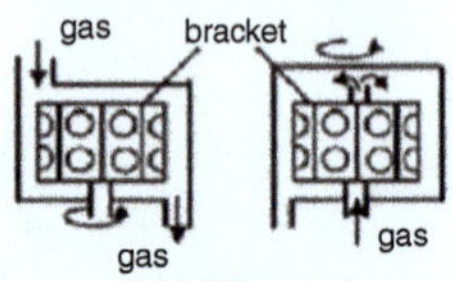
Series winding type	Plate heating mode(~1200) Infraread radiation heating mode(~1200)	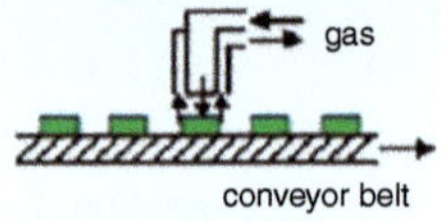

Figure 2.41 Types of chemical deposition.

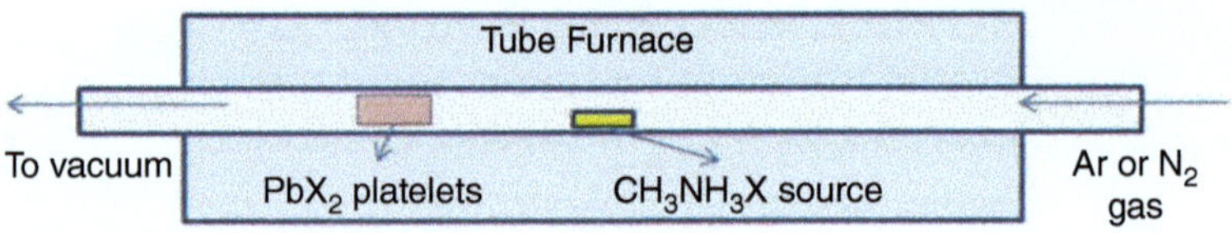

Figure 2.42 Preparation of methane-lead halide perovskite microsheets by vapor deposition. Source: Ha et al. [16] Copyright, 2014, Advanced Optical Materials.

2.4.5 Spraying Method

The spray coating process is an attractive option due to its wide use in industry and its ability to rapidly produce and efficiently utilize materials. It typically involves four processes: liquid generation, droplet transfer to the substrate, droplet aggregation into a wet film, and film drying. For instance, Meng et al. from the Institute of Physics at the Chinese Academy of Sciences used a two-step ultrasonic spraying method to produce centimeter-scale, uniform, and smooth perovskite $MAPbI_3$ films for large-area and high-efficiency solar cells. Compared to other deposition methods, this approach offers a larger area and higher smoothness, as well as better crystallinity, lower non-radiative complexation, and more efficient interfacial charge extraction performance. System optimization resulted in a highly efficient mesoscopic solar cell with a 16.03% efficiency (Figure 2.43).

The synthesis process of perovskite QDs involves combining spraying and supersaturated crystallization methods. Dai et al. developed a new method for synthesizing perovskite QDs using this approach. The process begins with the spraying of a 1 M lead bromide precursor dissolved in DMSO onto a titanium dioxide substrate heated to 60°C. Next, a solution of isopropanol dissolved in MAI is sprayed onto the PbI_2 wet film at 80°C. The film is then heat-treated to promote $MAPbI_3$ crystallization [55]. The resulting $MAPbBr_3$ QDs are highly crystalline and exhibit 100% absolute PLQY values for both colloidal solutions and QD films. The QD films

Figure 2.43 Schematic diagram of the spraying method. Source: Mohamad et al. [78] Copyright, 2022, American Chemical Society.

exhibit saturated green emission at 526 nm with a half-peak width of 20 nm and have been used as active light-emitting layers in QD-LEDs. These devices demonstrate good performance with an external quantum efficiency of 1.81% for EL.

2.4.6 Template Method

Kovalenko et al. developed the template method, a ligand-free approach for preparing inorganic PeNCs embedded in mesoporous silica [21]. The mesoporous silica acts as a template for the growth of inorganic PeNCs, where the precursors are grown into nanocrystals within the mesopores. The size and fluorescence color of the crystals can be controlled by the pore size and precursor ratio. This method is simple and convenient, and its lower organic solvent content compared to other methods makes it more suitable for efficient mass production (Figure 2.44).

2.4.7 Non-Template Method

Schmidt et al. synthesized $MAPbBr_3$ PeNCs without the use of a template method, resulting in a size of approximately 6 nm [79]. In this method, long-chain alkyl ammonium bromides such as octadecyl ammonium bromide ($CH_3(CH_2)_{17}NH_3Br$) were added to a solution of ODE containing OA at a temperature of 80°C to control the size of the nanocrystals. MABr and $PbBr_2$ were added sequentially to the solution, pre-dissolved in a small amount of DMF solvent to increase their solubility

Figure 2.44 Schematic diagram of the template method. Source: Dirin et al. [21] Copyright © 2022 American Chemical Society.

in ODE. The addition of acetone resulted in the formation of a yellow precipitate in the solution, which was separated by centrifugation to obtain the precipitate. The resulting $MAPbBr_3$ PeNCs were able to dissolve in toluene, indicating successful formation. The PLQY of these PeNCs was measured to be 20%.

Carrero's group and others conducted an in-depth study to improve the PLQY of $MAPbBr_3$ PeNCs using the non-template method. They were able to synthesize PeNCs with a PLQY of about 83% by regulating the ratio of MABr, $PbBr_2$, OABr, and ODE in the DMF solution [80].

2.5 Conclusion and Outlook

In conclusion, a wide range of methods have been demonstrated for the synthesis of perovskite QD materials for use in light-emitting devices. These include solution-based methods such as solvothermal, assisted ligand reprecipitation, anion exchange, spin coating, thermal injection, and templating techniques. Vapor deposition methods such as thermal vapor deposition, printing, spraying, and non-templating methods have also been used. However, there are still many challenges to be faced, such as the control of crystal growth direction and size, high defect densities, and toxicity issues.

Future research directions in this area should focus on developing more efficient and scalable synthesis methods that are cost-effective and environmentally friendly. Additionally, improving the stability and reproducibility of the synthesized materials is crucial for their commercialization. The development of new materials and structures, such as perovskite NWs and quantum rods, may also lead to new and improved light-emitting devices with enhanced performance. The research directions for perovskite QD synthesis may involve exploring new materials and methods to improve their stability and efficiency, as well as investigating their potential applications in various fields such as solar cells, displays, sensors, and biomedical imaging.

In addition, efforts can be made to further optimize the synthesis process and enhance the control over the size, morphology, and crystal structure of perovskite QDs. The use of new ligands, surfactants, and solvents can also be explored to improve their optical and electrical properties.

Finally, the integration of perovskite QD materials into other optoelectronic devices such as photodetectors, solar cells, and lasers should also be explored.

Overall, perovskite QDs have shown great potential in various fields due to their excellent electronic and optical properties. With the continuous advancement of synthesis methods and further research into their properties and applications, perovskite QDs are expected to become a key player in the next generation of optoelectronic materials.

References

1 Wells, H.L. (1893). Über die cäsium-und kalium-bleihalogenide. *Zeitschrift für anorganische Chemie* 3: 195–210.

2 Brivio, F., Walker, A.B., and Walsh, A. (2013). Structural and electronic properties of hybrid perovskites for high-efficiency thin-film photovoltaics from first-principles. *APL Materials* 1: 042111.

3 Chen, J., Zhou, S., Jin, S. et al. (2016). Crystal organometal halide perovskites with promising optoelectronic applications. *Journal of Materials Chemistry C* 4: 11–27.

4 Konstantakou, M. and Stergiopoulos, T. (2017). A critical review on tin halide perovskite solar cells. *Journal of Materials Chemistry A* 5: 11518–11549.

5 Le, Q.V., Shin, J.W., Jung, J.-H. et al. (2017). Control of the crystal growth shape in $CH_3NH_3PbBr_3$ perovskite materials. *Journal of Nanoscience and Nanotechnology* 17: 8169–8174.

6 Li, Z.J., Hofman, E., Li, J. et al. (2018). Photoelectrochemically active and environmentally stable $CsPbBr_3/TiO_2$ core/shell nanocrystals. *Advanced Functional Materials* 28: 1704288.

7 Yin, W.-J., Yang, J.-H., Kang, J. et al. (2015). Halide perovskite materials for solar cells: a theoretical review. *Journal of Materials Chemistry A* 3: 8926–8942.

8 Zhang, W., Eperon, G.E., and Snaith, H.J. (2016). Metal halide perovskites for energy applications. *Nature Energy* 1: 1–8.

9 Zhou, S., Tang, R., and Yin, L. (2017). Slow-photon-effect-induced photoelectrical-conversion efficiency enhancement for carbon-quantum-dot-sensitized inorganic $CsPbBr_3$ inverse opal perovskite solar cells. *Advanced Materials* 29: 1703682.

10 Van Le, Q., Jang, H.W., and Kim, S.Y. (2018). Recent advances toward high-efficiency halide perovskite light-emitting diodes: review and perspective. *Small Methods* 2: 1700419.

11 Chen, Q., De Marco, N., Yang, Y.M. et al. (2015). Under the spotlight: the organic–inorganic hybrid halide perovskite for optoelectronic applications. *Nano Today* 10: 355–396.

12 Mitzi, D. (2001). Templating and structural engineering in organic-inorganic perovskites. *J. Chem. Soc. Dalton Trans* 1: 1–12.

13 Mitzi, D.B., Feild, C., Harrison, W. et al. (1994). Conducting tin halides with a layered organic-based perovskite structure. *Nature* 369: 467–469.

14 Sichert, J.A., Tong, Y., Mutz, N. et al. (2015). Quantum size effect in organometal halide perovskite nanoplatelets. *Nano Letters* 15: 6521–6527.

15 Yuan, M., Quan, L.N., Comin, R. et al. (2016). Perovskite energy funnels for efficient light-emitting diodes. *Nature Nanotechnology* 11: 872–877.

16 Ha, S.T., Liu, X., Zhang, Q. et al. (2014). Synthesis of organic-inorganic lead halide perovskite nanoplatelets: towards high-performance perovskite solar cells and optoelectronic devices. *Advanced Optical Materials* 2: 838–844.

17 Tyagi, P., Arveson, S.M., and Tisdale, W.A. (2015). Colloidal organohalide perovskite nanoplatelets exhibiting quantum confinement. *The Journal of Physical Chemistry Letters* 6: 1911–1916.

18 Ling, Y., Yuan, Z., Tian, Y. et al. (2016). Bright light-emitting diodes based on organometal halide perovskite nanoplatelets. *Advanced Materials* 28: 305–311.

19 Kumar, S., Jagielski, J., Kallikounis, N. et al. (2017). Ultrapure green light-emitting diodes using two-dimensional formamidinium perovskites: achieving recommendation 2020 color coordinates. *Nano letters* 17: 5277–5284.

20 Kojima, A., Ikegami, M., Teshima, K. et al. (2012). Highly luminescent lead bromide perovskite nanoparticles synthesized with porous alumina media. *Chemistry Letters* 41: 397–399.

21 Dirin, D.N., Protesescu, L., Trummer, D. et al. (2016). Harnessing defect-tolerance at the nanoscale: Highly luminescent lead halide perovskite nanocrystals in mesoporous silica matrixes. *Nano Letters* 16: 5866–5874.

22 Protesescu, L., Yakunin, S., Bodnarchuk, M.I. et al. (2015). Nanocrystals of cesium lead halide perovskites ($CsPbX_3$, X = Cl, Br, and I): Novel optoelectronic materials showing bright emission with wide color gamut. *Nano Letters* 15: 3692–3696.

23 Zhang, F., Zhong, H., Chen, C. et al. (2015). Brightly luminescent and color-tunable colloidal $CH_3NH_3PbX_3$ (X= Br, I, Cl) quantum dots: potential alternatives for display technology. *ACS Nano* 9: 4533–4542.

24 Wang, D., Wu, D., Dong, D. et al. (2016). Polarized emission from $CsPbX_3$ perovskite quantum dots. *Nanoscale* 8: 11565–11570.

25 Chiba, T., Hoshi, K., Pu, Y.-J. et al. (2017). High-efficiency perovskite quantum-dot light-emitting devices by effective washing process and interfacial energy level alignment. *ACS applied materials & interfaces* 9: 18054–18060.

26 Li, G., Rivarola, F.W.R., Davis, N.J. et al. (2016). Highly efficient perovskite nanocrystal light-emitting diodes enabled by a universal crosslinking method. *Advanced Materials* 28: 3528–3534.

27 Im, J.H., Lee, C.R., Lee, J.W. et al. (2011). 6.5% efficient perovskite quantum-dot-sensitized solar cell. *Nanoscale* 3: 4088–4093.

28 Han, T.-H., Jang, K.Y., Dong, Y. et al. (2022). A roadmap for the commercialization of perovskite light emitters. *Nature Reviews Materials* 7: 757–777.

29 Manser, J.S., Christians, J.A., and Kamat, P.V. (2016). Intriguing optoelectronic properties of metal halide perovskites. *Chemical Reviews* 116: 12956–13008.

30 Travis, W., Glover, E.N.K., Bronstein, H. et al. (2016). On the application of the tolerance factor to inorganic and hybrid halide perovskites: a revised system. *Chemical Science* 7: 4548–4556.

31 Tan, D. and Garcia, F. (2019). Main group mechanochemistry: from curiosity to established protocols. *Chemical Society Reviews* 48: 2274–2292.

32 Palazon, F., El Ajjouri, Y., and Bolink, H.J. (2020). Making by grinding: mechanochemistry boosts the development of halide perovskites and other multinary metal halides. *Advanced Energy Materials* 10: 1902499.

33 Tong, Y., Bladt, E., Aygueler, M.F. et al. (2016). Highly luminescent cesium lead halide perovskite nanocrystals with tunable composition and thickness by ultrasonication. *Angewandte Chemie-International Edition* 55: 13887–13892.

34 Huang, H., Xue, Q., Chen, B. et al. (2017). Top-down fabrication of stable methylammonium lead halide perovskite nanocrystals by employing a mixture of ligands as coordinating solvents. *Angewandte Chemie-International Edition* 56: 9571–9576.

35 Li, Y., Huang, H., Xiong, Y. et al. (2018). Revealing the formation mechanism of $CsPbBr_3$ perovskite nanocrystals produced via a slowed-down microwave-assisted synthesis. *Angewandte Chemie-International Edition* 57: 5833–5837.

36 Pan, Q., Hu, H., Zou, Y. et al. (2017). Microwave-assisted synthesis of high-quality "all-inorganic" $CsPbX_3$ (X = Cl, Br, I) perovskite nanocrystals and their application in light emitting diodes. *Journal of Materials Chemistry C* 5: 10947–10954.

37 Chen, M., Zou, Y., Wu, L. et al. (2017). Solvothermal synthesis of high-quality all-inorganic cesium lead halide perovskite nanocrystals: from nanocube to ultrathin nanowire. *Advanced Functional Materials* 27: 1701121.

38 Chen, D., Fang, G., Chen, X. et al. (2018). Mn-doped $CsPbCl_3$ perovskite nanocrystals: solvothermal synthesis, dual-color luminescence and improved stability. *Journal of Materials Chemistry C* 6: 8990–8998.

39 Murray, C.B., Norris, D.J., and Bawendi, M.G. (1993). Synthesis and characterization of nearly monodisperse CdE (E = S, Se, Te) semiconductor nanocrystallites. *Journal of the American Chemical Society* 115: 8706–8715.

40 Lignos, I., Stavrakis, S., Nedelcu, G. et al. (2016). Synthesis of cesium lead halide perovskite nanocrystals in a droplet-based microfluidic platform: fast parametric space mapping. *Nano Letters* 16: 1869–1877.

41 Koolyk, M., Amgar, D., Aharon, S. et al. (2016). Kinetics of cesium lead halide perovskite nanoparticle growth; focusing and de-focusing of size distribution. *Nanoscale* 8: 6403–6409.

42 Liu, P., Chen, W., Wang, W. et al. (2017). Halide-rich synthesized cesium lead bromide perovskite nanocrystals for light-emitting diodes with improved performance. *Chemistry of Materials* 29: 5168–5173.

43 Yassitepe, E., Yang, Z., Voznyy, O. et al. (2016). Amine-free synthesis of cesium lead halide perovskite quantum dots for efficient light-emitting diodes. *Advanced Functional Materials* 26: 8757–8763.

44 Protesescu, L., Yakunin, S., Bodnarchuk, M.I. et al. (2016). Monodisperse formamidinium lead bromide nanocrystals with bright and stable green photoluminescence. *Journal of the American Chemical Society* 138: 14202–14205.

45 Pan, A., He, B., Fan, X. et al. (2016). Insight into the ligand-mediated synthesis of colloidal $CsPbBr_3$ perovskite nanocrystals: the role of organic acid, base, and cesium precursors. *ACS Nano* 10: 7943–7954.

46 Song, J., Xu, L., Li, J. et al. (2016). Monolayer and few-layer all-inorganic perovskites as a new family of two-dimensional semiconductors for printable optoelectronic devices. *Advanced Materials* 28: 4861–4869.

47 Zhang, D., Yu, Y., Bekenstein, Y. et al. (2016). Ultrathin colloidal cesium lead halide perovskite nanowires. *Journal of the American Chemical Society* 138: 13155–13158.

48 Papavassiliou, G.C., Pagona, G., Karousis, N. et al. (2012). Nanocrystalline/microcrystalline materials based on lead-halide units. *Journal of Materials Chemistry* 22: 8271–8280.

49 Huang, H., Susha, A.S., Kershaw, S.V. et al. (2015). Control of emission color of high quantum yield $CH_3NH_3PbBr_3$ perovskite quantum dots by precipitation temperature. *Advanced Science* 2: 1500194.

50 Arunkumar, P., Gil, K.H., Won, S. et al. (2017). Colloidal organolead halide perovskite with a high Mn solubility limit: A step toward Pb-free luminescent quantum dots. *The Journal of Physical Chemistry Letters* 8: 4161–4166.

51 Luo, B., Pu, Y.C., Lindley, S.A. et al. (2016). Organolead halide perovskite nanocrystals: branched capping ligands control crystal size and stability. *Angewandte Chemie-International Edition* 55: 8864–8868.

52 Veldhuis, S.A., Tay, Y.K., Eugene, B. et al. (2017). Benzyl alcohol-treated $CH_3NH_3PbBr_3$ nanocrystals exhibiting high luminescence, stability, and ultralow amplified spontaneous emission thresholds. *Nano Letters* 17: 7424–7432.

53 Luo, B., Naghadeh, S.B., Allen, A. et al. (2017). Peptide-passivated lead halide perovskite nanocrystals based on synergistic effect between amino and carboxylic functional groups. *Advanced Functional Materials* 27: 1604018.

54 Shamsi, J., Abdelhady, A.L., Accornero, S. et al. (2016). *N*-methylformamide as a source of methylammonium ions in the synthesis of lead halide perovskite nanocrystals and bulk crystals. *ACS Energy Letters* 1: 1042–1048.

55 Dai, S.W., Hsu, B.W., Chen, C.Y. et al. (2018). Perovskite quantum dots with near unity solution and neat-film photoluminescent quantum yield by novel spray synthesis. *Advanced Materials* 30: 1705532.

56 Kumar, S., Jagielski, J., Yakunin, S. et al. (2016). Efficient blue electroluminescence using quantum-confined two-dimensional perovskites. *ACS Nano* 10: 9720–9729.

57 Cho, J., Choi, Y.H., O'Loughlin, T.E. et al. (2016). Ligand-mediated modulation of layer thicknesses of perovskite methylammonium lead bromide nanoplatelets. *Chemistry of Materials* 28: 6909–6916.

58 Levchuk, I., Herre, P., Brandl, M. et al. (2017). Ligand-assisted thickness tailoring of highly luminescent colloidal $CH_3NH_3PbX_3$ (X = Br and I) perovskite nanoplatelets. *Chemical Communications* 53: 244–247.

59 Ahmed, G.H., Yin, J., Bose, R. et al. (2017). Pyridine-induced dimensionality change in hybrid perovskite nanocrystals. *Chemistry of Materials* 29: 4393–4400.

60 Kirakosyan, A., Kim, J., Lee, S.W. et al. (2017). Optical properties of colloidal $CH_3NH_3PbBr_3$ nanocrystals by controlled growth of lateral dimension. *Crystal Growth & Design* 17: 794–799.

61 Huang, H., Raith, J., Kershaw, S.V. et al. (2017). Growth mechanism of strongly emitting $CH_3NH_3PbBr_3$ perovskite nanocrystals with a tunable bandgap. *Nature Communications* 8: 996.

62 Weidman, M.C., Seitz, M., Stranks, S.D. et al. (2016). Highly tunable colloidal perovskite nanoplatelets through variable cation, metal, and halide composition. *Acs Nano* 10: 7830–7839.

63 Perumal, A., Shendre, S., Li, M. et al. (2016). High brightness formamidinium lead bromide perovskite nanocrystal light emitting devices. *Scientific Reports* 6: 36733.

64 Levchuk, I., Osvet, A., Tang, X. et al. (2018). Brightly luminescent and color-tunable formamidinium lead halide perovskite $FAPbX_3$ (X = CI, Br, I) colloidal nanocrystals. *Nano Letters* 18: 6106–6106.

65 Duong Nguyen, M., Kim, J., Hyon, J. et al. (2017). Room-temperature synthesis of widely tunable formamidinium lead halide perovskite nanocrystals. *Chemistry of Materials* 29: 5713–5719.

66 Li, X., Wu, Y., Zhang, S. et al. (2016). $CsPbX_3$ quantum dots for lighting and displays: Room-temperature synthesis, photoluminescence superiorities, underlying origins and white light-emitting diodes. *Advanced Functional Materials* 26: 2435–2445.

67 Seth, S. and Samanta, A. (2016). A facile methodology for engineering the morphology of $CsPbX_3$ perovskite nanocrystals under ambient condition. *Scientific Reports* 6: 37693.

68 Kostopoulou, A., Sygletou, M., Brintakis, K. et al. (2017). Low-temperature benchtop-synthesis of all-inorganic perovskite nanowires. *Nanoscale* 9: 18202–18207.

69 Zhang, X., Bai, X., Wu, H. et al. (2018). Water-assisted size and shape control of $CsPbBr_3$ perovskite nanocrystals. *Angewandte Chemie-International Edition* 57: 3337–3342.

70 Zhang, Z., Chen, Z., Yuan, L. et al. (2017). A new passivation route leading to over 8% efficient PbSe quantum-dot solar cells via direct ion exchange with perovskite nanocrystals. *Advanced Materials* 29: 1703214.

71 Wang, K., Lin, Z.-Y., Zhang, Z. et al. (2023). Suppressing phase disproportionation in quasi-2D perovskite light-emitting diodes. *Nature communications* 14: 397.

72 Huang, X., Guo, Q., Yang, D. et al. (2020). Reversible 3D laser printing of perovskite quantum dots inside a transparent medium. *Nature Photonics* 14: 82–88.

73 Sun, K., Tan, D., Fang, X. et al. (2022). Three-dimensional direct lithography of stable perovskite nanocrystals in glass. *Science* 375: 307–310.

74 Sun, C., Jiang, Y., Wei, K. et al. (2023). Perovskite light-emitting diodes toward commercial full-colour displays: progress and key technical obstacles. *Light: Advanced Manufacturing* 4: 15.

75 Kojima, A., Teshima, K., Shirai, Y. et al. (2009). Organometal halide perovskites as visible-light sensitizers for photovoltaic cells. *Journal of the American Chemical Society* 131: 6050.

76 Bade, S.G.R., Li, J., Shan, X. et al. (2016). Fully printed halide perovskite light-emitting diodes with silver nanowire electrodes. *ACS Nano* 10: 1795–1801.

77 Chen, J., Morrow, D.J., Fu, Y. et al. (2017). Single-crystal thin films of cesium lead bromide perovskite epitaxially grown on metal oxide perovskite ($SrTiO_3$). *Journal of the American Chemical Society* 139: 13525–13532.

78 Mohamad, D.K., Griffin, J., Bracher, C. et al. (2016). Spray-cast multilayer organometal perovskite solar cells fabricated in air. *Advanced Energy Materials* 6: 1600994.

79 Schmidt, L.C., Pertegas, A., Gonzalez-Carrero, S. et al. (2014). Nontemplate synthesis of $CH_3NH_3PbBr_3$ perovskite nanoparticles. *Journal of the American Chemical Society* 136: 850–853.

80 Gonzalez-Carrero, S., Schmidt, L.C., Rosa-Pardo, I. et al. (2018). Colloids of naked $CH_3NH_3PbBr_3$ perovskite nanoparticles: synthesis, stability, and thin solid film deposition. *ACS Omega* 3: 1298–1303.

3

Near-Infrared Perovskite Light-Emitting Devices

3.1 Introduction

Near-infrared (NIR) refers to the portion of the electromagnetic spectrum that lies between visible light and middle infrared (MIR). According to the American Society for Testing and Materials (ASTM) definition, NIR includes electromagnetic waves with wavelengths ranging from 780 to 2526 nm. NIR is commonly divided into two regions: NIR short-wave (780–1100 nm) and NIR long-wave (1100–2526 nm). As illustrated in Figure 3.1, NIR light sources have a wide range of applications, such as night vision devices, optical communication, sensing, biological imaging, facial recognition, and surveillance [1].

The key performance metrics of a near-infrared light-emitting diode (NIR LED) are the external quantum efficiency (EQE), optical output power (P), and radiance (R). The EQE is the product of four parameters, namely, the fraction of charge carriers passing through the electrodes that are injected into the emissive layer (γ), the fraction of excitons whose states have spin-allowed optical transitions (r), the photoluminescence quantum yield (PLQY or φ_{PL}) of the emitter, and the light outcoupling efficiency (η_{out}), which describes the probability that a photon escapes the device in the viewing direction. The parameters r and φ_{PL} are intrinsic properties of the selected emitters, γ depends on the device configuration and the charge-injection interlayers, and η_{out} depends on the refractive indices of the functional layers and the substrate, along with the orientation of the transition dipole moments of the emitter on the substrate [2].

At present, NIR LEDs are typically epitaxially grown from III to V semiconductors on sapphire substrates and require processing in a specialized clean room. However, this preparation method is expensive and restricts their use in integrated optoelectronic devices [3]. Consequently, there has been a persistent need to explore NIR-emitting materials that can effectively address these challenges and achieve high efficiency. Organic semiconductors and colloidal sulfur elemental lead quantum dots have received considerable attention in this regard. Nonetheless, the EQE of NIR-emitting devices based on these materials remains unsatisfactory due to certain inherent defects [4].

With respect to organic semiconductors, the energy gap law states that as the vibrational coupling increases, electronic energy can more readily convert to

Perovskite Light Emitting Diodes: Materials and Devices, First Edition. Hong Meng.

Figure 3.1 NIR application scenarios: (a) night vision. Source: Getmilitaryphotos/Adobe Stock Photos; (b) optical communication. Source: iconimage/Adobe Stock; (c) sensing.

vibrational energy. This can lead to the rapid formation of excited states [5, 6]. Consequently, as the emission wavelength increases, the luminescence quantum yield of organic semiconductors decreases. Moreover, in low-barrier materials, extending the conjugated system can result in tighter molecular packing, which promotes trilinear-state annihilation. Nevertheless, by employing Pt(II) complexes or thermally activated delayed fluorescence (TADF) materials, light-emitting diodes (LEDs) with peak emission wavelengths between 700 and 800 nm can attain external quantum efficiencies exceeding 10%. However, the device performance of organic LEDs deteriorates significantly at longer emission wavelengths (e.g. near 900 nm). Although quantum dot materials such as PbS and PbSe exhibit luminescence quantum yields greater than 50% in solution, their performance in devices is still limited [7]. Nevertheless, the lateral interdot coupling in quantum dot materials can facilitate transport-assisted capture and promote competition between exciton dissociation and radiative recombination processes, while this also renders them susceptible to strong luminescence annihilation in solid-state films [8].

3.2 Progress in Near-Infrared Perovskite Luminescence Materials

Metal halide perovskite materials have gained significant attention due to their solution processability, low cost, tunable bandgap, high optical absorption coefficient, long exciton diffusion length, high color purity, and high fluorescence quantum yield. Figure 3.2 illustrates some of the notable advances made in the EQE of NIR perovskite light-emitting devices in recent years.

In 2014, Friend's group demonstrated the use of perovskite materials as a light-emitting layer in the fabrication of LEDs. Although the device efficiency (EQE: 0.76%) was not high at that time, the low cost, tunability of the emission wavelength, and other advantages of perovskite materials sparked significant interest in research on solution-processed perovskite light-emitting materials [9]. Since then, numerous methods have been proposed to improve the performance of NIR perovskite materials. Leveraging the favorable exciton diffusion length of perovskite nanomaterials and the high fluorescence quantum efficiency of perovskite quantum dot materials (such as PbS, PbSe, and CdSe quantum dots), Sargent's group

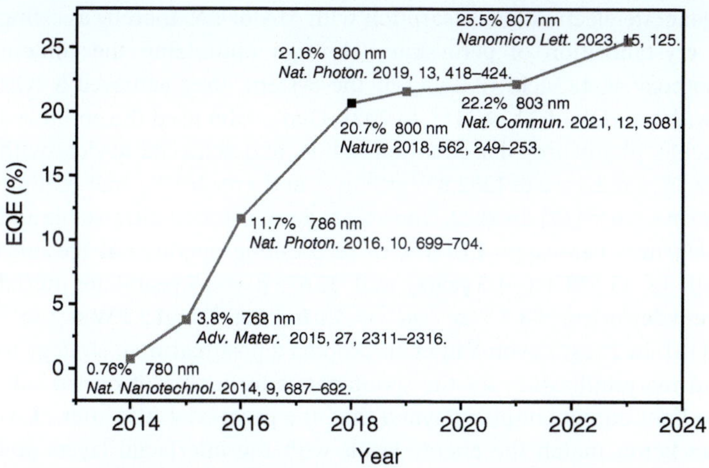

Figure 3.2 Progress in the study of near-infrared perovskite light-emitting devices.

introduced the concept of quantum dots in perovskite (QDiP) in 2015. This approach involves the heterogeneous epitaxial growth of perovskite with multiple types of quantum dots to create a structure in which the quantum dots are grown within a perovskite matrix. Additionally, calixarene-encapsulated quantum dots have been employed to further enhance the performance of QDiP devices [10]. In 2016, Gong and colleagues achieved a 5.2% EQE in NIR light-emitting devices by optimizing the halogen ion composition in the perovskite to achieve a better lattice match [11]. Subsequent researchers have explored the possibility of enhancing EQE by modifying the composition of quantum dots and the dimensionality of the perovskite structure in the QDiP system [12, 13]. Researchers have proposed various experimental approaches to improve the efficiency of simple perovskite NIR-emitting materials, such as size and dimensional tuning, defect passivation, and enhancing the light extraction efficiency. For example, in 2017, Rand et al. demonstrated the growth of 5.4 ± 0.8 nm nanocrystals in situ by incorporating organo-ammonium halide cations into perovskite precursors through size tuning. By using 0.8 nm nanocrystals, they obtained NIR light-emitting devices with an EQE of 7.9% [14]. In 2019, researchers were able to achieve significant improvements in light extraction efficiency and external quantum efficiency by adding 5-aminopentanoic acid (5AVA) and an excess of ammonium halide to the perovskite precursor, which resulted in the spontaneous formation of submicron-sized crystals in the thin film. This led to an EQE of 21.6% in the prepared NIR perovskite LEDs. In 2017, Rand and co-workers also achieved a high EQE of 7.9% by growing in situ nanocrystals with a particle size of 5.4 ± 0.8 nm through size tuning using organo-ammonium halide cations added to perovskite precursors. Other experimental measures to improve the efficiency of simple perovskite NIR-emitting materials include size and dimensional tuning, as well as defect passivation [15]. In 2021, Wang's team focused on improving the crystallization process of perovskite in thin films. They compared various amines, such as pentanamine and pentanoic acid, and found that amine additives could generate electrostatic adsorption with MA or FA, thereby assisting in the oriented crystallization of perovskite. Through optimizing the content of 2-(2-(2-aminoethoxy)acetic acid (AEAA) in the system, they achieved a NIR perovskite LED with an EQE of 22.2% [16]. In 2021, Guo's team used the optimized passivation molecule, phenylpropylammonium iodide, and achieved devices with an efficiency of 17.5%, a radiance of 1282.8 W sr^{-1} m^{-2}, and a record T_{50} half-lifetime of 130 h under 100 mA cm^{-2} [17]. In 2022, Guo's team demonstrated ultra-stable and efficient NIR (~800 nm) perovskite LEDs with record-long operational lifetimes (T_{50}, extrapolated) of 11 539 h (~1.3 years) and 32 675 h (~3.7 years) for initial radiance (or current densities) of 3.7 W sr^{-1} m^{-2} (~5.0 mA cm^{-2}) and 2.1 W sr^{-1} m^{-2} (~3.2 mA cm^{-2}) [18]. In 2023, Keyou Yan et al. adopted a posttreatment strategy to induce downward recrystallization for the spontaneous formation of buried sub-microcavities as a light output coupler. Beyond that, the perovskite film after PEAI posttreatment can better match the energy levels with the interfacial layers and reduce the carrier injection barrier, which provides advantages for the fabrication of highly efficient PeLEDs. Therefore, the PeLED achieved an unprecedented EQE of 25.5% with a radiance of 487 W sr^{-1} m^{-2} and low rolling-off. And the T_{75} stability exceeds 15 h at a current density of 62.5 mA cm^{-2}. This work provides a self-assembly method to integrate an output coupler for boosting the performance of PeLEDs [19].

This chapter provides an overview of NIR perovskite luminescent materials and explores methods for enhancing the EQE of devices. The content is divided into three main sections: (i) an introduction to the structure, properties, and research progress of NIR perovskite materials; (ii) a summary of enhancement schemes for improving the efficiency of these NIR perovskite devices; and (iii) a review of measures to improve the stability of NIR light-emitting devices, along with an outlook on future research directions for NIR perovskite materials.

3.3 Near-Infrared Perovskite Luminescent Materials

This section provides a brief overview of the common types of NIR perovskite luminescent materials and their current research progress. The perovskite layer is the most critical component that affects the performance of NIR light-emitting devices. For a detailed explanation of the structure and synthesis of metal halide perovskites, please refer to the review on perovskite nanocrystals [20].

3.3.1 Methylamine Lead Iodide ($MAPbI_3$)

Metal halide perovskites are highly advantageous compared to other types of light-emitting materials. They have a narrower luminescence spectrum than organic light-emitting devices, and they are easier to synthesize than colloidal quantum dots. Furthermore, they can be prepared on flexible substrates over larger areas than III–V semiconductors, making them a promising candidate for large-scale production of light-emitting devices [8]. Currently, NIR-perovskite materials can be classified into two categories: all-inorganic perovskites and organometallic perovskites. Among these, organometallic perovskites are the focus of more research on the NIR wavelength.

NIR-emitting organometallic perovskites mainly include $MAPbI_3$, $FAPbI_3$, and their dopants. Through years of research, perovskite structures have been explored in various forms in terms of scale and dimensionality. Currently, the most widely used perovskite structures for NIR-emitting devices are three-dimensional perovskites [14, 21]. Additionally, quasi-two-dimensional structures formed by the Ruddlesden–Popper phase have also been extensively studied in the context of NIR-perovskite materials, alongside three-dimensional perovskites. Over the years, perovskite structures have been explored in various forms, including different scales and dimensionalities [22, 23]. Among them, conventional 3D perovskites follow the chemical formula ABX_3, where A and B are cations and X is an anion. The high plasticity of the lattice in perovskite allows for the doping of different ions into the bulk, which can change the formation energy and improve stability, as well as tune the bandgap to change the luminescence wavelength of the perovskite material. However, not all atoms can be doped into the perovskite structure, and the doping must follow the Goldschmidt tolerance factor rule (τ) and octahedral coefficient (μ). In particular, other ions can be used to fully or partially replace the organic cation at position A, Pb^{2+} at position B, and I^- at position X. The following equations describe the Goldschmidt tolerance factor rule and octahedral coefficient:

$$\tau = \frac{r_X + r_A}{r_X + r_B} \tag{3.1}$$

$$\mu = \frac{r_B}{r_X} \tag{3.2}$$

where r_A, r_B, and r_X are the radii of ions A, B, and X, respectively. Perovskite structures typically exhibit a cubic structure when the value of τ is close to 1.0. However, when there is a significant difference between the 1.0 and τ values, the perovskite structure may take on a tetragonal or orthorhombic structure with lower symmetry, commonly known as the yellow phase [24]. Given the volatile nature of organic cations, some studies in the current literature suggest partially replacing them at the A-site with Cs^+ to enhance structural stability. This is particularly relevant due to the potential toxicity of Pb^{2+} and the limitation of the luminescence wavelength of lead-based halide perovskites to the first NIR window (NIR-I, 700–900 nm) [12]. To reduce potential toxicity, Pb^{2+} at the B-position in perovskite structures can be replaced with the less toxic, isochemically equivalent Sn^{2+} or with an Ag^+ and an In^{3+} (as discussed in the section on lead-free perovskites). Additionally, partial replacement of I^- with Cl^- at the X-position can enhance both efficiency and crystallinity [25]. However, this substitution can lead to a blueshift in the NIR wavelength [22]. While replacing I^- with Cl^- can improve efficiency and crystallinity, it may also cause a blueshift in the NIR wavelength. Furthermore, there are additional quasi-two-dimensional metal-organic perovskite structures, such as the multiple quantum well (MQW) structure, which are also described in the literature.

Organic ligands can aid in the synthesis of organometallic perovskites by promoting crystal growth in the nanometer size range and passivating surface defects,

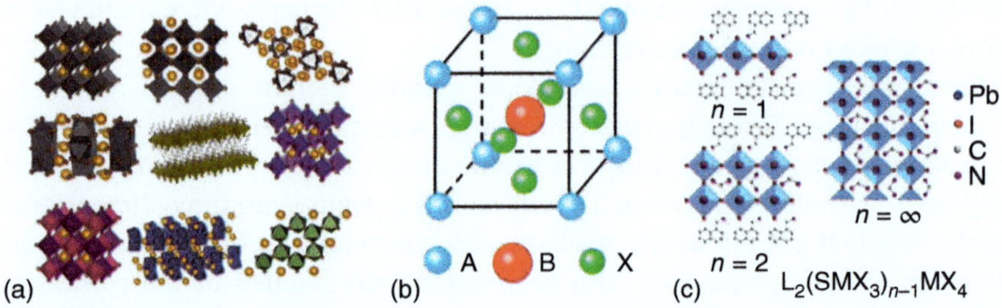

Figure 3.3 (a) Perovskite structure in different dimensions; (b) 3D perovskite structure; (c) quasi-2D perovskite structure. (a) Source: Shamsi et al. [20]/American Chemical Society. (b and c) Source: Wang et al. [22]/Springer Nature.

much like in conventional nanocrystal synthesis. This allows for precise tuning of the size and shape of nanocrystals, which can be either bulk-like (i.e. large enough to exhibit optical properties similar to those of bulk crystals or thin films) or nanoscale nanoplates, such as nanoplates (NPLs), nanosheets, nanowires, and quantum dots. The size of these nanostructures can be controlled down to a single perovskite layer with a thickness much smaller than the excitonic Bohr radius, resulting in strong quantum confinement effects. Moreover, the composition, structure, and size of nanocrystals can be modified not only during synthesis but also through post-synthesis processes such as ion exchange or exfoliation (Figure 3.3) [20]. Thin film materials for NIR applications are typically synthesized through low-temperature solution methods, which can be categorized into one-pot solution synthesis and poor solvent drop addition methods. The latter approach is particularly effective in producing perovskite films with minimal pinholes and cracks at the grain boundaries [26]. In situ synthesis of chalcocite through this method can yield uniform and dense films [14]. Currently, most synthesis methods rely on the use of poor solvent drop addition.

The first high-brightness infrared (IR) electroluminescent material synthesized in solution was the organometallic halide perovskite. Friend's group reported 13.2 W sr^{-1} m^{-2} of IR radiation from $MAPbI_{3-x}Cl_x$ at a current density of 363 mA cm^{-2}. The material exhibited maximum external and internal quantum efficiencies of 0.76% and 3.4%, respectively [9]. Feng Gao's group achieved excellent performance in NIR $FAPbI_3$-emitting devices through rational molecular passivation, resulting in a record EQE of 21.6% [27]. Metal halide perovskites are the most extensively studied NIR perovskite materials due to their simplicity in synthesis when compared to perovskite-coated quantum dot systems. However, the luminescence wavelengths of organo-lead-based halide perovskites are limited to the first NIR window due to fundamental bandgap restrictions [12]. The replacement or partial substitution of Pb^{2+} with Sn^{2+} can extend the IR wavelengths beyond NIR-I.

3.3.2 NIR-Emitting Materials Based on Perovskite

Colloidal quantum dots are emerging as promising IR light-emitting materials due to their tunable bandgap, high fluorescence quantum efficiency, and their solution preparation. However, quantum dots exhibit strong luminescence annihilation in the solid-state thin film state. To mitigate this issue, protective shell structures can be grown, insulating organic ligands can be encapsulated, or polymeric substrates can be introduced. However, these methods often require high voltages to inject sufficient current to achieve brighter luminescence, resulting in increased power loss. Therefore, current quantum dot films compromise between luminescence efficiency and charge transport, leading to unacceptably high power consumption [8]. To address the issue of luminescence annihilation in solid-state thin-film quantum dot materials, researchers have explored embedding the quantum dots into high-mobility perovskite substrates, which benefit from good diffusion lengths [10–13, 28]. One strategy to overcome the limitations of colloidal quantum dots in the NIR is to embed them into high-mobility perovskite substrates. This approach can improve the luminescence efficiency of quantum dots in the NIR and overcome the wavelength limitations of perovskite materials, such as the conventional $MAPbX_3$ luminescent materials that can only emit at a maximum wavelength of 780 nm [29].

To achieve epitaxial bonding between the perovskite substrate and the quantum dots, it is not sufficient for the two materials to have similar heterogeneous structures, as shown in Figure 3.4a. Energy considerations must also be taken into account. The epitaxial interface structure between the perovskite substrate and the quantum dots plays a crucial role in providing excellent passivation properties and improving fluorescence quantum efficiency. Currently, perovskite-coated quantum dots have a heterogeneous epitaxial structure with an I-type energy band arrangement (Figure 3.4b), which is suitable for luminescent materials by enhancing radiation complexation through spatially restricted interaction. The I-band arrangement allows for effective leakage of electrons and holes out of the matrix and their confinement in the quantum dots [11]. The synthesis of perovskite-encapsulated quantum dot structures involves two main steps: first, the preparation of colloidal quantum dots; second, the preparation of quantum dot and perovskite hybrid films using reprecipitation [10–13].

In 2015, Sargent's group proposed a structure for quantum dots embedded in a perovskite matrix through heterogeneous epitaxial growth [10]. In 2016, Sargent's group achieved an EQE of 5.2% by optimizing the composition of halogen ions in the perovskite to improve the lattice matching between the perovskite and quantum dots [11]. To overcome the challenges of fast and unbalanced carrier dynamics and large phase separation in metal halide perovskites, a perovskite-coated quantum dot material was developed using a dimensionally reduced perovskite matrix. This material achieved an EQE of 8.1% at 980 nm with radiation up to 7.4 W sr^{-1} m^{-2} [13]. Yusoff et al. incorporated $Ag_2S@SiO_2$ quantum dots into a perovskite matrix and optimized the device's charge injection balance, resulting in an EQE of 16.98% at 1397 nm and a power conversion efficiency of 11.28% [12].

The perovskite cladding quantum dot structure requires an additional step in the synthesis of luminescent materials compared to conventional NIR perovskite materials. Additionally, the sulfide PbS quantum dots used in the perovskite cladding structure have a large size, with a wave radius of 20 nm, and the emission spectra are

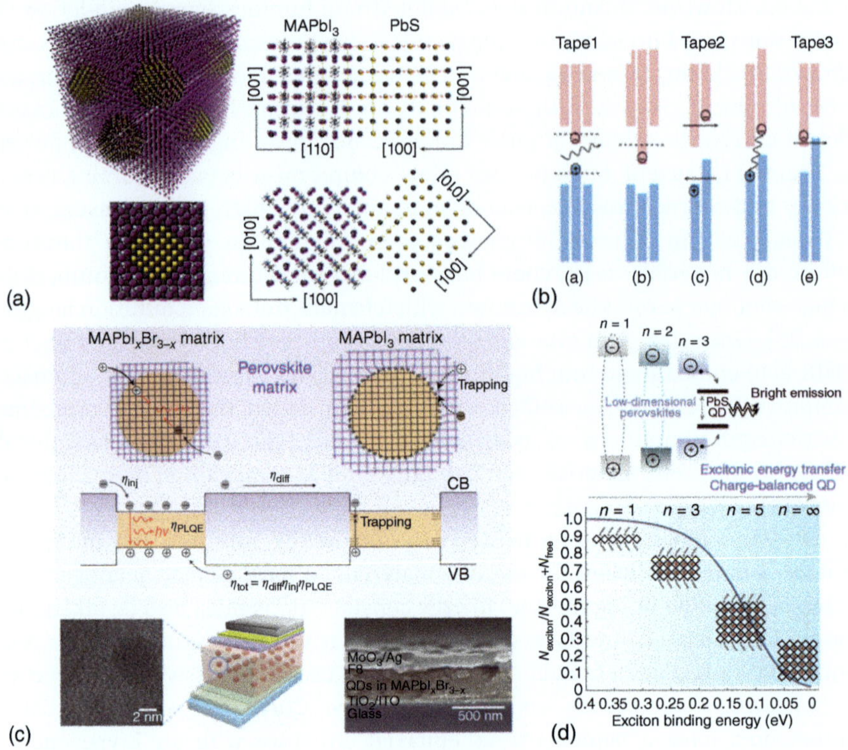

Figure 3.4 (a) Schematic diagram of the perovskite-covered quantum dot (QDiP) structure. (b) Heterogeneous epitaxial structures with different energy band arrangements; (c) QDiP structure with matched lattice parameters. (d) QDiP system with a quasi-dimensional perovskite substrate. (a) Source: Ning et al. [10]/Springer Nature. (b and c) Source: Gong et al. [11]/Reproduced from Springer Nature. (d) Source: Gao et al. [13]/Springer Nature.

closely related to their size. This increases the complexity, cost, and batch-to-batch variability of quantum dot synthesis [8]. The current perovskite-coated quantum dot devices have lower EQE compared to pure perovskite light-emitting devices. Another approach to achieve NIR emission is doping perovskites with lanthanides such as Yb^{3+} or Er^{3+}. This is similar to the perovskite cladding quantum dot system, where exciton energy is transferred from perovskite to lanthanides, resulting in emission at NIR wavelengths [28].

3.4 Strategies to Improve the Performance of NIR Perovskite Devices

Thanks to advancements in theoretical device physics, the development of advanced NIR materials, and the optimization of device structures, there has been a significant breakthrough in EQE from 0.76% to 25.5% in less than 10 years. This progress is the result of several strategies aimed at achieving better NIR optical performance and higher-efficiency light-emitting devices. The most representative strategies and their corresponding device performance parameters are summarized in Table 3.1.

3.4.1 NIR Perovskite Material Optimization

Based on previous research, the optimization pathway for perovskite materials can be categorized into four main strategies: regulation of particle size and dimensionality, preparation of uniform films, ligand engineering, and post-processing treatment [31]. These strategies aim to improve the purity, efficiency, and stability of the emitted light. In this subsection, we summarize how these strategies can be applied to optimize NIR perovskite materials. Specifically, we discuss how to broaden the range of NIR wavelengths and improve the radiative recombination efficiency of the material's excitons, taking into account both radiative and non-radiative recombination pathways.

3.4.1.1 Near-Infrared Wavelength Adjustment

Based on previous studies, the optical bandgap of halogenated perovskites can be adjusted by substituting A-site cations (e.g. Cs^+, MA^+, and FA^+) and halogen ions (Cl^-, Br^-, and I^-), but the emission from pure lead-based perovskite LEDs is limited to below 800 nm, which restricts their application in the NIR band [12, 29]. However, the approaches discussed in the chapters on doped Sn^{2+} and perovskite-based NIR emitting materials allow for red-shifting of the emitted NIR wavelengths.

Rand's group achieved tunable NIR light emission between 850 and 950 nm by utilizing mixed Pb-Sn perovskites. They obtained an efficient NIR device with a turn-on voltage of 1.65 V, an emission peak of 917 nm, and an EQE of 5.0% by using a Pb : Sn ratio of 0.6 : 0.45 hybrid perovskite [8]. Apart from substituting Pb^{2+} with Sn^{2+} to achieve red-shifted NIR wavelengths, there are other methods such as energy transfer from the perovskite substrate to lanthanide ions or forming quantum dot materials. Miyasaka and Ishii demonstrated this by doping $CsPbCl_3$ perovskite films with Yb^{3+} ions. Yb^{3+}:$CsPbCl_3$ films showed strong NIR emission due to efficient energy transfer from perovskite to Yb^{3+} ions, resulting in a high PLQY (over 60%) in thin film structures. Devices based on Yb^{3+}:$CsPbCl_3$ also exhibited bright electroluminescence near 1000 nm with an EQE of 5.9% [28]. It is important to note that the luminescence achieved by doping lanthanides into perovskites differs from that of previous methods, where the emission of NIR wavelengths is achieved through energy transfer to lanthanide ions, similar to the luminescence principle of the perovskite-coated quantum dot system. Due to the instability of perovskite in high-humidity and high-temperature environments, as well as the toxicity of lead ions, researchers have also explored double perovskites doped with Yb^{3+} or Er^{3+} [32, 33]. The double perovskite is a structure formed by replacing two Pb^{2+} ions with one monovalent and one trivalent ion. Due to its outstanding thermal stability, it has potential practical applications in the NIR. Lee et al. synthesized $Cs_2AgInCl_6$ double perovskite, which can withstand high temperatures up to 400 °C without decomposition. Lanthanide doping reduces the half-height width of the fluorescence spectrum at NIR wavelengths. However, the absorption wavelength is

Table 3.1 Enhancement strategies and corresponding device performance reported for NIR-emitting semiconductors based on perovskite materials.

Year [Ref.]	Strategies	Device structure	PLQY (%)	EL (nm)	EQE (%)	Radiance ($W\,sr^{-1}\,m^{-2}$)
2015 [21]	PEI improves surface topography, reducing non-radiative combination	ITO/ZnO/PEI/$MAPbI_{3-x}Cl_x$/TFB/MoO_x/Au	—	768	3.8	28
2016 [11]	QDiP structure enhances radiation combination and obtains longer NIR wavelengths	ITO/TiO_2/PbS QDs in $MAPbX_3$/F8/MoO_3/Ag	—	1390	5.2	2.6
2016 [23]	Adjusting the MA : PEA ratio to improve the surface topography degradation	ITO/TiO_2/$PEA_2(MA)_{n-1}PbI_{n3n+1}$/F8/$MoO_3$/Au	10.6	760	8.8	80
2016 [26]	Improve surface morphology and thus reduce non-radiative composites	ITO/PEDOT:PSS/$CsSnI_3$/PBD/LiF/Al	—	950	3.8	40
2016 [22]	Improve surface topography and reduce non-radiative recombination	ITO/ZnO/PEIE/$NMA_2FA_{n-1}Pb_nI_{3n+1}$/TFB/$MoO_x$/Au	60	800	9.6	55
2017 [14]	Modulation of grain size to improve radiation combination	ITO/poly-TPD/$MAPbI_3$/TPBi/LiF/Al	4.4	750	7.9	72
2018 [15]	Addition of amino acids improves the coupling output efficiency of light	ITO/ZnO/PEIE/$FAPbI_3$/TFB/MoO_x/Au	70	800	20.7	18.4
2018 [8]	Sn^{2+} doping to increase NIR wavelength	ITO/Poly-TPD/$MAPb_{0.6}SnI_{0.43}$/TPBi/LiF/Al	—	917	5.0	2.7
2019 [27]	Passivation to reduce the non-radiative compounds	ITO/ZnO/PEIE/$FAPbI_3$/TFB/MoO_3/Au	65	800	21.6	308
2019 [12]	QDiP structure enhances the radiation complex and increases the NIR wavelength	ITO/TiO_2/Ag_2S QDs in $Cs_{10}(MA_{0.17}FA_{0.83})Pb(Br_xI_{1-x})_3$/Porphyrin/$MoO_3$/Ag	84	1397	16.98	83.93
2019 [30]	Introduction of poly-TPD to improve carrier balance of devices	ITO/ZnO/PEIE/$FAPbI_3$/Poly-TPD/MoO_3/Al	—	799	20.2	57
2020 [13]	QDiP structure enhances the radiation complex and increases the NIR wavelength	ITO/PEDOT:PSS PbS QDs in $PEA_2Cs_2Pb_3Br_{10}$/TPBi/LiF/Al	38	980	8.1	7.4
2020 [28]	Doping of Yb^{3+} to $CsPbCl_3$ to increase the NIR wavelength	TCO/SnO_2/Liq/Yb^{3+} in $CsPbCl_3$/poly-TPD/PEDOT:PSS/Au	60	984	5.9	—
2021 [16]	Add AEAA to enhance the crystallization and PLQE	ITO/ZnO/PEIE/Perovskite/TFB/MoO_x/Au	—	803	22.2	250
2022 [18]	Introduce stabilizer SFB10 in the perovskite emissive layers to enhance lifetime	ITO/ZnO/PEIE/perovskite/TFB/ MoOx/Au	—	803	22.8	3.7
2023 [19]	Posttreatment to modify ZnO and enhance out-coupling efficiency	ITO/ZnO/Perovskite/Spiro-OMeTAD/MoO_x/Ag	—	807	25.5	487

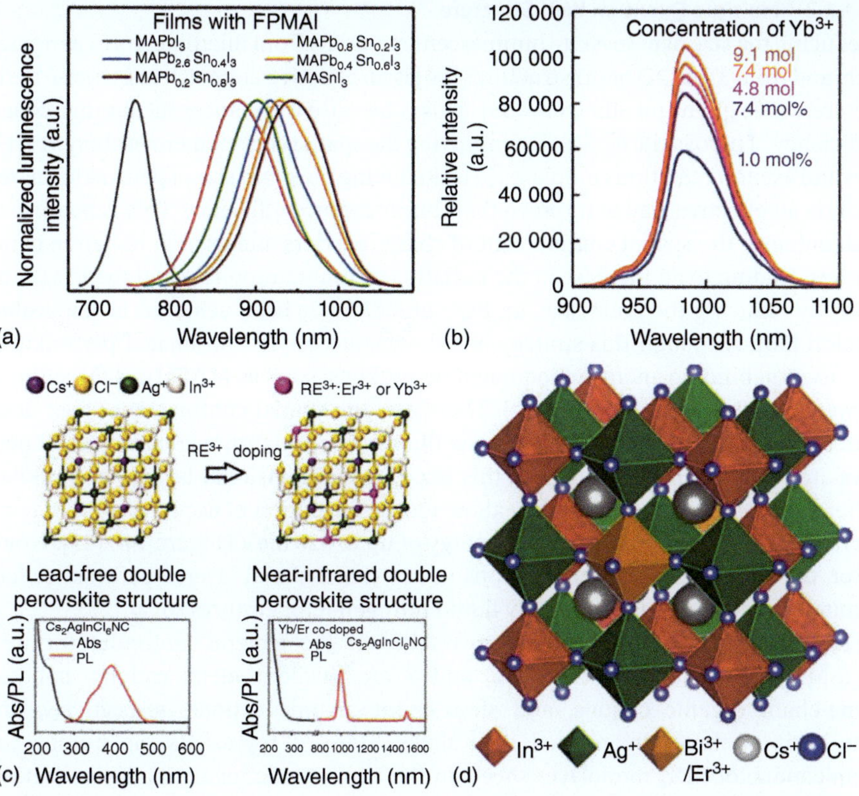

Figure 3.5 (a) EL spectra of tin–lead chalcocite. (b) PL spectra of Yb^{3+}-doped chalcocite and the variation of PLQY with Yb^{3+} concentration. (c) Double chalcocite structures with and without lanthanide doping and absorption emission spectra. (d) Bi^{3+}-Er^{3+} co-doped double chalcocite structures. (a) Source: Qiu et al. [8]/John Wiley & Sons. (b) Source: Ishii and Miyasaka [28]/John Wiley & Sons/CC BY 4.0. (c) Source: Lee et al. [33]/American Chemical Society. (d) Source: Arfin et al. [32]/John Wiley & Sons.

below 300 nm, which limits its practical application. To address this issue, Habibul Arfin et al. co-doped Bi^{3+}-Er^{3+} or Bi^{3+}-Yb^{3+} to increase the excitation wavelength to 400 nm. Bi^{3+} modifies the density of projection states at the band edge, creating new absorption channels at lower energies. The final NIR emission was obtained at 1537 and 996 nm for Bi^{3+}-Er^{3+} and Bi^{3+}-Yb^{3+}, respectively [32].

Quantum dots continue to be a promising NIR-emitting material for the NIR-II (1000–1700 nm) biological window, and they can be prepared using solution methods, as described in detail in the perovskite-based NIR-emitting materials subsection of this chapter. In addition, Yusof et al. developed a perovskite-encapsulated quantum dot system using silver sulfide ($Ag_2S@SiO_2$) quantum dots encapsulated by silica dispersed in $Cs_{0.01}(MA_{0.17}FA_{0.83})Pb(Br_xI_{1-x})_3$, which exhibited a peak emission at 1397 nm, an EQE of 16.98%, and an emissivity of 83.93 W sr^{-1} m^{-2} [12] (Figure 3.5).

3.4.1.2 Multiple Quantum Well Structure

Reducing the size of perovskite luminescent materials from micrometers to nanometers and from 3D to 2D or 0D structures [34] is an effective approach to increase their fluorescence quantum efficiency [35]. This is because it enhances the luminescence efficiency of perovskite devices by improving the spatial confinement of charge carriers and exciton radiation complexes [31]. Reducing the grain size of perovskite materials is an effective way to improve their luminescence efficiency. This is because it can enhance the spatial confinement of charge carriers and exciton radiation complexes, leading to an increase in the radiative recombination rate and the PLQY. In fact, by reducing the grain size, an EQE of 8.53% has been achieved in the visible region [36]. Although this strategy has shown success in brominated perovskites, the exciton binding energy of iodinated perovskites, such as $MAPbI_3$, is three times lower than that of the former [37]. Therefore, the spatial confinement effect does not result in a significant increase in the fluorescence quantum efficiency of IR perovskite materials [23]. To address this issue, two-dimensional layered perovskites offer a promising solution, as they allow for the formation of excitons at room temperature with a maximum binding energy of up to 320 meV (Figure 3.6) [23]. However, the strong quenching of excitons in two-dimensional layered perovskites has limited their light emission to only liquid nitrogen temperatures so far [22].

Layered organometallic halide perovskites of the general molecular formula $L_2(SMX_3)M_{n-1}X_4$, where M, X, L, and S are divalent metal cations, halides, long-chain organic cations, and short-chain organic cations, respectively, are the basis of three-dimensional perovskites. Here, n represents the number of semiconductor MX_4 monolayer sheets in the two organic insulating layers (cation

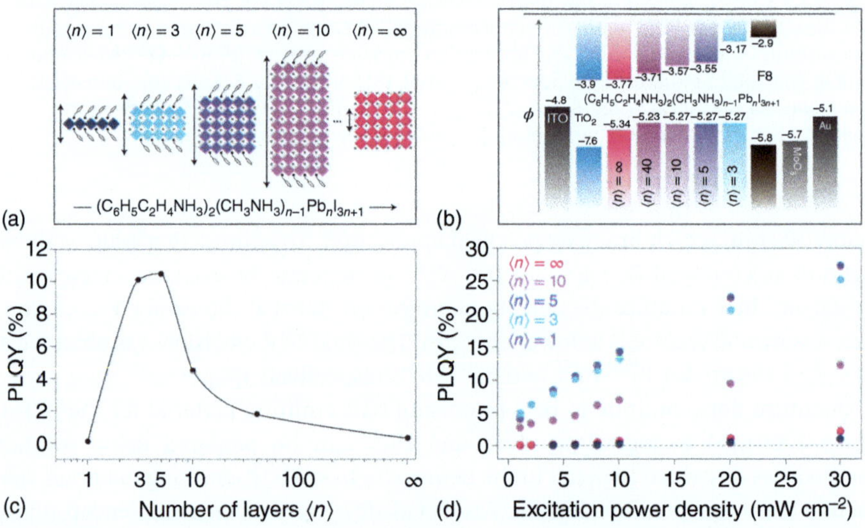

Figure 3.6 (a, b) Structure of multiple quantum wells (MQWs) and the schematic diagram of energy bands; (c) PLQY of perovskite materials with different lead-halide layer thicknesses; (d) PLQY of perovskite materials with different lead-halide layer thicknesses at different excitation intensities. Source: Yuan, et al. [23]/Springer Nature.

L), and $n = \infty$ corresponds to the structure of the three-dimensional perovskite SMX_3. The quantum domain-limiting effects, such as an increase in bandgap and exciton energy, can be significant as the number of MX_4 layers decreases, thus forming quantum well structures. In contrast, at $n = 1$, layered perovskites form a single-layer structure of two-dimensional perovskites L_2MX_4, which have excellent film-forming properties. Multilayered quasi-dimensional perovskites, combining the advantages of both two-dimensional and three-dimensional perovskites, have been synthesized by Sargent et al. through the doping of bulky cationic phenylethylamine (PEA = $C_8H_{19}NH_3^+$). PEA has a large ionic radius and cannot fit into the co-angular lead halide octahedral three-dimensional framework, resulting in the separation of the three-dimensional perovskite structure into layers. When the thickness is within quantum constraints, Ruddlesden–Popper phase layered perovskites naturally form a quantum well structure. However, despite their potential advantages, strong light emission has been observed from two-dimensional layered perovskites only at liquid nitrogen temperatures due to the rapid quenching of excitons [38]. The devices fabricated using this structured material exhibit an EQE of 8.8% and a radiance of 80 W sr^{-1} m^{-2} [23]. Huang Wei et al. incorporated iodinated 1-naphthylmethylamine into the perovskite precursor, resulting in an MQW-structured perovskite LED. This device demonstrated excellent stability and performance with an EQE of 11.7% and an energy conversion efficiency of 5.5% at a current density of 100 mA cm^{-2} [22].

3.4.1.3 Molecular Passivation

Efforts to achieve highly efficient perovskite LEDs have focused on enhancing the radiative recombination rate by limiting the spatial domains of electrons and holes. Various strategies have been explored, including using ultrathin light-emitting layers, fabricating nanoscale polycrystals, designing low-dimensional or MWQ structures, and synthesizing perovskite quantum dots. While these approaches have led to significant improvements in the EQE of devices, increasing it from less than 1%–14%, the achieved value is still much lower than what has been predicted by optical simulations [39]. Although the strategies mentioned above have shown improvements in the radiative recombination rate, they often come at the cost of charge injection efficiency and transport performance or the introduction of a large number of grain boundaries, which exacerbate non-radiative recombination. To achieve highly efficient perovskite LEDs, it is equally important to reduce non-radiative recombination. However, trap-mediated non-radiative losses remain a significant efficiency-limiting factor for perovskite devices. The trap state is typically associated with ionic defects, such as halide vacancies, and excessive addition of amine halide during synthesis can result in crystals with fewer defects. Thus, reducing defect density in perovskite semiconductors is crucial for improving device performance [40]. In addition, the use of molecular passivators (PAs) that can chemically bond to defects is an effective method of passivation to solve this problem [27]. Currently, thin 3D perovskite films consisting of well-passivated crystals with large particle sizes have shown promising potential as high-efficiency luminescent layers.

One reported method to greatly enhance the luminescence efficiency of perovskite films is by adding an excess of halogenated amines during the synthesis process. For instance, Huang Wei and Wang Jianpu's group achieved an EQE of 20.7% in NIR perovskite LEDs by preparing perovskite films with spontaneously formed isolated submicron grains using precursors that included 140% excess FAI and 70% 5AVA. Lin et al. also reported the formation of green perovskite LEDs by introducing a 100% additional methylammonium $CsPbBr_3/MABr$ quasi-core/shell structure, resulting in an EQE of 20% [41]. The shell layer was designed to passivate non-radiative defects and improve luminescence efficiency. However, the exact mechanism by which excess amine halide enhances the EQE is still unclear. To shed light on this, Yong-Heng Jia et al. used $Cs_{0.17}FA_{0.83}PbI_{2.5}Br_{0.5}$ as a model to investigate the role of excess amine halide (FAI in this model) in the perovskite crystallization process [40]. The mechanism by which excess amine halide enhances the EQE is not yet fully understood. However, Yong-Heng Jia et al. investigated the role of excess amine halide (FAI in this study) in the perovskite crystallization process using $Cs_{0.17}FA_{0.83}PbI_{2.5}Br_{0.5}$ as a model. The results suggest that excess FAI in the deposited state promotes the formation of dense, wide-bandgap mesophase films, which subsequently transform into isolated and highly crystalline perovskite grains during annealing. Excitation-related photoluminescence spectroscopy revealed that the excess FAI reduces the density of deep trapped states, leading to a decrease in non-radiative loss of the material.

Some functionalized groups such as $-NH_2$ and P=O are currently being used for molecular passivation in perovskite cells. Gao Feng's group used a diamine passivation molecule (TTDDA), which further suppressed non-radiative recombination in the perovskite film, leading to an improved fluorescence quantum efficiency of about 70% and an increased charge carrier lifetime [27]. However, their performance is highly dependent on the specific structure of the PA, even if they share the same functional groups. The lack of understanding regarding how the chemical structure of additives affects the passivation effect has hindered the rational design of molecular PAs to minimize non-radiative recombination losses. To address this issue, Huang Wei's team and collaborating groups used first principles to calculate that the iodide ion vacancies on the film surface are the main structural defects in 3D $FAPbI_3$ materials. They selected two similar amino-functionalized molecular PAs, 2,2′-(ethylenedioxy)diethylamine (EDEA) and hexamethylene diamine (HMDA) for comparison and found that HMDA is more effective in repairing defects due to its stronger hydrogen bonding with organic cations in perovskite. They then rationally designed the more weakly hydrogen-bonded 2,2′-[oxybis(ethoxy)]diethylamine (ODEA), which led to the development of an NIR perovskite semiconductor with an EQE of 21.6% [27] (Figure 3.7).

3.4.2 Device Structure Optimization

In a multilayer LED structure (e.g. anode/hole-injected layer/perovskite-emitting layer/electron-injected layer/cathode), the application of an electrical bias results in the injection of holes into the anode and electrons into the cathode. The holes then migrate through the highest occupied molecular orbital of the hole injection

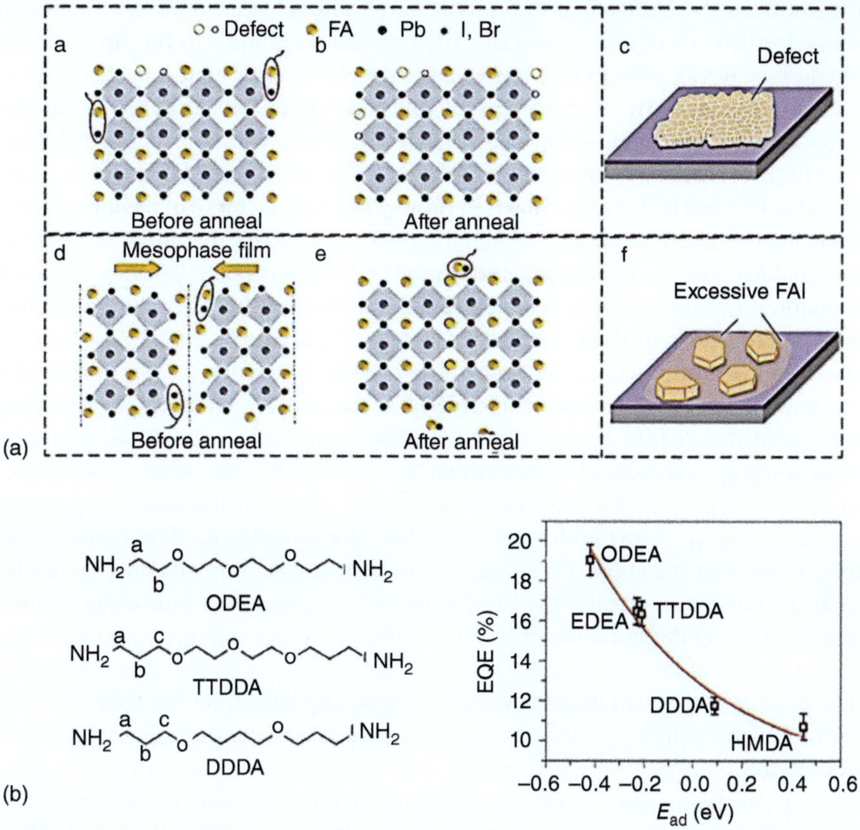

Figure 3.7 (a) Schematic diagram of the passivation of excess amine halide [40]; (b) structure of the passivated molecule and the corresponding EQE diagram of the perovskite light-emitting device. Source: Adapted from Xu et al. [27].

layer to the highest valence band of the perovskite, while the electrons travel through the lowest unoccupied molecular orbital of the electron injection layer to the lowest conduction band of the perovskite. These carriers combine within the perovskite layer to generate light with a bandgap energy similar to that of the perovskite, which is then emitted through the transparent electrode. The EQE of the device is primarily determined by the ratio of emitted photons to injected electrons and holes, which is calculated as

$$\mathrm{EQE} = \gamma\beta\varphi_L \tag{3.3}$$

where χ is the external coupling factor, γ is the charge balance factor, β is the probability of radiating light species, and φ_L is the quantum efficiency of luminescence. Therefore, the efficiency of NIR perovskite materials in this formulation is influenced by several factors, such as β and φ_L, which are affected by carrier-ion lattice coupling and defect states. Moreover, the inherent low fluorescence quantum efficiency of NIR perovskites, due to their low carrier radiative recombination rates, also plays a role. Additionally, rough morphology of NIR titanite, aggregation of colloidal perovskite quantum dots during film formation, operational instability, and metal diffusion from the electrode into the NIR perovskite all contribute to a reduction in the EL efficiency of the device [31]. The two factors χ and γ in the equation are related to the device structure. χ represents the poor external coupling efficiency of light, which is caused by optical loss at the interface. On the other hand, γ represents the low charge balance in the device, which results in electrical loss at the interface.

The EQE of a device is affected by two factors: the output coupling factor χ and the charge balance factor γ (as shown in Eq. (3.3)). These factors can be improved by optimizing the device structure through interlayer or interface treatment. Such treatment enables efficient charge injection into the NIR perovskite layer by reducing the charge injection barriers between the electrode and the perovskite and electron injection layers, leading to an enhanced charge balance factor γ. In a review on improving the luminescence efficiency of perovskite-based devices, optimizing the interfacial layers and interfaces is identified as an effective method for improving device performance [31]. A review on improving the luminescence efficiency of perovskite suggests that using an intermediate layer such as PFI can enhance hole injection, while using an intermediate layer such as PEI, a deposited atmospheric atomic layer ($Zn_{0.56}Mg_{0.44}O$ film), and interfacial solvent treatment (e.g. ethanolamine) can enhance electron injection. These treatments also reduce luminescence annihilation at the interface and alter the morphology of the perovskite film when applied before depositing the luminescent layer. Combining these intermediate layers has been shown to further improve device efficiency.

The interlayer plays an important role in extracting light from the light-emitting layer to the outer surface. For instance, in organic LEDs, controlling the refractive index of the charge injection/transport layer can improve the output coupling efficiency in perovskite LEDs. In addition, gold nanoparticles dispersed in poly(3,4-ethylenedioxythiophene):polystyrene sulfonate (PEDOT:PSS) and silver nanorods dispersed in N,N′-bis(1-naphthyl)-N,N′-bis(phenylbenzidine) (NPB) interlayers induce local surface plasmon resonance effects that increase the light extraction efficiency.

Current studies on perovskite LEDs focus on both the external coupling efficiency of the internal structure and the modification of the perovskite light-emitting layer. One such modification is the spontaneous formation of submicron-scale structures in solution-processed perovskites, as demonstrated by Wei Huang's team. This method leads to efficient, high-brightness electroluminescence that can extract light from the device and maintain wavelength- and viewing angle-independent electroluminescence. The addition of amino acids to the $FAPbI_3$ perovskite precursor solution resulted in a peak EQE of 20.7% (at a current density of 18 mA cm^{-2}) and an energy conversion efficiency of 12% (at a high current density of 100 mA cm^{-2}) [15] (Figure 3.8).

3.5 Conclusion and Outlook

This chapter provides an overview of current NIR perovskite materials, which can be classified into two categories: (i) metal halide perovskites and (ii) luminescent layers using perovskites as substrates. We discuss the structure, properties, and

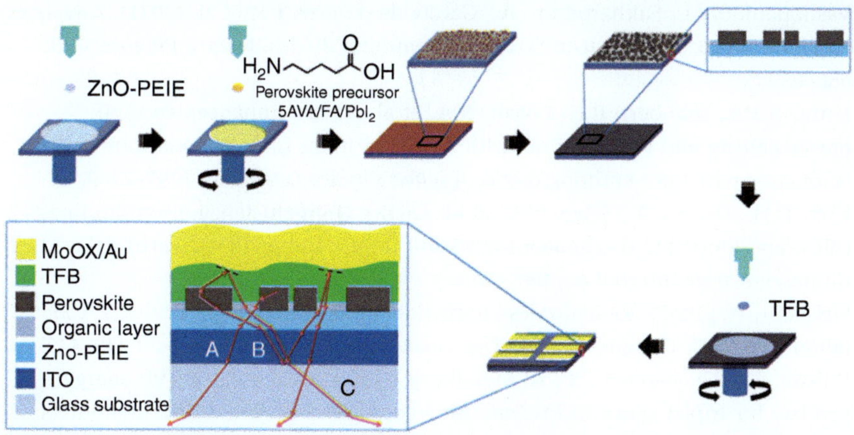

Figure 3.8 Light extraction device preparation as well as structure. Source: Cao et al. [15]/Springer Nature.

recent research progress of these materials. Based on previous studies, both metal halide perovskites and perovskite-based NIR LEDs have achieved EQEs exceeding 15% [27, 28]. Lead-based perovskite LEDs are limited to emitting light below 800 nm, and the synthesis of perovskite-coated quantum dots is relatively complex. Therefore, there is ample scope for enhancing the efficiency of lead-free perovskites. In this chapter, we review recent research progress in improving the electrical and optical properties of NIR perovskite materials and devices. This includes the tuning of NIR spectra, the use of MQW structures, and optimization of device structures.

Over the past six years, the EQE of perovskite LEDs has increased from 0.76% to 25.5%. However, the energy conversion efficiency and long-term stability of NIR perovskite LEDs still need to be improved for practical applications [31]. The future development of efficient NIR perovskite materials and devices that are compatible with industrial production, such as ink printing and transfer printing, is essential. In addition, the stability and lifetime of perovskite materials are crucial challenges that have been studied in the literature but still limit the practical application of NIR perovskite materials. Although NIR light sources have a wide range of applications, including night vision devices, optical communication, sensing, bio-imaging, facial recognition, and surveillance, there have been few reports on the application of NIR perovskites. Promising research topics include flexible, large-area, stretchable perovskite LEDs and multifunctional integration. Researchers can learn from the research history of organic LEDs in these areas, as organic LEDs have been studied for a longer time.

References

1 Xiang, H., Cheng, J., Ma, X. et al. (2013). Near-infrared phosphorescence: materials and applications. *Chemical Society Review* 42: 6128–6185.

2 Vasilopoulou, M., Fakharuddin, A., García de Arquer, F.P. et al. (2021). Advances in solution-processed near-infrared light-emitting diodes. *Nature Photonics* 15: 656–669.

3 Hunt, N.E.J., Schubert, E.F., Logan, R.A. et al. (1992). Enhanced spectral power density and reduced linewidth at 1.3 μm in an InGaAsP quantum well resonant-cavity light-emitting diode. *Applied Physics Letters* 61: 2287–2289.

4 Kim, D.H., D'Aléo, A., Chen, X.K. et al. (2018). High-efficiency electroluminescence and amplified spontaneous emission from a thermally activated delayed fluorescent near-infrared emitter. *Nature Photonics* 12: 98–104.

5 Siebrand, W. (1967). Radiationless transitions in polyatomic molecules. I. Calculation of Franck-Condon factors. *The Journal of Chemical Physics* 46: 440–447.

6 Wilson, J.S., Chawdhury, N., Al-Mandhary, M.R.A. et al. (2001). The energy gap law for triplet states in Pt-containing conjugated polymers and monomers. *Journal of the American Chemical Society* 123: 9412–9417.

7 Moroz, P., Liyanage, G., Kholmicheva, N.N. et al. (2014). Infrared emitting PbS nanocrystal solids through matrix encapsulation. *Chemistry of Materials* 26: 4256–4264.

8 Qiu, W., Xiao, Z., Roh, K. et al. (2019). Mixed lead-tin halide perovskites for efficient and wavelength-tunable near-infrared light-emitting diodes. *Advanced Materials* 31: 1806105.

9 Tan, Z.K., Moghaddam, R.S., Lai, M.L. et al. (2014). Bright light-emitting diodes based on organometal halide perovskite. *Nature Nanotechnology* 9: 687–692.

10 Ning, Z., Gong, X., Comin, R. et al. (2015). Quantum-dot-in-perovskite solids. *Nature* 523: 324–328.

11 Gong, X., Yang, Z., Walters, G. et al. (2016). Highly efficient quantum dot near-infrared light-emitting diodes. *Nature Photonics* 10: 253–257.

12 Vasilopoulou, M., Kim, H.P., Kim, B.S. et al. (2019). Efficient colloidal quantum dot light-emitting diodes operating in the second near-infrared biological window. *Nature Photonics* 14: 50–56.

13 Gao, L., Quan, L., García de Arquer, F.P. et al. (2020). Efficient near-infrared light-emitting diodes based on quantum dots in layered perovskite. *Nature Photonics* 14: 227–233.

14 Zhao, L., Yeh, Y.W., Tran, N.L. et al. (2017). In situ preparation of metal halide perovskite nanocrystal thin films for improved light-emitting devices. *ACS Nano* 11: 3957–3964.

15 Cao, Y., Wang, N., Tian, H. et al. (2018). Perovskite light-emitting diodes based on spontaneously formed submicrometre-scale structures. *Nature* 562: 249–253.

16 Zhu, L., Cao, H., Xue, C. et al. (2021). Unveiling the additive-assisted oriented growth of perovskite crystallite for high performance light-emitting diodes. *Nature Communications* 12: 5081.

17 Guo, Y., Apergi, S., Li, N. et al. (2021). Phenylalkylammonium passivation enables perovskite light emitting diodes with record high-radiance operational lifetime: the chain length matters. *Nature Communications* 12: 644.

18 Guo, B., Lai, R., Jiang, S. et al. (2022). Ultrastable near-infrared perovskite light-emitting diodes. *Nature Photonics* 16: 637–643.

19 Li, J., Duan, C., Zhang, Q. et al. (2023). Self-generated buried submicrocavities for high-performance near-infrared perovskite light-emitting diode. *Nano-Micro Letters* 15: 125.

20 Shamsi, J., Urban, A.S., Imran, M. et al. (2019). Metal halide perovskite nanocrystals: synthesis, post-synthesis modifications, and their optical properties. *Chemical Reviews* 119: 3296–3348.

21 Wang, J., Wang, N., Jin, Y. et al. (2015). Interfacial control toward efficient and low-voltage perovskite light-emitting diodes. *Advanced Materials* 27: 2311–2316.

22 Wang, N., Cheng, L., Ge, R. et al. (2016). Perovskite light-emitting diodes based on solution-processed self-organized multiple quantum wells. *Nature Photonics* 10: 699–704.

23 Yuan, M., Quan, L.N., Comin, R. et al. (2016). Perovskite energy funnels for efficient light-emitting diodes. *Nature Nanotechnology* 11: 872–877.

24 Li, W., Wang, Z., Deschler, F. et al. (2017). Chemically diverse and multifunctional hybrid organic-inorganic perovskites. *Nature Reviews Materials* 2: 16099.

25 Jeon, N.J., Noh, J.H., Yang, W.S. et al. (2015). Compositional engineering of perovskite materials for high-performance solar cells. *Nature* 517: 476–480.

26 Hong, W.L., Huang, Y.C., Chang, C.Y. et al. (2016). Efficient low temperature solution processed lead free perovskite infrared light emitting diodes. *Advanced Materials* 28: 8029–8036.

27 Xu, W., Hu, Q., Bai, S. et al. (2019). Rational molecular passivation for high-performance perovskite light-emitting diodes. *Nature Photonics* 13: 418–424.

28 Ishii, A. and Miyasaka, T. (2020). Sensitized Yb^{3+} luminescence in $CsPbCl_3$ film for highly efficient near-infrared light-emitting diodes. *Advanced Science* 7: 1903142.

29 Xing, G., Kumar, M.H., Chong, W.K. et al. (2016). Solution-processed tin-based perovskite for near-infrared lasing. *Advanced Materials* 28: 8191–8196.

30 Zhao, X. and Tan, Z.K. (2019). Large-area near-infrared perovskite light-emitting diodes. *Nature Photonics* 14: 215–218.

31 Kim, Y.H., Kim, J.S., and Lee, T.W. (2019). Strategies to improve luminescence efficiency of metal-halide perovskites and light-emitting diodes. *Advanced Materials* 31: 1804595.

32 Arfin, H., Kaur, J., Sheikh, T. et al. (2020). Bi^{3+}-Er^{3+} and Bi^{3+}-Yb^{3+} codoped $Cs_2AgInCl_6$ double perovskite near-infrared emitters. *Angewandte Chemie International Edition* 59: 11307–11311.

33 Lee, W., Hong, S., and Kim, S. (2019). Colloidal synthesis of lead-free silver-indium double-perovskite $Cs_2AgInCl_6$ nanocrystals and their doping with lanthanide ions. *The Journal of Physical Chemistry C* 123: 2665–2672.

34 Pietryga, J.M., Werder, D.J., Williams, D.J. et al. (2008). Utilizing the lability of lead selenide to produce heterostructured nanocrystals with bright, stable infrared emission. *Journal of the American Chemical Society* 130: 4879–4885.

35 Byun, J., Cho, H., Wolf, C. et al. (2016). Efficient visible quasi-2D perovskite light-emitting diodes. *Advanced Materials* 28: 7515–7520.

36 Cho, H., Jeong, S., Park, M. et al. (2015). Overcoming the electroluminescence efficiency limitations of perovskite light-emitting diodes. *Science* 350: 1222–1225.

37 Yang, Y., Yang, M., Li, Z. et al. (2015). Comparison of recombination dynamics in $CH_3NH_3PbBr_3$ and $CH_3NH_3PbI_3$ perovskite films: influence of exciton binding energy. *The Journal of Physical Chemistry Letters* 6: 4688–4692.

38 Si, J., Liu, Y., He, Z. et al. (2017). Efficient and high-color-purity light-emitting diodes based on in-situ grown films of $CsPbX_3$ (X = Br, I) nanoplates with controlled thicknesses. *ACS Nano* 11: 11100–11107.

39 Shi, X.B., Liu, Y., Yuan, Z. et al. (2018). Optical energy losses in organic inorganic hybrid perovskite light emitting diodes. *Advanced Optical Materials* 6: 1800667.

40 Jia, Y.H., Neutzner, S., Zhou, Y. et al. (2019). Role of excess FAI in formation of high-efficiency $FAPbI_3$-based light-emitting diodes. *Advanced Functional Materials* 30: 1906875.

41 Lin, K., Xing, J., Quan, L.N. et al. (2018). Perovskite light-emitting diodes with external quantum efficiency exceeding 20 per cent. *Nature* 562: 245–248.

4

Perovskite Red Light-Emitting Materials and Devices

4.1 The Development History of Perovskite Red Light-Emitting Diodes

Perovskite materials have gained significant interest for their exceptional optoelectronic properties, such as high carrier mobility, precise and tunable bandgap, high photoluminescence quantum yield, narrow half-peak width, excellent defect tolerance, and high color purity. However, their low formation energy makes them susceptible to poor stability, which limits their commercial application. For example, the red-light perovskite material $CsPbI_3$ is prone to phase transitions at room temperature, hindering its commercial viability.

Since the introduction of the first 3D perovskite-based light-emitting diodes (LEDs) utilizing perovskite materials in 2014 [1], researchers have made remarkable progress in enhancing their performance. Within just a few years, the external quantum efficiency has surpassed 20%, a substantial improvement from less than 1% in the initial stages. These advancements are a testament to the tireless efforts of researchers in this field. Perovskite LEDs offer several advantages over organic light-emitting diodes (OLEDs) and quantum dot LEDs, including their straightforward synthesis and the ability to easily and precisely tune their bandgap [2, 3]. The benefits of perovskite LEDs have led to a surge in research in this area. Researchers have been able to enhance the performance of these LEDs through a variety of techniques, including morphology tuning [4], ligand engineering [5], interface treatment [3], composition engineering [6], and device engineering [7, 8]. By optimizing the material and device structure, significant progress has been made in improving the overall performance of perovskite LEDs.

Perovskite-based red LEDs (620–760 nm) have significant potential as a component in lighting and display technology. However, their development has been constrained by limited stability compared to perovskite-based green LEDs [9]. Numerous optimization strategies have been developed to enhance the stability and optoelectronic properties of red perovskite materials [3, 5, 9–21]. Figure 4.1 displays the development of red perovskite LEDs since their initial introduction, showcasing the progress made in improving their performance, with the horizontal axis indicating the year and the vertical axis representing the external quantum efficiency (EQE). The circle size in the figure reflects the maximum brightness

Perovskite Light Emitting Diodes: Materials and Devices, First Edition. Hong Meng.

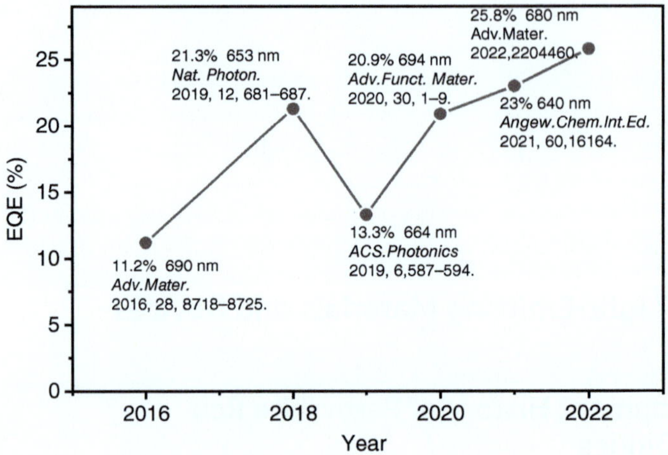

Figure 4.1 Schematic diagram of the development of the external quantum efficiency of red perovskite light-emitting diodes.

achieved by red perovskite LEDs in a given study, while the center point of the circle represents the EQE of the device and the year of publication. The data indicates a gradual improvement in the efficiency and brightness of red perovskite LEDs, with the maximum EQE above 25% in 2023, a substantial improvement from the initial less than 0.1% [22]. Jingbi You et al. achieved efficient red-light perovskite LEDs by mixing two different large organic cations to finely control the phase distribution of quasi-2D perovskites. This effectively reduced the number of smaller "*n*" phases while also achieving passivation of lead and halide defects in the perovskite film. As a result, the perovskite LED exhibited an EQE of 25.8% and a maximum brightness of 1300 cd m^{-2} at 680 nm, representing the highest performance for red-light perovskite LEDs to date [23].

This chapter provides a comprehensive overview of the materials and devices used in red perovskite LEDs. We will cover these topics from two main perspectives. Firstly, we will discuss the commonly used light-emitting materials for red perovskite diodes, their fundamental properties, material synthesis methods, and strategies for optimizing material properties. Secondly, we will examine the device structure of red perovskite LEDs, the materials commonly used in each functional layer, and the strategies for optimizing device performance. Finally, we will analyze the current research progress, summarize the challenges and opportunities in the field, and provide insight into the future of red perovskite LEDs.

Figure 4.1 presents a detailed decade-long journey, mapping the progress of external quantum efficiencies (EQEs) in red perovskite light-emitting devices and their performance within PeLEDs. Initially, red PeLEDs offered EQEs of a modest 11.2% with an emission peak at 690 nm, highlighting the hurdles and limitations encountered in the nascent stages of this technology. Nevertheless, a pivotal breakthrough in 2019, as reported in Nature Photonics, showcased an EQE of 21.3% at 653 nm. From this point onward, the timeline evidences a persistent surge in EQE values, such as an impressive 23%, signifying significant strides in this realm. The escalating trajectory of EQEs bears witness to ongoing technological improvements and the successful deployment of optimization strategies across the years. The recent apex in 2022, showcasing an EQE of 25.8%, serves to underline the effective application of cutting-edge methods in material processing, device fabrication, and design alterations. While the achievements so far are substantial, the most recent data points on the graph illustrate an unrelenting pursuit of further efficiency enhancements. Consequently, Figure 4.1 not only commemorates past accomplishments but also acts as a compass, directing future research efforts in the sphere of red PeLEDs.

4.2 Red Emission Perovskite Materials

Perovskite materials play a critical role in perovskite LEDs, and their properties largely determine the performance of the devices. As such, it is essential to introduce perovskite materials. This section provides a brief overview of commonly used red-light perovskite materials, their basic optoelectronic properties, synthesis methods, and optimization strategies.

4.2.1 Typical Red Emission Perovskite Material $CsPbI_3$

$CsPbI_3$ is a widely studied all-inorganic halide perovskite. Researchers have reported successful use of $CsPbI_3$ perovskite quantum dots in deep red perovskite LEDs with a wavelength of around 700 nm [24–26]. This is a significant achievement, as it allows perovskite materials to cover the entire visible spectrum.

$CsPbI_3$ exhibits four different crystalline phases, namely, α-cubic phase, β-quadratic phase, γ-rhombohedral phase, and δ-rhombohedral phase. Among these phases, the α-cubic phase is widely employed in red LEDs due to its appropriate bandgap (approximately 1.73 eV) [10, 14, 15, 27–31]. The α-$CsPbI_3$ perovskite has a typical cubic structure [32], as shown in Figure 4.2. The Cs ion occupies the top position of the cube (position A), the Pb ion occupies the center of the cube (position B), and the I ion occupies the face center of the cube (position X).

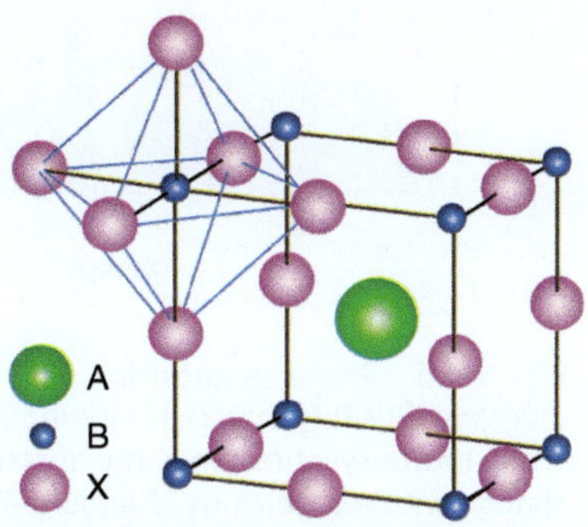

Figure 4.2 Crystal structure of perovskite materials.

The structure is characterized by an octahedral array of PbI_6 surrounded by Cs ions, which form a bandgap. The conduction band is made up of the 4P orbitals of the Pb atom [33]. In $CsPbI_3$, the I and Pb atoms are the main contributors to the material properties, while the Cs ions mainly serve to support the PbI_6 octahedral array and have little influence on the energy level structure of the material.

However, the α-$CsPbI_3$ phase is generally only stable at high temperatures. In a recent study, Even et al. used variable-temperature synchrotron X-ray diffraction to investigate the phase transformation behavior of $CsPbI_3$ at different temperatures, as shown in Figure 4.3b. The study found that the α-phase of $CsPbI_3$ is stable above 645 K. As the temperature decreases, the α-cubic phase transforms into the β-quadratic phase, then gradually into the γ-rhombohedral phase, and finally into the non-cubic δ-rhombohedral phase [34]. On the other hand, during the heating

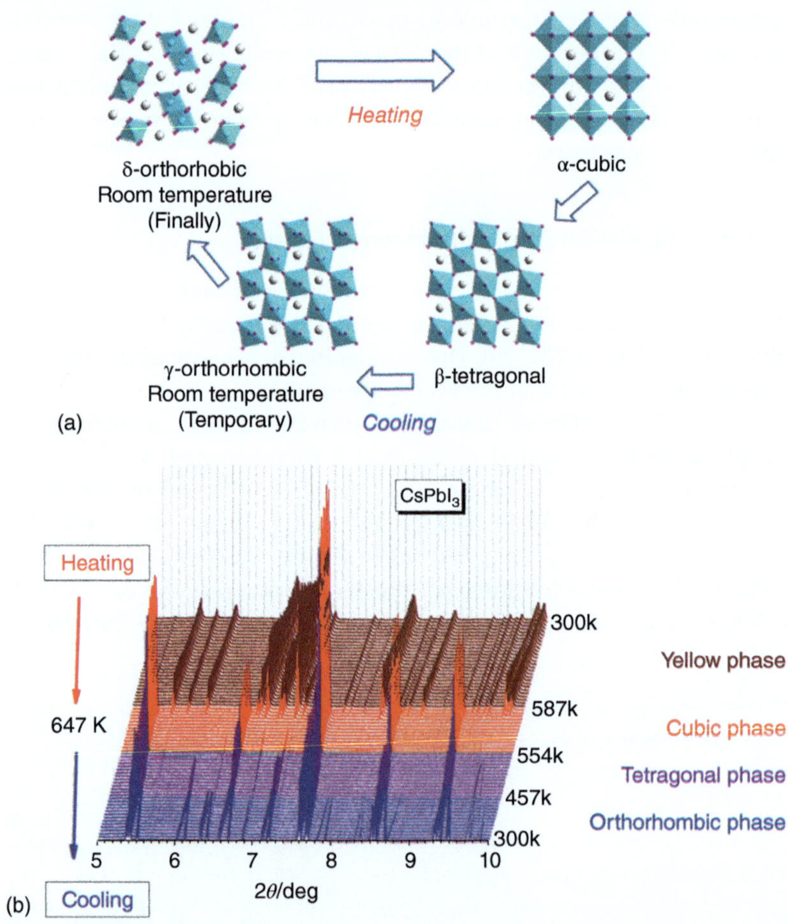

Figure 4.3 Phase composition of $CsPbI_3$ material at different temperatures (a) schematic diagram of $CsPbI_3$ material undergoing phase change with temperature change; (b) variable temperature synchrotron X-ray diffraction study of $CsPbI_3$ material undergoing phase change with temperature change [34].

process, the δ-phase transforms directly into the α-phase without passing through the intermediate phases. The schematic diagram of this phase transition is shown in Figure 4.3a.

Recent studies have found that the α-$CsPbI_3$ phase can be stabilized at room temperature when it is in the form of nanometer-sized particles [27, 35]. Recent studies have shown that nanometer-sized α-$CsPbI_3$ can be stable at room temperature, which has sparked further exploration into the stability of this material. However, it should be noted that α-$CsPbI_3$ is still prone to phase transition due to its thermodynamic instability, and its stability at room temperature is relatively poorer compared to $CsPbCl_3$ and $CsPbBr_3$. Hence, more research is needed to enhance the stability of $CsPbI_3$, such as developing new synthetic approaches and optimizing device structures.

4.2.2 Other Red Emission Perovskite Materials

In summary, α-$CsPbI_3$ is known to have relatively poor stability, and the presence of lead in the compound can be harmful to the environment and organisms. As a result, researchers have investigated alternative red-light perovskite materials, such as other ABX_3 types that could replace $CsPbI_3$, hybridized ABX_3 types, and double perovskite materials.

4.2.2.1 Other ABX_3 and Hybridized ABX_3-Type Materials

Perovskite materials typically follow the chemical formula ABX_3, where A and B represent monovalent and divalent cations, respectively, and X represents a monovalent anion. Due to the highly adaptable nature of the perovskite lattice, many different ions can be doped into the material. The introduction of these ions can alter the formation energies and modify the energy level structure of the material, thereby changing the emission wavelength. By replacing some or all of the Cs^+ ions at position A, Pb^{2+} ions at position B, and I^- ions at position X with other ions, it is possible to precisely adjust the stability and bandgap of the perovskite material [36]. However, it is important to note that not all atoms can be successfully doped into $CsPbI_3$, and doping/alloying must follow the rules of tolerance factor (τ) and octahedral factor (μ). These factors can be described as follows:

$$\tau = \frac{R_X + R_A}{\sqrt{2}(R_X + R_B)} \tag{4.1}$$

$$\mu = \frac{R_B}{R_X} \tag{4.2}$$

where R_A, R_B, and R_X denote the ionic radii of A, B, and X, respectively [3]. Typically, a perovskite material with a tolerance factor (τ) value close to 1.0 will exhibit a stable cubic structure. However, when the value of τ deviates from 1, the perovskite structure tends to become more square or orthorhombic in shape, exhibiting less symmetry [37]. Furthermore, in order to incorporate B-site ions into the octahedral lattice of the perovskite material, the octahedral factor must be taken into consideration. A straightforward mathematical calculation reveals that the BX_6 octahedra will

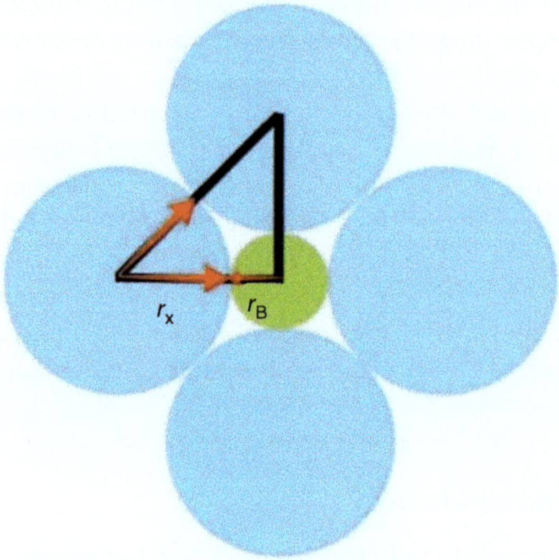

Figure 4.4 Schematic diagram of the crystal arrangement at octahedral factor $\mu = 0.41$.

be closely packed when the value of μ equals 0.41, as depicted in Figure 4.4. Typically, the tolerance factor (τ) falls within the range of 0.76–1.13 when the perovskite material satisfies the octahedral factor [38], a stable three-dimensional perovskite structure is formed as well as the octahedral factor (μ) between 0.442 and 0.895 is considered to be structurally stable [39]. Thus, with knowledge of the ionic radius, it is possible to make predictions about whether certain ions can be successfully doped into $CsPbI_3$. The ionic radii of commonly used ions are listed in Table 4.1 [40]. Anirban Dutta's group has created $\tau-\mu$ diagrams for each possible perovskite material by altering the radius parameter of one ion in ABX_3 while keeping the radii of the other two ions constant. The ion radii used in these calculations are listed in Table 4.1 (see Figure 4.5). Specifically, the A-site ion radius was varied, while the B and X sites were reserved for Pb^{2+} and I^-, respectively. These diagrams serve as a useful tool for understanding the crystal structure and stability of perovskite materials and provide a continuous perspective for discussion on the topic.

From Figure 4.5, it becomes apparent that many cations are capable of substituting for the A-site ions, but only a few can substitute for the B-site ions. Further details on this topic are discussed in the subsequent sections.

Other ABX_3 models have been explored, and it has been observed that a larger ion radius at position A leads to an increase in the tolerance factor. For $CsPbI_3$, the appropriate value of the tolerance factor is 0.893, which is already achieved by the Cs ion, the largest inorganic ion that can be utilized at this position. Therefore, there is no need to consider ions from the front of the periodic table of elements to further increase the tolerance factor. Since A-site ions primarily provide structural support to the perovskite lattice, they typically have little impact on the material's energy level structure. Thus, researchers have considered using larger organic cations to replace Cs. Table 4.1 lists the organic cations that can be used as substitutes for Cs. Currently, the most commonly used organic cations for perovskite materials are the methylammonium and formamidinium ions, which replace the A-site ions [9].

Table 4.1 Possible ionic radii for A, B, and X sites in perovskite materials.

A site	Size (Å)	B site	Size (Å)	I^- (Å)
Ammonium	1.46	Ni(II)	0.57	2.77
Rubidium	1.72	Hg(II)	0.61	2.80
Cesium	1.88	Ti(II)	0.66	2.86
Hydroxyammonium	2.16	V(II)	0.68	2.88
Methyl ammonium	2.17	Cr(II)	0.68	2.88
Hydrazine	2.17	Fe(II)	0.68	2.88
Nitroethylamine	2.50	Mn(II)	0.72	2.92
Formamide	2.53	Mg(II)	0.75	2.93
Imidazole	2.58	Ge(II)	0.77	2.97
Dimethyl ammonium	2.72	Cd(II)	0.81	3.01
3-Pyrrole	2.72	Ga(II)	0.92	3.12
Ethyl ammonium	2.74	Yb(II)	0.93	3.13
Guanidine	2.78	Tm(II)	0.95	3.15
Tetramethylammonium	2.92	Sn(II)	0.97	3.17
Thiazole	3.2	Dy(II)	0.97	2.20
Piperazine	3.22	Pb(II)	1.03	2.20
Cycloheptatriene	3.33	Sm(II)	1.11	3.31
Triethylenediamine	3.39	Sr(II)	1.18	2.22

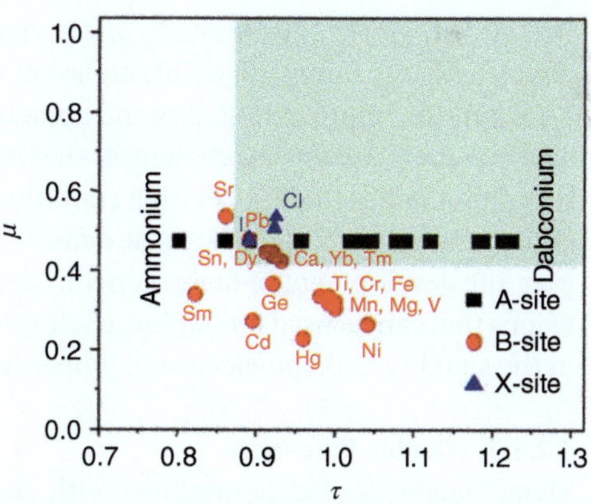

Figure 4.5 $\tau-\mu$ plots for various possible ABX_3 [35].

However, $FAPbI_3$ and $MAPbI_3$ perovskite materials, where these two organic cations are the main ions, are rarely reported. In perovskite materials, these organic cations are typically used as dopants rather than as main ions [29]. For position B, the choice of ions is more restrictive as both the tolerance factor and the octahedral factor need to be considered. Figure 4.5 shows the tolerance factor and octahedral factor for some ions after the substitution of Pb ions, where the red dots indicate the values for the substituted ions. It is evident that only a few ions can replace Pb ions. Recent reports have shown the successful synthesis of red-light perovskite materials that completely replace Pb ions, such as the $CsSnI_3$ perovskite material. However, it has a bandgap of 1.3 eV (in the near-infrared region) and much poorer stability than $CsPbI_3$ due to Sn^{2+} being easily oxidized to Sn^{4+} and having lower values for the tolerance factor and octahedral factor [31, 41]. Finally, with regard to the X-position, only a few halogens such as Cl^-, Br^-, and I^- can replace it. However, the choice of halogen can significantly affect the energy level structure of the material [9]. For example, $CsPbCl_3$ and $CsPbBr_3$ are examples of ABX_3-type materials that emit blue and green light, respectively. In general, there are relatively few ABX_3-type materials that emit red light in perovskites.

Heterogeneous perovskites are an area of active research in perovskite materials, as they offer advantages such as adjustable bandgaps and excellent stability. These materials are typically obtained by doping or other processes. In contrast to traditional perovskites, the doping ions in heterogeneous perovskites do not necessarily adhere to the tolerance factor and octahedral factor, resulting in a wider range of available ions for doping. In our discussion of heterogeneous perovskites, we will also consider the three positions, A, B, and X.

A position: As mentioned earlier, MA and FA can be doped into $CsPbI_3$ to form organic–inorganic hybrid perovskite materials. Recent studies have shown that by adding 10% FA (relative to Cs), a stable perovskite with a maintained stable emission wavelength can be obtained [29].

B position: Based on recent literature reports, the substitution of ions at the B position in red-light perovskites is limited, and only a few ions such as Sn, Mn, Sr, and Cu have been studied. They form cationic hybridized perovskites with Pb ions, primarily to enhance the stability and improve the optoelectronic properties of the material. In contrast, for the X position, hybridization is often used to fine-tune the emission wavelength and cover the visible emission spectrum [9].

Despite the improved stability and optoelectronic properties of heterogeneous perovskites, there are several challenges that need to be addressed. These include phase separation issues under operating conditions, miscibility of the two ions in the host lattice, and an increase in intrinsic defects, which can limit their applications. However, the development of heterogeneous perovskites has greatly expanded the range of ions that can be used for doping, leading to the potential for further improvements in the stability and optoelectronic properties of perovskite materials [42–45].

4.2.2.2 Double Perovskite

Metal halide double perovskites with the general formula $A_2M^IM^{III}X_6$ (where $A = Cs^+$; $M^I = Cu^+, Ag^+, Na^+$; $M^{III} = Bi^{3+}, Sb^{3+}, In^{3+}$; and $X = Cl^-, Br^-, I^-$) exhibit a hybrid structure composed of two perovskite crystals arranged in ordered cycles to maintain charge neutrality. Essentially, double perovskites are formed by replacing two Pb^{2+} ions in the $APbX_3$ structure with one M^+ and one M^{3+} metal ion. Figure 4.6 shows the typical bicalcite structure of $Cs_2AgInCl_6$. While double perovskites have the potential to overcome the stability issues of single perovskites, there are still challenges in achieving their commercial applications, such as the

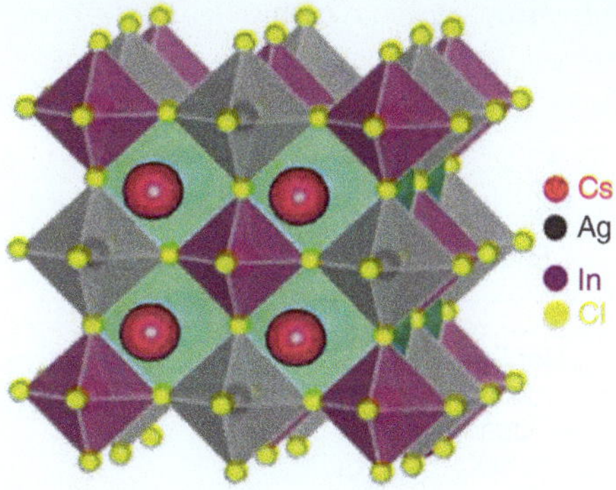

Figure 4.6 Schematic structure of a typical double perovskite ($Cs_2AgInCl_6$).

difficulty in synthesizing high-quality double perovskite materials and their limited optoelectronic properties compared to single perovskites [20].

Despite the potential advantages of metal halide double perovskite materials, such as maintaining charge neutrality and the replacement of Pb^{2+} ions with M^+ and M^{3+} metal ions, most of them suffer from issues such as large bandgaps, indirect bandgaps, poor stability, and poor carrier transport capability. Therefore, the development of red LEDs using double perovskite is still a significant challenge [20, 46–53]. Double perovskite materials doped with Mn^{2+} are generally found to emit red light among perovskite materials. This is mainly due to the spin of Mn, which prohibits the $4T^1$-6A [1] d-d transition, resulting in the emission of red light instead of yellow light [20, 21, 33].

4.2.3 Red Emission Perovskite Synthesis

4.2.3.1 Synthesis of Nanocrystals

Nanoparticles have been extensively studied due to their exceptional optical, thermal, electrical, magnetic, and mechanical properties [54, 55]. Perovskite nanocrystals have also been extensively studied due to their excellent optical, thermal, electrical, magnetic, and mechanical properties. These nanocrystals typically have a size range of 5–20 nm and exhibit a variety of morphologies, including cubic, flake, spherical, and linear shapes [56–59].

The current synthesis methods for red-light-emitting perovskite nanocrystals primarily rely on thermal injection methods. This method involves the rapid injection of a precursor solution into a hot solution that contains the remaining precursors, ligands, and a high-boiling-point solvent, as shown in Figure 4.7a [61–63]. The thermal injection method is typically used to synthesize small nanocrystals with a narrow size distribution by separating the nucleation and growth phases. The precursor solution is rapidly injected into a hot solution containing the remaining precursors, ligands, and a high-boiling-point solvent, as illustrated in Figure 4.7a. The thermal injection method enables the synthesis of small nanocrystals with a narrow size distribution by separating the nucleation and growth phases. Initially, rapid nucleation occurs, forming small nuclei. Subsequently, the rapid consumption of monomers terminates the nucleation phase, and the nuclei continue to grow without forming new

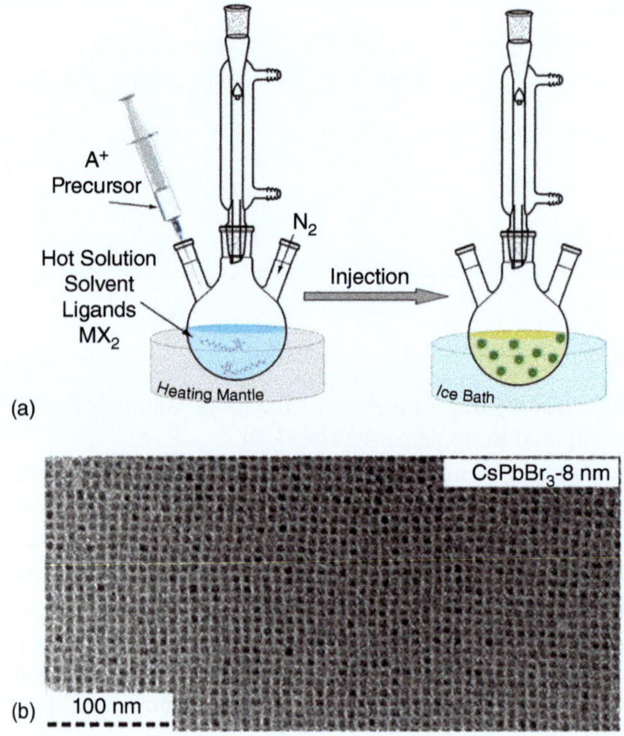

Figure 4.7 (a) Schematic diagram of a typical thermal injection method [2]. Source: Adapted from Tyagi et al. (b) Transmission electron microscopy image of nanocrystals obtained by thermal injection method. Source: Hong et al. [60]/Reproduced with permission from John Wiley & Sons.

nuclei. It is possible to dope the nanocrystals or alter their morphology by slightly adjusting the composition and ratio of the precursors or through ion exchange after the synthesis of the nanocrystals [10]. The entire synthesis process is very convenient, and the transmission electron microscopy images of the resulting nanocrystals generally show a uniform size distribution, as shown in Figure 4.7b. Nanocrystals synthesized by the thermal injection method typically have a size range of 5–20 nm and can form grains of about ten nanometers with excellent coverage after the formation of thin films [60].

4.2.3.2 Synthesis of Quasi-Two-Dimensional Films

Here, we are discussing quasi-2D films that are synthesized directly on the substrate as 2D films, rather than films where nanocrystals are transferred onto the substrate using some other method.

The chemical formula of chalcocite can also be expressed as $L_2A_{n-1}M_nX_{3n+1}$, where M and X represent short chains of divalent metal cations and halides, respectively, A represents monovalent metal cations or organic cations, L represents organic cations with long chains, and n represents layers of octahedral MX_6. The quasi-2D perovskite structure is illustrated in Figure 4.8b [64, 65]. As n increases

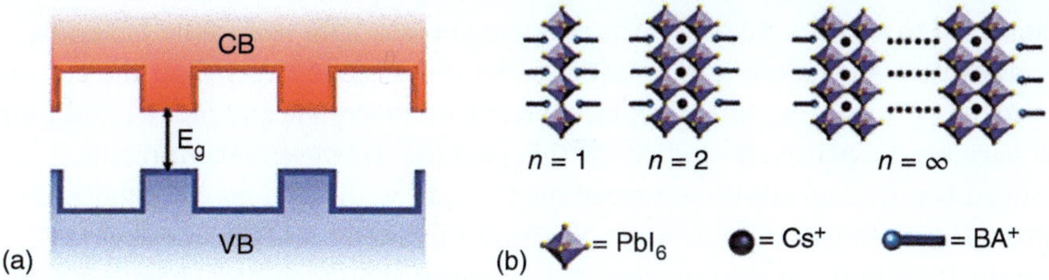

Figure 4.8 Structure and structural schematic of the quasi-two-dimensional perovskite energy band: (a) energy level structure; (b) structural schematic.

continuously, $L_2A_{n-1}M_nX_{3n+1}$ will eventually become $L_2(AMX_3)_n$, which is the structure of the three-dimensional perovskite AMX_3. Conversely, as n decreases to a small value, the material transforms into a two-dimensional perovskite. Many material properties undergo dramatic changes due to quantum-limited effects, such as increased exciton binding energy and bandgap [66, 67]. Quasi-two-dimensional perovskites are known for their natural quantum well structures, which facilitate efficient energy transfer and carrier radiative recombination. Consequently, many studies have focused on synthesizing quasi-two-dimensional films. The energy band diagram of the specific quantum well structure is depicted in Figure 4.8a.

For quasi-two-dimensional red-light perovskite materials, the most commonly used synthesis method is solution processing. Typically, the perovskite precursor reagent is first dissolved in a polar solvent. Then, the solution is spin-coated onto the substrate, and the polar solvent is removed by annealing to obtain quasi-two-dimensional perovskite thin films. Some recent studies on quasi-two-dimensional perovskite films have focused on multi-quantum well structures, which exhibit higher carrier injection efficiency. Such structures have the potential to improve the performance of red-light perovskite LEDs [12, 19].

4.2.4 Optimization Strategies of Red Perovskite Materials

As previously mentioned, various red-light perovskite materials have their drawbacks, such as instability or low fluorescence quantum yield, which greatly affect their application. In recent years, researchers have made significant efforts to optimize these material properties. The following five areas have been identified as potential strategies: morphology tuning, doping, surface passivation, multi-layer quantum well structures, and purification strategies.

4.2.4.1 Doping

In recent years, significant progress has been made in the doping of red perovskite materials. Doping can adjust the wavelength of light emission, reduce the content of Pb, and most importantly, increase the formation energy of perovskite, thereby improving its stability. As discussed earlier, several ions can be doped into red-light perovskite materials. However, not all doped ions in red-light perovskite materials will improve material stability and optoelectronic properties. Specific cases need to be analyzed to understand the effects of doping on red-light perovskite materials.

For position A, organic cations with spatial dimensions larger than the cesium ion are the only ones capable of replacing it [9]. Methacrylate and methanaminium ions have been extensively studied for this purpose. However, recent research has demonstrated that substituting some of the Cs^+ ions with FA^+ ions can significantly improve the material's stability and optoelectronic properties [29].

For the B position, current research is concentrated on replacing Pb with Mn, Sn, Sr, and Cu [11, 18, 30, 31, 41, 68, 69]. Notably, $CsMnI_3$, $CsSnI_3$, and $CsMnI_3$ cannot form stable perovskite structures or unstable perovskite structures. However, doping with Mn, Sn, and Cu can significantly enhance the stability and optoelectronic properties of red-light perovskite materials. Specifically, Mn doping can greatly improve the stability of $CsPbI_3$, allowing it to be preserved for several months. Moreover, the d state of Mn lies below the conduction band of $CsPbI_3$, which means that it does not alter the energy level structure of the material or affect the bandgap and optical properties of the red-light perovskite material [68]. Doping the perovskite material with Sn, another ion, leads to even better stability than Mn doping (up to 150 days). However, it does alter the bandgap of $CsPbI_3$ [31]. Zhang et al. recently discovered that Cu doping markedly enhances the fluorescence quantum yield and stability of $CsPbBrI_2$. However, it does not improve, and may even worsen, the stability of $CsPbI_3$. This is mainly because the octahedral factor of the Cu-doped $CsPbI_3$ material is too small, which hinders the maintenance of the crystal structure [18]. Apart from these dopants, Sr^{2+} ions have also been investigated for stabilizing $CsPbI_3$ nanocrystals [11].

For the X-position, only certain halides, such as Cl and Br, can be used for doping. The synthesis is typically achieved by initially using varying ratios of perovskite precursors or by later anion exchange to obtain X-position doped red-light perovskites [70–74]. Doping with different ratios of halides can effectively tune the luminescence wavelength, as halides have a significant impact on the conduction band of the perovskite system. Moreover, doping with various halides can also aid in stabilizing the new red-light perovskite material [9].

4.2.4.2 Surface Passivation

Perovskite nanocrystals possess a considerable specific surface area, and the existence of surface defects can significantly diminish the luminescence efficiency and stability of perovskite nanocrystals, thereby decreasing the LED device's performance [30, 76]. Consequently, finding ways to reduce surface defects has become a crucial area of research. Surface passivation, particularly through the use of surface ligands (typically organic amines or organic carboxylic acids), is one of the most effective methods to minimize surface defects. Surface ligands coordinate the hanging bonds produced by some uncoordinated anions/cations on the surface of perovskite nanocrystals, thus decreasing the number of defects on the surface of perovskite nanocrystals. This can be observed in Figure 4.9, where the black square represents the ligand layer formed by oleic acid (OA) and oleylamine (OAm). OAm and OA are the two most commonly used surface ligands in perovskite nanocrystal

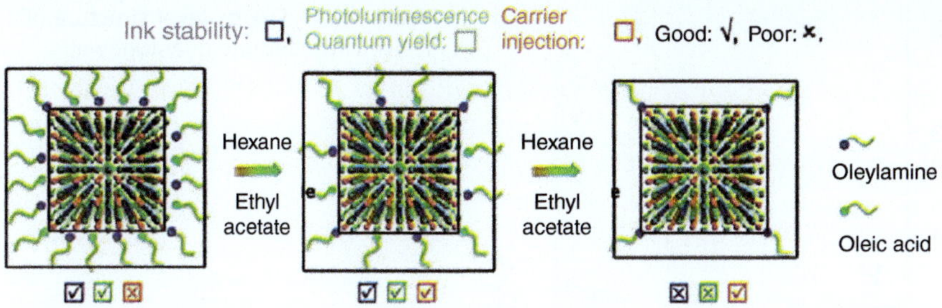

Figure 4.9 Ligand layers on the surface of perovskite nanocrystals and their effect on the photovoltaic properties of the nanocrystals [75].

synthesis [10, 11, 16, 18]. They aid in solution dispersion and colloidal stability and interact with the surface state to preserve certain material properties. Despite their benefits, OAm and OA are inherently insulating, which can limit carrier injection on the surface of the nanocrystals. Moreover, the dynamic and unstable nature of ligand binding can cause them to detach from the nanocrystal surface, resulting in an increased number of surface defects [77–79]. Pan's group used a bidentate ligand to passivate the surface of $CsPbI_3$, resulting in a significant improvement in stability and achieving a fluorescence quantum yield close to 100% [15]. Apart from the commonly used organic amines and organic carboxylic acids, inorganic ions like SCN^- and K^+ can also be employed for passivating the surfaces of red-light perovskite nanocrystals. For example, SCN^- can coordinate with $CsPbI_3$, while K^+ can coordinate with $CsPbI_{3-x}Br_x$ [16, 17].

4.2.4.3 Multiple Quantum Well Structure

While quasi-dimensional films have numerous advantages, such as improved film formation and higher exciton binding energy, their strong exciton–phonon interactions can result in a faster exciton decay rate at room temperature [66, 80, 81]. Consequently, the fluorescence quantum yield of quasi-dimensional perovskites tends to be very low. Jianpu Wang et al. demonstrated that a multi-quantum well structure can achieve a high EQE of 11.7%. This notable result is attributed to the efficient energy transfer that occurs within the structure [19]. The carrier injection process in the perovskite layer involves two stages. First, the carriers are injected into the perovskite layer, followed by their convergence at the quantum wells with larger n values ($L_2A_{n-1}M_nX_{3n+1}$), which are referred to as large n-MQWs due to the cascaded energy structure of the perovskite multi-quantum well films. The large n-MQWs layer then accumulates the electrons and holes, resulting in excellent electroluminescence performance. Figure 4.10 illustrates the energy level structure of the specific device. In 2023. Jingbi You et al. demonstrated a co-spacer cation quasi-2D $(PEA/m\text{-}F\text{-}PEA)_xNMA_{1-x})_2CsPb_2I_7$ film for developing efficient red PeLED. The low-dimensional phase is helpful for charge transfer and defect passivation. And the efficient red PeLED with a maximum EQE of 25.8% at 680 nm is achieved [23].

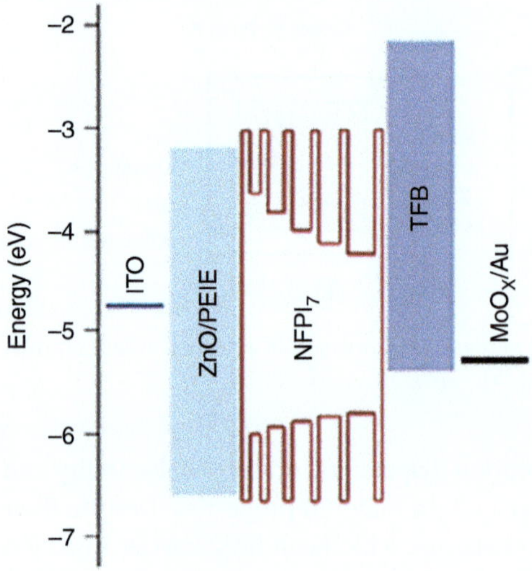

Figure 4.10 Energy level structure of a device with multiple quantum wells.

4.2.4.4 Ligand Engineering

In the previous section, we discussed the crucial role of surface passivation in achieving high fluorescence quantum efficiency in nanocrystals. Passivating ligands effectively reduce surface defects and are essential for achieving high quantum efficiency. However, these ligands often bind weakly to the nanocrystals and may detach from the surface during the purification process. Additionally, commonly used non-solvents such as methanol, ethanol, and acetone, which have high polarity, can alter the nanocrystal phases during overpurification.

To obtain nanocrystals with high fluorescence quantum efficiency, it is important to explore non-solvents with lower polarity and suitable purification paths that minimize phase alterations. This approach can help to overcome the challenges associated with weakly bound passivating ligands and prevent undesirable changes during the purification process [27, 82]. Luther et al. reported a successful method for purifying $CsPbI_3$ nanocrystals using low-polarity methyl acetate as a non-solvent, resulting in stable nanocrystals at room temperature. This purification approach also helped to elucidate the properties of the surface ligand layer [27]. The successful purification of nanocrystals such as $CsPbI_3$, $CsPbBr_3$, and $CsPbCl_3$ using methyl acetate has led to its widespread use as a non-solvent for this purpose.

Ligands play a crucial role in the synthesis of colloidal perovskites, such as in the dynamics of crystal growth, regulating the size and shape of the resulting crystals, and thus regulating the photoelectric properties. However, high dynamic binding between colloidal perovskite crystal surface and ligand is the source of chemical instability of colloidal perovskite. And because of that, the surface ligand is often desorbed with the colloidal nanocrystals (NC) surface of perovskite during aging, dilution, or washing purification, resulting in the formation of surface defects (mainly A-site and X-site vacancy defects), which greatly reduces PLQY and colloidal stability. Therefore, enhancing the coordination strength of the ligand on the crystal surface can have a positive effect on colloidal and chemical stability and thus on the conservation of its optical properties. In addition, surface ligands can be designed to prevent the aggregation of NCs and determine the extent of NC-solvent interactions, thus determining their dispersion in the medium and achieving uniform, stable luminescence. However, although long-chain ligands can passivate nanocrystal surface defects, the resulting decrease in electrical conductivity and charge transport difficulties both need to be traded off to achieve a relative balance and to achieve efficient and stable electroluminescence.

4.2.4.5 Additive Engineering

Colloidal perovskite is prone to ligand resolution during washing and purification, etc., resulting in A-site and X-site vacancy defects. And for 3D, quasi-2D, and 2D perovskites prepared by the solution method, there are many uncontrollable factors in the nucleation and growth processes, and there are still many trap states in the films. In general, the generation of trap states is related to the volatilization of the solvent during crystallization. In contrast, Lewis-base nonpolar solvents, such as DMSO, DMF, or NMP, readily form intermediate phases with metal halides and are widely used for the dissolution of perovskite precursors. Lewis base metal complexes decompose with the evaporation of the solvent during the annealing process, and metal dangling bonds and halogen vacancies are inevitably carried to the particle surface, i.e. vacancy defects are produced. These undesirable edge states are sensitive to moisture and oxygen, and once the oxygen atoms are adsorbed, they can provide additional low-energy orbitals that further serve as exciton capture sites. To suppress traps and improve perovskite stability and optoelectronic properties, vacancy defects should be effectively passivated.

4.3 Perovskite Red Light-Emitting Diodes

Despite the excellent stability and optoelectronic properties of red perovskite materials, their implementation in LED devices is hindered by several factors. These include the energy level mismatch between functional layers, photon self-absorption and reflection, and interface defects, all of which contribute to poor device performance [83]. Optimizing device structures, materials, and preparation processes is crucial for realizing the full luminescence potential of red perovskite materials.

4.3.1 Device Structure and Common Materials for Each Functional Layer

The device structure of perovskite LEDs closely resembles that of OLEDs, with electrodes, carrier transport layers (CTLs), and a light-emitting layer, as shown in Figure 4.11a. Typically, electrons are injected from the cathode, cross the electron injection layer, and reach the perovskite light-emitting layer. Simultaneously, holes from the anode injection layer pass through the hole injection layer and reach the perovskite layer. The holes and electrons then recombine and emit light from the perovskite layer.

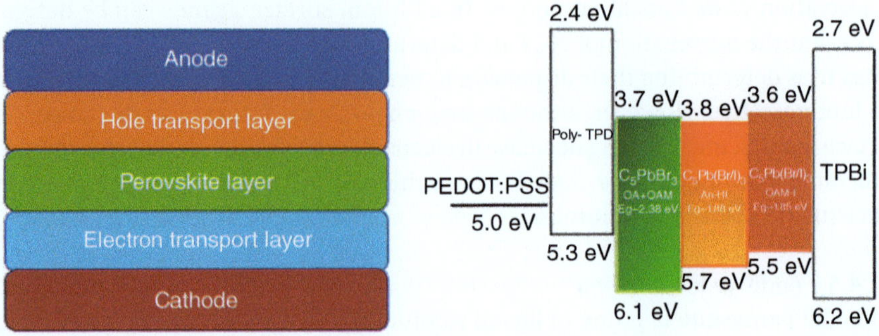

(a) (b)

Figure 4.11 Device structure of red light-emitting diodes of perovskite and energy level structure of typical devices: (a) device structure of common red light-emitting diodes of perovskite and (b) energy level diagram of current red light-emitting diodes of perovskite with the highest EQE.

Table 4.2 Common red perovskite light-emitting diode device structure in each layer of the material and the corresponding energy level position.

Anode	Energy level	HTL	Energy level	ETL	Energy level
ITO	−4.7 eV	PEDOT:PSS	−2.0 to 5.2 eV	ZnO	−3.7 to 7.3 eV
Cathode	Energy level	TCTA	−2.3 to 5.7 eV	TPBi	−2.7 to 6.2 eV
Ai/LiF	−3.0 eV	Poly-TPD	−2.0 to 5.4 eV	TCTA	−2.3 to 5.7 eV
Al/MoO_3	−5.6 eV	TFB	−2.3 to 5.3 eV	ZnO/PEIE	−3.2 to 6.5 eV
Au/MoO_3	−5.3 eV				
Ag/MoO_3	−5.3 eV				

Table 4.2 shows commonly used electrode materials and CTL materials for red perovskite LEDs, along with their respective energy level structures [10–19]. The current highest efficiency red perovskite LED exhibits an EQE of 21.3%. The device's energy level structure is presented in Figure 4.11b, utilizing Ploy-TPD and TPBi as the hole transport layer and electron transport layer, respectively [10].

4.3.2 Device Optimization Strategy

To enhance the performance of red perovskite LEDs, various processes involved in device preparation require further optimization. Fortunately, since the first perovskite LEDs were developed in 2014, several optimization methods for LEDs have been devised, including charge injection and transport enhancement, interfacial processing, and light extraction techniques. These are among the frequently employed optimization strategies for red perovskite LEDs based on perovskite.

4.3.2.1 Energy Level Regulation

For LED devices, the efficiency and brightness of the device are directly determined by the energy level alignment between the layers once the material of the light-emitting layer is established. Thus, it is crucial to compile the materials and common enhancement methods utilized for each functional layer.

It is widely recognized that in OLEDs, the carrier mobility and energy level structure of the CTL directly impact the efficiency, carrier injection balance, and blocking of reverse carriers, all of which are critical for achieving high luminous efficiency. Similarly, perovskite LEDs with CTLs commonly employ materials used in OLEDs. However, appropriate modifications are necessary to minimize the barrier between the CTL and the light-emitting layer in perovskite devices. For instance, Rogach et al. introduced a perfluorinated ionomer layer between the poly-TPD hole injection layer and the perovskite luminescent layer, which increased the valence band of the hole injection layer by 0.34 eV, promoting hole injection and resulting in a maximum luminance increase from 340 to 1377 cd m^{-2} [84]. Similar operations are performed to modify ZnO with PEI [85], PEIE-modified ZnO, PEI-modified PEDOT:PSS [6], PVK:PBD complex [86], and PEDOT:MoO_3 [87]. All of these methods serve to decrease the barrier between the CTL and the luminescent layer, thus improving carrier injection. It should be noted that commonly used carrier transport materials are inorganic oxides, mainly due to their lower electron affinity, which can significantly enhance carrier injection [88–90].

On the other hand, electrode modification is another approach that can be employed. A recent report showed that incorporating multi-walled carbon nanotubes into the electrode increased the carrier injection capacity [91, 92]. Another approach is to use defect engineering to enhance carrier transport capability [93].

4.3.2.2 Light Extraction Technology

Despite the high internal quantum efficiency of perovskite materials, not all emitted photons from the perovskite layers can escape the device layers and enter free space [83]. This is primarily due to inherent device structure limitations, including light self-absorption and reflection by the layers. Numerous methods exist to enhance perovskite LED performance. For instance, Sun's group employed silver nanoshell plasmonically coupled quantum dots to utilize excitons [94]. Another approach is to use special textures and silica microspheres to enhance the optical coupling output of perovskite LEDs [83]. Figure 4.11a demonstrates the impact of particular substrate microstructures on the EQE of the LED.

4.3.2.3 Interface Treatment Method

Gas–solid interactions. In the preparation of intercalated perovskite LEDs, it is crucial to consider solvent orthogonality to prevent solvent cleaning of the previous layer. However, only a few solvents are suitable for this purpose, which increases the difficulty of preparation. Tan et al. addressed this issue by utilizing a cross-linking method based on trimethyl aluminum vapor. They deposited TMA on a thin film of perovskite quantum dots using atomic layer deposition to immobilize and passivate the nanocrystals. This prevented the solvent used in the subsequent layer from cleaning off the red perovskite quantum dots, resulting in a red perovskite LED with an EQE of 5.7%. Figure 4.11b schematically illustrates the surface state of the nanocrystals after the specific TMA treatment [14].

Thermal annealing. Heat treatment is known to have a significant impact on the performance of perovskite LEDs. Proper annealing processes can help to reduce defects and improve the crystallinity of each layer, thereby reducing the leakage current and contact resistance of the device.

Interfacial passivation. Both devices utilize a p-i-n architecture, where the CTL made of metal oxide is in direct contact with the quantum dot layer. However, traps present in the metal oxide can result in exciton bursts at the interface between the CTL and quantum dots, leading to device inefficiencies. Therefore, it is crucial to passivate any surface defects present in the metal oxide layer.

Polymer-assisted films. When perovskite films are synthesized directly on the substrate, the film formation is often poor. To improve the quality of the films, perovskite precursors are typically mixed with polymers before film preparation. This results in better-quality films and increases the printability of the films when using polymers as aids [1, 13, 95]. Zhang et al. (2016) utilized PEI as an assistant to synthesize perovskite films with low perovskite defect passivation, resulting in red LEDs with an EQE of 7.25%. Similarly, in 2018, Tian et al. added PEO directly to the perovskite precursor to prepare a composite film of PEO and perovskite, resulting in a red LED with a maximum brightness of 1392 cd m^{-2} and a maximum EQE of 6.23%, as well as an increase in stability [13].

4.4 Conclusion and Outlook

Recent years have seen significant progress in the development of red-light perovskite LEDs. This article discusses the crystal structure, energy level structure, and optoelectronic properties of a typical red-light perovskite material, CsPbI3. However, the toxicity and phase sensitivity of $CsPbI_3$ have prompted researchers to explore alternative materials, such as doping with other elements or using double perovskite structures. Two main methods for synthesizing red-light perovskite materials are thermal injection and solution methods, with the former being more mature due to extensive studies on nanocrystals. Optimization techniques, including doping, surface passivation, multiple quantum wells, and purification strategies, have been developed to enhance the stability and optoelectronic properties of perovskite materials.

Regarding red-light perovskite LEDs, this article covers the basic structure and commonly used functional layer materials, as well as optimization techniques such as charge injection and transport enhancement, interface treatment, and light extraction. Despite an impressive EQE of 25.8%, issues with stability, emission wavelength tuning, and reproducibility still require further refinement to expand the application of red perovskite LEDs. Specific issues include the lack of a low-temperature synthesis method for red-light perovskite nanocrystals, limited anion exchange methods, and low maximum brightness of eye-red perovskite LEDs. However, with continued efforts and multidisciplinary collaboration, researchers are optimistic about the future development of red-light perovskite LEDs in terms of stability, photovoltaic properties, and synthesis methods.

Overall, the development of perovskite LEDs, especially red-light perovskite LEDs, has been progressing rapidly in recent years. The crystal structure, energy level structure, and optoelectronic properties of red-light perovskite materials have been well studied. However, issues such as stability, toxicity, and reproducibility still need to be addressed in order to improve the application of red perovskite devices.

In terms of synthesis methods, the thermal injection method for nanocrystals is relatively mature, while the solution method for quasi-two-dimensional thin films is still under development. Various optimization methods have been explored, including doping, surface passivation, multiple quantum wells, and purification strategies. These methods have shown promising results in improving the stability and optoelectronic properties of red perovskite devices.

Moving forward, further research is needed to address the specific issues related to red perovskite devices, such as the need for a low-temperature synthesis method for nanocrystals, the need for more efficient anion exchange methods, and the need to increase the maximum brightness of red-light perovskite LEDs. With continued efforts and multi-disciplinary collaboration, it is expected that these issues will be addressed and red-light perovskite LEDs will continue to advance in terms of stability, photovoltaic properties, and synthesis methods in the coming years.

References

1 Tan, Z., Moghaddam, R.S., Lai, M.L. et al. (2014). Bright light-emitting diodes based on organometal halide perovskite. *Nature Nanotechnology* 9: 687–692.

2 Shamsi, J., Urban, A.S., Imran, M. et al. (2019). Metal halide perovskite nanocrystals: synthesis, post-synthesis modifications, and their optical properties. *Chemical Reviews* 119: 3296–3348.

3 Shan, Q., Song, J., Zou, Y. et al. (2017). High performance metal halide perovskite light-emitting diode: from material design to device optimization. *Small* 13: 1701770.

4 Yu, J.C., Kim, D.B., Jung, E.D. et al. (2016). High-performance perovskite light-emitting diodes via morphological control of perovskite films. *Nanoscale* 8: 7036–7042.

5 Kim, Y.H., Kim, J.S., and Lee, T.W. (2019). Strategies to improve luminescence efficiency of metal-halide perovskites and light-emitting diodes. *Advanced Materials* 31: 1804595.

6 Cho, H., Jeong, S., Park, M. et al. (2015). Overcoming the electroluminescence efficiency limitations of perovskite light-emitting diodes. *Science* 350: 1222–1225.

7 Lin, K., Xing, J., Quan, L.N. et al. (2018). Perovskite light-emitting diodes with external quantum efficiency exceeding 20 percent. *Nature* 562: 245–248.

8 Wang, H., Yu, H., Xu, W. et al. (2018). Efficient perovskite light-emitting diodes based on a solution-processed tin dioxide electron transport layer. *Journal of Materials Chemistry C* 6: 6996–7002.

9 Dutta, A. and Pradhan, N. (2019). Phase-stable red-emitting $CsPbI_3$ nanocrystals: successes and challenges. *ACS Energy Letters* 4: 709–719.

10 Chiba, T., Hayashi, Y., Ebe, H. et al. (2018). Anion-exchange red perovskite quantum dots with ammonium iodine salts for highly efficient light-emitting devices. *Nature Photonics* 12: 681–687.

11 Yao, J., Ge, J., Wang, K. et al. (2019). Few-nanometer-sized α-$CsPbI_3$ quantum dots enabled by strontium substitution and iodide passivation for efficient red-light emitting diodes. *Journal of the American Chemical Society* 141: 2069–2079.

12 Zhang, S., Yi, C., Wang, N. et al. (2017). Efficient red perovskite light-emitting diodes based on solution-processed multiple quantum wells. *Advanced Materials* 29: 1606600.

13 Tian, Y., Zhou, C., Worku, M. et al. (2018). Highly efficient spectrally stable red perovskite light-emitting diodes. *Advanced Materials* 30: 1707093.

14 Li, G., Rivarola, F.W.R., Davis, N.J.L.K. et al. (2016). Highly efficient perovskite nanocrystal light-emitting diodes enabled by a universal crosslinking method. *Advanced Materials* 28: 3528–3534.

15 Pan, J., Shang, Y., Yin, J. et al. (2018). Bidentate ligand-passivated $CsPbl_3$ perovskite nanocrystals for stable near-unity photoluminescence quantum yield and efficient red light-emitting diodes. *Journal of the American Chemical Society* 140: 562–565.

16 Yang, J., Song, Y., Yao, J. et al. (2020). Potassium-bromide surface passivation on $CsPbI_{3-x}Br_x$ nanocrystals for efficient and stable pure red perovskite light emitting diodes. *Journal of the American Chemical Society* 142: 2956–2967.

17 Lu, M., Guo, J., Lu, P. et al. (2019). Ammonium thiocyanate-passivated $CsPbI_3$ perovskite nanocrystals for efficient red light-emitting diodes. *The Journal of Physical Chemistry C* 123: 22787–22792.

18 Zhang, J., Zhang, L., Cai, P. et al. (2019). Enhancing stability of red perovskite nanocrystals through copper substitution for efficient light-emitting diodes. *Nano Energy* 62: 434–441.

19 Wang, N., Cheng, L., Ge, R. et al. (2016). Perovskite light-emitting diodes based on solution-processed self-organized multiple quantum wells. *Nature Photonics* 10: 699–704.

20 Nandha, K.N. and Nag, A. (2018). Synthesis and luminescence of Mn-doped $Cs_2AgInCl_6$ double perovskites. *Chemical Communications* 54: 5205–5208.

21 Majher, J.D., Gray, M.B., Strom, T.A. et al. (2019). $Cs_2NaBiCl_6$:Mn^{2+} – a new orange-red halide double perovskite phosphor. *Chemistry of Materials* 31: 1738–1744.

22 Wang, Y.K., Yuan, F.L., Dong, Y.T. et al. (2021). All-inorganic quantum-dot LEDs based on a phase-stabilized alpha-$CsPbI_3$ perovskite. *Angewandte Chemie-International Edition* 60: 16164–16170.

23 Jiang, J., Chu, Z., Yin, Z. et al. (2022). Red perovskite light-emitting diodes with efficiency exceeding 25% realized by co-spacer cations. *Advanced Materials* 34: 2204460.

24 Yassitepe, E., Yang, Z., Voznyy, O. et al. (2016). Amine-free synthesis of cesium lead halide perovskite quantum dots for efficient light-emitting diodes. *Advanced Functional Materials* 26: 8757–8763.

25 Davis, N.J.L.K., La Pena, F.D., Tabachnyk, M. et al. (2017). Photon reabsorption in mixed $CsPbCl_3$:$CsPbI_3$ perovskite nanocrystal films for light-emitting diodes. *The Journal of Physical Chemistry C* 121: 3790–3796.

26 Zou, C., Huang, C., Sanehira, E.M. et al. (2017). Highly stable cesium lead iodide perovskite quantum dot light-emitting diodes. *Nanotechnology* 28: 455201.

27 Swarnkar, A., Marshall, A.R., Sanehira, E.M. et al. (2016). Quantum dot-induced phase stabilization of α-CsPbI3 perovskite for high-efficiency photovoltaics. *Science* 354: 92–95.

28 Liu, F., Ding, C., Zhang, Y. et al. (2017). Colloidal synthesis of air-stable alloyed $CsSn_{1-x}Pb_xI_3$ perovskite nanocrystals for use in solar cells. *Journal of the American Chemical Society* 139: 16708–16719.

29 Protesescu, L., Yakunin, S., Kumar, S. et al. (2017). Dismantling the "red wall" of colloidal perovskites: highly luminescent formamidinium and formamidinium-cesium lead iodide nanocrystals. *ACS Nano* 11: 3119–3134.

30 Lu, M., Zhang, X., Zhang, Y. et al. (2018). Simultaneous strontium doping and chlorine surface passivation improve luminescence intensity and stability of $CsPbI_3$ nanocrystals enabling efficient light-emitting devices. *Advanced Materials* 30: 1804691.

31 Liu, F., Zhang, Y., Ding, C. et al. (2017). Highly luminescent phase-stable $CsPbI_3$ perovskite quantum dots achieving near 100% absolute photoluminescence quantum yield. *ACS Nano* 11: 10373–10383.

32 Park, N.-G. (2015). Perovskite solar cells: an emerging photovoltaic technology. *Materials Today* 18: 65–72. https://doi.org/10.1016/j.mattod.2014.07.007.

33 Locardi, F., Cirignano, M., Baranov, D. et al. (2018). Colloidal synthesis of double perovskite $Cs_2AgInCl_6$ and Mn-doped $Cs_2AgInCl_6$ nanocrystals. *Journal of the American Chemical Society* 140: 12989–12995.

34 Marronnier, A., Roma, G., Boyer-Richard, S. et al. (2018). Anharmonicity and disorder in the black phases of cesium lead iodide used for stable inorganic perovskite solar cells. *ACS Nano* 12: 3477–3486.

35 Protesescu, L., Yakunin, S., Bodnarchuk, M.I. et al. (2015). Nanocrystals of cesium lead halide perovskites ($CsPbX_3$, X = Cl, Br, and I): novel optoelectronic materials showing bright emission with wide color gamut. *Nano Letters* 15: 3692–3696.

36 Akkerman, Q.A., Raino, G., Kovalenko, M.V. et al. (2018). Genesis, challenges and opportunities for colloidal lead halide perovskite nanocrystals. *Nature Materials* 17: 394–405.

37 Wang, Z., Shi, Z., Li, T. et al. (2017). Stability of perovskite solar cells: a prospective on the substitution of the A cation and X anion. *Angewandte Chemie* 56: 1190–1212.

38 Li, W., Wang, Z.M., Deschler, F. et al. (2017). Chemically diverse and multifunctional hybrid organic-inorganic perovskites. *Nature Reviews Materials* 2: 16099.

39 Travis, W., Glover, E.N.K., Bronstein, H. et al. (2016). On the application of the tolerance factor to inorganic and hybrid halide perovskites: a revised system. *Chemical Science* 7: 4548–4556.

40 Saparov, B. and Mitzi, D.B. (2016). Organic-inorganic perovskites: structural versatility for functional materials design. *Chemical Reviews* 116: 4558–4596.

41 Lei, T., Lai, M.L., Kong, Q. et al. (2018). Electrical and optical tunability in all-inorganic halide perovskite alloy nanowires. *Nano Letters* 18: 3538–3542.

42 Li, Z., Yang, M.J., Park, J.S. et al. (2016). Stabilizing perovskite structures by tuning tolerance factor: formation of formamidinium and cesium lead iodide solid-state alloys. *Chemistry of Materials* 28: 284–292.

43 Hazarika, A., Zhao, Q., Gaulding, E.A. et al. (2018). Perovskite quantum dot photovoltaic materials beyond the reach of thin films: full-range tuning of A-site cation composition. *ACS Nano* 12: 10327–10337.

44 Jong, U.G., Yu, C.J., Kye, Y.H. et al. (2018). A first-principles study on the chemical stability of inorganic perovskite solid solutions $Cs_{1-x}Rb_xPbI_3$ at finite temperature and pressure. *Journal of Materials Chemistry A* 6: 17994–18002.

45 Brennan, M.C., Draguta, S., Kamat, P.V. et al. (2018). Light-induced anion phase segregation in mixed halide perovskites. *ACS Energy Letters* 3: 204–213.

46 Volonakis, G., Filip, M.R., Haghighirad, A.A. et al. (2016). Lead-free halide double perovskites via heterovalent substitution of noble metals. *The Journal of Physical Chemistry Letters* 7: 1254–1259.

47 Mcclure, E.T., Ball, M.R., Windl, W. et al. (2016). Cs_2AgBiX_6 (X = Br, Cl): new visible light absorbing, lead-free halide perovskite semiconductors. *Chemistry of Materials* 28: 1354.

48 Slavney, A.H., Hu, T., Lindenberg, A.M. et al. (2016). A bismuth-halide double perovskite with long carrier recombination lifetime for photovoltaic applications. *Journal of the American Chemical Society* 138: 2138–2141.

49 Slavney, A.H., Leppert, L., Bartesaghi, D. et al. (2017). Defect-induced band-edge reconstruction of a bismuth-halide double perovskite for visible-light absorption. *Journal of the American Chemical Society* 139: 5015–5018.

50 Creutz, S.E., Crites, E.N., De Siena, M.C. et al. (2018). Colloidal nanocrystals of lead-free double-perovskite (elpasolite) semiconductors: synthesis and anion exchange to access new materials. *Nano Letters* 18: 1118–1123.

51 Du, K., Meng, W., Wang, X. et al. (2017). Bandgap engineering of lead-free double perovskite $Cs_2AgBiBr_6$ through trivalent metal alloying. *Angewandte Chemie* 56: 8158–8162.

52 Volonakis, G., Haghighirad, A.A., Milot, R.L. et al. (2017). $Cs_2InAgCl_6$: a new lead-free halide double perovskite with direct band gap. *The Journal of Physical Chemistry Letters* 8: 772–778.

53 Zhou, J., Xia, Z., Molokeev, M.S. et al. (2017). Composition design, optical gap and stability investigations of lead-free halide double perovskite $Cs_2AgInCl_6$. *Journal of Materials Chemistry* 5: 15031–15037.

54 Rossetti, R., Nakahara, S., and Brus, L.E. (1983). Quantum size effects in the redox potentials, resonance Raman spectra, and electronic spectra of CdS crystallites in aqueous solution. *Journal of Chemical Physics* 79: 1086–1088.

55 Brus, L.E. (1986). Electronic wave functions in semiconductor clusters: experiment and theory. *The Journal of Physical Chemistry* 90: 2555–2560.

56 Pan, A.Z., He, B., Fan, X.Y. et al. (2016). Insight into the ligand-mediated synthesis of colloidal $CsPbBr_3$ perovskite nanocrystals: the role of organic acid, base, and cesium precursors. *ACS Nano* 10: 7943–7954.

57 Parobek, D., Dong, Y.T., Qiao, T. et al. (2018). Direct hot-injection synthesis of Mn-doped $CsPbBr_3$ nanocrystals. *Chemistry of Materials* 30: 2939–2944.

58 Wang, A.F., Guo, Y.Y., Muhammad, F. et al. (2017). Controlled synthesis of lead-free cesium tin halide perovskite cubic nanocages with high stability. *Chemistry of Materials* 29: 6493–6501.

59 Seth, S. and Samanta, A. (2016). A facile methodology for engineering the morphology of $CsPbX_3$ perovskite nanocrystals under ambient condition. *Scientific Reports* 6: 37693.

60 Li, X., Zhang, K., Li, J. et al. (2018). Heterogeneous nucleation toward polar-solvent-free, fast, and one-pot synthesis of highly uniform perovskite quantum dots for wider color gamut display. *Advanced Materials Interfaces* 5: 1800010.

61 Yu, W.W. and Peng, X. (2002). Formation of high-quality CdS and other II–VI semiconductor nanocrystals in noncoordinating solvents: tunable reactivity of monomers. *Angewandte Chemie* 41: 2368–2371.

62 Bullen, C. and Mulvaney, P. (2004). Nucleation and growth kinetics of CdSe nanocrystals in octadecene. *Nano Letters* 4: 2303–2307.

63 Deka, S., Genovese, A., Zhang, Y. et al. (2010). Phosphine-free synthesis of p-type copper(I) selenide nanocrystals in hot coordinating solvents. *Journal of the American Chemical Society* 132: 8912–8914.

64 Tyagi, P., Arveson, S.M., and Tisdale, W.A. (2015). Colloidal organohalide perovskite nanoplatelets exhibiting quantum confinement. *The Journal of Physical Chemistry Letters* 6: 1911–1916.

65 Calabrese, J.C., Jones, N.L., Harlow, R.L. et al. (1991). Preparation and characterization of layered lead halide compounds. *Journal of the American Chemical Society* 113: 2328–2330.

66 Hong, X., Ishihara, T., and Nurmikko, A.V. (1992). Dielectric confinement effect on excitons in PbI_4-based layered semiconductors. *Physical Review B* 45: 6961–6964.

67 Ishihara, T. (1994). Optical properties of PbI-based perovskite structures. *Journal of Luminescence* 60: 269–274.

68 Akkerman, Q.A., Meggiolaro, D., Dang, Z. et al. (2017). Fluorescent alloy $CsPb_xMn_{1-x}I_3$ perovskite nanocrystals with high structural and optical stability. *ACS Energy Letters* 2: 2183–2186.

69 Zou, S., Liu, Y., Li, J. et al. (2017). Stabilizing cesium lead halide perovskite lattice through Mn(II) substitution for air-stable light-emitting diodes. *Journal of the American Chemical Society* 139: 11443–11450.

70 Jing, Q., Zhang, M., Huang, X. et al. (2017). Surface passivation of mixed-halide perovskite $CsPb(Br_xI_{1-x})_3$ nanocrystals by selective etching for improved stability. *Nanoscale* 9: 7391–7396.

71 Dastidar, S., Egger, D.A., Tan, L.Z. et al. (2016). High chloride doping levels stabilize the perovskite phase of cesium lead iodide. *Nano Letters* 16: 3563–3570.

72 Chen, X., Peng, L., Huang, K. et al. (2016). Non-injection gram-scale synthesis of cesium lead halide perovskite quantum dots with controllable size and composition. *Nano Research* 9: 1994–2006.

73 He, Y., Gong, J., Zhu, Y. et al. (2018). Highly pure yellow light emission of perovskite $CsPb(Br_xI_{1-x})_3$ quantum dots and their application for yellow light-emitting diodes. *Optical Materials* 80: 1–6.

74 Gualdron-Reyes, A.F., Yoon, S.J., Barea, E.M. et al. (2019). Controlling the phase segregation in mixed halide perovskites through nanocrystal size. *ACS Energy Letters* 4: 54–62.

75 Li, J., Xu, L., Wang, T. et al. (2017). 50-Fold EQE improvement up to 6.27% of solution-processed all-inorganic perovskite CsPbBr3 QLEDs via surface ligand density control. *Advanced Materials* 29: 1603885. https://doi.org/10.1002/adma.201603885.

76 Zhang, X., Lu, M., Zhang, Y. et al. (2018). PbS capped $CsPbI_3$ nanocrystals for efficient and stable light-emitting devices using p-i-n structures. *ACS Central Science* 4: 1352–1359.

77 De Roo, J., Ibanez, M., Geiregat, P. et al. (2016). Highly dynamic ligand binding and light absorption coefficient of cesium lead bromide perovskite nanocrystals. *ACS Nano* 10: 2071–2081.

78 Shao, H., Bai, X., Cui, H. et al. (2018). White light emission in Bi^{3+}/Mn^{2+} ion co-doped $CsPbCl_3$ perovskite nanocrystals. *Nanoscale* 10: 1023–1029.

79 Pan, G., Bai, X., Yang, D. et al. (2017). Doping lanthanide into perovskite nanocrystals: highly improved and expanded optical properties. *Nano Letters* 17: 8005–8011.

80 Gauthron, K., Lauret, J.S., Doyennette, L. et al. (2010). Optical spectroscopy of two-dimensional layered $(C_6H_5C_2H_4\text{-}NH_3)_2\text{-}PbI_4$ perovskite. *Optics Express* 18: 5912–5919.

81 Li, R.Z., Yi, C., Ge, R. et al. (2016). Room-temperature electroluminescence from two-dimensional lead halide perovskites. *Applied Physics Letters* 109: 151101.

82 Behera, R.K., Adhikari, S.D., Dutta, S.K. et al. (2018). Blue-emitting $CsPbCl_3$ nanocrystals: impact of surface passivation for unprecedented enhancement and loss of optical emission. *The Journal of Physical Chemistry Letters* 9: 6884–6891.

83 Richter, J.M., Abdijalebi, M., Sadhanala, A. et al. (2016). Enhancing photoluminescence yields in lead halide perovskites by photon recycling and light out-coupling. *Nature Communications* 7: 13941.

84 Zhang, X., Lin, H., Huang, H. et al. (2016). Enhancing the brightness of cesium lead halide perovskite nanocrystal based green light-emitting devices through the interface engineering with perfluorinated ionomer. *Nano Letters* 16: 1415–1420.

85 Wang, J., Wang, N., Jin, Y. et al. (2015). Interfacial control toward efficient and low-voltage perovskite light-emitting diodes. *Advanced Materials* 27: 2311–2316.

86 Ling, Y., Yuan, Z., Tian, Y. et al. (2016). Bright light-emitting diodes based on organometal halide perovskite nanoplatelets. *Advanced Materials* 28: 305–311.

87 Kim, D.B., Yu, J.C., Nam, Y.S. et al. (2016). Improved performance of perovskite light-emitting diodes using a PEDOT:PSS and MoO_3 composite layer. *Journal of Materials Chemistry C* 4: 8161–8165.

88 Hoye, R.L.Z., Chua, M.R., Musselman, K.P. et al. (2015). Enhanced performance in fluorene-free organometal halide perovskite light-emitting diodes using tunable, low electron affinity oxide electron injectors. *Advanced Materials* 27: 1414–1419.

89 Shan, Q., Li, J., Song, J. et al. (2017). All-inorganic quantum-dot light-emitting diodes based on perovskite emitters with low turn-on voltage and high humidity stability. *Journal of Materials Chemistry C* 5: 4565–4570.

90 Shi, Z., Li, Y., Zhang, Y. et al. (2017). High-efficiency and air-stable perovskite quantum dots light-emitting diodes with an all-inorganic heterostructure. *Nano Letters* 17: 313–321.

91 Jiang, Y., Liu, H., Gong, X. et al. (2016). Enhanced performance of planar perovskite solar cells via incorporation of Bphen/Cs_2CO_3-MoO_3 double interlayers. *Journal of Power Sources* 331: 240–246.

92 Cheng, N., Liu, P., Qi, F. et al. (2016). Multi-walled carbon nanotubes act as charge transport channel to boost the efficiency of hole transport material free perovskite solar cells. *Journal of Power Sources* 332: 24–29.

93 Peng, B., Yu, G., Zhao, Y. et al. (2016). Achieving ultrafast hole transfer at the monolayer MoS_2 and $CH_3NH_3PbI_3$ perovskite interface by defect engineering. *ACS Nano* 10: 6383–6391.

94 Yang, X., Hernandezmartinez, P.L., Dang, C. et al. (2015). Electroluminescence efficiency enhancement in quantum dot light-emitting diodes by embedding a silver nanoisland layer. *Advanced Optical Materials* 3: 1439–1445.

95 Bade, S.G.R., Li, J., Shan, X. et al. (2016). Fully printed halide perovskite light-emitting diodes with silver nanowire electrodes. *ACS Nano* 10: 1795–1801.

5

Perovskite Green Light-Emitting Materials and Devices

5.1 History of Green Perovskite Light-Emitting Diodes

The exceptional photovoltaic properties of perovskites have generated significant interest, leading to their use in a variety of applications such as solar cells, light-emitting diodes (LEDs), and photodetectors, among others [1–4]. Compared to conventional displays such as liquid crystal displays (LCDs) and organic light-emitting diodes (OLEDs), perovskite light-emitting diodes (PeLEDs) offer significant advantages, such as a narrow spectral half-peak width, high color saturation, good defect tolerance, and color and brightness tunability in the visible light range. Additionally, they offer high carrier mobility and low cost, making them an attractive option for various applications [5–7]. The advantages of PeLEDs are evident from research findings. Studies have indicated that red and green PeLEDs demonstrate superior electroluminescence (EL) efficiency and brightness when compared to blue PeLEDs. Currently, both red and green PeLEDs achieve an external quantum efficiency (EQE) exceeding 30%, while blue PeLEDs achieve approximately 19% EQE. Given the human eye's high sensitivity to green light, there is significant interest in exploring the application of green PeLEDs in various fields, such as lighting and displays.

The chemical structure of a three-dimensional perovskite material can be represented by the formula ABX_3, where A and B are cations and X is an anion. In contrast, quasi-two-dimensional perovskite materials have a general formula of $L_2[A_{n-1}B_nX_{3n+1}]$, where L is an organic spacer cation, A and B are different cations, X is an anion, and n represents the number of inorganic layers. The quasi-two-dimensional structure is achieved through a self-assembly process of the organic cations and the inorganic layers, resulting in the formation of alternating organic and inorganic layers with a thickness of a few nanometers. This structure is of great interest for optoelectronic applications due to its unique electronic and optical properties, such as high exciton binding energy and tunable bandgap [7, 8]. The A-site in perovskite materials is typically occupied by an organic or metal cation, such as the methylamine ion ($CH_3\ NH_3^+$, MA^+), formamidine ion ($CH(NH)_2^{2+}$, FA^+), or cesium ion (Cs^+). The B-site is usually a divalent transition metal cation, with lead ion (Pb^{2+}) being commonly used. The C-site is occupied by a halogen ion, typically Br for green perovskite materials. In quasi-two-dimensional perovskite materials, the L represents

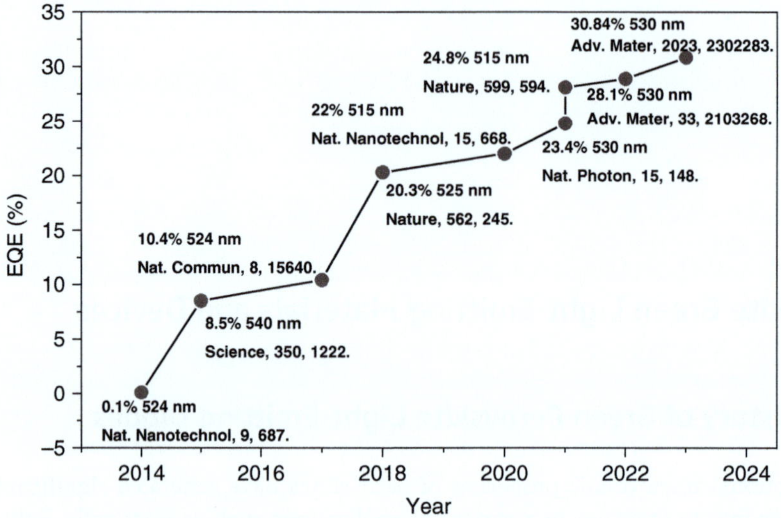

Figure 5.1 Diagram of the development of green perovskite light-emitting diodes.

long-chain organic ammonium cations with large ionic radii, such as butylamine (BA) or phenylethylamine (PEA), among others.

Figure 5.1 shows the progress in the EQE of green perovskite light-emitting devices over time. Friend's group created the first green PeLED that could operate at room temperature. They used $MAPbBr_3$ as the light-emitting material and produced a device with an EQE of 0.1% [9]. Since then, various strategies have been developed to enhance the performance of green PeLEDs, including surface passivation, defect engineering, and composition optimization. These strategies have led to significant improvements in the efficiency and stability of green PeLEDs, making them promising candidates for next-generation lighting and display applications. In 2015, Zeng Haibo's group reported the first inorganic perovskite-based LEDs, using $CsPbBr_3$ quantum dots as the light-emitting layer material. This approach addressed the instability issues associated with $MAPbBr_3$ as the light-emitting material and achieved an EQE of 0.12% [10]. In the same year, Tae Woo Lee et al. made significant progress by achieving an EQE of 8.53% through precise adjustment of the MABr to $PbBr_2$ ratio in the precursor components and the introduction of additives to the hole injection layer (HIL) to regulate the energy levels of each functional layer. This marked a significant improvement in the performance of green PeLEDs and opened up new possibilities for their use in commercial applications [11]. The synthesis process has also been adapted to improve the fluorescence efficiency. In 2017, a considerable number of research articles were published on the use of crystal engineering and optimization of device structure to enhance the performance of PeLED devices [12–16]. Jingbi You's group achieved an EQE of 10.4% by using the mixed cation $Cs_{0.87}MA_{0.13}PbBr_3$ as the luminescent material [17]. In March 2018, Mhaisallar et al. utilized material structure design to construct a multilayered structure consisting of three-dimensional $FAPbBr_3$ combined with a two-dimensional octylamine PbBr microsheet. This resulted in a fluorescence quantum efficiency of over 80% and an EQE of 13.4% [18]. In September 2018, Haibo Zeng's team passivated the surface of $CsPbBr_3$ quantum dots with organic–inorganic hybrid ligands, resulting in a device with an EQE of 16.48% [19]. In October 2018, Wei Zhanhua's group further improved the EQE by preparing a $CsPbBr_3$/MABr core–shell structure to passivate the surface defect state of $CsPbBr_3$. Poly(methyl methacrylate) (PMMA) was also introduced to balance the carrier transport equilibrium, resulting in an EQE increase of 20.3% [3]. In 2019, several research groups achieved remarkable results by improving the performance of green PeLEDs through surface modification to passivate defects [2, 20–22]. Zhang's group made significant progress in improving the performance of green PeLEDs through surface modification to passivate defects. They also synthesized quasi-two-dimensional $FAPbBr_3$ nanosheets with 1–4 layers of Pb by controlling the type and content of large cations. This resulted in the highest-purity green PeLEDs at the time, with 97% coverage compared to the 2020 standard color gamut [23]. In 2020, Dong et al. developed a method to treat the surface of $CsPbBr_3$ perovskite quantum dots and prepare bipolar shell $CsPbBr_3$ quantum dots, consisting of an internal anionic shell and an external cationic and polar solvent molecular shell, through electrostatic adsorption. This method reduced the surface defect density and increased carrier mobility, resulting in an EQE of 22% [24]. In 2021, Lee et al. utilized a single-doping alloying strategy to dope guanidinium salts into $FAPbBr_3$ nanocrystals, limiting their solubility and filling in ligand defects with excess guanidinium salts. The resulting perovskite nanocrystals had uniform grain size and high stability, leading to excellent EQEs of up to 23.4% in the resulting devices [25]. In 2021, Sargent's group improved the stability and performance of green perovskite thin films by passivating PEA-based perovskite with tris(4-fluorophenyl)phosphine oxide (TFPPO) in the counter-solvent. TFPPO interacts with lead in the system to passivate defects, while fluorine atoms bridge the PEA cations through hydrogen bonds, enhancing film and device stability. This resulted in green perovskite thin films with an EQE of up to 24.8%. The stability was also greatly improved, with T_{50} at an initial brightness of 7200 cd cm^{-1} for up to two hours [26]. The stability of the LEDs was also greatly improved, maintaining an initial brightness of 7200 cd cm^{-1} for up to two hours. In a similar timeframe, Su's group at Southeast University published a paper in which they added 18-crown-6 ether and polyethylene glycol methyl ether acrylate (polyethylene glycol methyl ether acrylate [MPEG-MAA]) as additives to the perovskite system. The ether bond (C—O—C) in 18-crown-6 ether not only effectively passivates defects but also prevents self-aggregation of organic ligands in the system, inducing the formation of quasi-core/shell structure perovskite nanocrystals. After thermal annealing, MPEG-MAA and its carbon-carbon double bond (C=C) can polymerize to give comb-like polymers, further protecting the perovskite nanocrystals from water and oxygen attack. The result is a green PeLED with an EQE of 28.1%, which is the highest ever achieved for a green PeLED, and a 10-fold increase in operating stability compared to a device without additives [27]. In 2022, an astounding world record was set for the performance of green PeLEDs, leveraging an innovative in situ core/shell structure. The results demonstrated a peak current efficiency of 151 cd A^{-1}, coupled with an exceptional maximum EQE of 28.9%. The maximum brightness recorded was around 470 000 cd m^{-2}, with a remarkably low-efficiency roll-off of about 5% even at an intensity of 400 000 cd m^{-2}. Further, the half-life spanned 520 hours at an initial brightness of 1000 cd m^{-2}, with projections indicating an estimated half-life surpassing 30 000 hours at 100 cd m^{-2}. It's noteworthy that these devices emitted green light at an EL peak of 540 nm. Impressively, this high-stability, high-efficiency green perovskite material was synthesized using a simple in situ reaction. This reaction involved benzylphosphonic acid (BPA) as an additive with three-dimensional (3D) polycrystalline perovskite films, thereby eliminating the need for separate synthesis processes. This achievement marks a significant leap forward in the field of optoelectronics [28]. Rong-Jun Xie et al. demonstrated a modified hole transport layer, $Ni_{1-y}Mg_yO_x$, to increase the hole mobility, while the addition of the polyethylene glyco (PEG) layer between the HTLs and the perovskite EMLs blocks electron leakage. This strategy not only enables negligible electron leakage but also effectively regulates the near-field light distribution in the functional layers, increasing the light outcoupling efficiency from 30.02% to 41.82%. As a result, an ultrahigh-efficiency green PeLED with a maximum EQE of 30.84% at a luminance of 6514 cd m^{-2}, presents the most efficient PeLED to date, which is comparable to commercial OLEDs [29]. Despite the exceptional optoelectronic properties exhibited by PeLEDs, there are several challenges they face, such as a tendency to roll-off at high current densities, sensitivity to light and moisture, and a short device lifetime due to ion migration. These issues can significantly affect device performance and even lead to device failure [30–33]. The EQE is a crucial parameter for evaluating the performance of PeLEDs, and it can be affected by various factors, such as non-radiative recombination, unbalanced carrier transport, and leakage current in the device structure. To make PeLEDs commercially viable, researchers have proposed several strategies, including defect state passivation, nanostructure optimization, and ligand engineering, to enhance the EL efficiency of the devices and mitigate the efficiency roll-off issue [12, 30, 34, 35]. Environmental concerns have become increasingly important in recent years. One of the main concerns is the use of lead, a toxic element in PeLEDs, which could limit their widespread use in everyday applications. To address this issue, researchers are exploring lead-free perovskite materials to meet the demands of society.

This chapter provides an overview of the development of green PeLEDs, including materials, synthesis, and device assembly. The main focus is on improving the EQE of green PeLEDs and the factors influencing it, as well as summarizing and generalizing strategies to enhance the EQE from the literature. Additionally, this chapter discusses the stability and lifetime of green PeLEDs and aims to provide insights into improving their performance.

5.2 Green Light Perovskite Materials

The chemical structure of a typical green halogen perovskite is represented by the general formula ABX_3. The A-site is commonly an organic amine cation or metal

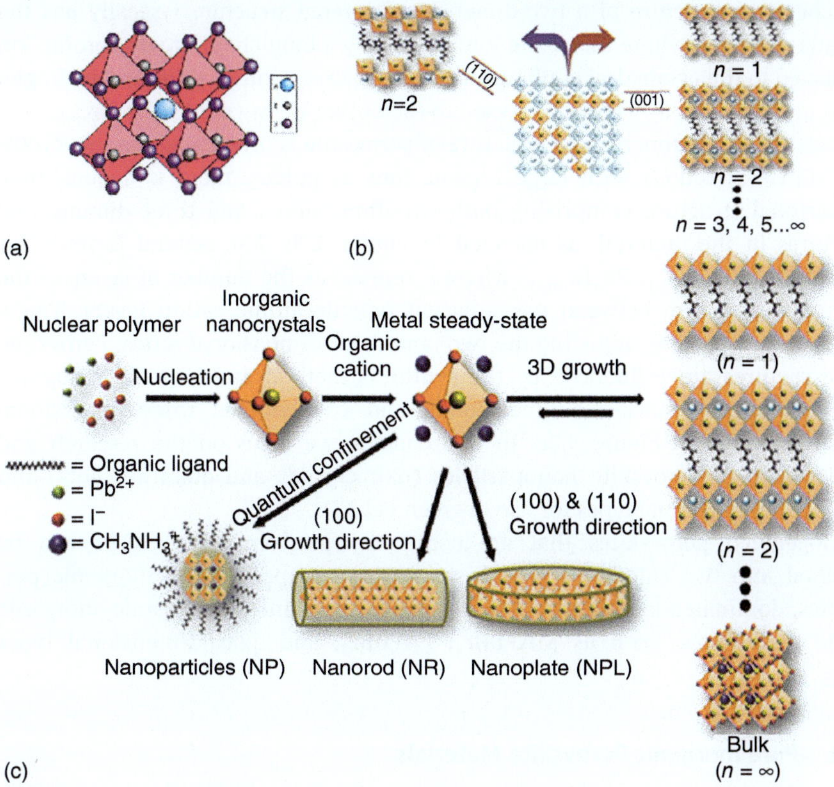

Figure 5.2 (a) Cubic ABX_3 perovskite crystal structure [36]; (b) quasi-dimensional perovskite structure; (c) schematic diagram of the relationship between the crystalline phase and morphology formation of heterogeneous perovskites [53].

cesium ion, while the B-site is typically the divalent transition metal Pb^{2+}, and X represents the halogen atom Br. Figure 5.2a shows that the B-site Pb^{2+} is typically surrounded by six halogen ions, Br^-, resulting in an octahedral configuration $[PbBr_6]^{4-}$ [36]. In three dimensions, the octahedra are linked by vertices to create a spatial network structure. These octahedral voids are filled by A-site cations. The perovskite structure is formed only when the A-site ions precisely occupy the octahedral gaps, which necessitates that the ionic radii of the A, B, and X site ions are satisfied (Eq. (5.1)).

$$t = \frac{R_A + R_B}{\sqrt{2}(R_A + R_B)} \tag{5.1}$$

where t denotes the tolerance factor, roughly ranging from 0.81 to 1.11 [37]. When the value of t falls outside this range, the mutual repulsion between ions causes distortion of the cubic crystal configuration of perovskite. If the radius of the A-site ion is too large to fit the gaps between the $[PbBr_6]^{4-}$ octahedra, the vertex connection between layers in the octahedra breaks, and the structure transforms from a conventional three-dimensional configuration to a layered two-dimensional structure.

The chemical structure of a two-dimensional layered structure typically has the formula A_2PbBr_4, where the A-site ion is typically a long-chain organic amine ion connected to the octahedral configuration via ion-to-ion interactions and hydrogen bonding, while the layers are connected to each other by van der Waals forces.

When the three-dimensional structure of perovskite is oriented along the <100> and <001> directions with large organic ions as guides, there is a quasi-two-dimensional structure comprising both two-dimensional and three-dimensional structures in the material, as depicted in Figure 5.2b. The general formula for this structure is $L_2A_{n-1}Pb_nBr_{3n+1}$, where n represents the number of layers of the octahedral structure between two adjacent organic amine cation layers. The n value can be varied by adjusting the two- and three-dimensional ratios. Perovskite nanostructures are influenced by the synthesis method and include nanoparticles, nanosheets, quasi-two-dimensional, and conventional three-dimensional types, as shown in Figure 5.2c. In this chapter, we focus on the research and development of perovskite nanoparticles (nanocrystals and quantum dots) and quasi-two-dimensional types for use in green PeLEDs.

Halogenated perovskites that are commonly used in green LEDs can be classified into two categories. The first category comprises pure inorganic perovskites, dominated by $CsPbBr_3$. The second category includes organic–inorganic hybrid perovskites, such as $MAPbBr_3$, $FAPbBr_3$, and quasi-dimensional types $L_2A_{n-1}Pb_nBr_{3n+1}$.

5.2.1 Pure Inorganic Perovskite Materials

Zeng Haibo's group reported the first study that employed $CsPbBr_3$ as a luminescent layer in 2015 [10]. Using the thermal injection method, they synthesized $CsPbBr_3$ quantum dots with a uniform size dispersion and an average diameter of 8 nm, as illustrated in Figure 5.3. The device structure is: ITO/PEDOT:PSS/Perovskite/QDs/TPBi/LiF/Al.

The device showed an EQE of 0.12% and a luminescence of 946 cd m^{-2}. In comparison to organic–inorganic hybrid perovskite materials, inorganic perovskite materials have several advantages, such as good stability and high photoluminescence

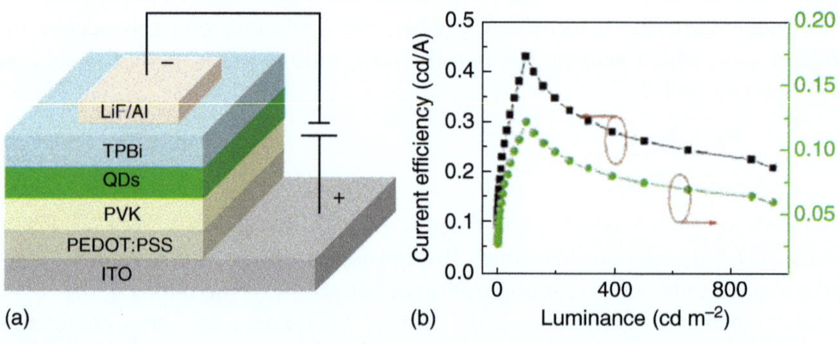

Figure 5.3 (a) Schematic diagram of the device structure. (b) Current efficiency–brightness–external quantum efficiency (CE–L–EQE) curve. Source: Song et al. [10]/John Wiley & Sons.

(PL) quantum efficiency (>90%). However, they also face some challenges, such as the low exciton binding energy (40 meV) of $CsPbBr_3$, which leads to the easy dissociation of excitons into free carriers and the formation of non-radiative composites. Furthermore, quantum dots have a large specific surface area, which makes it easy for them to aggregate and reduce the surface morphology and quality of the film, thus leading to a degradation in device performance. Researchers have proposed various strategies to address these issues, such as altering the ratio of synthetic raw materials and introducing long organic alkyl chains as ligands [22, 38]. As a result, many research studies have been conducted, and different strategies have been proposed to address these challenges. These strategies include altering the ratio of synthetic raw materials and introducing long organic alkyl chains as ligands.

Mathews et al. first introduced phenylethylamine bromide (PEABr) into $CsPbBr_3$ to transform the three-dimensional $CsPbBr_3$ perovskite structure into a quasi-two-dimensional lamellar structure [13]. By adjusting the molar ratio of PEABr and $CsPbBr_3$, the surface morphology of the film was obtained as shown in Figure 5.4. When the molar ratio of PEABr to $CsPbBr_3$ was 0.8:1, the film surface was dense with small grain sizes, small pores, and uniform coverage. Using this quasi-dimensional $CsPbBr_3$ perovskite material as the light-emitting layer, the EQE of the device is 1.97%, which is 50 times higher than that of a pure $CsPbBr_3$ PeLED (Figure 5.5).

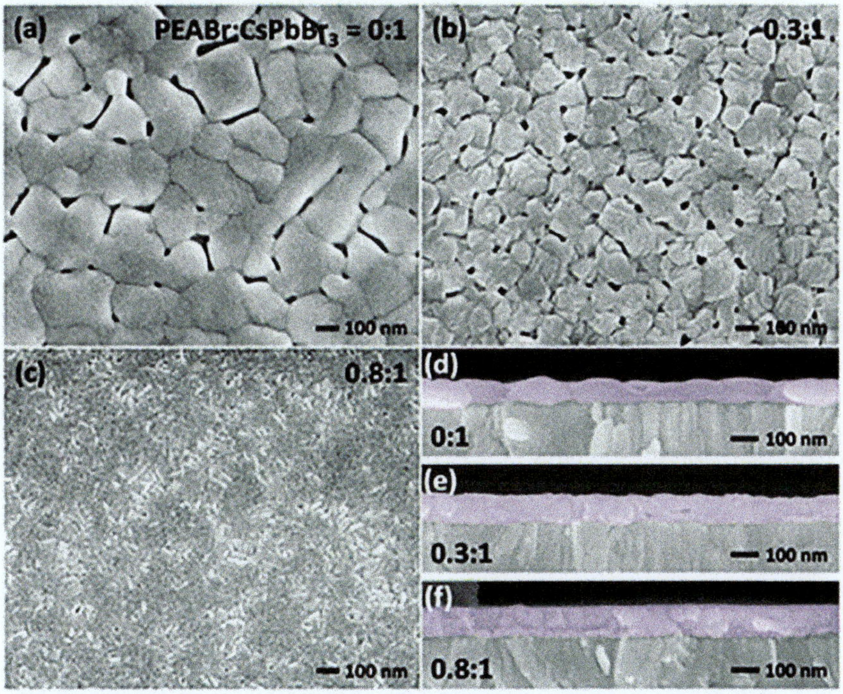

Figure 5.4 FESEM images of $CsPbBr_3$ perovskite films prepared with different PEABr:$CsPbBr_3$ molar ratios (a–c) film surface morphology (d–f) film cross-sectional morphology. Source: Ng et al. [13]/Reproduced from Royal Society of Chemistry.

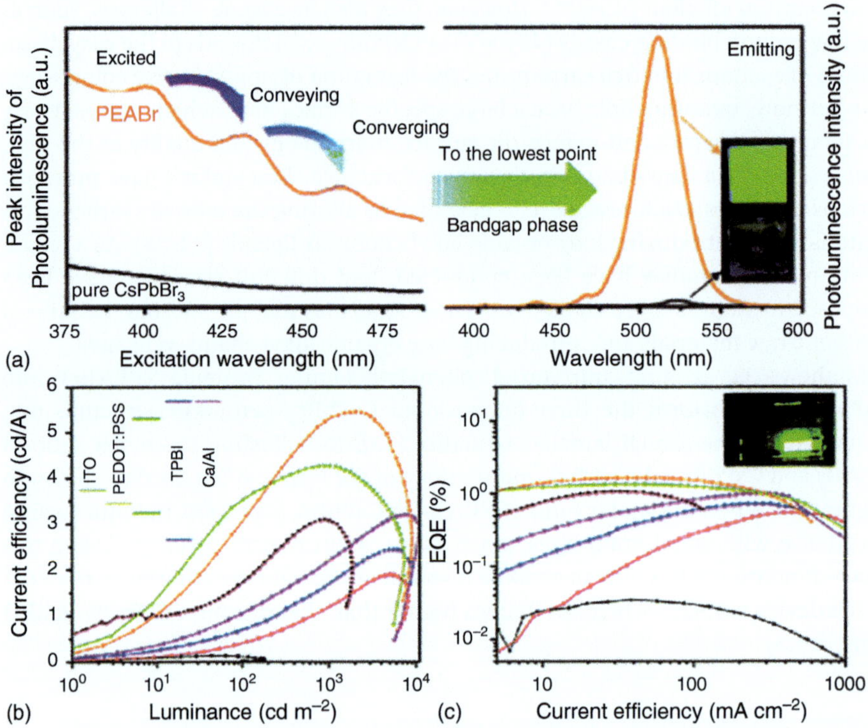

Figure 5.5 (a) Schematic diagram of the role of PEABr in devices assembled from $CsPbBr_3$ films prepared with different PEABr:$CsPbBr_3$ molar ratios (b) CE–L curves; (c) EQE–J curves. Source: Ng et al. [13]/Reproduced with permission from Royal Society of Chemistry.

5.2.2 Organic–Inorganic Hybrid Perovskite Materials

Organic–inorganic hybrid perovskite materials can be classified into two categories – those with a typical perovskite cubic configuration where cesium ions are replaced by methylamine and formamidine ions, and those with quasi-dimensional perovskites of structural formula $L_2A_{n-1}Pb_nBr_{3n+1}$. These materials exhibit excellent performance due to the presence of insulating organic alkyl chains. These chains enable the nanoparticles to be uniformly dispersed while passivating surface defects, resulting in thin films with uniform surface coverage.

$FAPbBr_3$ outperforms $MAPbBr_3$ due to the larger ion radius of FA^+ compared to MA^+, which leads to a smaller bandgap and faster carrier transport [39]. In 2014, the first PeLED with $MAPbBr_3$ as the luminescent layer was reported with an EQE of 0.1% [11]. In 2017, a team led by Tae-Woo Lee in Korea improved the EQE of PeLEDs to 8.79% by incorporating nanocrystals (A-NCP) to reduce the grain size of $MAPbBr_3$ and decrease the surface defect density of the film, thereby minimizing non-radiative recombination at the grain boundaries [40]. The main challenge for $MAPbBr_3$ is its tendency to decompose and produce methylamine gas, which hinders its development. On the other hand, $FAPbBr_3$ has shown better performance and stability. In 2016, $FAPbBr_3$ was used as a light-emitting layer in LED devices,

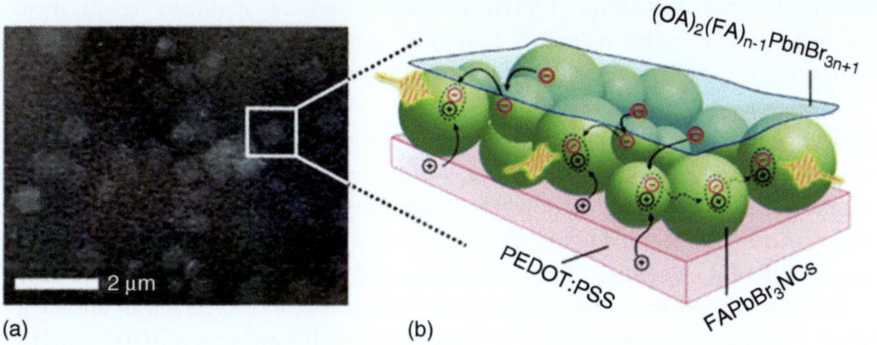

Figure 5.6 (a) Thin film FESEM diagram. (b) Schematic representation of the multilayered structure and energy transfer of $FAPbBr_3$ nanocrystals. Source: Chin et al. [18]/Reproduced from Royal Society of Chemistry.

resulting in pure green luminescence, 85% quantum efficiency, and a maximum PL emission peak at 530 nm [41]. In 2018, Mhaisalkar and colleagues constructed a multilayered structure by self-assembling $FAPbBr_3$ nanocrystals and 2D octylamine lead bromide microsheets to form energy-level strings, as shown in Figure 5.6. This approach improved the exciton complex efficiency and resulted in a PeLED with an EQE of 13.4%, a current efficiency of 57.6 cd A^{-1}, and a power efficiency exceeding 58 Im W^{-1} [18].

Although PeLEDs constructed with $MAPbBr_3$ and $FAPbBr_3$ as light-emitting layer materials have shown superior photovoltaic properties compared to conventional three-dimensional bulk material PeLEDs, they are still susceptible to environmental factors such as moisture. To address this issue, long-chain organic cations have been introduced into perovskite materials to form a quasi-two-dimensional perovskite material with the structure $L_2A_{n-1}Pb_nBr_{3n+1}$, as shown in Figure 5.2b. These materials are mostly in the Ruddlesden–Popper phase, with $PbBr_6$ octahedral layers surrounded by large organic cations at the top and bottom to form a "sandwich" structure. The long organic alkyl hydrocarbon chains are cut along the <100> crystal orientation of the 3D chalcocite, resulting in a more stable structure that is less sensitive to environmental factors [42]. Quasi-two-dimensional perovskite materials have unique properties that make them highly versatile. Compared to three-dimensional bulk perovskites, quasi-two-dimensional perovskites have larger exciton binding energies, fewer surface defects, higher photoluminescence quantum yield (PLQY), and higher stability. They also exhibit distinct absorption and PL spectra with significant Stoke shifts [43].

5.2.3 Synthesis of Perovskite Green Light-Emitting Materials

Various methods have been reported in the literature for synthesizing perovskite materials with different nanostructures. For example, ligand-assisted redeposition (LARP) and thermal injection methods are commonly used for synthesizing perovskite nanoparticles, while spin coating and solution vacuum vapor deposition (chemical vapor deposition [CVD]) are often used for synthesizing quasi-two-dimensional perovskite materials [44]. The synthesis of quasi-two-dimensional perovskites is mostly done using spin coating and solution vacuum vapor deposition (CVD) methods. Ligand-assisted redeposition (LARP) is a method for synthesizing perovskite nanoparticles, as shown in Figure 5.8a, by injecting precursor solutions dissolved in good solvents such as *N*,*N*-dimethylformamide (DMF) or dimethyl sulfoxide (DMSO) into poor solvents containing ligands (e.g. toluene) at room temperature. This method takes advantage of the difference in solubility of precursors in different solvents to form supersaturated solutions. Han et al. used LARP to synthesize $FAPbBr_3$ nanocrystals with a particle size of 5–20 nm by dissolving 3,3-bromodiphenylpropylamine (3,3-bromodiphenylpropylamine [DDPA-Br]) in toluene as an antisolvent. The resulting thin film had a quantum efficiency of 78%, as shown in Figure 5.7b [46]. Another method used for the synthesis of quasi-two-dimensional perovskites is spin-coating and solution vacuum vapor deposition (CVD). Ligand-assisted redeposition (LARP) is a method used for synthesizing perovskite nanoparticles. As shown in Figure 5.7a, precursor solutions dissolved in good solvents such as DMF and DMSO are injected into poor solvents containing ligands (e.g. toluene) at room temperature to form supersaturated solutions. Han et al. synthesized $FAPbBr_3$ nanocrystals with a particle size of 5–20 nm using LARP and 3,3-bromodiphenylpropylamine (DDPA-Br) as an antisolvent, achieving a thin film quantum efficiency of 78% (as shown in Figure 5.7b). They also optimized the film quality by controlling the titration time of the antisolvent and the amount of ligand added, distinguishing between the nanocrystal peg formation process and the in situ LARP process. Thermal injection is another method proposed by Loredana for the synthesis of homogeneous $FAPbBr_3$ colloidal nanocrystals [41]. There are various methods for synthesizing perovskite materials that form different nanostructures, as reported in the literature. For instance, spin-coating and solution vacuum vapor deposition (CVD) methods are commonly used for the synthesis of quasi-two-dimensional perovskites. Ligand-assisted redeposition (LARP) and thermal injection methods, on the other hand, are frequently used for the synthesis of perovskite nanoparticles.

LARP involves injecting precursor solutions, dissolved in good solvents such as DMF and DMSO, into poor solvents containing ligands (e.g. toluene) at room temperature, taking advantage of the difference in solubility of precursors in different solvents to form supersaturated solutions. $FAPbBr_3$ nanocrystals with a particle size of 5–20 nm were synthesized using LARP by dissolving 3,3-bromodiphenylpropylamine (DDPA-Br) in toluene as an antisolvent, resulting in a thin film quantum efficiency of 78%. The nanocrystal peg formation process and the in situ LARP process were distinguished by controlling the titration time of the antisolvent and the amount of ligand added to optimize the film quality.

Thermal injection is another method proposed for the synthesis of homogeneous $FAPbBr_3$ colloidal nanocrystals. In this method, the oleylamine (OAm) solution is injected into the FA-Pb precursor solution under heated conditions and then rapidly cooled down to room temperature, where the crystals form and grow. The nanocrystals synthesized by this method have a crystal size of 50 nm, a thin film

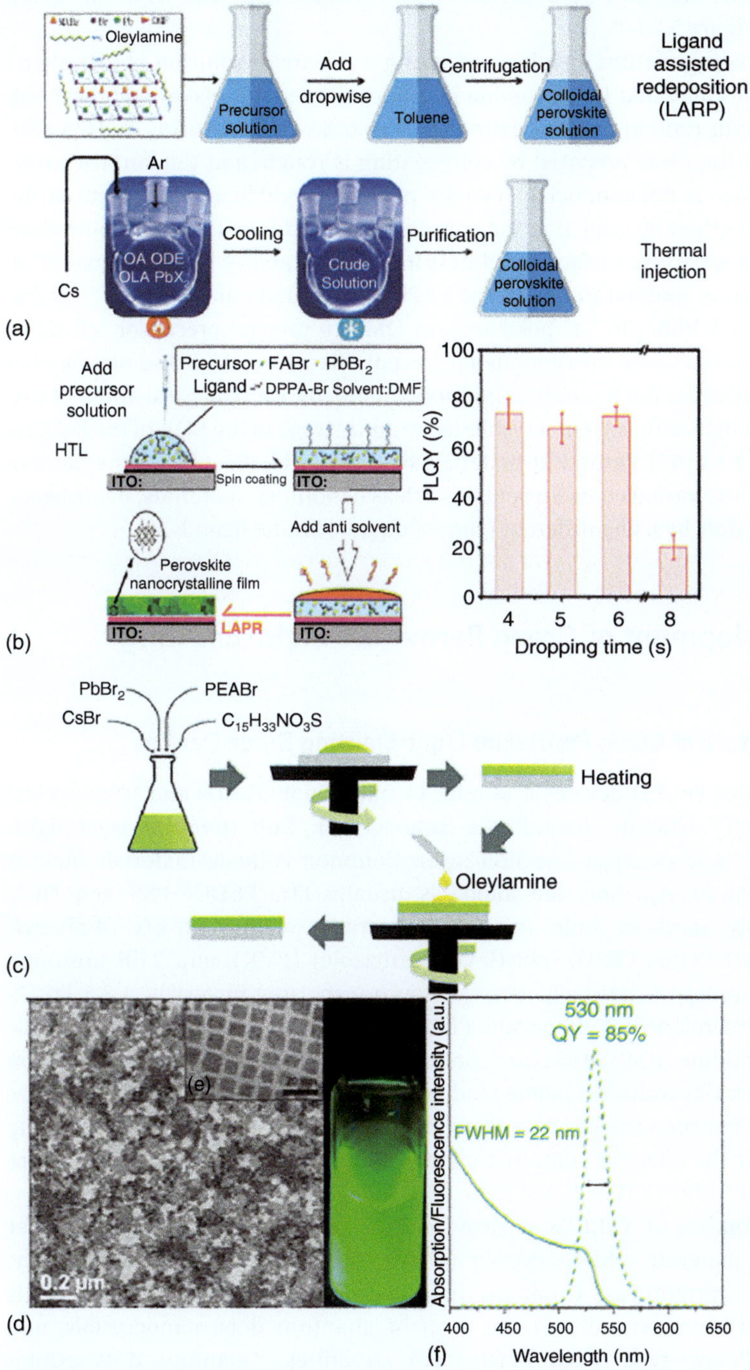

Figure 5.7 (a) Synthesis flow diagram of the ligand-assisted redeposition (LARP) and thermal injection methods [12]; (b) flow diagram of the LARP synthesis of $FAPbBr_3$ and PLQYs of $FAPbBr_3$ films obtained by adjusting the anti-solvent drop addition time. (c) schematic diagram of the spin-coating and solution methods; (d,e) TEM images of $CsPbBr_3$ 12 nm nanocrystals prepared by the thermal injection method; (f) $CsPbBr_3$ absorption and PL spectra of nanocrystals [55].

PLQY of 85%, an emission wavelength of 530 nm, and a half-peak width of 22 nm, as shown in Figure 5.7d–f.

The spin-coating method involves preparing a precursor solution by dissolving ammonium bromide and lead bromide in a nonvolatile, non-plasmonic polar solvent in a certain ratio and then coating the substrate and annealing it. However, the surface of the films prepared by spin coating is rough, and the surface coverage is low, which is not conducive to device performance. Shen et al. improved the spin coating method by adjusting the concentration and type of ligand to produce $CsPbBr_3$ films with low surface roughness and good stability [44]. Perumal et al. utilized a solution method to synthesize $FAPbBr_3$ colloidal nanocrystals. They dissolved FABr and $PbBr_2$ in the polar solvent DMF to prepare precursor solutions, which were then dropped into a nonpolar solvent mixture of OAm and oleic acid in toluene to synthesize the nanocrystals at room temperature. The passivation of surface defects in the resulting nanocrystals led to an increase in the EQE of the PeLEDs from 7.0% to 11.1% [47]. Perovskite materials synthesized by the solution method can be structured into various nanostructures such as nanotubes, nanorods, nanosheets, and quantum dots by using different types of organic amine ligands.

5.3 Development of Green Perovskite Light-Emitting Diodes

5.3.1 Structure of Green Perovskite Light-Emitting Diode Devices

The structure of PeLED devices is similar to traditional OLEDs and is composed of a "sandwich" structure, including a cathode, HIL, hole transport layer, light-emitting layer, and electron injection layer. Common cathode materials include metals like Al or Ag, and the anode is usually ITO/PEDOT:PSS and NiO_x are commonly used as hole injection materials, while poly-bis (4-phenyl) (4-butylphenyl) amine (TPD), poly(9-vinylcarbazole) (PVK), and TFB are often used as hole transport materials. For electron-transporting materials, 1,3,5-Tris(1-phenyl-1H-benzimidazol-2-yl)benzene (TPBI) and 4,6-Bis(3,5-di-3-pyridylphenyl)-2-methylpyrimidine (B3PYMPM) are commonly used. To facilitate carrier injection and balance carrier transport, some studies have added a thin layer of polymer as a buffer layer between the electron injection layer (or HIL) and the light-emitting layer, enabling the effective combination of excitons. The first green PeLEDs were structured as ITO/PEDOT:PSS/$MAPbBr_3$/F8/Ca/Al [32].

The performance of PeLEDs is heavily influenced by the luminescent layer of perovskite material, which affects properties such as EQE, current efficiency, and luminous brightness. Common light-emitting layer perovskite materials for green LEDs are typically in the form of quantum dots, nanocrystals, and Ruddlesden–Popper-type quasi-dimensional structures. Quantum dots exhibit a quantum-limited effect, while nanocrystals and quasi-dimensional structures have a quantum well structure that facilitates effective radiative recombination of excitons [48].

5.3.2 Quantum Dot Green Perovskite Light-Emitting Diodes

In comparison to OLEDs, perovskite quantum dot diodes offer advantages such as higher brightness, shorter response time, and narrower emission spectra [10, 15]. In 2015, Haibo Zeng's team published the first $CsPbBr_3$ quantum dot LED device, which was synthesized using the thermal injection method. The LED devices showed an EQE of 0.12% and a luminescence of 946 cd m^{-2}. In 2017, Kido's group investigated the effect of solvents such as butanol (BuOH), ethyl acetate (AcOEt), and butyl acetate (AcOBu) on the performance of $CsPbBr_3$ quantum dots after washing away excess ligands. Figure 5.8g shows the specific posttreatment flow [45]. The effect of various solvents such as BuOH, AcOEt, and AcOBu on the

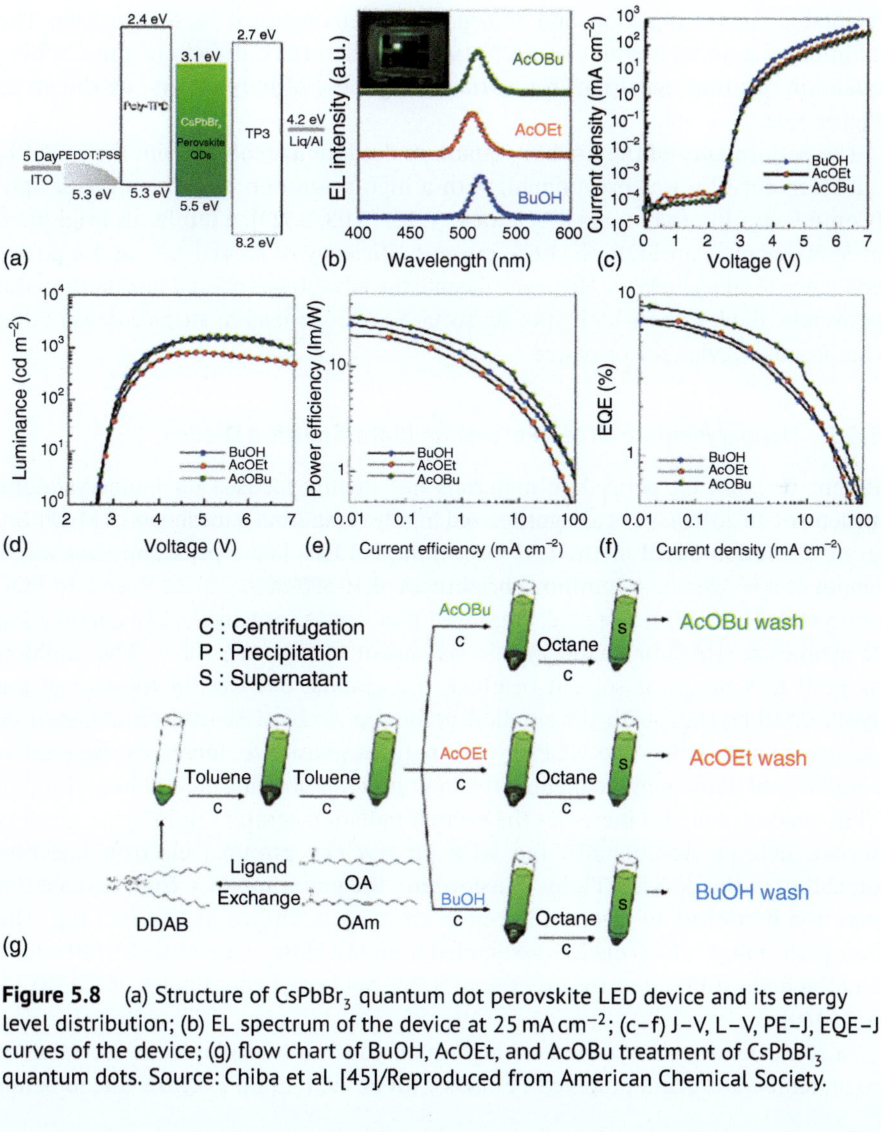

Figure 5.8 (a) Structure of $CsPbBr_3$ quantum dot perovskite LED device and its energy level distribution; (b) EL spectrum of the device at 25 mA cm^{-2}; (c–f) J–V, L–V, PE–J, EQE–J curves of the device; (g) flow chart of BuOH, AcOEt, and AcOBu treatment of $CsPbBr_3$ quantum dots. Source: Chiba et al. [45]/Reproduced from American Chemical Society.

performance of $CsPbBr_3$ quantum dot LED devices was investigated by Kido's group in 2017. The $CsPbBr_3$ quantum dots were treated with each solvent to wash away excess ligands, and the AcOBu-treated quantum dots demonstrated better performance. Figure 5.8b–f shows that the devices exhibited a lower driving voltage ($V_{turn\text{-}on}$ = 2.9 eV), a narrow luminescence spectrum (FWHM = 17 nm), an improved EQE of 8.73%, and a power efficiency of 31.7 Im W^{-1}.

In 2018, Haibo Zeng's group utilized three short-chain ligands, namely tetraoctylammonium bromide (TOAB), didodecyldimethylammonium bromide (DDAB), and octanoic acid (OTAc), along with a small amount of FA^+ to synthesize $CsPbBr_3$ quantum dots, as depicted in Figure 5.9a [49]. TOAB was used to optimize the synthesis process for colloidal stabilization, DDAB helped prevent agglomeration of quantum dots, thereby improving the PL properties of $CsPbBr_3$, and OTAc facilitated carrier injection and transport, as demonstrated in Figure 5.9b. The addition of a small amount of FA^+ also reduced surface defects in the $CsPbBr_3$ quantum dot film, resulting in a surface roughness of only 2.3 nm, as shown in Figure 5.9c.

The performance of the $CsPbBr_3$ quantum dot film assembled from the $CsPbBr_3$ quantum dot film was exceptional, with a high saturation of emitted green light [Commission Internationale d'Eclairage (CIE) (0.05, 0.71)], a luminous brightness of 55 800 cd m^{-2}, an EQE of 11.6%, current efficiency of 45.4 cd A^{-1}, and a power efficiency of 44.65 lm W^{-1}. However, despite the advantages offered by quantum dot perovskite diodes, their high specific surface area can lead to surface defects that compromise device performance.

5.3.3 Nanocrystalline Green Perovskite Light-Emitting Diodes

Recent research on perovskite materials has mainly focused on nanocrystalline structures. In 2016, Gao et al. synthesized highly crystalline nanosheets of $MAPbBr_3$ using a solution-based method. The resulting PeLEDs had a peak emission wavelength of λ = 529 nm, a luminous brightness of 10 590 cd m^{-2} at 12 V, and an EQE of 0.54% [47]. In 2017, Lee et al. employed nanocrystal pegging (NCP) engineering to synthesize $MAPbBr_3$ nanocrystals, as shown in Figure 5.10b–f. The addition of TPBI to a nonpolar solvent resulted in a gradual decrease in the size of the synthesized crystals, with the smallest grain size range of 50–200 nm obtained by adding 0.1 wt% TPBI. The NCP process reduces grain size, increases the exciton confinement domain by weakening the intergranular bonding of $MAPbBr_3$ through TPBI coating, and also increases the exciton radiation complex ratio by passivating surface defects. Additionally, the NCP process can promote electron injection capability in the device ETL by transforming the grain interface from an electron injection barrier to an electron injection channel, as shown in Figure 5.10g. The best-performing PeLED devices assembled from $MAPbBr_3$ with 0.1 wt% TPBI added had the highest PL intensity, an EQE of 8.79%, and color coordinates of CIE (0.27, 0.71) [40].

In 2018, Demir et al. synthesized high-quality MAPbBr3 colloidal nanocrystals at room temperature and prepared PeLEDs with an EQE of up to 12.9% due to better

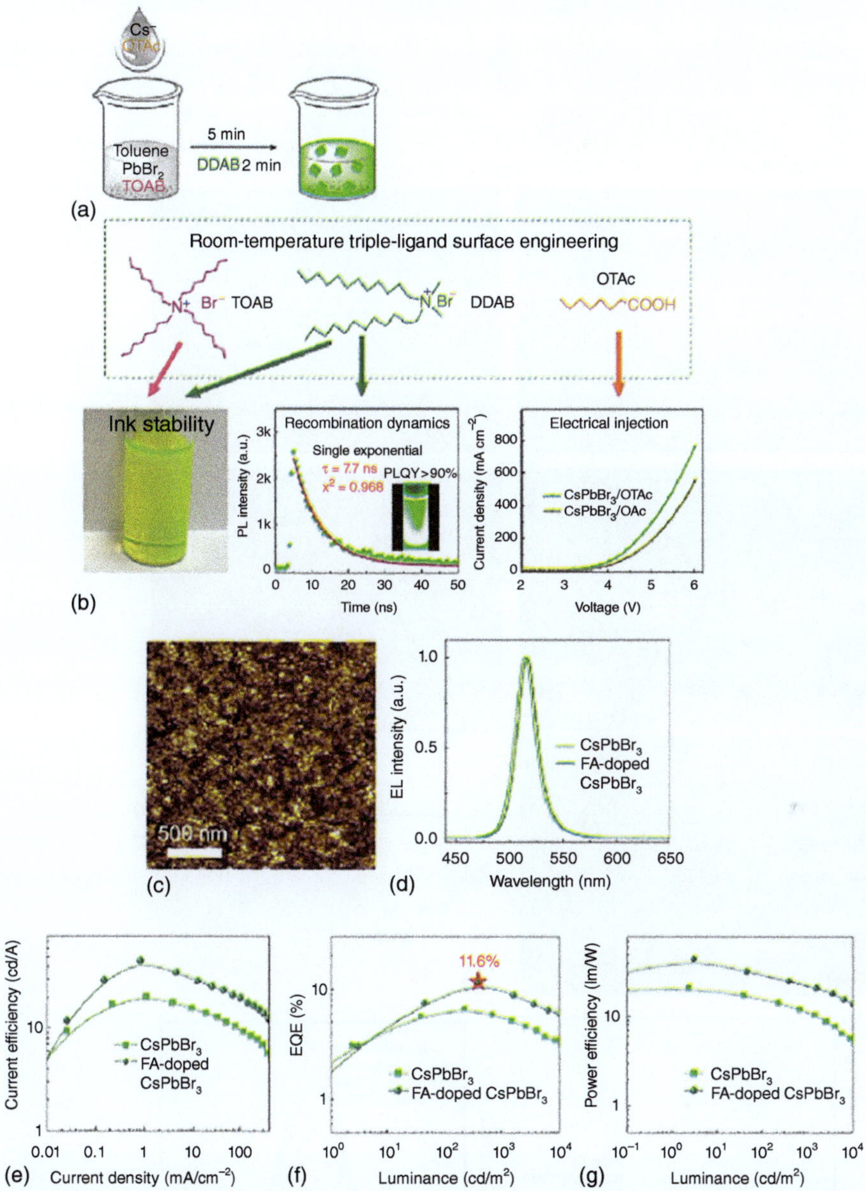

Figure 5.9 (a) Schematic diagram of the synthesis of $CsPbBr_3$ quantum dots; (b) structural formulae of TOAB, DDAB, and OTAc and their roles in FA-doped $CsPbBr_3$ quantum dots; (c) AFM plots of FA-doped $CsPbBr_3$ quantum dot films; (d) normalized EL spectra of the devices at 3 V; (e–g) EQE–L, CE–J, PE–J plots. Source: Song et al. [49]/Reproduced from John Wiley & Sons, Inc.

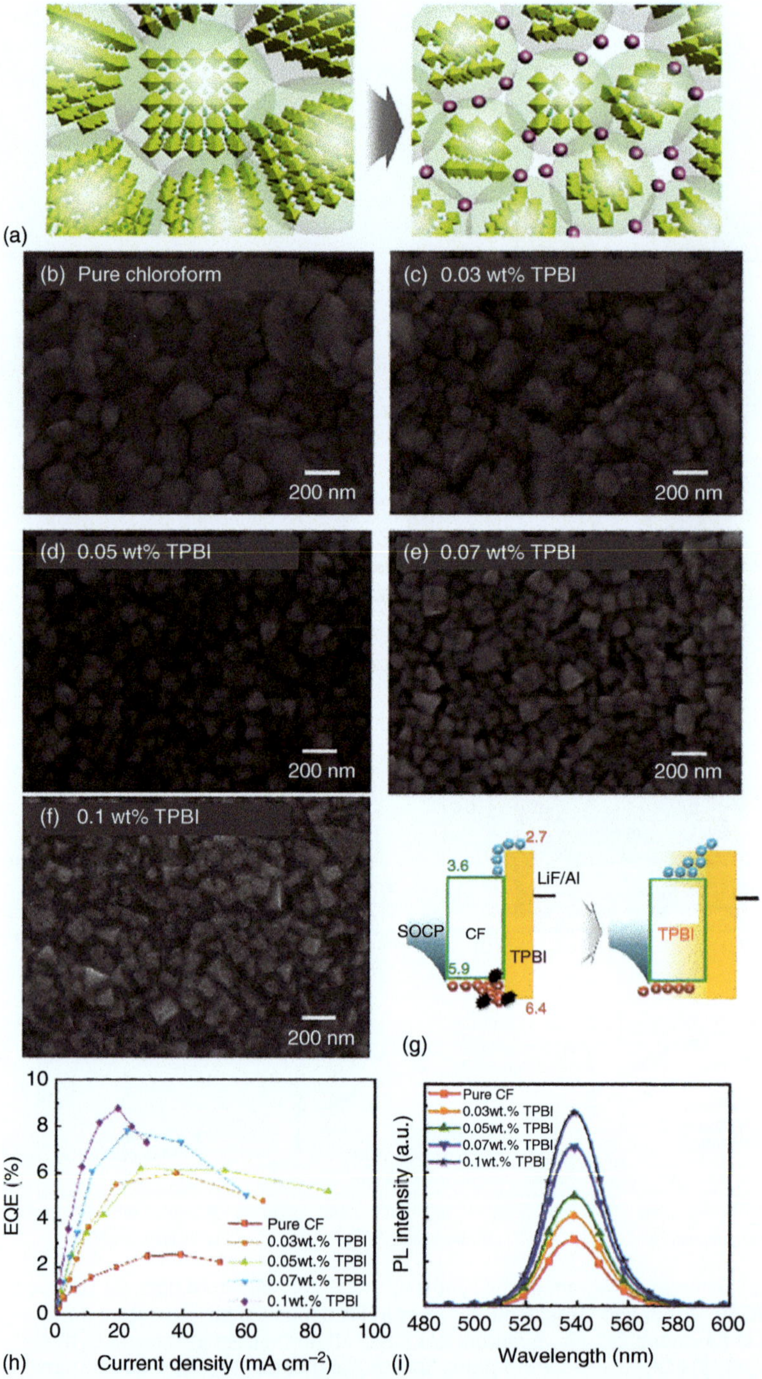

Figure 5.10 (a) Schematic diagram of intergranular repulsion induced by the A-NCP process; (b–f) SEM images of $MAPbBr_3$ films with different amounts of TPBI added; (g) schematic diagram of electron and hole transport equilibrium in a perovskite light-emitting diode; (h) steady-state PL spectra of $MAPbBr_3$ with 0–0.1 wt% TPBI added; (i) device EQE–V curve. Source: Park et al. [40]/Reproduced with permission from Elsevier.

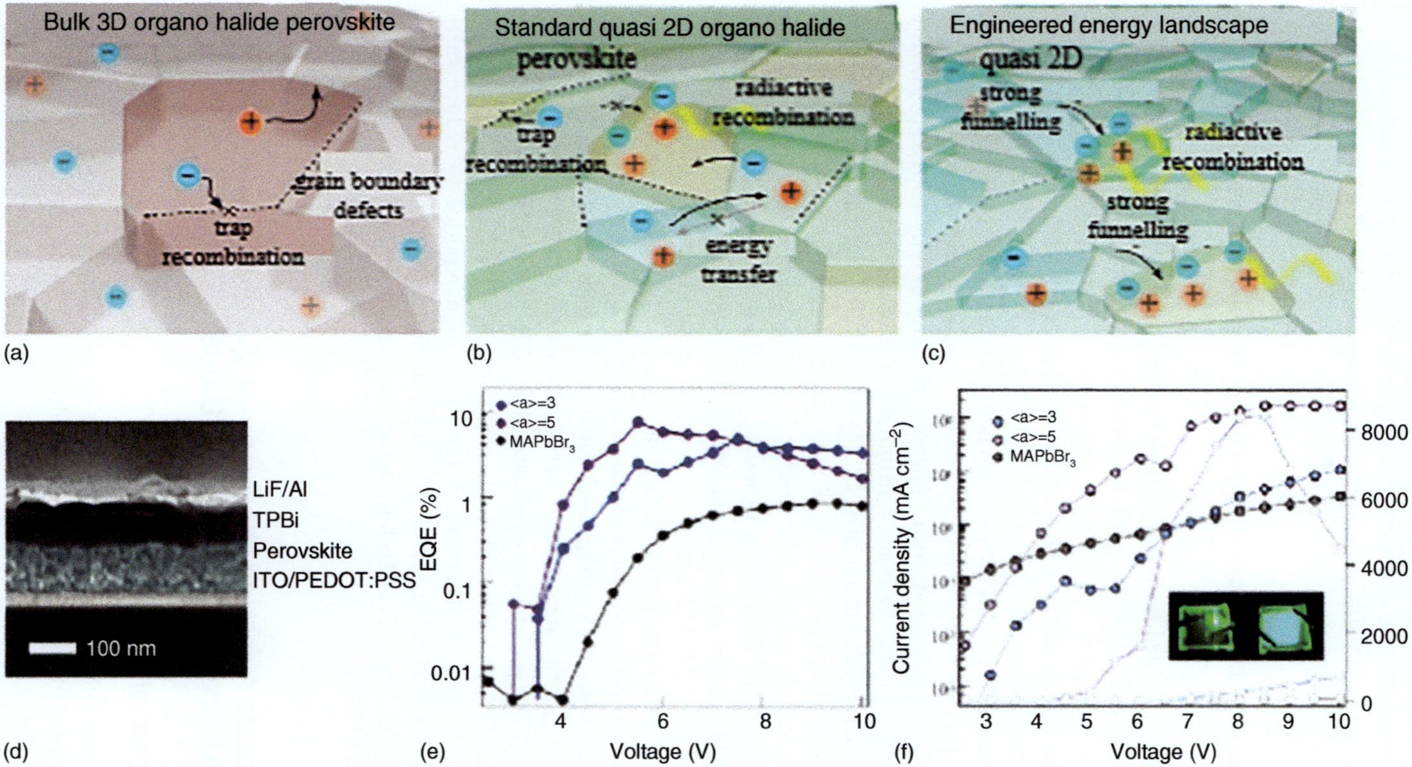

Figure 5.11 (a) Schematic diagram of a conventional 3D perovskite structure with the presence of carriers with high diffusion coefficients that are not conducive to radiative combination; (b) schematic diagram of a common quasi-2D perovskite structure with the coexistence of free carriers and excitons; (c) schematic diagram of a quasi-2D perovskite structure with faster energy transfer than non-radiative combination with energy level engineering optimization; (d) SEM image of the device structure cross-section; *n* values (e) EQE–V diagram; (f) J–V–L diagram for different material assemblies of perovskite light-emitting diodes. Source: Quan et al. [14]/Reproduced with permission from American Chemical Society.

carrier transport due to the small size of the grains [50]. They also observed that the Auger process, which typically causes rapid luminescence decay, can be effectively suppressed at low current densities. There have been several other studies exploring similar strategies with similarly promising results, which will not be discussed in detail here.

5.3.4 Quasi-2D Ruddlesden–Popper Green Perovskite Light-Emitting Diodes

LED devices assembled from conventional polycrystalline perovskite films tend to have low exciton binding energies, leading to the decomposition of excitons into free electron-hole pairs. Additionally, the surface defect states cause poor surface quality of the film, leading to the degradation of device performance. In 2016, Byun et al. synthesized quasi-2D structured $PEA_2(MA)_{n-1}Pb_nBr_{3n+1}$ perovskite materials by introducing PEA to $MAPbBr_3$ and adjusting the ratio of MA to PEA, as shown in Figure 5.11 [11]. However, the PeLED devices prepared from this material did not perform as well as expected, with a current efficiency of only 4.90 cd A^{-1}. In 2017, Sargent et al. adjusted the main distribution of n values in quasi-2D to enhance energy transfer in the material [14]. Figure 5.11a–c illustrates the differences between conventional 3D perovskite structures, general quasi-2D structures, and quasi-2D structures treated with energy level regulation engineering. Quasi-2D perovskite materials were first synthesized in 2016 by Byun et al. by introducing PEA to $MAPbBr_3$ and adjusting the MA to PEA ratio. Figure 5.11d shows that quasi-2D assembled PeLEDs with $n = 5$ exhibit good PL performance, including a luminance of 8400 cd m^{-2} and an EQE of 7.4%. The faster energy transfer rate in this structure outpaces the non-radiative composite rate.

In 2018, Zhanao Tan's group used BA instead of PEA to synthesize quasi-2D $BA_2(CsPbBr)_{3n-1}PbBr_4$ perovskite materials. They coated the surface with polyethylene oxide (PEO) polymer to increase the EQE of PeLEDs to 8.42%. These devices exhibited an extremely high brightness of 33 533 cd m^{-2} [4]. Quasi-2D $CsPbBr_3$/MABr core/shell structures have also been reported to reduce carrier injection and transport barriers and improve device performance, as discussed in previous studies. These PeLEDs have achieved a current efficiency of 78 cd A^{-1} and an EQE of 20.3% [3].

5.4 Factors Affecting the External Quantum Efficiency of Perovskite Green Light-Emitting Diodes

5.4.1 Aspects of Materials

One of the main challenges with PeLEDs is efficiency roll-off at high current densities, which is primarily caused by non-radiative recombination processes. These processes include defect-assisted non-radiative recombination and intermittent non-radiative recombination, with the latter being more likely to occur at high current densities [51]. The occurrence of non-radiative combinations in perovskite materials is one of the main reasons for the efficiency roll-off in PeLEDs, particularly at high current densities. Surface defects, which are inherent in perovskite films and can be generated during synthesis and post-processing, contribute to non-radiative recombination. These defects can arise from the decomposition of unstable ligands, large crystal sizes, and grain interfaces resulting from grain size reduction processes. When there are too many defect states on the surface of the film, carriers will flow along the defects without forming excitons, leading to non-radiative recombination. The Auger process, which can affect the open-circuit and driving voltages of PeLEDs, is related to the small number of carriers at the surface of the film. Liao's group has provided a detailed mechanism for the Auger process [52]. In addition to the aforementioned factors, Mhaisalkar et al. have also suggested that the loss of current at the interface can have a detrimental effect on device efficiency [53].

5.4.2 Aspects of the Device Structure

Another major reason for the efficiency roll-off in PeLEDs is the energy level mismatch between the functional layer materials, which results in carrier transport imbalance and the occurrence of leakage current. Due to the much smaller hole mobility compared to electron mobility, the exciton complex region may not be exactly in the middle of the light-emitting layer, causing some carriers to be trapped at the interface and leading to low external coupling efficiency [34]. Hence, to overcome the challenges related to carrier transport imbalance and leakage current, it is crucial to regulate the energy levels of the functional layer material or include a carrier blocking or buffer layer, which can facilitate and balance carrier injection and transport.

5.5 Strategies for Improving the External Quantum Efficiency of Green Perovskite Light-Emitting Diodes

5.5.1 Ligand Engineering

The surface morphology of perovskite materials has a significant impact on device performance, and the addition of organic ligands can improve film surface quality. However, the addition of ligands has both advantages and disadvantages for the device. While a certain amount of ligands on the surface of perovskite material can passivate surface defects and improve fluorescence quantum efficiency and stability, excessive addition can form an insulating layer and hinder carrier injection, damaging the optoelectronic performance of the LED device. Commonly used organic ligands are long-chain alkyl amines, such as OAm and BA, and aromatic amines, such as PEA and 3,3-diphenylpropylamine (DDPA). The alkyl chain length and ligand concentration have a significant impact on device properties such as brightness and EQE. Moreover, the molecular and dimensional dimensions of perovskite materials can also be affected by the addition of organic ligands. Some studies have used posttreatment methods, such as solvent washing, to remove excess ligands and weaken their negative effects on perovskite materials.

Li et al. utilized a cyclic solvent treatment process with a hexane/ethyl acetate solvent mixture to regulate the surface ligand concentration of synthesized $CsPbBr_3$ QDs. This improved the EQE of the PeLEDs to 6.27%, which was a 50-fold increase compared to the EQE of $CsPbBr_3$ QDs PeLEDs post-treated with acetone [15]. In 2016, Bakr et al. synthesized $CsPbBr_3$ QDs using short-chain organic amines

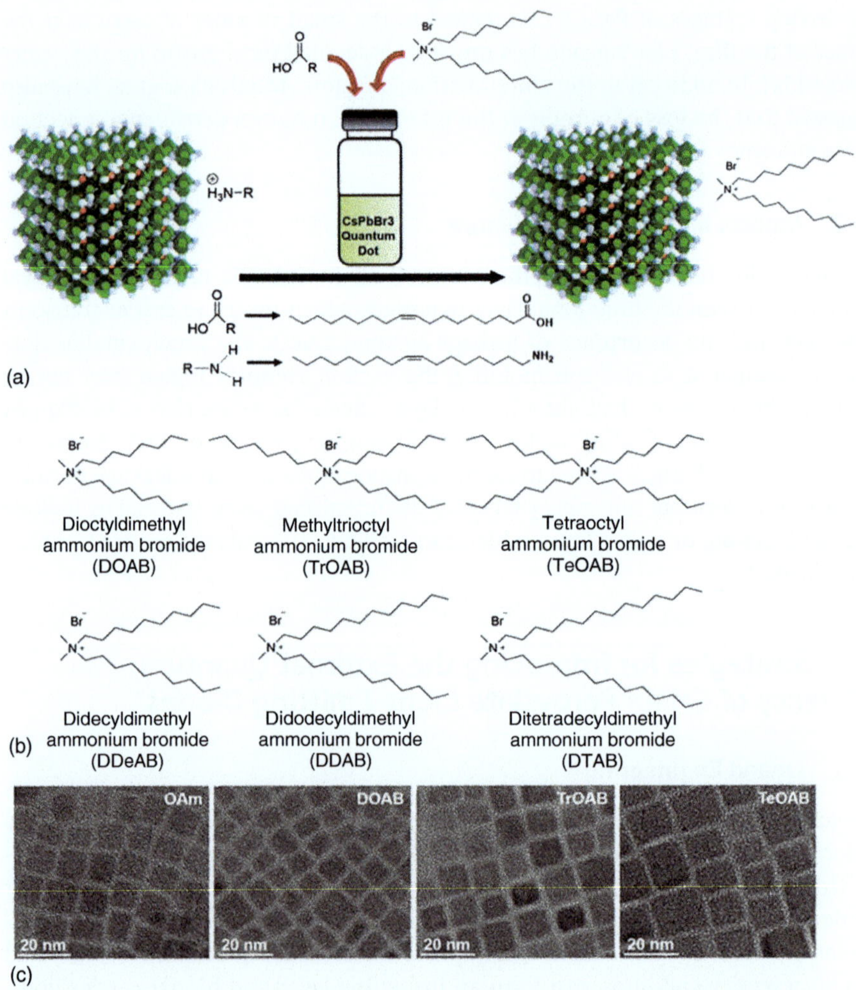

Figure 5.12 (a) Schematic diagram of the synthesis process of $CsPbBr_3$ nanocrystals; (b) chemical structural formulae of brominated quaternary ammonium salt ligands with different sizes and alkyl chain lengths; (c) TEM images of $CsPbBr_3$ nanocrystals with different sizes of ligands; (d,e) PLQY and PL spectra of $CsPbBr_3$ nanocrystals under normal conditions; (f) $CsPbBr_3$ nanocrystalline perovskite light-emitting diode device structure; (g–j) J-V, CE-J, L-V, EQE-J plots of the devices. Source: Park et al. [56]/American Chemical Society.

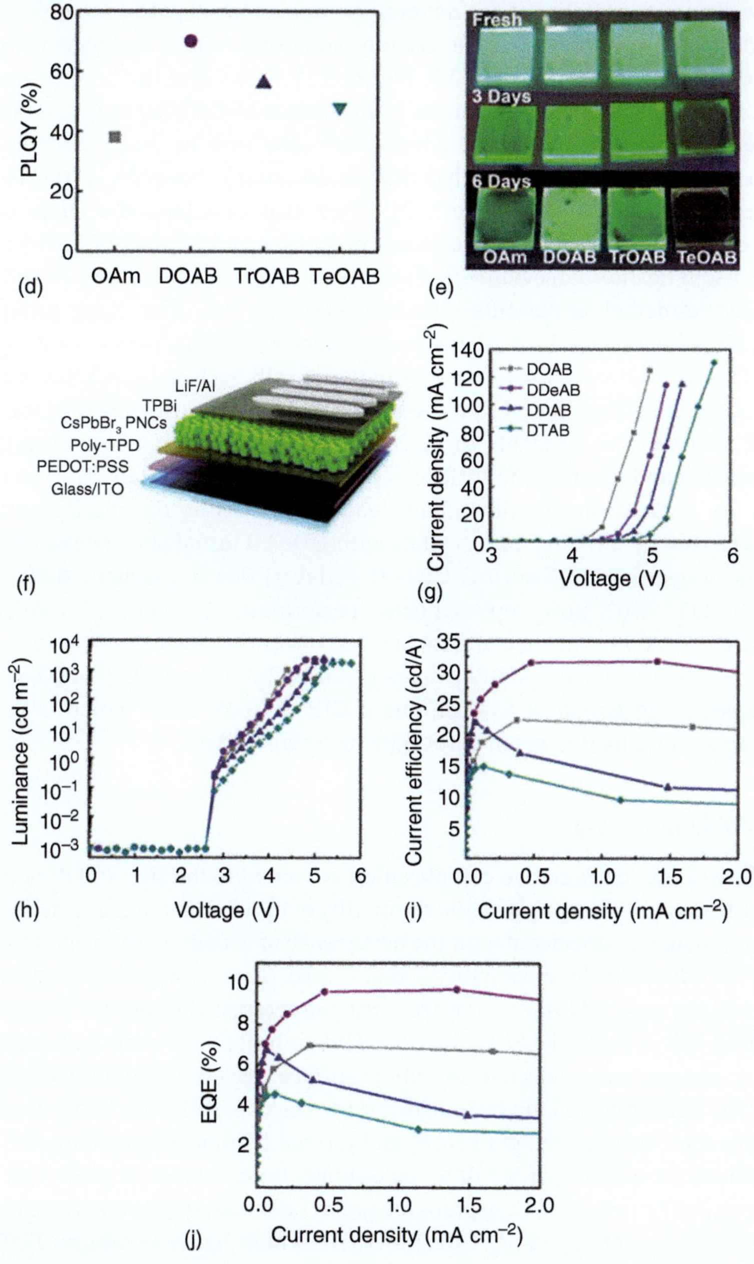

Figure 5.12 (*Continued*)

(DDAB) instead of long-chain ligands (OAm and oleic acid), which facilitated carrier injection and reduced film surface roughness. The resulting PeLED exhibited an EQE of 3.0% and a narrow luminescence peak (FWHM = 19 nm) [54]. In 2019, Liao's group investigated the effects of both the type and concentration of ligand amines on $CsPbBr_3$ and device performance [55]. The addition of ligands improves the structural stability of the material by weakening negative ion defects and grain interfaces, thus enhancing the exciton radiation complex and increasing the EQE of the device from 7.0% to 11.1%. Figure 5.12 shows that the Song group investigated the effect of ligand size on the performance of $CsPbBr_3$ nanocrystals using different brominated quaternary ammonium salts such as dioctyldimethylammonium bromide (DOAB), methyltrioctylammonium bromide (TrOAB), and tetraoctylammonium bromide (TeOAB). They also examined the effect of ligand alkyl chain length on the optoelectronic properties and stability of $CsPbBr_3$ nanocrystals using ligands didecyldimethyl ammonium bromide (DDeAB), DDAB, and ditetradecyldimethyl ammonium bromide (DTAB) [56]. The Song group investigated the effect of ligand size on the performance of $CsPbBr_3$ nanocrystals by selecting different brominated quaternary ammonium salts such as DOAB, TrOAB, and TeOAB as ligands. They found that $CsPbBr_3$ nanocrystals with DOAB ligands had high PL and were more stable to moisture, mainly due to the appropriate concentration of ligands coating the surface of the $CsPbBr_3$ nanocrystals. They also investigated the effect of ligand alkyl chain length by selecting ligands DDeAB, DDAB, and DTAB. The $CsPbBr_3$ nanocrystals with DDeAB ligand showed the best optoelectronic properties (EQE = 9.7%, CE = 31.7 cd A^{-1}) due to excellent surface passivation. In 2020, Wu's group reported better performance of colloidal $FAPbBr_3$ using betaine (SBE-18) as a ligand compared to the conventional ligand OAm /oleic acid [57]. The SBE-18-coated $FAPbBr_3$ nanocrystals exhibited a PLQY of 90.6%, an emission peak wavelength of 534 nm, and a CIE of (0.25, 0.70), which is the "greenest" background light of green perovskite materials to date.

5.5.2 Crystal Engineering

The probability of exciton radiative complexation is related to the size and dimensionality of perovskite nanocrystals, while the quality of film formation is affected by their grain size, which is associated with the occurrence of leakage currents and the loss of EQE. To reduce the dimensionality and grain size of perovskite nanocrystals, two approaches are used: additive and restructuring processes. The additive process involves adding halogen additives (e.g. MABr, CsBr) or functional small molecules (PIP, TPBI) to the precursor solution to reduce surface defects and grain size. In Figure 5.13, You's group coated the ZnO surface with a layer of PVP, which increased the nucleation sites, reduced the grain size, and formed a homogeneous film [17]. They also utilized the addition of MABr to the $CsPbBr_3$ lattice to reduce grain interfaces in the perovskite material. They prepared green LEDs using a device structure of ITO/ZnO/PVP/$Cs_{0.87}MA_{0.13}PbBr_3$/CBP/MoO_3/Al, which had a maximum EQE of 10.4% and a current efficiency of 33.9 cd A^{-1}. In 2015, Greenham's team used the polar functional group and hydrophilicity of PIP to prevent crystal growth and block carrier outflow [58].

The restructuring process involves methods such as NCP and the introduction of quasi-two-dimensional nanostructures to reduce the dimensionality and grain size of the perovskite structure. These techniques prevent rapid crystal growth by removing excess solvent during the spin-coating of the precursor solution,

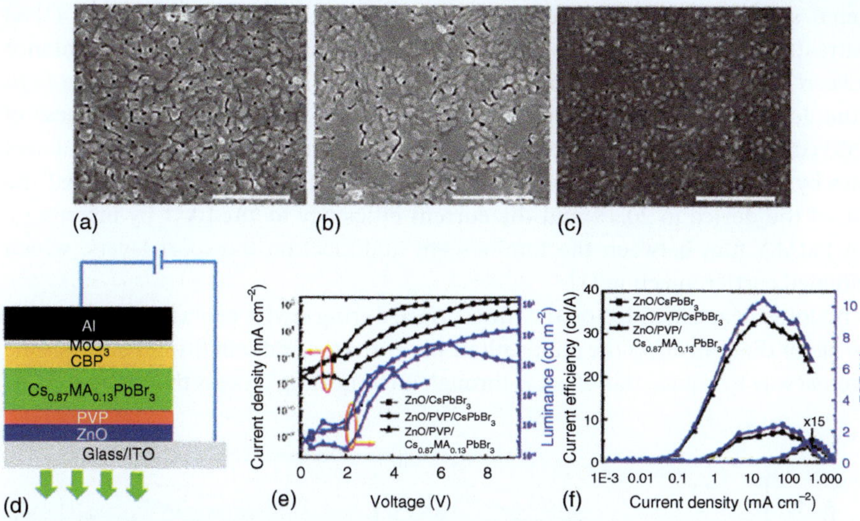

Figure 5.13 (a–c) Planar SEM images of $CsPbBr_3$, $Cs_{0.87}MA_{0.13}PbBr_3$ deposited on ZnO, ZnO/PVP, respectively; (d) device structure; (e,f) J–V–L, CE–J–EQE plots of the devices. Source: Zhang et al. [17]/Reproduced with permission from Springer Nature / CC BY 4.0.

resulting in a homogeneous perovskite material with excellent EL properties. Quasi-dimensional perovskite materials have enhanced optoelectronic properties due to exciton-limiting domains when compared to conventional two-dimensional materials. These materials have a quantum well structure that allows rapid energy transfer from wide bandgaps (small n values) to narrow bandgaps (large n values) and avoids exciton bursts. Moreover, the quasi-dimensional structure prevents the growth of large crystals, which reduces the number of defect states at grain boundaries. Surface passivation of organic amines further reduces the surface roughness of the film.

5.5.3 Surface Engineering

The quality of the surface of perovskite films plays a crucial role in determining the luminescence efficiency of the device, and poor film quality can lead to significant current leakage. Various methods have been proposed in the literature to improve the film quality and obtain dense, uniformly covered films over a large area. These methods include adding polymer films, annealing treatment, anti-solvent vapor treatment, and two-step deposition [59, 60]. The rapid crystallization process can result in the formation of cavities on the film surface, which can negatively impact the efficiency of PeLEDs. To prepare highly efficient devices, methods such as the addition of polymeric films, annealing treatment, anti-solvent vapor treatment, and two-step deposition have been proposed to improve the film quality, leading to dense, uniformly covered films. For instance, the addition of PEO acts as a barrier to excessive crystal growth and maintains the stability of the quasi-2D layered structure. Zhanao Tan's group showed that an isopropyl alcohol (IPA) posttreatment can wash away excess insulating alkyl chains and achieve a balance between high fluorescence quantum efficiency and fast carrier injection. The EQE of the device was 8.42% without PEO but increased to a luminous brightness of 33 533 cd m^{-2} with PEO (Figure 5.14b) due to the reduction of non-radiative defect states by surface passivation and the IPA treatment [4]. Lin's team increased the EQE of the device to 20.3% and the current efficiency to 78 cd A^{-1} by inserting a thin PMMA film between the luminescent and electron transport layers, which facilitated carrier injection [3].

The Joule heat generated on the film surface during device operation is an important factor that can affect the photovoltaic performance and stability of PeLEDs. One effective way to reduce this effect is through an annealing process that increases the

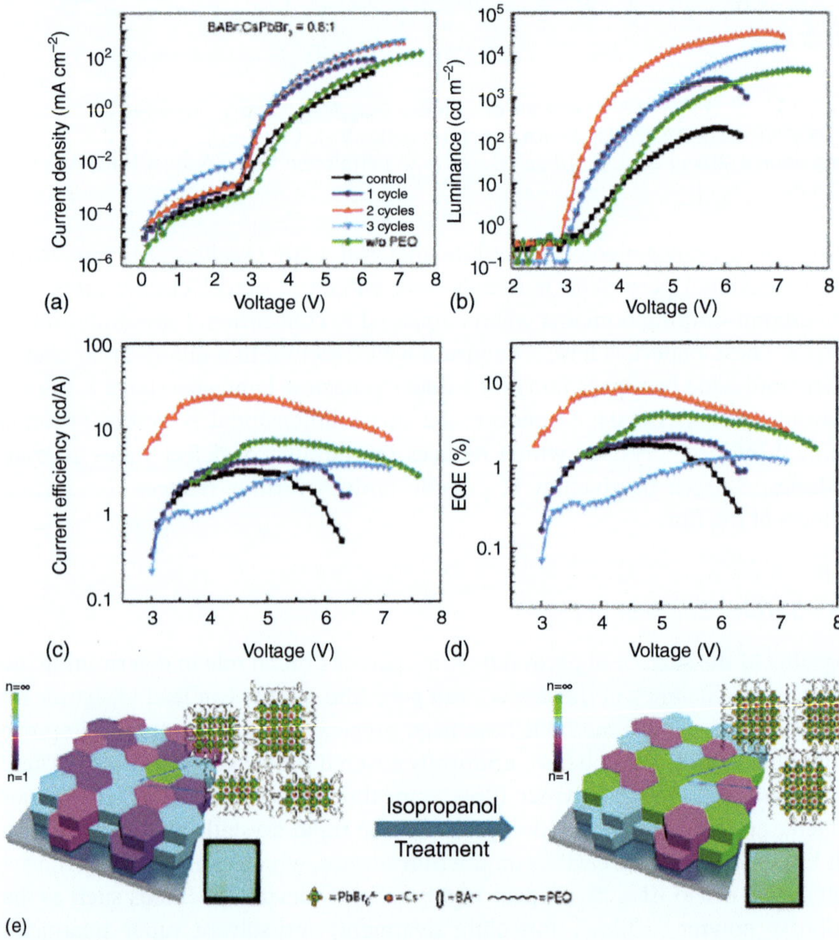

Figure 5.14 (a–d) J–V, L–V, CE–V, EQE–V diagrams of $BA_2(CsPbBr)_{3n-1}PbBr_4$-PEO perovskite light-emitting diodes, respectively; (e) schematic diagram of the structure of quasi-2D $BA_2(CsPbBr)_{3n-1}PbBr_4$-PEO thin film before/after IPA treatment [1].

crystallinity of the film and results in a more homogeneous film that is conducive to radiative recombination. This, in turn, can lead to an overall improvement in the uniformity of the films [61]. In 2019, a significant reduction of defects at grain boundaries was achieved by Wu's group through a simple and rapid chemical sintering treatment of $MAPbBr_3$ films [21]. Annealing treatment is a common method to improve the surface quality of perovskite films and enhance device performance. However, excessive thermal annealing treatment may lead to the loss of constituent elements in the perovskite, which can destroy the perovskite structure and cause device performance degradation. Therefore, careful optimization of the annealing process is necessary to balance the improvement of film quality with the preservation of perovskite structure.

5.5.4 Passivation Engineering

During the synthesis of perovskite materials, structural traps are formed that tend to trap carriers, resulting in non-radiative recombination of excitons and efficiency roll-off in the device. To mitigate this issue, several studies have reported the use of organic additives, Lewis acid additives, and amine additives to fill defects at grain boundaries and perform surface passivation. Lin and colleagues added MABr to the precursor solution for the synthesis of $CsPbBr_3$ to form a core/shell structure, reducing the non-radiative defect sites. The device showed a remarkable EQE of 20.3% and an increase in current efficiency from 0.36 to 22.81 cd A^{-1} [3]. Wang et al. employed cesium trifluoroacetate (CsTFA) as a cesium source to regulate the crystal growth and passivate the grain interface of $CsPbBr_3$ [20]. The bonding of TFA^- with Pb^{2+} led to a reduction of the grain size from 300 to 60 nm, resulting in a device with an EQE of 10.5%. However, in the process of reducing grain size, defects can be generated at the surface or grain boundaries of quasi-two-dimensional perovskites. To address this issue, You's group employed trioctylphosphine oxide (TOPO) coated with $PEA_2(FAPbBr)_{3n-1}PbBr_4$ to passivate trap states. As shown in Figure 5.15, TOPO-treated $PEA_2(FAPbBr)_{3n-1}PbBr_4$ (at $n = 3$) exhibited optimal PL luminescence properties due to the effective passivation of surface defects by TOPO. Additionally, they added methylammonium cloride (MACl) to the precursor solution to slow down the crystal growth rate and fill the intergranular boundaries with Cl. This resulted in a PeLED with an EQE of 14.36% and a current efficiency of 62.4 cd A^{-1}.

In 2021, Wei's group utilized fluorinated TFPPO as a passivating agent to enhance the stability and efficiency of green light-emitting devices based on PEA-based perovskites [26]. As depicted in Figure 5.16a, the dimensional distribution of TFPPO was controlled by inducing crystallization in PEA-based perovskites. During the process, $[PbBr]_6^{4-}$ atomic nuclei, MA^+ and Cs^+ ions combined to form a perovskite sheet, enabling PEA^+ ions to diffuse into the perovskite lattice and form a multi-quantum well structure. In the absence of TFPPO, the fast diffusion of PEA led to polydispersity of quantum wells. However, the addition of TFPPO in the antisolvent limited the diffusion of PEA by binding its fluorine atoms through hydrogen bonding, thereby promoting the formation of monodisperse quantum

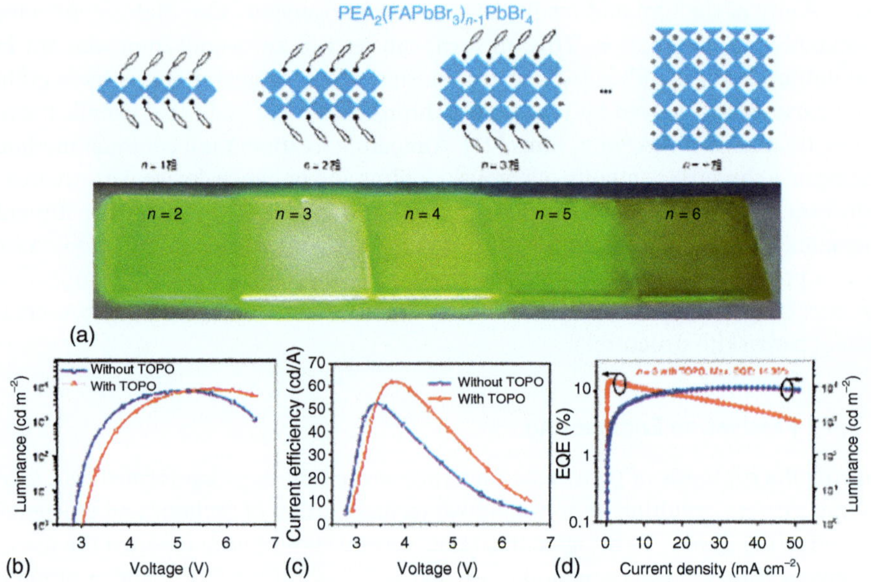

Figure 5.15 (a) Schematic diagram of the structure of $PEA_2(FAPbBr)_{3n-1}PbBr_4$ and PL diagram of $PEA_2(FAPbBr)_{3n-1}PbBr_4$ under UV irradiation. Source: Yang et al./Reproduced with permission from Springer Nature; (b,c) L–V, CE–V plots of $PEA_2(FAPbBr)_{3n-1}PbBr_4$ ($n = 3$) devices with/without TOPO cladding; (d) TOPO cladding of $PEA_2(FAPbBr)_{3n-1}PbBr_4$ EQE–J–L curve plots of perovskite light-emitting diodes with/without TOPO coating [4].

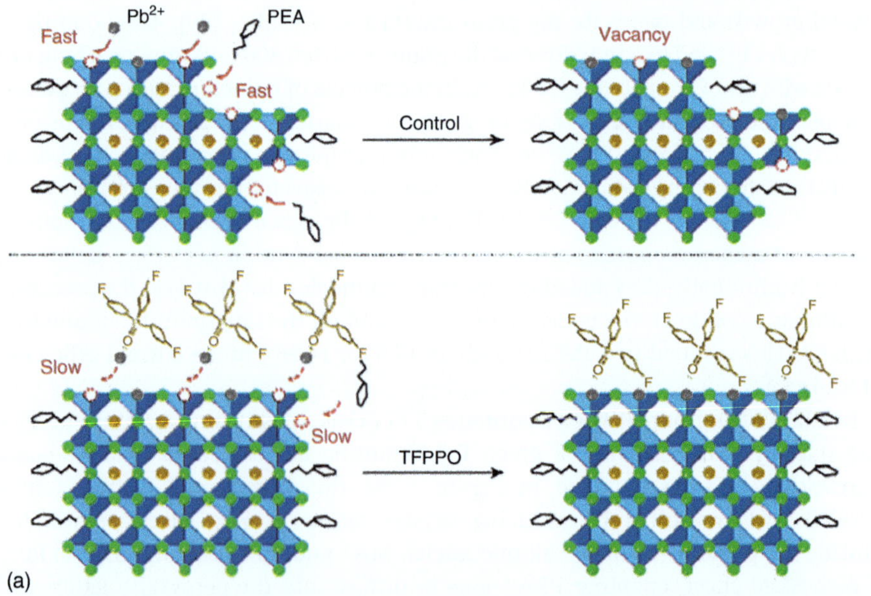

Figure 5.16 (a) Schematic diagram of TFPPO control dimension distribution; (b) device structure (left), cross-sectional TEM image (middle) and energy band diagram (right); (c) current density versus voltage (left), luminance versus voltage (middle) and EQE versus luminance (right) curves; (d) EQE histogram of 40 devices based on control, TPPO-treated and TFPPO-treated RDPs; (e) current density versus voltage curves of hole-only (h) and electron-only (e) devices based on control, TPPO-treated and TFPPO-treated RDPs; (f) operating stability of LEDs based on TFPPO-treated RDPs. Source: Ma et al. [26]/Reproduced from Springer Nature.

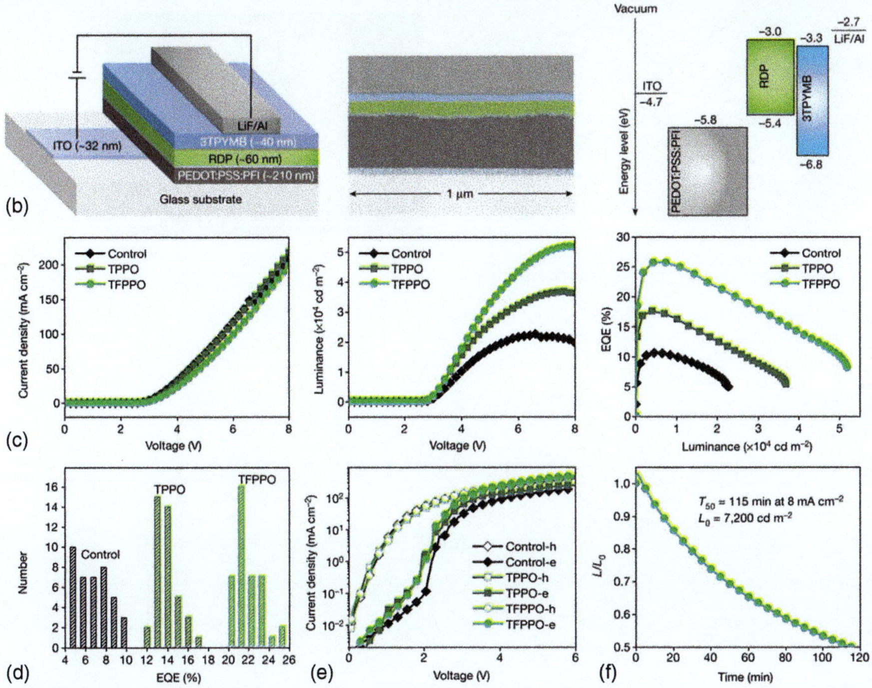

Figure 5.16 (*Continued*)

wells. Additionally, the P=O portion of TFPPO passivated the grain boundaries of the perovskite. Figure 5.16b illustrates the device performance data, where the introduction of TFPPO reduced the device leakage current, increased the device operating brightness, and improved the EQE to a maximum EQE value of 25.6%. Moreover, the device exhibited excellent stability due to the effective passivation of defects, with a T_{50} lifetime of up to two hours at an initial brightness of 7200 cd m^{-2}.

5.5.5 Optimization of the Device Structure

The optimization of device structure involves two aspects: efficient carrier injection and balancing carrier transport. Polymers or inorganic materials can be added to the charge transport layer to regulate the energy level distribution for efficient carrier injection. It is also important to balance the transport of holes and electrons into the emitting layer, ensuring that excitons are efficiently compounded in the emitting layer and that exciton escape is prevented.

As illustrated in Figure 5.17, Zhang et al. utilized a perfluorinated ionic cross-linked polymer (PFI) as an additive in the HIL to modify the energy level alignment of the HIL. This adjustment increased the valence band maximum (VBM) of the poly-TPD, enabling it to match the energy level of the adjacent functional layer, which facilitated efficient hole injection (Figure 5.17b,c) [62]. Additionally, the polyetherimide (PFI) layer contributed to the structural stability of the CsPbBr3 nanocrystal. The resulting device exhibited a low drive voltage of only

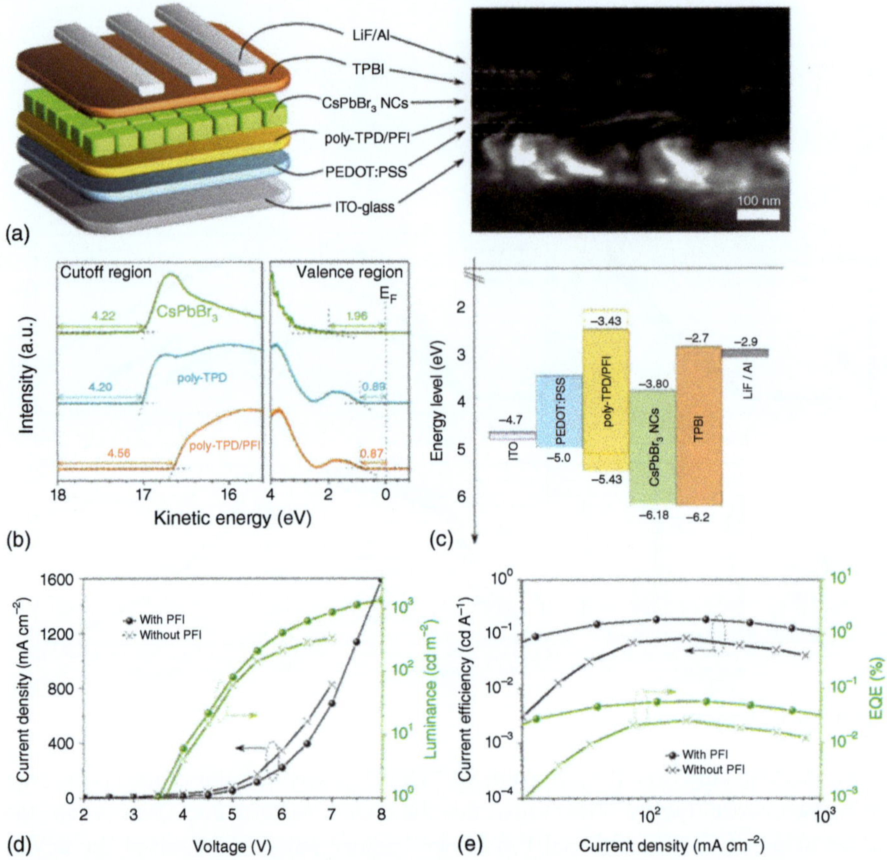

Figure 5.17 (a) Device structures; (b) UPS spectra of $CsPbBr_3$ nanocrystalline films, poly-TPD films, poly-TPD/PFI films deposited on ITO substrates; (c) overall energy level distribution of the perovskite light-emitting diodes; (d,e) J–V–L, CE–J–EQE profiles of the devices with/without PFI addition layer. Source: Zhang et al. [62]/American Chemical Society.

2.5 V, a narrow full width at half maximum (FWHM) of 18 nm, and a significant increase in EQE from 0.026% to 0.6%.

In 2017, Tae-Woo Lee's team employed a mixture of PEDOT:PSS and PFI as a buffer HIL for $CsPbBr_3$ PeLEDs. The film surface roughness was as low as 3.46 nm, and the energy levels were well matched among each layer, effectively preventing current shunting and leakage currents in the device [63]. This significantly reduces the occurrence of current shunting and leakage currents in the device.

In 2018, Wu's team developed a novel structure for PeLEDs known as an "insulator-perovskite-insulator" (IPI) structure, as illustrated in Figure 5.18 [64]. They added ultrathin LiF insulating films at the interface between the two ends of the light-emitting layer, resulting in the device structure ITO/LiF/Perovskite/LiF/Bphen (short for Bathophenanthroline)/LiF/Al (device C). Three common materials for green light perovskites ($MAPbBr_3$, $FAPbBr_3$, $CsPbBr_3$) were used as light-emitting layers, while the common device structures PEDOT:PSS/

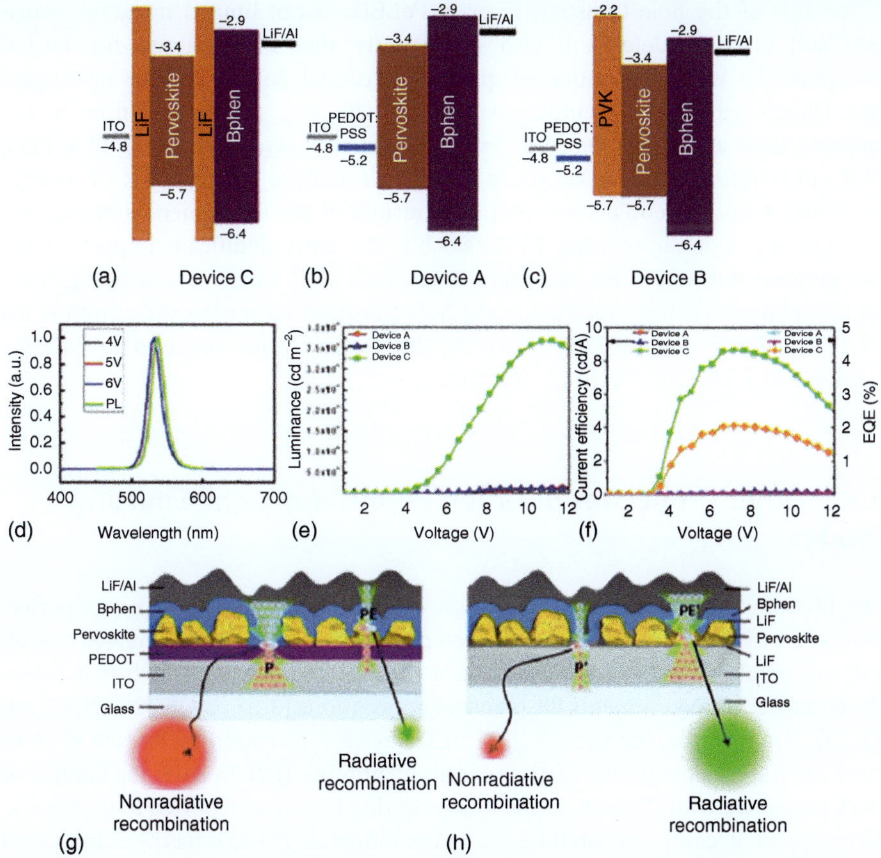

Figure 5.18 (a–c) Device structures; (b) spectrograms; (e,f) L–V and CE–V–EQE plots for devices with different structures; (g,h) schematic diagrams of cross-sections of the perovskite light-emitting diode device structures [64].

Perovskites/Bphen (device A) and PEDOT:PSS/PVK/Perovskites/Bphen (device B) were used for comparison (Figure 5.18a–c). The EL peak of the device with $MAPbBr_3$ as the luminescent layer is slightly blueshifted compared to the PL peak, with a wavelength of 535 nm (Figure 5.18d). Among the three devices with $MAPbBr_3$ as the emitting layer, device C demonstrated the best EL performance, with the highest brightness (36 854 cd m^{-2}), current efficiency (8.67 cd A^{-1}), and EQE (2.36%) (Figure 5.18e,f). The IPI structure with its very thin LiF insulating layer allows most of the carriers to be transported directly through the perovskite crystal without flowing through its defects, which minimizes leakage currents. In 2020, Richard's group also deposited a very thin LiF layer of approximately 1 nm between the light-emitting layer and the hole transport layer and doped the light-emitting layer with 5% tetraphenylphosphonium chloride (TPPCl) to control the crystallinity and size of the perovskite, improving the crystal quality and carrier lifetime of the perovskite film. This led to an increase in the photoluminescence quantum efficiency (PLQE) of the thin film to 65% and the EQE of the device to 19.1% [65].

The role of the hole transport layer in PeLEDs is not limited to carrier transport and interface defects. It also significantly affects the energy distribution and potential well distribution of quasi-dimensional perovskites. To investigate this, Chen's team in 2021 employed $PEA_2Cs_{n-1}Pb_nBr_{3n+1}$ perovskite films as the emitting layer and tested four different hole transport layers, including PVK:PEO, PVK, PEDOT:PSS, and nickel oxide (NiO_x), to analyze their impact on energy distribution and potential traps [66]. The results of the experiment indicate that the hole transport layer using PVK:PEO has the most significant impact on the energy potential well of the perovskite layer, with PEO having a passivating effect on it. The device using PVK:PEO as the hole transport layer achieved a luminance of 23 110 cd m^{-2} and an EQE of 11.5%, the highest value reported to date for unmodified $PEA_2Cs_{n-1}Pb_nBr_{3n+1}$ PeLEDs.

5.6 Other Properties of Green Perovskite Light-Emitting Diodes

The lifetime and stability of a PeLED are crucial factors in evaluating its performance. However, the EL intensity of green perovskite diodes often decreases rapidly and can reach 50% of its initial value in a short period of time, which falls below the commercial requirements for practical applications [3, 16, 20, 23, 28, 29, 62, 64, 67–78]. For example, the $CsPbBr_3$/PFI LED device constructed by Rogach et al. in 2016 had a short operational lifetime of only 10 minutes [62]. In 2018, the lifetime of the LED device with $CsPbBr_3$ as the light-emitting layer was reported to increase to 120 minutes. Recently, the lifetime of all pure inorganic PeLEDs has been improved significantly, with T_{50} values reaching up to 436 hours [22]. The stability and lifetime of a perovskite device are influenced by various factors, including the stability of the perovskite material itself and the device's operating conditions. The crystal structure of perovskite material can be altered by exposure to high temperatures and environmental factors, and the stability of the nanoparticles in solution and thin film states must also be considered. Furthermore, bias pressure during device operation can cause ion migration in the perovskite material and diffusion of electrode metals, leading to the formation of p-i-n junctions [59]. The stability of perovskite devices is affected by various factors such as the crystal structure of the material, stability of nanoparticles in solution and thin film states, ion migration, electrode metal diffusion, and Joule heating during operation. Ion migration, particularly of halogen ions, is the main cause of instability in PeLEDs. This occurs because the perovskite material is an ionic crystal with weak lattice binding energy, and ions can easily escape and migrate along defect sites. Additionally, the electric field formed during device operation accelerates ion migration. The thermal degradation of the material due to Joule heating at high current densities can also contribute to reduced device performance [79]. A current strategy to effectively inhibit ion migration in perovskite materials is to use "all-inorganic" approaches, which involve using all-inorganic perovskite materials and all-inorganic device structures (Table 5.1).

Table 5.1 Lifetime of green perovskite light-emitting diodes.

Year	Emitting materials	EQE (%)	PL peak (nm)	Working time
2016 [73]	$MAPbBr_3$ QDs	1.06	518	$T_{50} = 230$ h
2018 [4]	$PEA_2(FAPbBr)_{3n-1}PbBr_4$	14.36	532	$T_{50} = 65$ min
2018 [81]	$PEA_2(CsPbBr)_{3n-1}PbBr_4$-crown	15.5	520	$T_{50} = 1.5$ h
2018 [3]	$CsPbBr_3$/MABr PCs	20.3	525	$T_{50} = 105$ h
2019 [23]	2D $FAPbBr_3$ NPs	3.53	532	$T_{50} = 32$ min
2019 [20]	$FA_{0.11}MA_{0.10}Cs_{0.79}PbBr_3$	10.5	518	$T_{50} = 250$ h
2019 [22]	PDB-$CsPbBr_3$ QDs	—	520	$T_{82} = 436$ h
2021 [69]	PEABr-$CsPbBr_3$	16.8	520	$T_{50} = 208$ h
2021 [68]	$PEA_2PbBr_4(CsPbBr_3)_4$	22.49	512	$T_{50} = 52$ min
2021 [27]	PEABr-$CsPbBr_3$	28.1	514	$T_{50} = 4.04$ h
2022 [71]	$CsPbBr_3$ film	15.4	520	$T_{50} = 664$ h
2022 [28]	$(FA_{0.7}MA_{0.1}GA_{0.2})_{0.87}Cs_{0.13}PbBr_3$	28.9	540	$T_{50} = 520$ h
2023 [29]	$PEA_2FA_2Pb_3Br_{10}$	30.8	530	$T_{50} = 101$ min

Lee's team has proposed three effective strategies to enhance the stability of PeLEDs: A-site cation engineering, utilization of Ruddlesden–Popper crystalline phases, and incorporation of additives or interfacial layers to suppress ion migration [59]. One effective way to improve the stability of PeLEDs is through A-site cation engineering, which replaces the amidine ion with a more stable cation in green PeLEDs. Table 5.2 shows that $CsPbBr_3$, a pure inorganic perovskite material, is the most stable of the polycrystalline green perovskite materials. The stability of PeLEDs is dependent on dimension and is unevenly developed, with three-dimensional structures (single or polycrystalline) developing the fastest, zero-dimensional structures (quantum dots or nanocrystals) developing second, and quasi-two-dimensional structures developing the slowest. In 2017, Sun et al. added PEO to $CsPbBr_3$ and treated it with anti-solvent chloroform (CF) vapor after spin-coating to form a film with perovskite polycrystals of 5 μm size, which had few defects and a high binding energy of 128.4 meV. The device had an initial luminance value of 1000 cd m^{-2} and only decayed by 18% after 80 hours

Table 5.2 Some thermal stability parameters of green light perovskite materials.

Perovskite materials	Thermal decomposition temperature (°C)	Carrier mobility ($cm^2\ V^{-1}\ S^{-1}$)	Trap density (cm^3)
$CsPbBr_3$	500	4500	1.1×10^{10}
$FAPbBr_3$	200	62	9.6×10^9
$MAPbBr_3$	200	24	3×10^{10}

of continuous operation. The use of Ruddlesden–Popper crystalline phases and the addition of additives or interfacial layers to inhibit ion migration are other effective ways to improve PeLED stability proposed by Lee's team [75]. In 2016, Zhang et al. synthesized $MAPbBr_3$ quantum dots using ligand-assisted redeposition and assembled them into devices, resulting in a lifetime (T_{50}) of 230 hours. Li's group developed a new device in 2018 with a similar lifetime of 230 hours [73]. Li's group proposed an all-inorganic perovskite device structure in 2018, consisting of ITO/NiO/$CsPbBr_3$ quantum dots/$Zn_{0.8}Mg_{0.2}O$ nanoparticles/Al, to improve device stability [35]. In solution state, $CsPbBr_3$ quantum dots have a high PLQY of 83.6% and show only a minor decrease of 6.2% after 30 days. Assembled devices based on $CsPbBr_3$ quantum dots exhibit an EQE of 3.79% and a long lifetime (T_{80}) of over 12 hours at 20°C, 30–40% humidity, and a constant voltage of 8 V. The devices were subjected to multiple cycles of temperature changes, from 330 to 393 K and back to 300 K, with only a 15.1% reduction in the EL strength. In 2018, Mohite et al. reported that quasi-dimensional systems with more constituents tend to exhibit higher stability [80]. Therefore, they theorized that the presence of a three-dimensional composition in the quasi-two-dimensional system could have a negative effect on device stability. They also found that the reduction in device EL was related to erosion and degradation of the interface rather than the perovskite material itself. Hee Ju An et al. introduced a novel co-hole transport layer (HTL) of 4,4′-bis(*N*-carbazolyl)-1,1′-biphenyl (CBP) and PVK into PeLEDs. By optimizing the composition ratio of CBP and PVK, the performance of CsPbBr3-based PeLEDs is significantly improved via efficient Förster resonant energy transfer and enhanced charge transfer owing to the well-aligned energy levels of the HTLs with the emission layers [71]. In 2017, Deschler et al. added crown ether to the precursor solution of $CsPbBr_3$ and PEABr and assembled devices with the structure ITO/poly-TPD/Perovskites/TPBi/LiF/Al, which achieved an EQE of 15.5% [81]. Deschler et al. conducted a study where crown ethers were added to the precursor solution of $CsPbBr_3$ and PEABr. They concluded that the addition of crown ethers benefits the carrier solution by facilitating carrier domain limiting, reducing non-radiative combination, and suppressing ion migration. The study also found that the inherently thermally stable nature of the perovskite material cannot tolerate the Joule heating generated by device operation, which limits device stability. Priyabrata Sadhukhan et al. In the experiment, 1-butyl-3-methylimidazole tetrafluoroboric acid ionic liquid (IL) was introduced with a potential additive of the $CsPbBr_3$ perovskite precursor, resulting in an ultrabright green PeLED achieving a peak brightness of $3.28 \times 105\,cm^{-2}$ only at a bias voltage of 6 V and a peak EQE of 13.75%. This achievement is the outcome of multirole support from IL that simultaneously enables superior control over the perovskite-film morphology, passivates defects, modifies the band energy levels, and prevents ion migration. Hence, this work demonstrates IL as a novel alternative additive with the potential to outperform conventional long-chain ligands in high-performance PeLED device fabrication [70]. However, most studies have not explored the underlying mechanisms by which the proposed measures improve device stability.

In summary, there are three main methods for improving the stability of PeLEDs: (1) incorporating appropriate ligands or polymers into the perovskite material to create a homogeneous microstructure; (2) developing new perovskite materials with high thermal stability; and (3) optimizing the device structure to facilitate carrier injection and transport and reducing the driving voltage of the device.

Future research on device stability could focus on the development of quasi-two-dimensional green perovskite materials and understanding ion migration at room temperature. It is also important to investigate the underlying mechanisms by which these methods improve device stability, which have not been thoroughly explored in most studies.

5.7 Conclusion and Outlook

The field of green PeLEDs has seen rapid progress in recent years due to their many advantages over conventional LEDs, such as their direct bandgap, narrow emission spectra, high brightness, tunable colors, and low cost. This chapter provides a comprehensive overview of the development of green PeLEDs in terms of perovskite structure, synthesis methods, device assembly, and their performance. Specifically, the chapter focuses on the development of device efficiency and factors that affect EQE. Practical measures to address the efficiency roll-off of PeLEDs are also discussed, including material improvement and device optimization.

Factors that contribute to the loss of EQE in perovskite light-emitting devices include non-radiative combination processes in the perovskite material and carrier transport imbalances in the device. These issues can be addressed by crystal engineering, surface passivation, and ligand and device engineering. As a result, the current EQE of green PeLEDs has reached 30.84%.

However, the stability and lifetime of these devices remain significant challenges. Green PeLEDs are highly unstable when exposed to light and water vapor, and ions in the material tend to migrate under bias pressure, leading to short lifetimes. Relevant literature suggests three approaches to improve device stability: (1) using suitable ligands or polymers to cover the perovskite material to prevent material agglomeration, ion migration, and moisture oxidation; (2) actively developing new perovskite materials with high thermal stability; and (3) optimizing the device structure by adding a buffer layer or exciton blocking layer to achieve carrier injection and transmission balance and improve exciton combination efficiency.

References

1 Wang, Z., Wang, F., Sun, W. et al. (2018). Manipulating the trade-off between quantum yield and electrical conductivity for high-brightness quasi-2D perovskite light-emitting diodes. *Advanced Functional Materials* 28: 1804187.

2 Krieg, F., Ong, Q.K., Burian, M. et al. (2019). Stable ultraconcentrated and ultradilute colloids of $CsPbX_3$ (X = Cl, Br) nanocrystals using natural lecithin as a capping ligand. *Journal of the American Chemical Society* 141: 19839–19849.

3 Lin, K., Xing, J., Quan, L.N. et al. (2018). Perovskite light-emitting diodes with external quantum efficiency exceeding 20 per cent. *Nature* 562: 245–248.

4 Yang, X., Zhang, X., Deng, J. et al. (2018). Efficient green light-emitting diodes based on quasi-two-dimensional composition and phase engineered perovskite with surface passivation. *Nature Communications* 9: 570.

5 Fang, Z., Chen, W., Shi, Y. et al. (2020). Dual passivation of perovskite defects for light-emitting diodes with external quantum efficiency exceeding 20%. *Advanced Functional Materials* 30: 1909754.

6 Li, C., Zang, Z., Chen, W. et al. (2016). Highly pure green light emission of perovskite $CsPbBr_3$ quantum dots and their application for green light-emitting diodes. *Optics Express* 24: 15071–15078.

7 Sadhanala, A., Ahmad, S., Zhao, B. et al. (2015). Blue-green color tunable solution processable organolead chloride-bromide mixed halide perovskites for optoelectronic applications. *Nano Letters* 15: 6095–6101.

8 Levchuk, I., Osvet, A., Tang, X. et al. (2017). Brightly luminescent and color-tunable formamidinium lead halide perovskite $FAPbX_3$ (X = Cl, Br, I) colloidal nanocrystals. *Nano Letters* 17: 2765–2770.

9 Tan, Z.-K., Moghaddam, R.S., Lai, M.L. et al. (2014). Bright light-emitting diodes based on organometal halide perovskite. *Nature Nanotechnology* 9: 687–692.

10 Song, J., Li, J., Li, X. et al. (2015). Quantum dot light-emitting diodes based on inorganic perovskite cesium lead halides ($CsPbX_3$). *Advanced Materials* 27: 7162–7167.

11 Byun, J., Cho, H., Wolf, C. et al. (2016). Efficient visible quasi-2D perovskite light-emitting diodes. *Advanced Materials* 28: 7515–7520.

12 Shan, Q., Song, J., Zou, Y. et al. (2017). High performance metal halide perovskite light-emitting diode: from material design to device optimization. *Small* 13: 1701770.

13 Ng, Y.F., Kulkarni, S.A., Parida, S. et al. (2017). Highly efficient Cs-based perovskite light-emitting diodes enabled by energy funnelling. *Chemical Communications* 53: 12004–12007.

14 Quan, L.N., Zhao, Y., Garcia de Arquer, F.P. et al. (2017). Tailoring the energy landscape in quasi-2D halide perovskites enables efficient green-light emission. *Nano Letters* 17: 3701–3709.

15 Li, J., Xu, L., Wang, T. et al. (2017). 50-fold EQE improvement up to 6.27% of solution-processed all-inorganic perovskite $CsPbBr_3$ QLEDs via surface ligand density control. *Advanced Materials* 29: 1603885.

16 Shi, Z., Li, Y., Zhang, Y. et al. (2017). High-efficiency and air-stable perovskite quantum dots light-emitting diodes with an all-inorganic heterostructure. *Nano Letters* 17: 313–321.

17 Zhang, L., Yang, X., Jiang, Q. et al. (2017). Ultra-bright and highly efficient inorganic based perovskite light-emitting diodes. *Nature Communications* 8: 15640.

18 Chin, X.Y., Perumal, A., Bruno, A. et al. (2018). Self-assembled hierarchical nanostructured perovskites enable highly efficient LEDs via an energy cascade. *Energy & Environmental Science* 11: 1770–1778.

19 Song, J., Fang, T., Li, J. et al. (2018). Organic-inorganic hybrid passivation enables perovskite QLEDs with an EQE of 16.48%. *Advanced Materials* 30: 1805409.

20 Wang, H., Zhang, X., Wu, Q. et al. (2019). Trifluoroacetate induced small-grained $CsPbBr_3$ perovskite films result in efficient and stable light-emitting devices. *Nature Communications* 10: 665.

21 Xi, J., Xi, K., Sadhanala, A. et al. (2019). Chemical sintering reduced grain boundary defects for stable planar perovskite solar cells. *Nano Energy* 56: 741–750.

22 Xie, Q., Wu, D., Wang, X. et al. (2019). Branched capping ligands improve the stability of cesium lead halide ($CsPbBr_3$) perovskite quantum dots. *Journal of Materials Chemistry C* 7: 11251–11257.

23 Fang, H., Deng, W., Zhang, X. et al. (2018). Few-layer formamidinium lead bromide nanoplatelets for ultrapure-green and high-efficiency light-emitting diodes. *Nano Research* 12: 171–176.

24 Dong, Y., Wang, Y.K., Yuan, F. et al. (2020). Bipolar-shell resurfacing for blue LEDs based on strongly confined perovskite quantum dots. *Nature Nanotechnology* 15: 668–674.

25 Kim, Y.H., Kim, S., Kakekhani, A. et al. (2021). Comprehensive defect suppression in perovskite nanocrystals for high-efficiency light-emitting diodes. *Nature Photonics* 15: 148–155.

26 Ma, D., Lin, K., Dong, Y. et al. (2021). Distribution control enables efficient reduced-dimensional perovskite LEDs. *Nature* 599: 594–598.

27 Liu, Z., Qiu, W., Peng, X. et al. (2021). Perovskite light-emitting diodes with EQE exceeding 28% through a synergetic dual-additive strategy for defect passivation and nanostructure regulation. *Advanced Materials* 33: 2103268.

28 Kim, J.S., Heo, J.M., Park, G.S. et al. (2022). Ultra-bright, efficient and stable perovskite light-emitting diodes. *Nature* 611: 688–694.

29 Bai, W., Xuan, T., Zhao, H. et al. (2023, https://doi.org/10.1002/adma.202302283). Perovskite light-emitting diodes with an external quantum efficiency exceeding 30%. *Advanced Materials* 2302283.

30 Liu, X.K. and Gao, F. (2018). Organic-inorganic hybrid ruddlesden-popper perovskites: an emerging paradigm for high-performance light-emitting diodes. *The Journal of Physical Chemistry Letters* 9: 2251–2258.

31 Akkerman, Q.A., Raino, G., Kovalenko, M.V. et al. (2018). Genesis, challenges and opportunities for colloidal lead halide perovskite nanocrystals. *Nature Materials* 17: 394–405.

32 Lee, H.D., Kim, H., Cho, H. et al. (2019). Efficient ruddlesden-popper perovskite light-emitting diodes with randomly oriented nanocrystals. *Advanced Functional Materials* 29: 1901225.

33 Yang, S., Wang, Y., Liu, P. et al. (2016). Functionalization of perovskite thin films with moisture-tolerant molecules. *Nature Energy* 1: 15016.

34 Kim, Y.H., Kim, J.S., and Lee, T.W. (2019). Strategies to improve luminescence efficiency of metal-halide perovskites and light-emitting diodes. *Advanced Materials* 31: 1804595.

35 Shi, Z., Li, S., Li, Y. et al. (2018). Strategy of solution-processed all-inorganic heterostructure for humidity/temperature-stable perovskite quantum dot light-emitting diodes. *ACS Nano* 12: 1462–1472.

36 Green, M.A., Ho-Baillie, A., and Snaith, H.J. (2014). The emergence of perovskite solar cells. *Nature Photonics* 8: 506–514.

37 Manser, J.S., Christians, J.A., and Kamat, P.V. (2016). Intriguing optoelectronic properties of metal halide perovskites. *Chemical Reviews* 116: 12956–13008.

38 Jin, F., Zhao, B., Chu, B. et al. (2018). Morphology control towards bright and stable inorganic halide perovskite light-emitting diodes. *Journal of Materials Chemistry C* 6: 1573–1578.

39 Amat, A., Mosconi, E., Ronca, E. et al. (2014). Cation-induced band-gap tuning in organohalide perovskites: interplay of spin-orbit coupling and octahedra tilting. *Nano Letters* 14: 3608–3616.

40 Park, M.H., Jeong, S.H., Seo, H.K. et al. (2017). Unravelling additive-based nanocrystal pinning for high efficiency organic-inorganic halide perovskite light-emitting diodes. *Nano Energy* 42: 157–165.

41 Protesescu, L., Yakunin, S., Bodnarchuk, M.I. et al. (2016). Monodisperse formamidinium lead bromide nanocrystals with bright and stable green photoluminescence. *Journal of the American Chemical Society* 138: 14202–14205.

42 Ortiz-Cervantes, C., Carmona-Monroy, P., and Solis-Ibarra, D. (2019). Two-dimensional halide perovskites in solar cells: 2D or not 2D? *ChemSusChem* 12: 1560–1575.

43 Kim, H., Huynh, K.A., Kim, S.Y. et al. (2019). 2D and quasi-2D halide perovskites: applications and progress. *Physica Status Solidi* 14: 1900435.

44 Cho, H., Jeong, S., Park, M. et al. (2015). Overcoming the electroluminescence efficiency limitations of perovskite light-emitting diodes. *Science* 350: 1222–1225.

45 Chiba, T., Hoshi, K., Pu, Y.J. et al. (2017). High-efficiency perovskite quantum-dot light-emitting devices by effective washing process and interfacial energy level alignment. *ACS Applied Materials & Interfaces* 9: 18054–18060.

46 Shang, Y., Li, G., Liu, W. et al. (2018). Quasi-2D inorganic $CsPbBr_3$ perovskite for efficient and stable light-emitting diodes. *Advanced Functional Materials* 28: 1801193.

47 Ling, Y., Yuan, Z., Tian, Y. et al. (2016). Bright light-emitting diodes based on organometal halide perovskite nanoplatelets. *Advanced Materials* 28: 305–311.

48 Xing, J., Yan, F., Zhao, Y. et al. (2016). High-efficiency light-emitting diodes of organometal halide perovskite amorphous nanoparticles. *ACS Nano* 10: 6623–6630.

49 Song, J., Li, J., Xu, L. et al. (2018). Room-temperature triple-ligand surface engineering synergistically boosts ink stability, recombination dynamics, and charge injection toward EQE-11.6% perovskite QLEDs. *Advanced Materials* 30: 1800764.

50 Yan, F., Xing, J., Xing, G. et al. (2018). Highly efficient visible colloidal lead-halide perovskite nanocrystal light-emitting diodes. *Nano Letters* 18: 3157–3164.

51 Yan, F., Tan, S.T., Li, X. et al. (2019). Light generation in lead halide perovskite nanocrystals: LEDs, color converters, lasers, and other applications. *Small* 15: 1902079.

52 Yuan, S., Liu, Q., Tian, Q. et al. (2020). Auger effect assisted perovskite electroluminescence modulated by interfacial minority carriers. *Advanced Functional Materials* 30: 1909222.

53 Veldhuis, S.A., Boix, P.P., Yantara, N. et al. (2016). Perovskite materials for light-emitting diodes and lasers. *Advanced Materials* 28: 6804–6834.

54 Pan, J., Quan, L.N., Zhao, Y. et al. (2016). Highly efficient perovskite-quantum-dot light-emitting diodes by surface engineering. *Advanced Materials* 28: 8718–8725.

55 Shen, W.S., Yuan, S., Tian, Q.S. et al. (2019). Surfacial ligand management of a perovskite film for efficient and stable light-emitting diodes. *Journal of Materials Chemistry C* 7: 14725–14730.

56 Park, J.H., Lee, A.Y., Yu, J.C. et al. (2019). Surface ligand engineering for efficient perovskite nanocrystal-based light-emitting diodes. *ACS Applied Materials & Interfaces* 11: 8428–8435.

57 Zu, Y., Xi, J., Li, L. et al. (2020). High-brightness and color-tunable $FAPbBr_3$ Perovskite nanocrystals 2.0 enable ultrapure green luminescence for achieving recommendation 2020 displays. *ACS Applied Materials & Interfaces* 12: 2835–2841.

58 Li, G., Tan, Z.K., Di, D. et al. (2015). Efficient light-emitting diodes based on nanocrystalline perovskite in a dielectric polymer matrix. *Nano Letters* 15: 2640–2644.

59 Cho, H., Kim, Y.H., Wolf, C. et al. (2018). Improving the stability of metal halide perovskite materials and light-emitting diodes. *Advanced Materials* 30: 1704587.

60 Almeida, G., Goldoni, L., Akkerman, Q. et al. (2018). Role of acid-base equilibria in the size, shape, and phase control of cesium lead bromide nanocrystals. *ACS Nano* 12: 1704–1711.

61 Chih, Y.K., Wang, J.C., Yang, R.T. et al. (2016). NiO_x electrode interlayer and $CH_3NH_2/CH_3NH_3PbBr_3$ interface treatment to markedly advance hybrid perovskite-based light-emitting diodes. *Advanced Materials* 28: 8687–8694.

62 Zhang, X., Lin, H., Huang, H. et al. (2016). Enhancing the brightness of cesium lead halide perovskite nanocrystal based green light-emitting devices through the interface engineering with perfluorinated ionomer. *Nano Letters* 16: 1415–1420.

63 Kim, Y.H., Wolf, C., Kim, Y.T. et al. (2017). Highly efficient light-emitting diodes of colloidal metal-halide perovskite nanocrystals beyond quantum size. *ACS Nano* 11: 6586–6593.

64 Shi, Y., Wu, W., Dong, H. et al. (2018). A strategy for architecture design of crystalline perovskite light-emitting diodes with high performance. *Advanced Materials* 30: 1800251.

65 Zhao, B., Lian, Y., Cui, L. et al. (2020). Efficient light-emitting diodes from mixed-dimensional perovskites on a fluoride interface. *Nature Electronics* 3: 704–710.

66 Li, T., Xiang, T., Wang, M. et al. (2021). Unraveling the energy landscape and energy funneling modulated by hole transport layer for highly efficient perovskite LEDs. *Laser & Photonics Reviews* 15: 2000495.

67 Liu, Z., Qiu, W., Peng, X. et al. (2021). Perovskite light-emitting diodes with EQE exceeding 28% through a synergetic dual-additive strategy for defect passivation and nanostructure. *Regulation* 33: 2103268.

68 Chu, Z., Ye, Q., Zhao, Y. et al. (2021). Perovskite light-emitting diodes with external quantum efficiency exceeding 22% via small-molecule passivation. *Advanced Materials* 33: 2007169.

69 Han, B., Yuan, S., Cai, B. et al. (2021). Green perovskite light-emitting diodes with 200 hours stability and 16% efficiency: cross-linking strategy and mechanism. *Advanced Functional Materials* 31: 2011003.

70 Sadhukhan, P., Kim, M.S., Baek, S.-D. et al. (2023). Super-bright green perovskite light-emitting diodes using ionic liquid additives. *Small Methods* 7: 2201407.

71 An, H.J., Baek, S.D., Kim, D.H. et al. (2022). Energy and charge dual transfer engineering for high-performance green perovskite light-emitting diodes. *Advanced Functional Materials* 32: 2112849.

72 Li, J., Bade, S.G., Shan, X. et al. (2015). Single-layer light-emitting diodes using organometal halide perovskite/poly (ethylene oxide) composite thin films. *Advanced Materials* 27: 5196–5202.

73 Deng, W., Xu, X., Zhang, X. et al. (2016). Organometal halide perovskite quantum dot light-emitting diodes. *Advanced Functional Materials* 26: 4797–4802.

74 Wei, Z., Perumal, A., Su, R. et al. (2016). Solution-processed highly bright and durable cesium lead halide perovskite light-emitting diodes. *Nanoscale* 8: 18021–18026.

75 Wu, C., Zou, Y., Wu, T. et al. (2017). Improved performance and stability of all-inorganic perovskite light-emitting diodes by antisolvent vapor treatment. *Advanced Functional Materials* 27: 1700338.

76 Chen, F., Boopathi, K.M., Imran, M. et al. (2020). Thiocyanate-treated perovskite-nanocrystal-based light-emitting diodes with insight in efficiency roll-off. *Materials* 13: 367.

77 Aygüler, M.F., Weber, M.D., Puscher, B.M.D. et al. (2015). Light-emitting electrochemical cells based on hybrid lead halide perovskite nanoparticles. *The Journal of Physical Chemistry C* 119: 12047–12054.

78 Yu, J.C., Kim, D.B., Baek, G. et al. (2015). High-performance planar perovskite optoelectronic devices: a morphological and interfacial control by polar solvent treatment. *Advanced Materials* 27: 3492–3500.

79 Chen, S., Wen, X., Sheng, R. et al. (2016). Mobile ion induced slow carrier dynamics in organic-inorganic perovskite $CH_3NH_3PbBr_3$. *ACS Applied Materials & Interfaces* 8: 5351–5357.

80 Tsai, H., Nie, W., Blancon, J.C. et al. (2018). Stable light-emitting diodes using phase-pure Ruddlesden-Popper layered perovskites. *Advanced Materials* 30: 1704217.

81 Ban, M., Zou, Y., Rivett, J.P.H. et al. (2018). Solution-processed perovskite light emitting diodes with efficiency exceeding 15% through additive-controlled nanostructure tailoring. *Nature Communications* 9: 3892.

6

Blue Perovskite Light-emitting Materials and Devices

6.1 Technology Development of Blue Perovskite Light-emitting Diodes

As we delve into the realm of emerging technology, our focus narrows down to a significant stride in materials science – the advent of blue perovskite materials and devices. This chapter, titled “Blue Perovskite Light-emitting Materials and Devices,” meticulously explores the transformative journey of blue perovskite technology from its inception to its current state-of-the-art application and potential future pathways.

Perovskite materials, particularly their blue-emitting variants, have gained substantial attention in the field of optoelectronics due to their unique properties, such as high photoluminescence quantum yields, a tunable bandgap, and simplified solution processing techniques. These distinctive attributes open a world of opportunities, including enhancing the performance and energy efficiency of light-emitting diodes (LEDs), revolutionizing display technologies, and potentially even enabling new solar energy harvesting techniques. However, developing stable and efficient blue perovskite materials and devices presents significant challenges given their intricate structure, operational instability, and the enigmatic nature of their degradation mechanisms.

This chapter will systematically guide you through the critical milestones in blue perovskite technology. It will shed light on the fundamental concepts underpinning these materials and devices, their synthesis techniques, device fabrication, performance optimization strategies, and prevailing challenges. Furthermore, it will present current research advancements and prospective areas of exploration in the pursuit of achieving stability, enhanced efficiency, and broader applicability of blue perovskite materials and devices. By the end of this chapter, readers will gain comprehensive insights into the blue perovskite technology landscape, arming themselves with the knowledge necessary to further contribute to this dynamic and promising field [1, 2].

Since 2018, significant improvements in the performance parameters of perovskite blue light materials and devices have been achieved, with highly efficient devices prepared using a variety of methods reported in recent years [3, 4]. The focus of scientific attention has shifted from simply improving device efficiency to enhancing device lifetime, increasing brightness, reducing efficiency roll-off, and widening the

Perovskite Light Emitting Diodes: Materials and Devices, First Edition. Hong Meng.

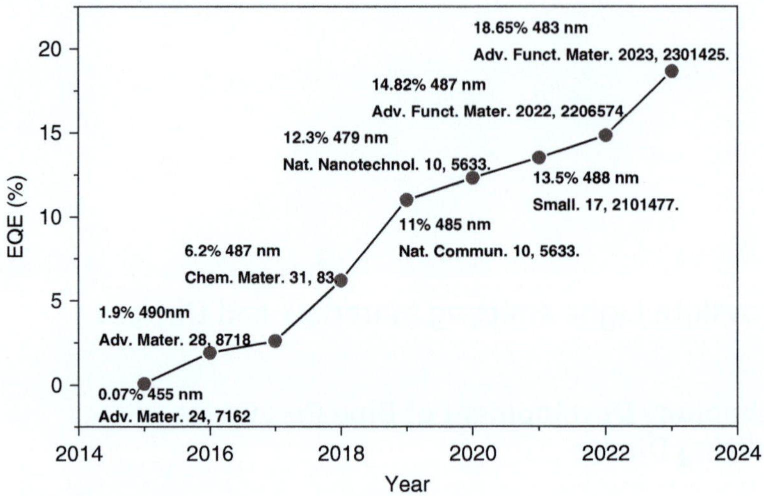

Figure 6.1 The development in EQEs of blue PeLEDs [5–14].

bandgap further to achieve a blueshift in the emitted light. Blue light is defined as visible light with wavelengths between 400 and 480 nm, while the wavelength range of sky-blue light is 480–490 nm, and cyan light falls between blue and green light with wavelengths between 490 and 510 nm. Although sky-blue and cyan light do not meet the requirements of display applications due to their wavelength ranges, they can be easily blended with other light colors to form white light for solid-state lighting applications, making them important to consider. Figure 6.1 indicates the development in EQEs (external quantum efficiencies) of blue PeLEDs in the last 10 years.

Figure 6.1 presents a decade-long overview of the advancements in the external quantum efficiencies (EQEs) of blue perovskite light-emitting diodes (PeLEDs). It captures a clear trend of progressive enhancement in the efficiency values, illustrating the significant strides made in optimizing the performance of these devices. The x-axis represents the chronological progression of years, while the y-axis plots the corresponding EQE values, demonstrating how they have evolved over the given timeframe. From a cursory glance, it's evident that there have been substantial improvements in the EQEs of blue PeLEDs. Early data points reflect a period of modest EQEs, indicating the challenges faced during the initial stages of blue perovskite technology development. However, the graph shows a consistent upward trend, a testament to the relentless research efforts and optimization strategies implemented over the years. In the latter part of the timeline, a steep increase in EQE values signifies the successful incorporation of breakthrough techniques in material processing, device fabrication, and overall design improvements. This reflects the maturation of the technology and its journey toward achieving the goals of higher efficiency and performance. Nevertheless, the graph also underscores the ongoing nature of this work as researchers strive to push the boundaries of EQE values even further, thereby enhancing the feasibility of blue PeLEDs in a wide range of applications. Thus, Figure 6.1 serves as both a historical record of achievement and a guidepost for future innovation in the field of blue PeLEDs.

6.2 Blueshift Strategy

In recent years, PeLEDs have advanced rapidly, achieving EQEs near to 30% in near-infrared, red, and green emissions. However, realizing high-performance deep-blue perovskite LEDs, which meet the Rec. 2020 blue standard with color coordinates of (0.131, 0.046), continues to pose a significant challenge.

Typically, blue perovskite emitters are classified into three categories (Figure 6.2): 3D perovskites, which are produced by spin-coating mixed-halide (Cl/Br) precursors followed by an annealing process; 2D/quasi-2D perovskites, which are generated by combining large amounts of organic cations to segregate the inorganic octahedra $[PbX_6]^{4-}$ layers; and 0D perovskite quantum dots (PQDs), which are composed of isolated metal halide octahedral anions, metal halide clusters, or perovskite nanocrystals (NCs).

Over the years, perovskite green light-emitting devices have achieved relative maturity in terms of both efficiency and stability. Therefore, most strategies for achieving high-efficiency blue perovskite devices have been based on blueshift strategies applied to the existing strategies for high-efficiency green light devices.

Energy band theory suggests that the periodic parameters of the lattice of a semiconductor material, such as the lattice constant and symmetry, play a crucial role in determining the bandgap. In the case of perovskite materials, the lattice is composed of a lead-halide octahedron as the "backbone" and organic or inorganic cations that fill the gaps between the octahedrons. This unique structure implies that the bandgap of a perovskite material is mainly determined by its lead-halide octahedron backbone, while the cations in the A-site play an indirect role in this process.

The comparative properties of 3D, (quasi-) 2D, corrugated-2D, 1D, and 0D perovskites (excluding Cs_4PbBr_6) are systematically detailed in Table 6.1 [16]. Upon comparison, it is noteworthy that 0D perovskites outshine their 3D, 2D, and 1D counterparts in terms of stability. This superior stability of 0D perovskites positions them favorably for commercial applications.

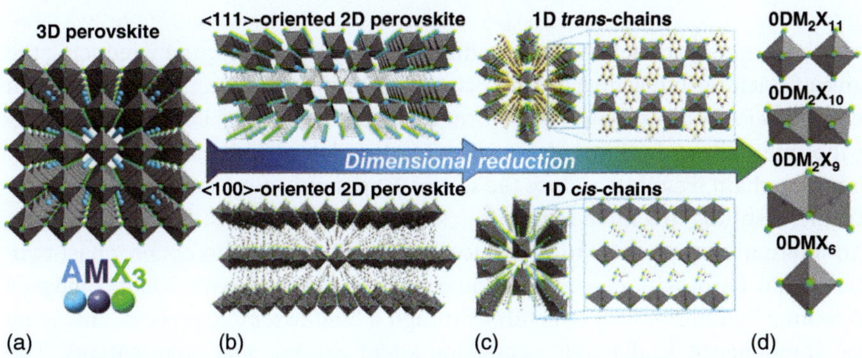

Figure 6.2 Crystal structures of (a) cubic 3D perovskite, (b) 2D perovskite, (c) 1D perovskite, and (d) 0D perovskite at molecular levels. Source: McCall et al. [15]/American Chemical Society.

Table 6.1 Property comparisons of 3D, (quasi-) 2D, corrugated-2D, 1D, and 0D perovskites (except Cs_4PbBr_6).

Parameter	3D	(Quasi-) 2D	Corrugated-2D	1D	0D (except Cs_4PbBr_6)
Stokes shift (nm)	10–20	10–20	40–200	100–200	100–350
FWHM (nm)	12–40	10–30	90–250	100–200	100–200
PLQY (%)	≈100	≈80	≈20	≈30	≈100
E_b (meV)	20–60	100–400	100–400	100–500	150–600
PL mechanism	FEs	FEs	FEs and STEs	FEs and STEs	STEs
Stability	Low	Medium	Medium	Medium	High

FWHM = full width at half maximum; E_b = exciton binding energy.
Source: Sun et al. [16]/John Wiley & Sons/CC BY 4.0.

There are three main strategies for achieving a blueshift based on highly efficient green-light perovskite materials.

One approach to achieving blueshift is to adjust the internal composition of lead halide octahedra to form 3D perovskites, which can change the lattice parameters and structure. This can be accomplished by replacing or doping the basic ions of the "skeleton," Pb^{2+}, with X^-, and by replacing a lattice consisting purely of Br^- with Br^-&Cl^- doping, which is the most common form of blueshift. For instance, Richard H. Friend's group reported research results on this approach in 2015 [17]. They achieved this by replacing or doping the basic ions of the "skeleton" (Pb^{2+}) with X^- and replacing a lattice consisting purely of Br^- with Br^-&Cl^-. To prepare perovskite films with different Br/Cl values, they mixed methylamine bromide and lead acetate with methylamine chloride and lead acetate, respectively, to create precursors consisting of two anions. By adjusting the mixing ratio of the two precursors, they were able to prepare 3D perovskite films, which were then used to create perovskite LEDs. The EL peaks showed a significant blueshift with increasing doping of chloride ions due to the change in the internal structure of the lead halide octahedra (Figure 6.3).

Another strategy for achieving blueshift is to widen the bandgap by reducing the degree of stacking and aggregation of octahedra through the quantum confinement effect of semiconductor materials to form 2D/quasi-2D perovskites without changing the internal composition of the octahedral skeleton. This can be achieved by using long-chain ligands to reduce the size of 3D perovskite crystals to the quantum dot scale, resulting in a finite blueshift.

An alternative strategy is to introduce bulk A-site cations to obtain quasi-two-dimensional perovskite crystals (2D/quasi-2D perovskites) with wider bandgaps and multiple quantum well structures through a "self-assembly process" involving three components: lead halide octahedra, small cations, and large cations. The reduction in size of perovskite crystals through this approach can produce a more pronounced blueshift when the reduction is only in one dimension, according to theoretical calculations. However, reducing the crystal structure to a fully 2D

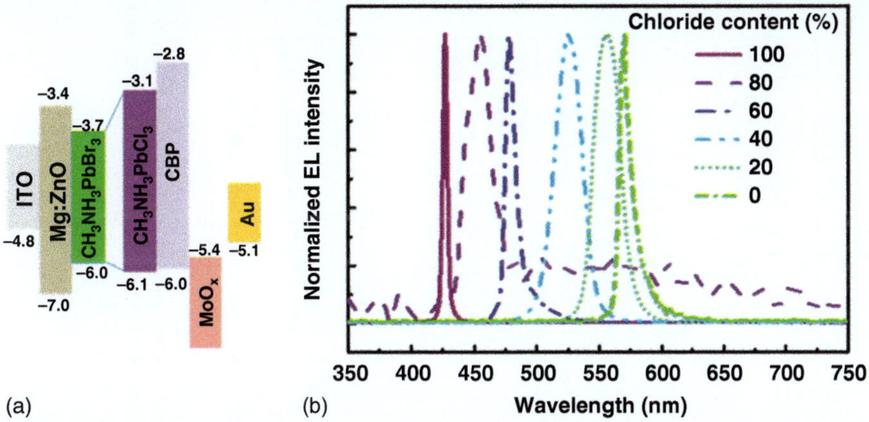

Figure 6.3 (a) Device structure of a perovskite light-emitting diode and (b) normalized electroluminescence spectra of devices prepared from perovskite films at different bromine to chloride ratios (measured at 77 K). Source: Sadhanala et al. [17]/American Chemical Society/CC BY 4.0.

structure means that one dimension of the crystal structure will only contain one layer of lead-halide octahedra, at which point the blueshift of the emitted light will reach its limit. For example, in 2016, Cao Yong's group reported on the preparation of quasi-2D perovskite crystals with wide bandgaps and multiple quantum well structures using this approach [18]. The researchers added the long-chain ammonium salt, 2-phenoxyethylamine (POEA), as an additive to the perovskite precursor. The amine cation in POEA replaced the original methylammonium (MA) cation on the surface of the perovskite, which led to the formation of π–π stacks of the phenyl groups in POEA, resulting in the orientation of the perovskite during crystallization, as shown in Figure 6.4a. The morphology of the film changed from massive to rod-like to flaky as the amount of POEA added increased, as observed from the scanning electron microscopy results. X-ray diffraction results provided direct evidence of the formation of a low-dimensional phase. As the amount of low-dimensional perovskite phase increased, there was a significant blueshift in the emission peaks, as previously described. However, the low purity of the low-dimensional phase on the films resulted in multiple peaks in both the fluorescence and electroluminescence spectra of the films, as shown in Figure 6.4b. The paper also discusses the effect of passivating defects on the surface of the films on the position of the emission peaks.

In addition to the previous two strategies, the influence of small cations in the A-site on the perovskite lattice should also be considered. Studies have shown that partial substitution of Cs^+ in the lattice for smaller cations such as Rb^+ or K^+ from the same main group can result in a blueshift in the emission peak. It is expected that the experimental methods used to adjust these "small cations" to achieve blueshift will become more advanced in the future.

It is important to note that each of these blueshift strategies, whether used individually or in combination, can lead to degradation in the performance of the

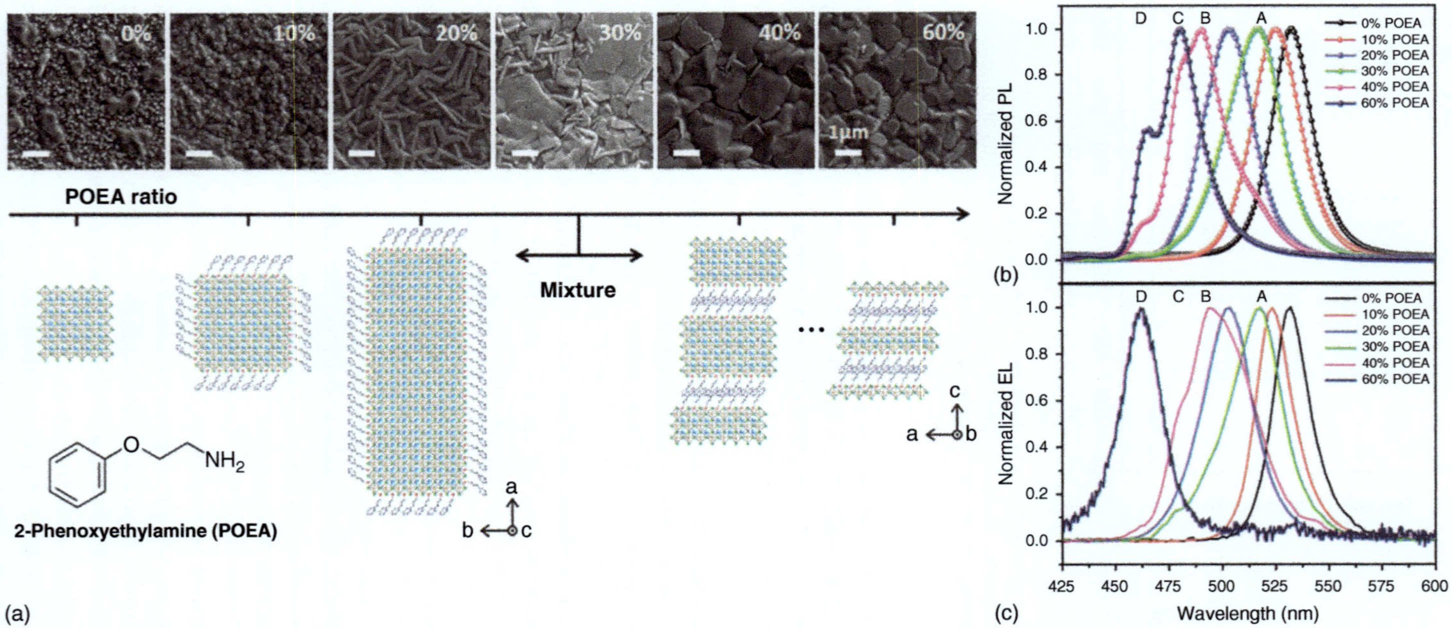

Figure 6.4 (a) SEM images of perovskite films made from $MAPbBr_3$ precursor solution with increasing POEA doping concentration, a schematic diagram of the transformation of the perovskite structure from a bulk film to a layered structure; (b) PL spectra of perovskite films with different doping ratios; and (c) EL spectra of luminescent devices with different doping ratios. Source: Chen et al. [18]/Reproduced from John Wiley & Sons, Inc.

perovskite layer [13]. Therefore, designing high-performance blue LED preparation methods requires more than just modifying the well-established process for green LEDs to achieve a blueshift. In recent years, several effective methods have been reported to optimize materials and devices for blue perovskite LEDs, which will be discussed in greater detail later.

6.3 Perovskite Blue Light-emitting Materials

6.3.1 Perovskite Blue Light-emitting Materials with a Quasi-two-dimensional Structure

In the initial preparation of 3D perovskite crystals, solvent evaporation often leads to the formation of crystalline films with a large number of defects that can be categorized depending on their scale. One typical approach for passivating these defects and improving film formation is the addition of various long-chain ammonium salts. However, adding more long-chain ammonium salts can alter the crystallization of the perovskite, resulting in chalcocite crystals that appear as flat flakes in scanning electron microscope (SEM) images. A study of this phenomenon revealed that in the chalcocite precursor solution (a colloid), the long-chain ammonium cations have a strong affinity for the lead halide octahedral aggregates, which cluster around them. When the long-chain ammonium cations contain a large volume of branched groups, the strong repulsive forces between the high spatial resistance groups cause the intermediate phase aggregates in the solvent to prefer forming intermediate layers with large resistance ammonium cations on both sides, creating a "sandwich" structure with several layers of stacked lead-halide octahedra that is not uniformly packed as a spherical core–shell structure.

The quasi-2D perovskite crystals with a sandwich structure, known as Ruddlesden–Popper phases (RP phases), are a type of pseudo-cubic cell. They have a more complex gap layer structure between the "sandwiches." Compared to 3D perovskite crystals, quasi-2D crystals have a different chemical structure, where the n value represents the number of layers of lead-halide octahedra stacked in the sandwich. In the interstices between the lead-halide octahedra in the sandwich, there is a minor cation. This unique structure provides advantages as a luminescent layer material, which can be summarized as follows:

(1) The preparation process is simple, controllable, and highly reproducible.
(2) Quasi-2D crystals have a low defect density and result in a smooth and flat surface, making them ideal for use as light-emitting layers in devices.
(3) Due to the higher dielectric constant of the ligand portion on both sides of the quasi-2D perovskite compared to the lead-halide octahedral lattice in the middle, a dielectric effect occurs, resulting in a wider bandgap. This effect enables the excitons generated to be tightly bound in the crystal with a large exciton binding energy. This promotes the combination of injected electrons and holes into excitons and inhibits the dissociation of excitons, thereby enhancing the efficiency of the device.

(4) The *n*-value of quasi-dimensional perovskites is often not fixed in the prepared films, indicating that the bandgap and energy level structure of the perovskite crystals in quasi-dimensional phase films are distributed in a discrete range. With a constant injection of charge, charge and energy transfer can occur between phases with different energy level structures. These transfer processes can enhance exciton utilization and suppress non-radiative exciton decay.

As previously discussed, quasi-2D perovskite crystals with lower n values result in a significant blueshift in the wavelength of emitted light (both photoluminescence and electroluminescence) from the perovskite films. In 2019, significant breakthroughs were achieved in the sky-blue region (480–490 nm) and blue region (450–480 nm) for blue perovskite light-emitting materials with quasi-2D structures, with EQEs of 9.5% and 2.6%, respectively. These breakthroughs demonstrate the great potential of this low-dimensional material for the preparation of blue LEDs [3, 19]. This highlights the immense potential of using low-dimensional perovskite crystals for the development of blue LEDs [17, 20].

In 2016, Jin's group successfully synthesized discoidal 2D perovskite crystals with the structural formula PEA_2PbBr_4 using a solvent atmosphere treatment method [21]. The method used by Jin's group to prepare quasi-2D perovskite crystals resulted in emission peaks that can be considered to reach the theoretical limit. This is because a 2D phase crystal with a single layer of lead-halide octahedra in one dimension was formed due to the large size of the PEA^+ ions, which could not fit in the interstices between the octahedra. However, the electroluminescence emission peak of this film was in the deep-blue region, and the device performance was unsatisfactory. The main reasons for this are twofold. Firstly, the 2D phase crystals consisting of a single layer of lead-halide octahedra are weakly conductive and do not allow for effective charge injection at a given voltage. Secondly, the low-dimensional phase with $n = 1$ is a thermodynamically unstable state, making it difficult to prepare perovskite films and devices in the usual way.

In current research on efficient blue light-emitting perovskite devices, quasi-2D phases with n values between 2 and 4 have become a hot topic, as this series of perovskite crystals can achieve both blue light emission and desired light-emitting properties through optimization. However, obtaining pure phases with a defined n value can be challenging, and prepared perovskite films are often mixtures of quasi-dimensional crystals with different *n* values, resulting in multiple absorption or emission peaks when measured. To obtain as monodisperse a quasi-dimensional phase as possible and avoid the generation of a large number of phases with high *n* values leading to red-shifted emission peaks, researchers have offered the following strategies to optimize the luminescence properties of such perovskite crystals when determining the *n* value of a phase.

6.3.1.1 Development of New Bulky Cations

(1) Figure 6.5 depicts recent literature regarding the fabrication of blue perovskite LEDs using "large cations." In 2018, Hu Bin's group employed propylamine bromide (PABr) as a long-chain ligand to create the low-dimensional phase, which lacks bulky groups that cause spatial site resistance effects. This

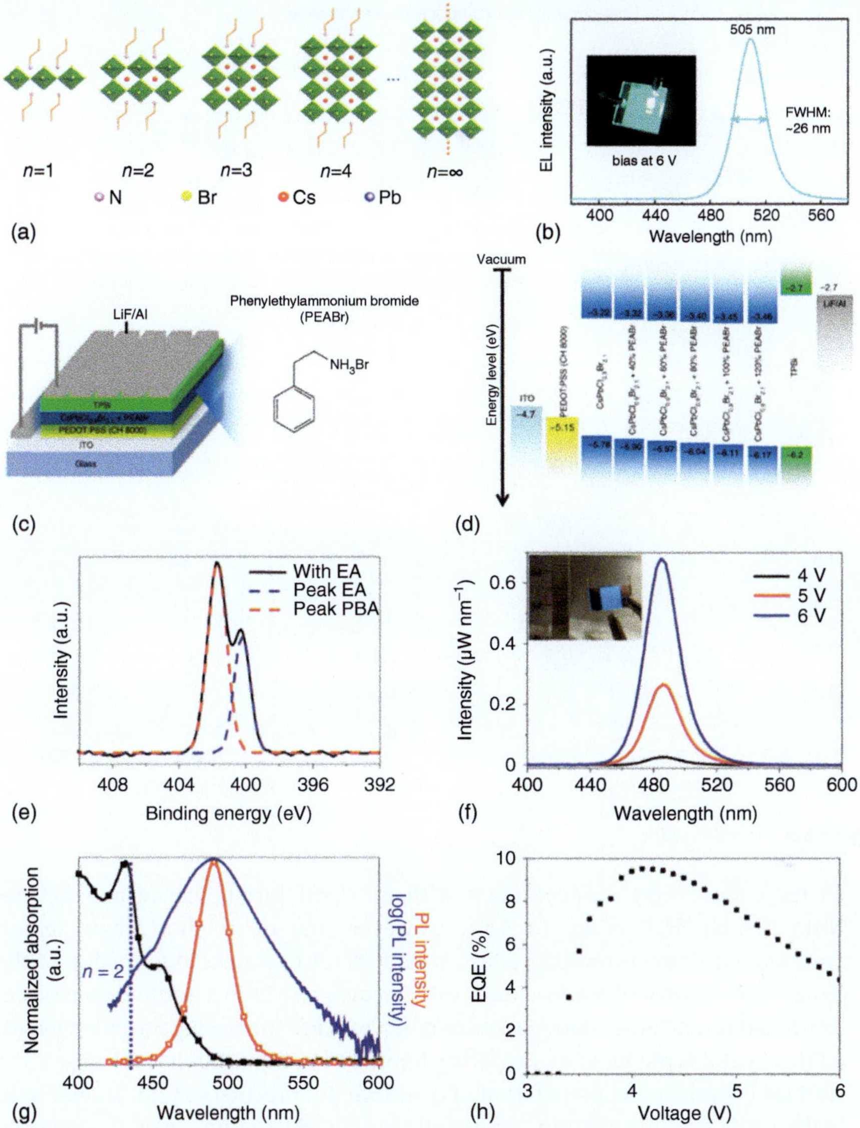

Figure 6.5 (a) Schematic structure of low-dimensional perovskite crystals with $n = 0$ to positive infinity; (b) PLQY diagram of the excitation intensity of the films since (excitation light wavelength of 365 nm). Source: Chen et al. [22]/Reproduced with permission from Elsevier; (c and d) structure and energy level structure diagram of the PeLED device. Source: Li et al. [23]/Springer Nature/CC BY 4.0; (e) XPS spectra of the film components after anti-solvent treatment; (f) EL spectra of the device; (g) absorption and emission spectra of the films; (h) the corresponding EQE–voltage relationship. Source: Yang et al. [24]/Reproduced from John Wiley & Sons, Inc; (i) schematic diagram of the altered component structure due to the new ligand. Source: Yuan et al. [19]/John Wiley & Sons; (j–l) physical diagram of low-dimensional perovskite single crystals with $n = 1$, 2, and 3; and (m and n) photoluminescence absorption and emission spectra of single crystals. Source: Adapted from Yang et al. [24].

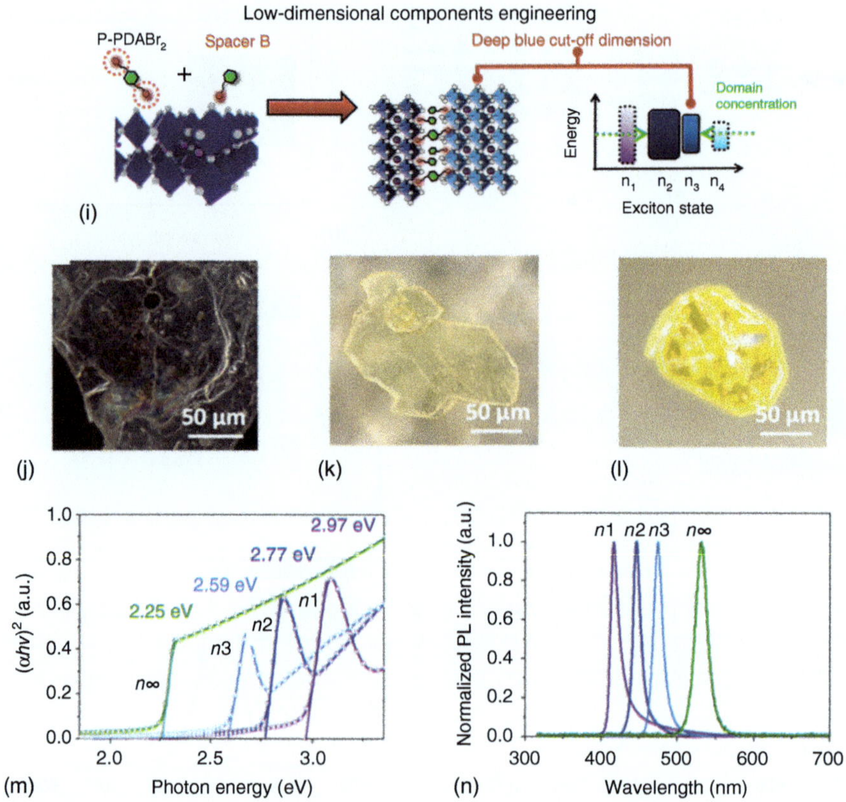

Figure 6.5 *(Continued)*

"large cation" can produce films with excellent luminescence and devices with a high EQE of up to 3.6%. However, due to its short chain length and limited domain-limiting effect, it is difficult to reduce the n value of the quasi-dimensional phase to a lower value, which results in a luminescence spectrum that is not "blue" enough (quasi-green region). Increasing the chain length of this ligand is not ideal for preparing luminescent films, but it can be used as a surface ligand for the preparation of quantum dot precursors [22]. However, in 2016, Chih-Jen Shih's group successfully produced low-dimensional perovskite films with quantum wells using the quantum dot preparation method, blurring the classification boundary between low-dimensional perovskite films and quantum dot microcrystalline films [25].

In 2018, Yong Cao's group employed two strategies to blueshift the perovskite emission simultaneously [23]. The group successfully achieved an EQE of up to 5% for perovskite blue LED devices by implementing two blueshift strategies. The first strategy involved varying the molar ratio of two anions (Br^- and Cl^-) to obtain blue light films with good film-forming properties. In addition, the group added PEABr to passivate the surface and assist in the formation of the low-dimensional quasi-dimensional phase. This led to an improvement in the luminescence efficiency and color stability of the device, along with a significant blueshift in the position of the emitted light as the amount of PEABr addition increased.

The second strategy involved indirectly controlling the longitudinal distribution of the grains of the perovskite layer by changing the preparation process of the hole injection layer PEDOT:PSS, which further increased the device's efficiency. The appropriate carbon chain length and bulky branched groups of PEA^+ make it a suitable "large cation" for the preparation of quasi-dimensional blue perovskite films. There are many other reports of similar findings using this "large cation," but they are not included in this discussion [26].

The performance of a luminescent device can be seriously affected by the electron–phonon coupling effect (EPI) of the luminescent layer material during carrier injection. In 2018, Edward H. Sargent's group prepared luminescent devices using three different large cationic ligands: BA^+ (butylamine ion), PhA^+ (aniline ion), and PBA^+ (phenylbutylamine ion) via a controlled ternary solvent crystallization method to produce quasi-dimensional perovskite discoidal crystals [27]. Figure 6.6 displays crystal structures determined by a combination of experiments and calculations, showing the results of tests using the ternary solvent crystallization control method. The researchers compared the strength of the electro–acoustic coupling effect in various crystals using several methods, including transient fluorescence spectroscopy at different temperatures, to obtain the temperature-dependent radiative and non-radiative complex rate constants k_{rad} and k_{non}. Higher k_{rad} values were observed in systems with smaller Bohr radii. Variable-temperature fluorescence spectroscopy was used to obtain the temperature dependence of the half-peak width of the emission peaks, which allowed the researchers to calculate the value of the lattice deformation potential D (depicted in the figure). Resonance Raman spectra were also used to gauge the strength of the phonon coupling during electron energy transfer. Additionally, experimental and computational data related to the energy band structure and lattice relaxation (depicted in the figure) were analyzed.

In 2019, Jianpu Wang's group employed a phenylbutylamine cation (PBA^+) as a large cation to prepare thin films with components $PBABr_{1.1}Cs_{0.7}FA_{0.3}PbBr_3$ in a quasi-dimensional perovskite structure. The resulting devices exhibited the highest EQE (9.5% at 483 nm) among all the tested films. This achievement not only demonstrated the effectiveness of the reduced-dimensional blueshift strategy in the preparation of sky-blue perovskite LEDs but also revealed a promising bulk cationic ligand that can efficiently prepare monodisperse quasi-2D perovskite phases. Compared to phenylethylamine, phenylbutylamine possesses a longer carbon chain, which can better promote the formation of monodisperse, low-n-value quasi-dimensional phases, facilitate the orientation of perovskite thin film crystals, and result in more complex interlayer structures between the layers. By combining this ligand with the corresponding anti-solvent process, the researchers successfully produced films with only monodisperse phase distributions of $n = 2$ and 3 [3].

In 2019, Liang-Sheng Liao's group proposed a novel double cationic ligand with positive charges at both ends, namely PDA, for the preparation of stable low-dimensional perovskite films. p-PDA as a ligand results in more stable low-dimensional films. The dispersion of domains and the kinetics of film formation

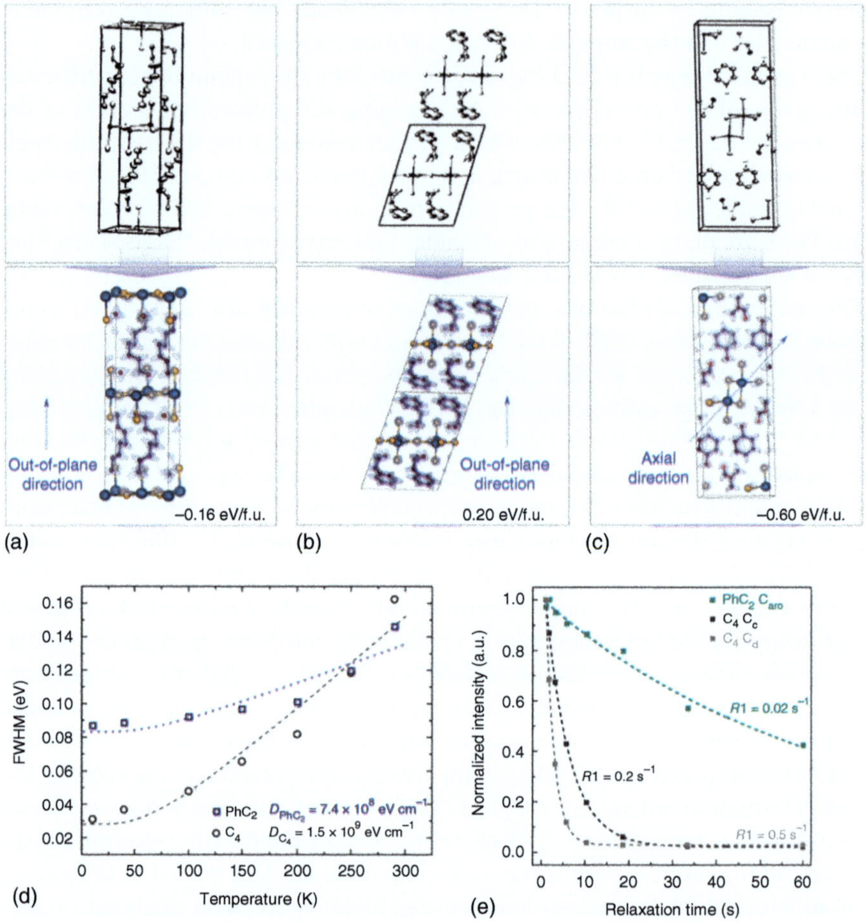

Figure 6.6 (a–c) Atomic structures of the C_4, PhC_2, and Ph crystals obtained from single crystal XRD analysis and the total electron charge density of each crystal simulated using the experimental crystal structure; (d) estimated formation potentials of PhC_2 and C_4 by fitting the FWHM of the PL profiles at different temperatures; the dashed lines are obtained by fitting the model discussed in the "Methods" section; and (e) spin–lattice relaxation rates (R1) fitted to the terminal aromatic carbon of the PhC_2 crystal and to the last two carbon atoms of the methyl terminal of the C_4 crystal. Source: Gong et al. [27]/Springer Nature.

for the quasi-dimensional films were strongly influenced by the cation used. The group mixed p-PDABr with PEABr to modulate the composition of the films. By optimizing the composition, it was found that the addition of a certain amount of PEA^+ reduced the content of the $n = 1$ phase and increased the content of the $n = 2$ and 3 phases in the films. This helped to avoid the reduction of device efficiency due to long-distance energy transfer and improved the device's performance, resulting in deep-blue light emission [19].

In 2020, Peidong Yang's group conducted an extensive investigation of low-dimensional perovskite crystals utilizing PBA^+ as the ligand cation. The direct characterization of single crystals is a more effective method for investigating the properties of materials than thin films. Yang Peidong's group prepared quasi-dimensional perovskite single crystals with $n = 1, 2$, and 3 using a solution method. The crystal bandgap, temperature dependence of the lattice, and device properties were all affected by the different n values. The final $n = 3$ single-crystal device achieved an EQE with a maximum value exceeding 1, making it the first successful blue single-crystal perovskite LED. However, the instability of the device's performance was mainly due to nonuniform surfaces generating Joule heat during operation, leading to thermodynamic disorder in the lattice. This study provides new insights for future researchers on how to use PBA as a ligand to create highly efficient devices [28].

6.3.1.2 Mixing of Bulky Cations

(2) To obtain the desired film components, a mixed ligand strategy is often used, as the number of bulk ligands that can directly contribute is limited. Chen Jun's group discovered that the mixing ratio of two ligands, PEA and isopropylamine (IPA), significantly affected the emitted light color, fluorescence quantum efficiency, and exciton lifetime of the films [29]. The disruption of the π–π stacking between the components of the phenylethylamine ligand by the shorter chain IPA is the main reason for the change in the emitted light color, fluorescence quantum efficiency, and exciton lifetime of the films. Notably, the researchers also investigated the impact of film thickness on device performance and found a balance between brightness and efficiency.

(3) In 2019, Liang-Sheng Liao's group investigated the effect of mixing two ligands, phenethylamine cation and double cation N-(2-bromoethyl)-1,3-propane-diamine dihydrobromide (NPA^{2+}), on the preparation of quasi-dimensional perovskite films. The addition of PEA to the NPA film not only maintains the blue light emission but also enhances the fluorescence quantum efficiency, as shown in Figure 6.7. The TA characterization results indicate that the film mainly consists of $n = 2$ and 3 low-dimensional phases, and this distribution leads to high device performance, with the maximum EQE exceeding 2% [30].

6.3.1.3 Cationic Doping

Edward H. Sargent's group utilized two blueshift techniques to produce perovskite LEDs with emission wavelengths of 465 and 475 nm, respectively. They employed dimensionality reduction using PEA to form quasi-2D perovskites, and cation exchange. Cs^+ and Rb^+ are homologous cations, but Rb^+ has a smaller ionic radius. Substituting Cs^+ with Rb^+ leads to lattice contraction, resulting in blueshift emission. Moreover, Rb^+ has been reported to have a passivating effect on perovskite films. The research group initially replaced Cs^+ with Rb^+ to obtain blue-emitting raw films and subsequently passivated the films by introducing a large amount of Rb^+. The passivation led to a redshift of the emission from 465 to 475 nm, and the resulting perovskite LED exhibited an EQE of 1.35% (Figure 6.8) [29].

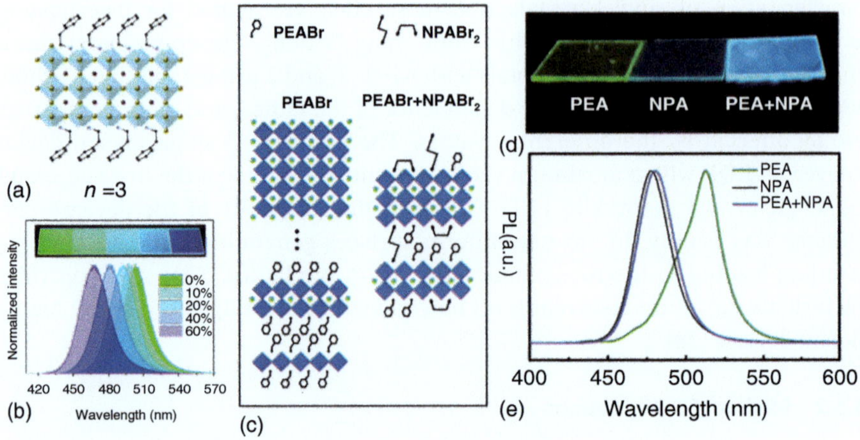

Figure 6.7 (a) Atomic model of PEA/IPA hybrid quasi-dimensional perovskite with $n = 3$; (b) PL spectra of perovskite films with different IPABr additions, insets are physical maps of the corresponding films under 365 nm irradiation. Source: Jiang et al. [29]/Springer Nature/CC BY 4.0; (c) schematic diagram of the dimensional distribution of the multi-ligand modulated low-dimensional perovskite phase; and (d and e) physical maps and PL maps under UV irradiation with different ligand additions. Source: Jin et al. [30]/Reproduced from John Wiley & Sons, Inc.

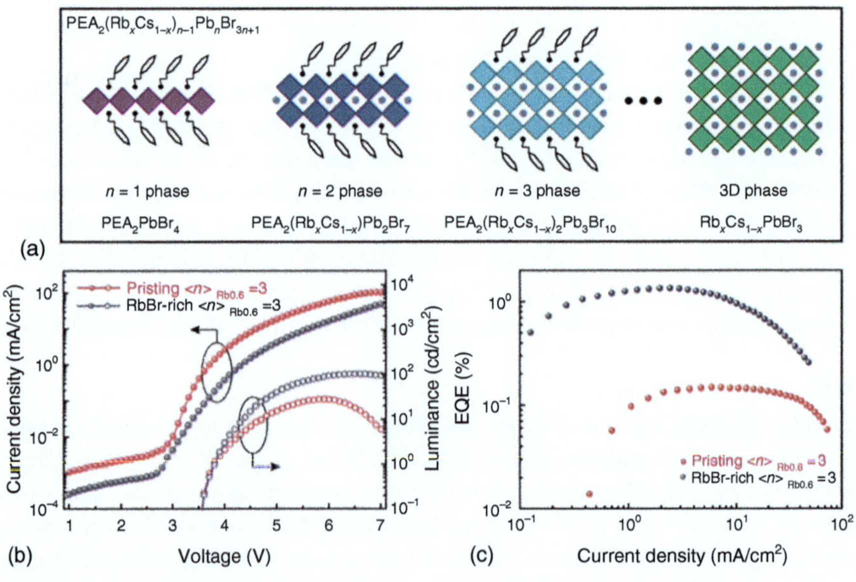

Figure 6.8 (a) Schematic structure of the Cs-Rb mixed quasi-2D perovskite phase and (b and c) performance curves of the prepared devices at different Rb^+ additions. Source: Jiang et al. [29]/Springer Nature/CC BY 4.0.

6.3.2 Blue Light Perovskite Nanocrystals or Quantum Dot Materials

Quantum dots are semiconductor nanostructures that confine excitons in all three dimensions [31]. They are typically on the nanoscale and can range in size from 2 to 10 nm (equivalent to 10–50 atoms), although self-assembled quantum dots can be larger, typically ranging from 10 to 50 nm. By controlling the chemical composition and optimizing the synthesis method, it is possible to produce quantum dots of varying sizes and shapes, which in turn affects their luminescence properties [32].

One strategy for preparing blue light-emitting quantum dot materials is to achieve a blueshift by modifying the structure of the lead halide octahedra. Co-doping Br and Cl anions into the lattice is the most common approach, and the higher the content of Cl^-, the bluer the light emission peak of the nanocrystal. However, doping chloride ions can lead to instability of the perovskite lattice due to the small size of the ion, which can cause ion migration under the effect of an applied electric field. This can limit device lifetime and efficiency at high voltages, and cause a significant shift in the wavelength of the emitted light over time or voltage (often exhibiting a redshift in blue light devices). To achieve deep-blue light emission in 3D or quasi-2D perovskite crystals requires a high doping of chloride ions, but too high a doping can result in very unstable light colors in the prepared devices, which diminishes their practical application. However, perovskite quantum dots or nano-microcrystalline materials, with long-chain ligand molecules wrapped around their surface, are more tolerant to chloride ion doping, resulting in more photostable devices. In summary, perovskite quantum dots (nano-microcrystalline) have great potential for use in deep-blue LEDs.

There are several common strategies used to optimize blue perovskite quantum dot (nano-microcrystalline) materials.

6.4 Synthesis and Use of New Long-Chain Ligands

The choice of ligands and the preparation process are critical factors affecting the performance of perovskite nanocrystal films. With the right ligand and preparation method, the stability of the internally wrapped perovskite nanocrystals can be significantly improved, leading to an increase in the fluorescence quantum efficiency of the solution. In 2019, Maksym V. Kovalenko's group introduced a short-chain ligand that is electrically neutral, which enhances the stability of Br/Cl-doped perovskite crystals while also improving the electrical properties of the films by washing away excess ligands. In order to ensure the electrical properties of the film, the structure of the device and the thickness of the crystal stack were optimized, and polystyrene (PS) was added as an additive to reduce the leakage current. The concentration of polystyrene in the precursor solution was also optimized to produce a highly efficient blue perovskite LED with a peak at 463 nm and a maximum EQE of 1.4% [33] (Figure 6.9).

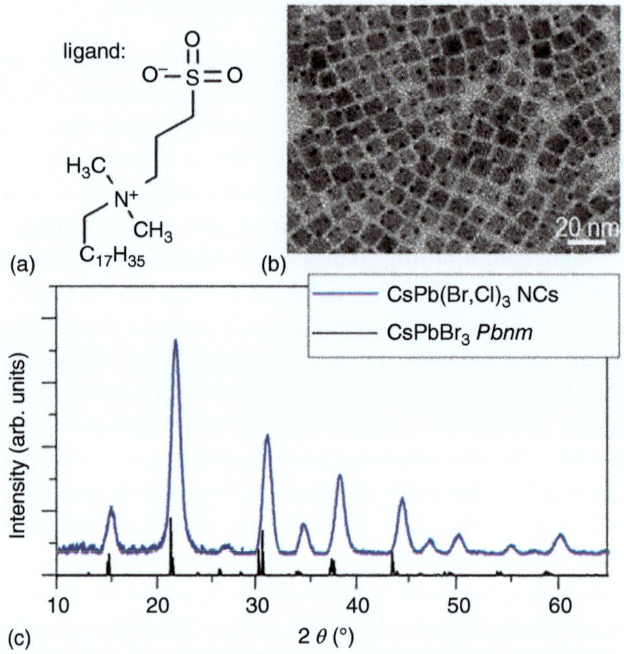

Figure 6.9 (a) Structure of the long chain amphiphilic sulfobetaine molecule used as a capping ligand for the nanocrystals; (b) TEM image of the nanocrystals; and (c) XRD pattern of the nanocrystals and reference pattern of $CsPbBr_3$ in the orthogonal galena *Pbmn* space group. Source: Ochsenbein et al. [33]/American Chemical Society.

6.5 Surface Modification of Nanostructures

As mentioned earlier, the high fluorescence quantum efficiency of perovskite quantum dot precursors is challenging to demonstrate on thin films. Therefore, surface modification plays a critical role in enhancing the performance of perovskite nanocrystals. The surface of nanocrystals is covered by numerous long-chain organic ligands, and surface modification techniques typically involve modifying the ligands.

The surface modification of nanocrystals is crucial in enhancing the performance of perovskite nanocrystals. Gao Feng's group proposed a new method using K^+ for surface modification, as shown in Figure 6.10. The group prepared blue precursors using mixed ligands, octanoic acid (OTAc) and didodecyl dimethyl ammonium bromide (DDAB), which wrapped the surfaces of the nanocrystals. After introducing K^+ as an additive, the crystallization kinetics in the precursor solution were affected, resulting in smaller spacing, larger grain size, and a wider distribution of size on the modified crystalline surfaces. The chemical environment of Br^-, Cl^-, and Pb^{2+} in the formation of ionic crystals was also altered. The addition of K^+ tends to bind to the halogen anion in the lattice, reducing the number of long-chain ligands on the surface, particularly OTAc , which reduces the defect density on the surface of the nanocrystals. This change improves the fluorescence quantum efficiency values of

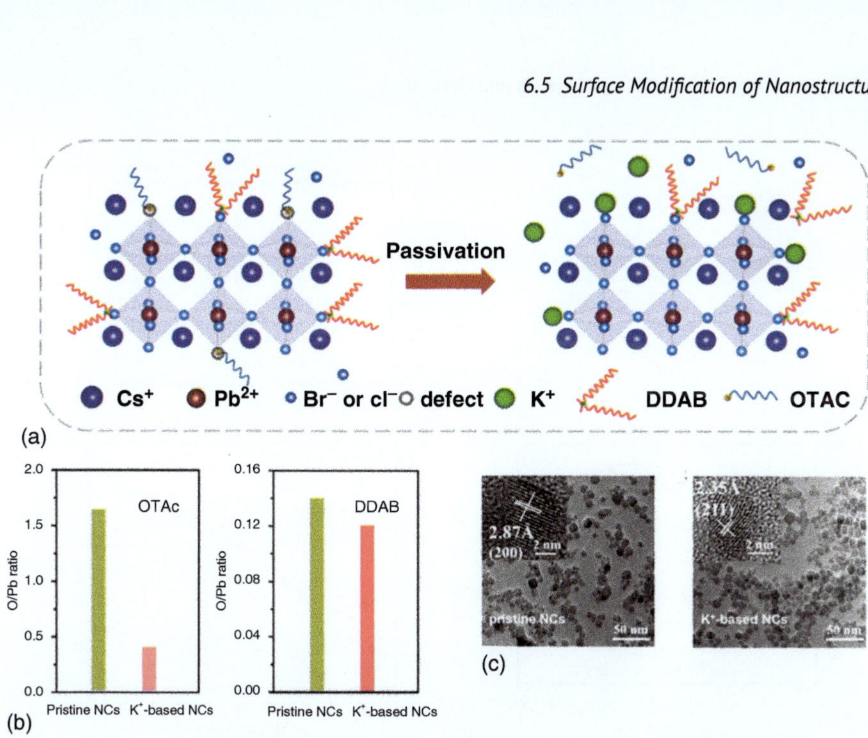

Figure 6.10 (a) Mechanistic diagram of the passivation effect on the potassium ion surface; (b) O/Pb and N/Pb values before and after treatment; and (c) TEM images before and after treatment, with insets showing the corresponding high-resolution TEM images. Source: Yang et al. [24]/Reproduced from John Wiley & Sons, Inc.

the precursors and the exciton lifetime on the surface of the films. By optimizing the amount of K^+ added and the material of the device transmission layer, a deep-blue PeLED with excellent performance was produced. The device emits light at a wavelength of 477 nm, with a maximum brightness of 89.65 cd m^{-2} and a peak EQE of 1.96% [24].

Surface modifiers can also be added to the precursor solution after the crystallization process to optimize the surface of the nanocrystals. Jin Young Kim's group proposed a posttreatment method called the LMPT method, which involves using ligand salts containing different halide anions. In this method, DDAB and DDAC were added as ion-pair ligands to the $CsPbBr_xCl_{3-x}$ nanocrystal precursors that were already prepared, and the resulting solution was treated to optimize the surface of the nanocrystals. To optimize performance, the ratio of the two long-chain salts used in the posttreatment process is modulated. The dodecyl dimethyl ammonium ion (DDA^+) is attracted to the halide anion in the lattice, and the Br^- and Cl^- ions in the salt are incorporated into the lattice, resulting in a blueshift in luminescence color as confirmed by high-resolution transmission electron microscopy (HRTEM) and X-ray diffraction spectroscopy. The surface structure of the nanocrystals is also altered, leading to an enhancement in fluorescence quantum efficiency. After careful tuning of the ratio of the two modified salts, the group successfully produced two PeLEDs emitting light at wavelengths of 470 and 480 nm, with maximum EQEs of 0.44% and 0.86%, respectively [34] (Figure 6.11).

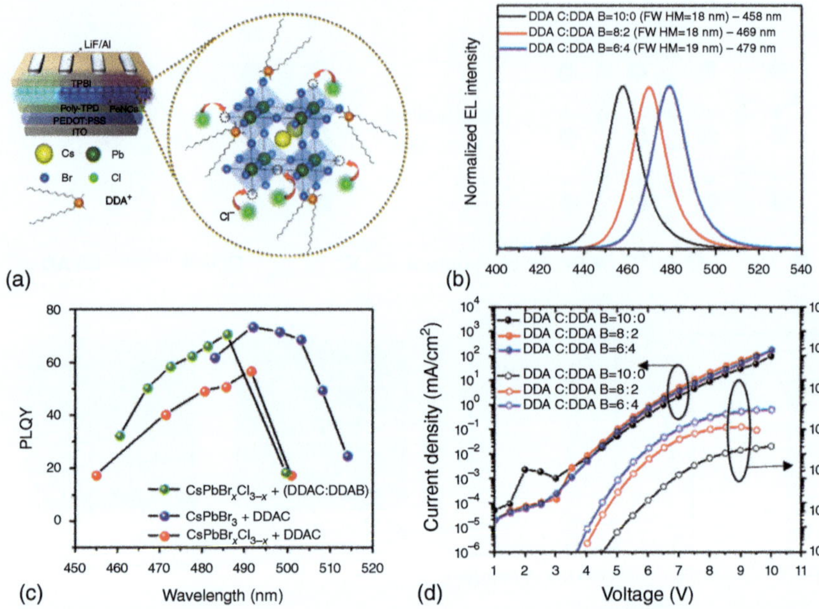

Figure 6.11 (a) Schematic diagram of the ligand action on the surface of the nanocrystals and the structure of the light-emitting device; (b) EL diagram of the device; (c) values of PLQY of the nanocrystals under different treatments; and (d) *J-V-L* curve of the electrical properties of the device. Source: Adapted from Shin et al. [34].

6.6 Optimization of the Internal Structure

In addition to synthesizing new ligands and surface modifications, exploring new internal structures of nanocrystals is a fundamental approach to optimization. This approach typically involves pure inorganic types (such as $CsPbX_3$) and organic–inorganic hybrid types (such as $B_xC_{1-x}PbX_3$, where B and C can be any organic or inorganic perovskite A-site cation). Internal structure modification can be achieved through four main methods: first, by doping other ions into the crystal structure; second, by using new preparation methods to create new structures based on the original elemental composition; third, by utilizing surface modifications to control the shape or size of the nanocrystals; and fourth, by designing new types of perovskite-structured crystals with different elemental compositions.

It has been reported that doping Mn^{2+} can significantly improve the fluorescence quantum efficiency of perovskite nanocrystals. As a magnetic dopant ion, Mn^{2+} can partially replace the central Pb^{2+} in the lattice, significantly reduce the defect density of the microcrystal, and form long-lived, stable emission energy states in the lattice. However, due to the all-bromine system's unfavorable conditions for manganese ion doping, Daniel N. Congreve's group developed a new two-step doping method for preparing highly efficient deep-blue-light nanocrystals. The group's method involved synthesizing chloride nanocrystals with high manganese content, followed by introducing bromine ions for ion exchange, resulting in up to

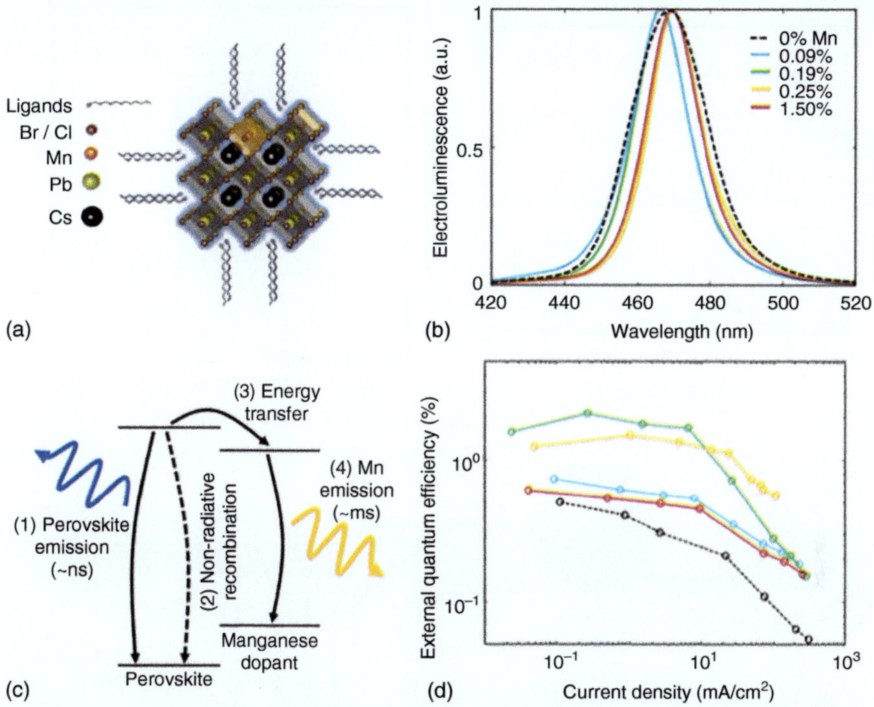

Figure 6.12 (a) Perovskite crystal structure under manganese ion doping. Source: Hou et al. [35]/with permission of Elsevier; (b) EL performance diagram of the device under different Mn doping. Source: Hou et al. [35]/with permission of Elsevier; (c) energy diagram of the doped nanocrystal system. Where (1) band-edge emission competes with (2) non-radiative complexation and (3) energy transfer to the manganese ion, which can then emit for a long time (4). Emission from perovskite occurs on nanosecond time scales, while long-lived Mn emission occurs on microsecond to millisecond time scales. Source: Hou et al. [35]/with permission of Elsevier; (d) device performance plots for different amounts of manganese doping. Source: Hou et al. [35]/with permission of Elsevier; (e) XRD patterns of $CsPbCl_3$ nanocrystals and Cs_4PbBr_6 nanocrystals. Source: Sun et al. [36]/American Chemical Society; (f) fluorescence spectra of different ratios of $Cs_4PbBr_6/CsPbCl_3$. Source: Sun et al. [36]/American Chemical Society; (g) silver exchange and surface reconstruction schematic. Source: Sun et al. [36]/Reproduced from John Wiley & Sons, Inc.; and (h) carrier dynamics model in ultra-small perovskite quantum dots. Source: Li et al. [37]/American Chemical Society. The extended state (green region) and the localized tail state are divided by the mobility edge (Eme). For simplicity, the tail state is assumed to be close to the 1Se state. The localized states have an exponential density of states, as shown by the orange shading. The light carriers produced by above-gap excitation (process II) can either recombine directly (process IV) or relax into the tail state and undergo radiative recombination (process III). Sub-energy gap excitation (process I) produces light carriers trapped in the tail state, which may jump into the extended state. As shown by the arrows, carriers trapped in the tail state can transition between lower and higher energy levels. The capture of the tail state and subsequent radiative recombination suppress the non-radiative recombination (process V) and thus increase PLQY.

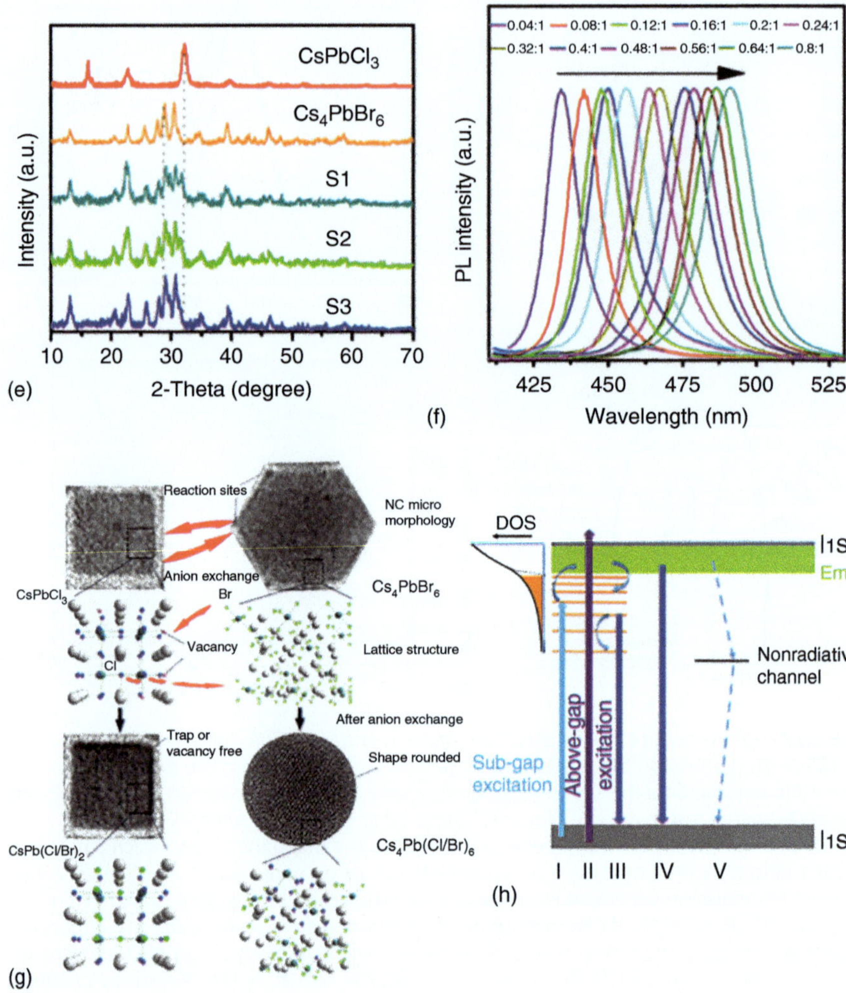

Figure 6.12 (*Continued*)

1.5% manganese ion doping. The group utilized transient fluorescence spectroscopy, UV–vis absorption spectroscopy, X-ray diffraction spectroscopy, and other characterization methods to investigate the effect of Mn^{2+} doping content on the internal structure of the nanocrystals. They found that the optimal device performance was achieved with 0.19% Mn ion doping, resulting in an emission peak at 466 nm and a maximum EQE of 2.12% [35].

W.G. Bi's group has proposed a novel strategy for synthesizing blue-light nanocrystals. They mixed $CsPbCl_3$ solution with Cs_4PbBr_6 solution and observed changes in their absorption and emission spectra based on the mixing ratios, as depicted in Figure 6.12. The researchers characterized the reaction in situ at different stages of the mixed precursor and presented the reaction mechanism in Figure 6.12. During the mixing process, the square-shaped $CsPbCl_3$ crystals and the hexagonal

Cs_4PbBr_6 crystals undergo anion exchange, resulting in rhombic-shaped crystals that eventually become circular. Transient fluorescence spectroscopy revealed that this ion exchange process effectively passivates the surface defects of the crystals, resulting in a blue backplane material with excellent luminescence and a pure light color. The research group utilized this material to create highly efficient PeLEDs. However, it is expected that this series of preparation methods and reaction mechanisms can provide novel approaches for preparing blue perovskite light-emitting layers [36].

The impact of size on the properties of perovskite nanocrystals is widely acknowledged, although no model exists for correlating size with properties beyond the quantum confinement effect. Zhizhen Ye's group synthesized $CsPbBr_3$ quantum dot crystals with a size of only 3 nm and discovered that the band-tailed state structure of these small-sized quantum dot crystals is quite unique, as determined by characterization techniques. The surface ligands have a significant impact on the detailed structure of perovskite nanocrystals, as demonstrated in Figure 6.13, which includes both continuous and discrete states. At higher excitation intensities, energy transfer results in relaxed luminescence in the band-tailed state, while at lower excitation intensities, the discrete region of the band-tailed state can

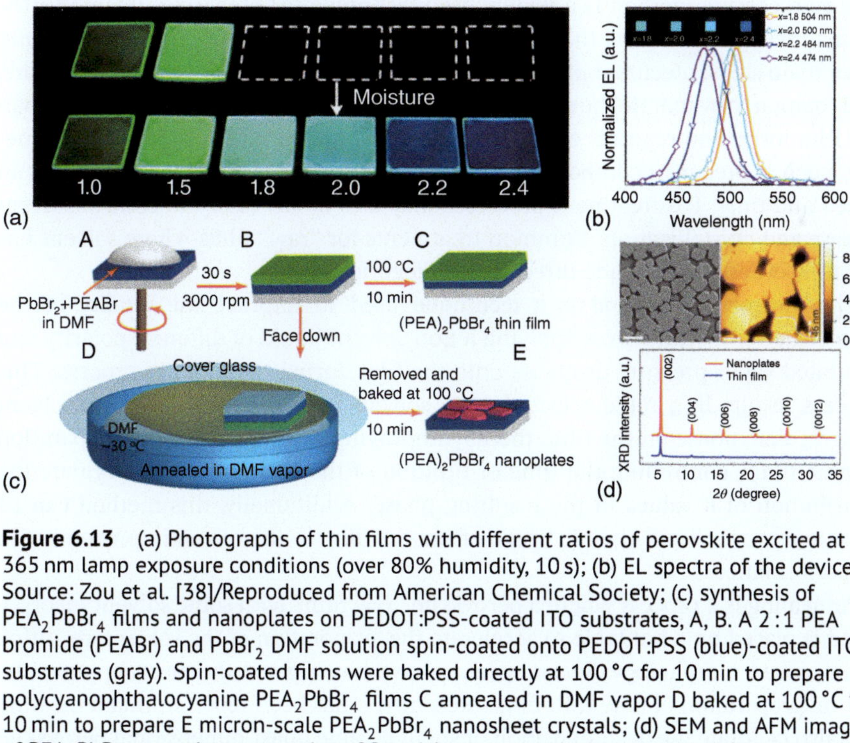

Figure 6.13 (a) Photographs of thin films with different ratios of perovskite excited at 365 nm lamp exposure conditions (over 80% humidity, 10 s); (b) EL spectra of the devices. Source: Zou et al. [38]/Reproduced from American Chemical Society; (c) synthesis of PEA_2PbBr_4 films and nanoplates on PEDOT:PSS-coated ITO substrates, A, B. A 2 : 1 PEA bromide (PEABr) and $PbBr_2$ DMF solution spin-coated onto PEDOT:PSS (blue)-coated ITO substrates (gray). Spin-coated films were baked directly at 100 °C for 10 min to prepare polycyanophthalocyanine PEA_2PbBr_4 films C annealed in DMF vapor D baked at 100 °C for 10 min to prepare E micron-scale PEA_2PbBr_4 nanosheet crystals; (d) SEM and AFM images of PEA_2PbBr_4 nanosheets at scales of 2 and 1 μm, respectively. Inserted white contour curves show the thickness of the nanoplate at 46 nm, with XRD patterns of the polycrystalline film (blue) and nanoplate (red) below. Source: Liang et al. [21]/Reproduced from American Chemical Society.

produce relaxed luminescence. This effect indicates that smaller-sized quantum dot microcrystals with corresponding surface ligand cladding can prevent excitons or carriers with specific energies from being captured by the defect state [37].

In addition to altering the composition of ions within nanocrystals to achieve a blue color, the substitution of Pb^{2+} in the B-position can also be explored. Previous studies have shown that replacing Pb^{2+} with Bi^{3+}, Sn^{2+}, Cd^{2+}, and Zn^{2+} can lead to the blue coloration of crystals. However, such substitutions can significantly alter the structure of perovskite crystals, leading to decreased stability and optoelectronic properties. Recent research in the field of blue perovskite nanocrystals has focused on replacing Pb with Bi, and further studies are necessary to explore the potential of this approach.

6.7 Process for the Preparation of Blue Light-Emitting Layers

6.7.1 Preparation of Three-Dimensional and Quasi-Two-Dimensional Perovskite Films

The spin coating method is a widely used technique for preparing organic LEDs, with a homogenizer being the primary equipment. The process can be divided into three main steps: precursor preparation, coating of the precursor onto the substrate, and rotation to evaporate the solvent and form the film. Achieving optimal spin coating film formation requires extensive experimentation and adjustment of parameters, such as precursor composition and concentration, substrate selection, and spin speed and time. Posttreatment processes may also be necessary to control solvent release and crystallization. Common treatments for "raw" films where solvent has not fully evaporated include three broad categories.

The antisolvent method is a technique used to improve the quality of the spin-coated film. It involves applying a non-solvent liquid of different polarity and saturated vapor pressure dropwise onto the film during the rotation process. This process results in a rapid release of the solvent from the uniformly flat solvent intermediate phase on the film, thereby improving its quality. For the preparation of blue films, controlling the time of titration of the antisolvent can regulate the distribution of n values in the resulting phase. Additionally, this method can be used to remove excess organic ligands from the film, leading to an improvement in its performance.

Annealing is a process where a perovskite "raw film" with some solvent incorporated is placed on a hot table to accelerate the evaporation of the solvent, resulting in improved morphology. Several groups have investigated the annealing process for the preparation of blue thin films. Recently, it has been reported that annealing can also facilitate the decomposition of small n-value quasi-dimensional perovskite phases into larger n-value phases [39].

Solvent atmosphere treatment is a method that involves treating the "raw film" just after spin coating using a solvent atmosphere. During the treatment, the film surface undergoes a small-scale dissolution-recrystallization process, which effectively reduces film defects. Sun Baoquan's group has successfully used this method to form low-dimensional perovskite discoidal crystals in the raw film during the treatment process, resulting in the emission of deep-blue light [38]. Jin Song's group employed cesium bromide, lead bromide, and benzyl bromide as primary raw materials to fabricate blue-light perovskite films. The addition of excess CsBr resulted in the formation of Cs_4PbBr_6 phase on the films. Treatment with water vapor for a specific time period facilitated the decomposition of Cs_4PbBr_6, which then combined with phenylethylamine salt to form discoidal low-dimensional crystals in the films. After optimization of the component content and water vapor treatment time, the films demonstrated more efficient blue light emission.

6.7.2 Preparation of Nano-Microcrystalline Precursors

Ligand-assisted precipitation is one of the earliest and simplest methods used for the preparation of perovskite quantum dots. The method involves first preparing the perovskite precursor in a polar solvent, which is then injected into a nonpolar solvent with a different structure and stirred. The nonpolar solvent acts as a ligand to coat the outside of the intermediate phase structure of the precursor, resulting in the formation of nanocrystals with different structures (blocks, flakes, spheres, layers, etc.) and shapes (spheres, wires, flakes).

The thermal injection method is commonly used for the synthesis of cesium-containing quantum dot nano-microcrystals. Initially, PbX_2 salt is dissolved in a mixture of oleic acid, oleylamine, and octadecene at a temperature range of 140–200 °C. Subsequently, cesium oleate and the organic salt Ph-COX are injected into the mixture for a few seconds, followed by rapid cooling in an ice bath. The chain length and structure of alkylamines and alkyl-carboxylic acids can be modulated to obtain quantum dots with different shapes.

Additional methods such as ultrasonication or microwave treatment during the precursor reaction have been shown to impact the resulting nanocrystals. Due to the wide variety of structures found in nanocrystals, instability within the internal structure can be compensated for by the use of long ligand chains for encapsulation. As a result, this luminescent layer material has unique advantages in the preparation of deep-blue PeLEDs.

6.8 Device Performance Optimization and Interface Engineering

6.8.1 Passivation of Film Defects

Perovskite materials are characterized by an ionic lattice with large bond lengths and weak bond energies, which can have three negative effects [40]. Firstly, the material is less durable and can be eroded by oxygen and moisture in the air, leading to the destruction of its lattice structure. Secondly, at higher temperatures, the ions in the lattice have a greater tendency to deviate from their original positions. Finally, the process of lattice structure formation is often accompanied by the formation of a large number of microscopic defects. These three factors can seriously affect or even cause the loss of photovoltaic properties of the lattice.

Surface defects in perovskite film layers can be classified into three categories based on their size. The first category consists of macroscopic defects at the micron level, such as pinholes, roughness of the film surface, and poor film coverage. The second category comprises mesoscopic defects such as grain boundaries and phase boundaries, which range in size from a few to several tens of nanometers. The third category includes internal lattice defects, which are the smallest defects that cannot be observed by scanning electron microscopy. These internal lattice defects consist of lattice vacancies, intra-lattice gaps, and lattice mismatches.

6.8.2 Selection and Optimization of Hole and Electron Injection Layers

The energy level structure of the transport layer material is a crucial factor that needs to be considered while selecting the material. The HOMO and LUMO levels of the material are the primary consideration to ensure efficient carrier injection during device operation. The energy level structure of commonly used transport layers is depicted in Figure 6.14 [41].

When selecting a transport layer material, the energy level structure is crucial to ensure proper carrier injection during device operation. The HOMO and LUMO values of the material should be considered first. Additionally, carrier mobility is an important factor for device operation under high voltage or high current. The balance of electron and hole injection also affects device efficiency. For instance, Cao Yong's group discovered that using TPD\PVK and PO-T2T as the hole and electron injection layers, respectively, greatly improved device performance compared to the TPD\Perovskite\TPBi structure. The EQE increased from 1.1% to 1.96%.

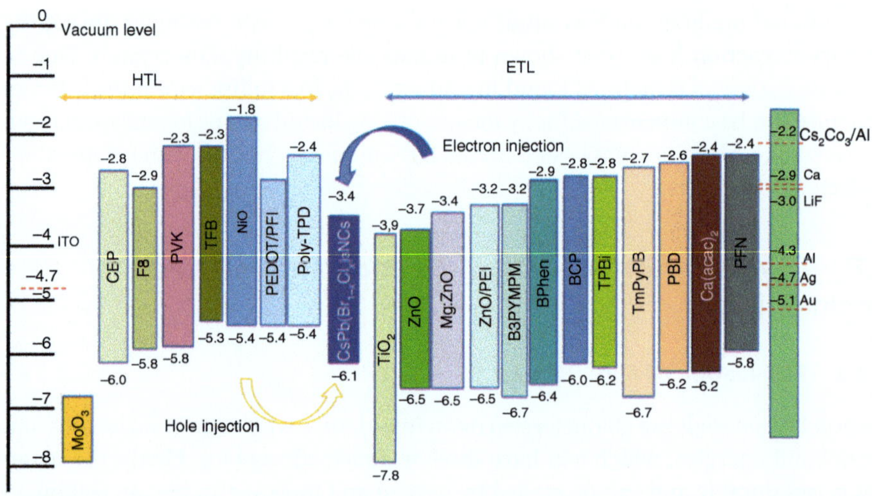

Figure 6.14 Energy level diagram for common transport layer materials. Source: Kumawat et al. [41]/Royal Society of Chemistry.

This improvement was due to TPD's hole mobility of 1×10^{-4} and TPBi's electron mobility of 1×10^{-6}, which differ by two orders of magnitude. In contrast, PO-T2T has an electron mobility of 1.1×10^{-4}, one order of magnitude higher than TPD's. The balanced injection of electrons and holes contributed to the device's efficiency increase. In addition, the hole injection layer is often used as a substrate material for the perovskite light-emitting layer, making the selection of the transport layer critical in determining the crystallization process and film quality of the perovskite film. Conversely, in inverted-structured light-emitting devices, the electron transport material is used as a substrate for the perovskite layer, and careful selection and optimization of the surface structure are necessary [24].

As there are not many reports on perovskite blue light-emitting devices, the choice of transport layer is relatively limited, for the hole injection layer is mainly divided into two types of metal oxides and organic polymers, while the electron injection layer mostly uses small molecules of organic semiconductors.

6.8.3 Interface Engineering

One approach to improving device performance is by enhancing the interlayer structure to prevent exciton quenching. Hongbin Yao's team found that NiO_x, when used as the hole injection layer in blue-light perovskite devices, led to subpar

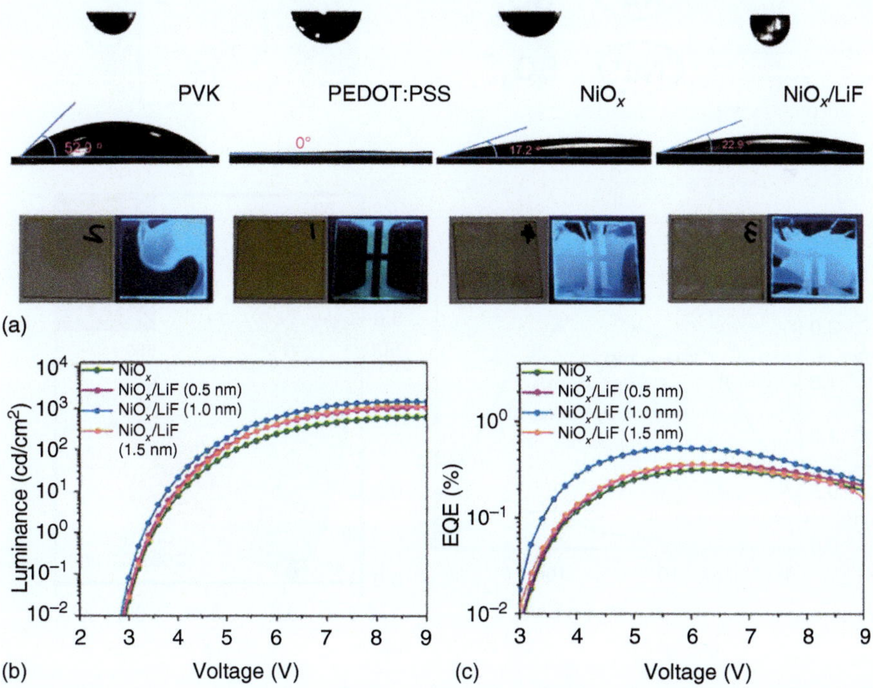

Figure 6.15 (a) Wettability of different substrates to the perovskite precursor solution; and (b and c) device performance plots. Source: Wang et al. [42]/Reproduced from American Chemical Society.

performance. This was primarily due to NiO_x's poor surface morphology, creating a plethora of defects and consequentially trapping energy level states. Such states cause excitons to quench through non-radiative transitions, trapping carriers in the process. To mitigate this problem, the group discovered that a 1 nm layer of LiF, deposited on NiO_x via vapor deposition, considerably reduced quenching effects and enhanced the device's performance [42] (Figure 6.15).

The second aspect is adjusting the energy level of the transport layer. The blue light-emitting layer material has a deep bandgap, which results in a deeper HOMO energy level, increasing the potential barrier for hole injection during device operation. Wallace C.H. Choy's group reported that modifying the NiO_x transport

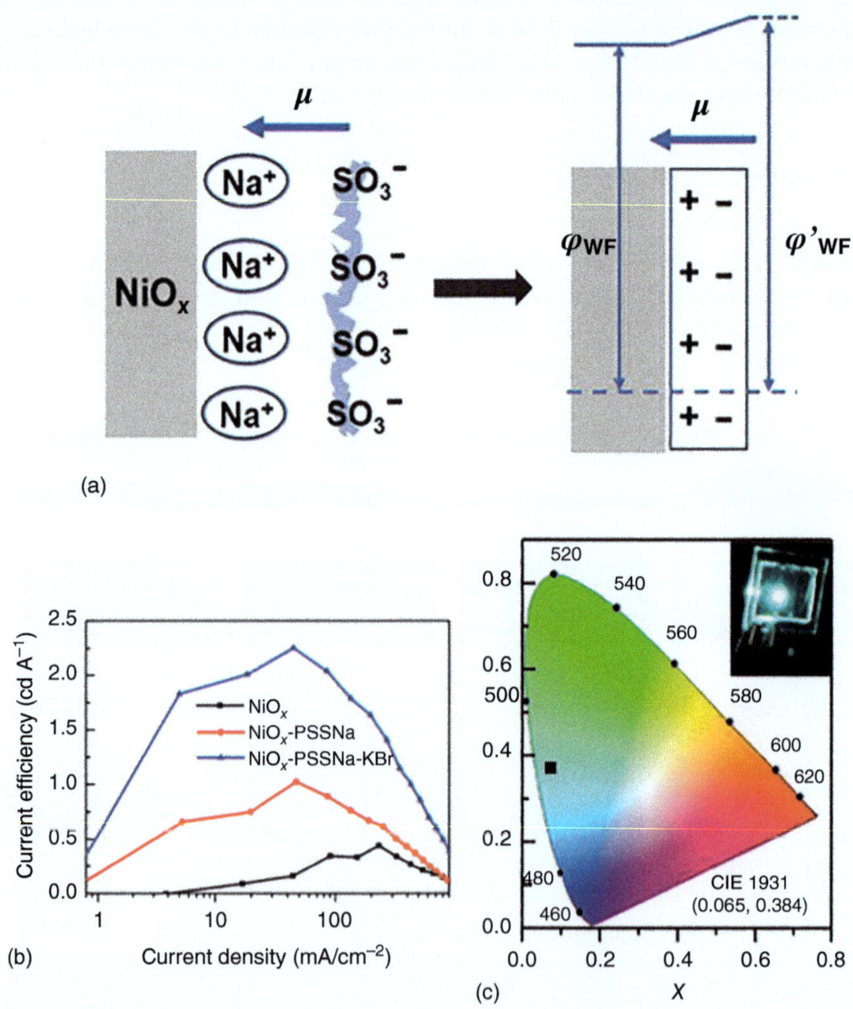

Figure 6.16 (a) PSS-Na forms a dipole moment on NiO_x, and the dipole moment increases the function of the NiO_x-PSS-Na film; (b) device efficiency plot; and (c) CIE plot of the emitted light from the device. Source: Ren et al. [43]/Reproduced from John Wiley & Sons, Inc.

layer with PSS-Na can form a molecular dipole moment on its surface, which can effectively adjust the energy level structure of the transport layer and improve the device's performance [43] (Figure 6.16).

6.9 Optimization of Device Stability

Despite the rapid development of PeLEDs, stability remains a major challenge that hinders their practical applications. This challenge can be attributed to two main factors: materials and devices. On the material side, the crystal structure of the perovskite, environmental factors, and the stability of the perovskite nanoparticles in solution and thin film states can affect stability. On the device side, ion migration, p-i-n junction formation, and the diffusion of metal electrode atoms can cause instability. These challenges are particularly pronounced for blue perovskite diodes due to the difficulty in preparing and constructing materials and devices. Currently, there are three major stability issues that need to be addressed for blue perovskite diodes: improving their lifetime, reducing efficiency roll-off, and maintaining stable emission color over a long period.

Device stability is a crucial factor for the commercialization of perovskite-based devices. One of the main challenges in improving device stability is to prevent the degradation of the perovskite material due to exposure to moisture and oxygen in the air.

One approach to improving device stability is to use encapsulation materials to protect the perovskite layer from environmental factors. For example, researchers have used various encapsulation materials such as metal oxides, polymers, and inorganic materials to protect perovskite layers. However, the encapsulation process can add complexity and cost to device fabrication.

Another approach is to modify the perovskite layer itself to improve its stability. For instance, researchers have developed various strategies to improve the stability of perovskite materials, including doping with metal ions, incorporating small molecules, and using additives to control crystal growth.

Additionally, the choice of charge transport materials can also affect device stability. For example, the use of stable and low-hygroscopic transport materials such as PCBM and C60 can improve device stability. Moreover, modifying the interface between the perovskite layer and the transport layer can also enhance device stability. For instance, introducing self-assembled monolayers or interfacial layers can effectively improve the stability of perovskite-based devices.

Overall, optimizing device stability is critical for the commercialization of perovskite-based devices, and a combination of approaches can be used to achieve this goal.

6.9.1 Lifetime of Perovskite Blue Light-emitting Diodes

Currently, the operational lifetime of PeLEDs, regardless of their emission color, is significantly lower than that of organic LEDs. This represents a major limitation to the commercialization of perovskite-based technology. For blue perovskite LEDs, the stability issue is particularly challenging due to the difficulty in preparing and fabricating the relevant materials and devices. Table 6.2 summarizes the reported operational lifetimes of blue PeLEDs to date.

As shown in Table 6.2, as of 2019, the longest reported lifetime for a sky-blue PeLED was 220 min (at a luminance of $150\,cd\,m^{-2}$). In the blue-light region, the longest lifetime was 10 min (at a drive voltage of 4.4 V), and in the deep-blue region, the longest lifetime was 13.5 min (at a drive current of $0.35\,mA\,cm^{-2}$). The lifetime of blue PeLEDs is significantly shorter than that of green devices, and there is a noticeable drop in device lifetime as the wavelength of the emitted light shifts toward the blue region. It is worth noting that, unlike green perovskite diodes, the long-lived blue devices have a quasi-dimensional perovskite structure. These stability issues are a major barrier to the practical application of perovskite LEDs.

6.9.2 Optimization of Efficiency Stability in Perovskite Light-emitting Diodes

The stability of device efficiency is crucial for the practical application of PLEDs. The efficiency roll-off during device operation is caused by complex factors. Firstly, stability issues related to the light-emitting material can cause a reduction in device efficiency at high currents, and optimizing material stability and efficiency stability often occur simultaneously. Secondly, the change in charge injection at high voltages can also have a significant impact, which may result in unbalanced charge injection and shift the compound region, affecting device performance. Finally, at high carrier injection, the carrier density at the light-emitting layer of the device dramatically increases, leading to the generation of three-carrier combinations, which reduces the device's efficiency.

Efficiency stability is a crucial parameter for the practical application of PeLEDs. Efficiency roll-off, which is the decrease in efficiency at high current densities, is a major challenge for the stability of PeLEDs. To improve the efficiency and stability of PeLEDs, researchers have explored various approaches, including material design, device structure optimization, and encapsulation methods.

Material design strategies include the use of mixed halide perovskites, which can improve the stability of the perovskite structure by reducing the formation of defect states. Another strategy is to incorporate small molecules, such as phenylethylamine, into the perovskite layer, which can help to reduce non-radiative recombination and improve the stability of the devices.

Device structure optimization is another approach to improve the efficiency and stability of PeLEDs. This includes the use of interfacial layers, such as TFB (2,2′,2″-((1,3,5-benzenetriyl)tris(1-phenylethyl))tris(1-phenyl-1*H*-benzimidazole)), which can improve the injection and transport of charge carriers, leading to more stable device performance. Another approach is to use a graded perovskite structure, where the halide composition of the perovskite layer is gradually changed, resulting in improved efficiency and stability.

Table 6.2 Reported operational lifetimes and related parameters of blue PeLEDs.

Time [Ref.]	Luminous layer	Device structure	EL (nm)	EQE (%)	L_0 or drive voltage	T_{50}
2019 [38]	$PEA_2(CsPbBr_3)_nPbBr_4$	ITO/PVK/perovskite/TPBi/LiF/Al	484	0.13	4 V	3.5 min
2019 [24]	$CsPbBr_xCl_{3-x}$NCs	ITO/PEDOT:PSS/TPD/perovskite/MoO_3/Al	476	1.19	—	4.5 min
2019 [19]	$PPDA_2(CsPbBr_3)PbBr_4$	ITO/PVK/PFI/perovskite/3TPYMB/Liq/Al	465	2.6	0.35 mA cm^{-2}	13.5 min
2019 [34]	$CsPbBr_xCl_{3-x}$NCs	ITO/PEDOT:PSS/TPD/perovskite/TPBi/LiF/Al	479	0.85	4 V	50 s
2019 [23]	$PEA_2Cs_{n-1}Pb_n(Cl_xBr_{1-x})_{3n+1}$	ITO/PEDOT:PSS/perovskite/TPBi/LiF/Al	480	5.7	4.4 V	10 min
2018 [44]	$PEA_2IPA_{1.5}Pb_{2.5}Br_{8.5}$	ITO/PEODT:PSS/NiO_x/PVK/perovskite/TPBi/LiF/Al	490	1.5	10 cd m^{-2} 20 cd m^{-2} 210 cd m^{-2}	10 min 4 min 0.5 min
2019 [3]	$PBABr_{1.1}(Cs_{0.7}FA_{0.3}PbBr)_3$	ITO/NiO/TFB/PVK/perovskite/MoO_x/Au	483	9.5	100 cd m^{-2}	100 s
2017 [7]	$MAPbBr_3$:$EAPbBr_3$	ITO/PEDOT:PSS/perovskite/TmPyPB/CsF/Al	473 485	2.6	100 cd m^{-2}	10 min
2017 [45]	$Cs_{10}(MA_{0.17}FA_{0.83})_{100-x}Pb(Br_xI_{100-x})_3$	ITO/ZnO/perovskite/NPD/MoO_3/Al	475	1.7	374.5 cd m^{-2}	4 min
2019 [43]	$PA_2(CsPbBr)_{3n-1}PbBr_4$	ITO/NiO_x-PSSNa/perovskite/TPBi/LiF/Al	492	1.45	150 cd m^{-2} 415 cd m^{-2}	220 min 120 min
2019 [38]	$PEA_2(Rb_{0.6}Cs_{0.4})_{n-1}Pb_nBr_{3n+1}$	ITO/PEDOT:PSS/perovskite/TmPyPB/LiF/Al	475	1.35	4.5 V	14.5 min
2021 [26]	$PEA_2CsPb_2Cl_yBr_{(7-y)}$	ITO/PVK/perovskite/PO-T2T/Liq/Al	485	11.2	2.55 V	17.5 min
2022 [20]	p-F-$PEACsPbBr_3Cl_{0.5}$	ITO/PVK/perovskite/TPBi/Liq/Al	457	3.41	—	6.5 min
2023 [46]	p-F-$PEA_2Cs_{n-1}Pb_n(Br_{0.55}Cl_{0.45})_{3n+1}$	ITO/TFB&PVK/perovskite/TPBi/LiF/Al	468	11.87	3.2 V	23 min
2023 [47]	$PEA_2(Cs_xEA_{1-x}PbBr_3)_2PbBr_4$	ITO/PEDOT:PSS/perovskite/TPBi/LiF/Al	486	16.07	3.0 V	5 min

Encapsulation methods, such as the use of thin-film barrier layers, can also improve the efficiency and stability of PeLEDs by reducing the ingress of oxygen and moisture into the device, which can degrade the perovskite layer and reduce the device's performance.

In summary, the optimization of efficiency stability in PeLEDs is crucial for their practical application. Various approaches, including material design, device structure optimization, and encapsulation methods, have been explored to improve the efficiency and stability of PeLEDs.

6.9.3 Light Color Stability Optimization

Achieving long-term stability in the color of the emitted light is another critical aspect that needs to be addressed for practical applications of perovskite light-emitting devices. The instability of the perovskite material itself, as well as the device structure, can cause changes in the emitted color over time, which can be a significant drawback for commercial use. Therefore, there is a need to optimize the material and device design to maintain color stability over extended periods.

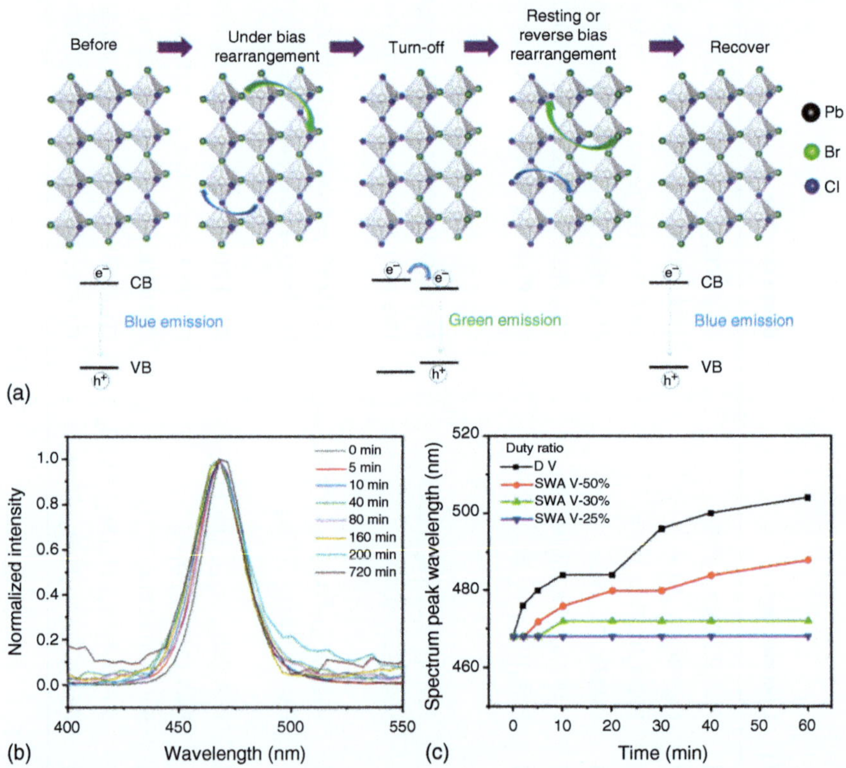

Figure 6.17 (a) Variation within the chalcocite lattice under square wave drive; (b) spectral stability under square wave drive; and (c) effect of different square wave drive parameters on the EL spectrum. Source: Tan et al. [48]/John Wiley & Sons.

Light-emitting devices with unstable light colors are not practical. The color stability of perovskite devices is impacted by high-temperature degradation, ion migration, and lattice chirality of the perovskite materials under electric fields. In blue perovskite devices, redshift is often observed during operation. To address this issue, structural design of the perovskite material is required. The most common approach is to use quasi-2D-structured Br-based perovskite films to prevent anion doping and ion migration during device operation. However, for deep-blue light emission, a blueshift strategy involving doping with chloride ions is used with careful control of the doping amount and material structure design using long-chain-wrapped quantum dot structures. Recently, new blueshift strategies have emerged that involve replacing small cations or ions in the Pb position to produce stable deep-blue light devices.

In 2019, Tang Jiang's group proposed a method to address the luminescence redshift issue at the device level. The researchers prepared quasi-2D-structured perovskite films with a large amount of chloride ion doping as a blueshift strategy, using POEA as a ligand. Although the large amount of doping initially resulted in blue emission, the operation of the device caused severe redshift. However, when a square wave driver was applied, the original blue emission of the device was restored after a period of time with the application of a voltage in the opposite direction [48] (Figure 6.17).

6.10 Conclusion and Outlook

The development of blue PeLEDs has been slower than that of near-infrared, red, and green light-emitting diodes, and further improvements in efficiency and stability are necessary. The current approach to preparing blue perovskite devices involves reducing the dimensionality of lead–bromine perovskite to form a quasi-2D structure through a domain-limiting effect, which requires a significant amount of "large" organic ammonium cations to modulate the luminescence. However, the presence of excessive organic matter can negatively impact charge transport performance, resulting in low brightness for the blue perovskite devices prepared by this method, despite their acceptable efficiency. Moving forward, the focus should shift toward developing 3D Br/Cl hybrid systems with additive modifications to stabilize the luminescence spectrum and reduce defects, enabling the production of highly efficient and bright blue perovskite devices.

In conclusion, perovskite blue light-emitting materials and devices have shown great potential in the field of optoelectronics due to their high color purity, tunability, and low cost. However, the development of efficient and stable blue perovskite devices still faces many challenges, such as efficiency roll-off, color stability, and material stability. Researchers have made significant progress in addressing these issues, including the use of quasi-dimensional structures, blueshift strategies, and square wave drivers.

Despite these advancements, there is still a long way to go before perovskite blue light-emitting devices can be widely used in industry. Future research efforts should focus on exploring new materials and device architectures to achieve high-brightness, efficient, and stable blue emission. By addressing these challenges, perovskite blue light-emitting devices have the potential to revolutionize the field of optoelectronics and enable the development of new technologies in areas such as solid-state lighting, displays, and sensing.

However, it is worth noting that, despite the challenges, significant progress has been made in the development of perovskite blue light-emitting materials and devices in recent years. Researchers are continuously exploring new strategies for improving efficiency, stability, and color stability. The development of new blueshift strategies, such as the use of large cations or hybrid Br/Cl systems, also shows promise for achieving efficient, high-brightness blue perovskite devices. As the field continues to evolve and more research is conducted, it is expected that perovskite blue light-emitting materials and devices will become increasingly competitive with other types of light-emitting technologies, ultimately leading to their widespread industrial application.

Therefore, there is still a need for further research and development in the area of perovskite blue light-emitting materials and devices. Efforts are being made to address the challenges of improving the efficiency, stability, and color stability of perovskite blue light-emitting diodes. Some promising strategies include the use of new materials, such as 3D Br/Cl hybrid systems with additive modifications, as well as innovative device design and engineering techniques. With continued research and development, perovskite blue light-emitting materials and devices have the potential to become a competitive alternative to traditional OLED technology in a wide range of applications, from consumer electronics to lighting and displays.

References

1 Zhao, Q.Q., Chen, F.T., Li, C.Q. et al. (2023). Challenges and developments for the blue perovskite nanocrystal light-emitting diodes. *Dalton Transactions* 52: 3921–3941.

2 Ren, Z.W., Wang, K., Sun, X.W. et al. (2021). Strategies toward efficient blue perovskite light-emitting diodes. *Advanced Functional Materials* 31: 2100516.

3 Liu, Y., Cui, J.Y., Du, K. et al. (2019). Efficient blue light-emitting diodes based on quantum-confined bromide perovskite nanostructures. *Nature Photonics* 13: 760−+.

4 Ge, C.Y., Fang, Q.R., Lin, H.R. et al. (2021). Review on blue perovskite light-emitting diodes: recent advances and future prospects. *Frontiers in Materials* 8: 7.

5 Song, J.Z., Li, J.H., Li, X.M. et al. (2015). Quantum dot light-emitting diodes based on inorganic perovskite cesium lead halides ($CsPbX_3$). *Advanced Materials* 27: 7162−+.

6 Pan, J., Quan, L.N., Zhao, Y.B. et al. (2016). Highly efficient perovskite-quantum-dot light-emitting diodes by surface engineering. *Advanced Materials* 28: 8718–8725.

7 Wang, Q., Ren, J., Peng, X.F. et al. (2017). Efficient sky-blue perovskite light-emitting devices based on ethylammonium bromide induced layered perovskites. *ACS Applied Materials & Interfaces* 9: 29901–29906.

8 Vashishtha, P., Ng, M., Shivarudraiah, S.B. et al. (2019). High efficiency blue and green light-emitting diodes using Ruddlesden-Popper inorganic mixed halide perovskites with butylammonium interlayers. *Chemistry of Materials* 31: 83–89.

9 Wang, Q., Wang, X.M., Yang, Z. et al. (2019). Efficient sky-blue perovskite light-emitting diodes via photoluminescence enhancement. *Nature Communications* 10: 8.

10 Dong, Y.T., Wang, Y.K., Yuan, F.L. et al. (2020). Bipolar-shell resurfacing for blue LEDs based on strongly confined perovskite quantum dots. *Nature Nanotechnology* 15: 668−+.

11 Liu, Y.L., Zhang, L.Z., Chen, S. et al. (2021). Water-soluble conjugated polyelectrolyte hole transporting layer for efficient sky-blue perovskite light-emitting diodes. *Small* 17: 8.

12 Shen, Y., Li, Y.Q., Zhang, K. et al. (2022). Multifunctional crystal regulation enables efficient and stable sky-blue perovskite light-emitting diodes. *Advanced Functional Materials* 32: 2206574.

13 Zhou, W., Shen, Y., Cao, L.X. et al. (2023). Manipulating ionic behavior with bifunctional additives for efficient sky-blue perovskite light-emitting diodes. *Advanced Functional Materials* 33: 2301425.

14 Yu, Y., Tang, Y., Wang, B. et al. (2023). Red perovskite light-emitting diodes: recent advances and perspectives. *Laser & Photonics Reviews* 17: 2200608.

15 McCall, K.M., Morad, V., Benin, B.M. et al. (2020). Efficient lone-pair-driven luminescence: structure-property relationships in emissive $5s^2$ metal halides. *ACS Materials Letters* 2: 1218–1232.

16 Sun, S., Lu, M., Gao, X. et al. (2021). 0D perovskites: unique properties, synthesis, and their applications. *Advanced Science* 8: 2102689.

17 Sadhanala, A., Ahmad, S., Zhao, B.D. et al. (2015). Blue-green color tunable solution processable organolead chloride-bromide mixed halide perovskites for optoelectronic applications. *Nano Letters* 15: 6095–6101.

18 Chen, Z.M., Zhang, C.Y., Jiang, X.F. et al. (2017). High-performance color-tunable perovskite light emitting devices through structural modulation from bulk to layered film. *Advanced Materials* 29: 8.

19 Yuan, S., Wang, Z.K., Xiao, L.X. et al. (2019). Optimization of low-dimensional components of quasi-2D perovskite films for deep-blue light-emitting diodes. *Advanced Materials* 31: 9.

20 Zhou, Y.H., Wang, C.Y., Yuan, S. et al. (2022). Stabilized low-dimensional species for deep-blue perovskite light-emitting diodes with EQE approaching 3.4%. *Journal of the American Chemical Society* 144: 18470–18478.

21 Liang, D., Peng, Y.L., Fu, Y.P. et al. (2016). Color-pure violet-light-emitting diodes based on layered lead halide perovskite nanoplates. *ACS Nano* 10: 6897–6904.

22 Chen, P., Meng, Y., Ahmadi, M. et al. (2018). Charge-transfer versus energy-transfer in quasi-2D perovskite light-emitting diodes. *Nano Energy* 50: 615–622.

23 Li, Z.C., Chen, Z.M., Yang, Y.C. et al. (2019). Modulation of recombination zone position for quasi-two-dimensional blue perovskite light-emitting diodes with efficiency exceeding 5%. *Nature Communications* 10: 10.

24 Yang, F., Chen, H.T., Zhang, R. et al. (2020). Efficient and spectrally stable blue perovskite light-emitting diodes based on potassium passivated nanocrystals. *Advanced Functional Materials* 30: 1908760.

25 Kumar, S., Jagielski, J., Yakunin, S. et al. (2016). Efficient blue electroluminescence using quantum-confined two-dimensional perovskites. *ACS Nano* 10: 9720–9729.

26 Yuan, S., Cui, L.S., Dai, L.J. et al. (2021). Efficient and spectrally stable blue perovskite light-emitting diodes employing a cationic π-conjugated polymer. *Advanced Materials* 33: 2103640.

27 Gong, X.W., Voznyy, O., Jain, A. et al. (2018). Electron-phonon interaction in efficient perovskite blue emitters. *Nature Materials* 17: 550–+.

28 Chen, H., Lin, J., Kang, J. et al. (2020). Structural and spectral dynamics of single-crystalline Ruddlesden-Popper phase halide perovskite blue light-emitting diodes. *Science Advances* 6: 8.

29 Jiang, Y.Z., Qin, C.C., Cui, M.H. et al. (2019). Spectra stable blue perovskite light-emitting diodes. *Nature Communications* 10: 9.

30 Jin, Y., Wang, Z.K., Yuan, S. et al. (2020). Synergistic effect of dual ligands on stable blue quasi-2D perovskite light-emitting diodes. *Advanced Functional Materials* 30: 1908339.

31 He, H.Y., Mei, S.L., Wen, Z.Q. et al. (2022). Recent advances in blue perovskite quantum dots for light-emitting diodes. *Small* 18: 29.

32 Hu, Y.Y., Cao, S.J., Qiu, P. et al. (2022). All-inorganic perovskite quantum dot-based blue light-emitting diodes: recent advances and strategies. *Nanomaterials* 12: 18.

33 Ochsenbein, S.T., Krieg, F., Shynkarenko, Y. et al. (2019). Engineering color-stable blue light-emitting diodes with lead halide perovskite nanocrystals. *ACS Applied Materials & Interfaces* 11: 21655–21660.

34 Shin, Y.S., Yoon, Y.J., Lee, K.T. et al. (2019). Vivid and fully saturated blue light-emitting diodes based on ligand-modified halide perovskite nanocrystals. *ACS Applied Materials & Interfaces* 11: 23401–23409.

35 Hou, S.C., Gangishetty, M.K., Quan, Q.M. et al. (2018). Efficient blue and white perovskite light-emitting diodes via manganese doping. *Joule* 2: 2421–2433.

36 Sun, C., Gao, Z.Y., Liu, H.X. et al. (2019). One stone, two birds: high-efficiency blue-emitting perovskite nanocrystals for LED and security ink applications. *Chemistry of Materials* 31: 5116–5123.

37 Li, J., Gan, L., Fang, Z.S. et al. (2017). Bright tail states in blue-emitting ultrasmall perovskite quantum dots. *Journal of Physical Chemistry Letters* 8: 6002–6008.

38 Zou, Y.T., Xu, H., Li, S.Y. et al. (2019). Spectral-stable blue emission from moisture-treated low-dimensional lead bromide-based perovskite films. *ACS Photonics* 6: 1728–1735.

39 Cheng, L., Jiang, T., Coo, Y. et al. (2020). Multiple-quantum-well perovskites for high-performance light-emitting diodes. *Advanced Materials* 32: 9.

40 Liu, Z., Qiu, W.D., Peng, X.M. et al. (2021). Perovskite light-emitting diodes with EQE exceeding 28% through a synergetic dual-additive strategy for defect passivation and nanostructure regulation. *Advanced Materials* 33: 2103268.

41 Kumawat, N.K., Liu, X.K., Kabra, D. et al. (2019). Blue perovskite light-emitting diodes: progress, challenges and future directions. *Nanoscale* 11: 2109–2120.

42 Wang, K.H., Peng, Y.D., Ge, J. et al. (2019). Efficient and color-tunable quasi-2D $CsPbBr_xCl_{3-x}$ perovskite blue light-emitting diodes. *ACS Photonics* 6: 667–676.

43 Ren, Z.W., Xiao, X.T., Ma, R.M. et al. (2019). Hole transport bilayer structure for quasi-2D perovskite based blue light-emitting diodes with high brightness and good spectral stability. *Advanced Functional Materials* 29: 9.

44 Xing, J., Zhao, Y.B., Askerka, M. et al. (2018). Color-stable highly luminescent sky-blue perovskite light-emitting diodes. *Nature Communications* 9: 8.

45 Kim, H.P., Kim, J., Kim, B.S. et al. (2017). High-efficiency, blue, green, and near-infrared light-emitting diodes based on triple cation perovskite. *Advanced Optical Materials* 5: 1600920.

46 Liu, Y., Wang, S., Yu, Z. et al. (2023). A multi-functional additive strategy enables efficient pure-blue perovskite light-emitting diodes. *Advanced Materials* e2302161.

47 Chu, Z., Zhang, W., Jiang, J. et al. (2023). Blue light-emitting diodes based on quasi-two-dimensional perovskite with efficient charge injection and optimized phase distribution via an alkali metal salt. *Nature Electronics* 6: 360–369.

48 Tan, Z.F., Luo, J.J., Yang, L.B. et al. (2020). Spectrally stable ultra-pure blue perovskite light-emitting diodes boosted by square-wave alternating voltage. *Advanced Optical Materials* 8: 8.

7

Effect of Metal Ion Doping on Perovskite Light-Emitting Materials

Currently, the development of perovskite light-emitting diodes is still facing certain challenges. One of the most significant challenges is improving the device efficiency, particularly the luminous efficiency of blue perovskite light-emitting diodes [1]. Secondly, there is an urgent need to address the issue of crystal phase transition and poor stability to humidity, high temperature, and light during device operation [2–4]. Impurity doping has been a well-established method for modifying the properties of substrates since the 1940s. This technique is commonly used in microelectronics and optoelectronics to alter the electrical and optical characteristics of materials [5–7]. The unique ionic crystal structure of halogenated perovskite materials provides more options for doping with metal ions compared to conventional semiconductor materials [8]. Scientists have investigated the substitution of various A- or B-site elements, including but not limited to K^+, Rb^+, Ag^+, Sr^{2+}, Mn^{2+}, Bi^{3+}, and rare earth metal ions, in halogenated perovskite materials due to their special ionic crystal structure [9–15]. Recent studies have demonstrated that doping can improve the stability of perovskite materials and enhance the efficiency and lifetime of devices. Figure 7.1 illustrates the progress made in the field of perovskite light-emitting diodes through metal ion doping and surface passivation modifications. For near-infrared perovskite light-emitting devices, Bi-doping of $MAPbI_3$ perovskite materials was initially reported but showed unsatisfactory performance. Rare earth metal elements such as Yb and Er have since been used for doping, resulting in higher efficiency. In the field of red perovskite light-emitting devices, Mn-doping was the primary method initially, while Sr, Cu, K, and other elements showed excellent performance when used for doping. In green light perovskite light-emitting devices, A-site and B-site doping have been reported, where Li, La, K, Zn, Cd, Sn, and Ce plasma are used as doping elements. Metal ion doping had significantly improved the efficiency of green-light perovskite. However, the study of blue-light-doped perovskite is relatively lagging behind, and although studies on A-site and B-site doping with Cs, Mn, and K have been conducted, the efficiency is still very low.

In this chapter, we provide a comprehensive summary of the effects of metal ion doping on perovskite light-emitting diodes, with a focus on the impacts of doping on crystal structure, exciton dynamics, wavelength tuning, and device performance. We also discuss recent developments in metal ion-doped perovskite light-emitting

Perovskite Light Emitting Diodes: Materials and Devices, First Edition. Hong Meng.

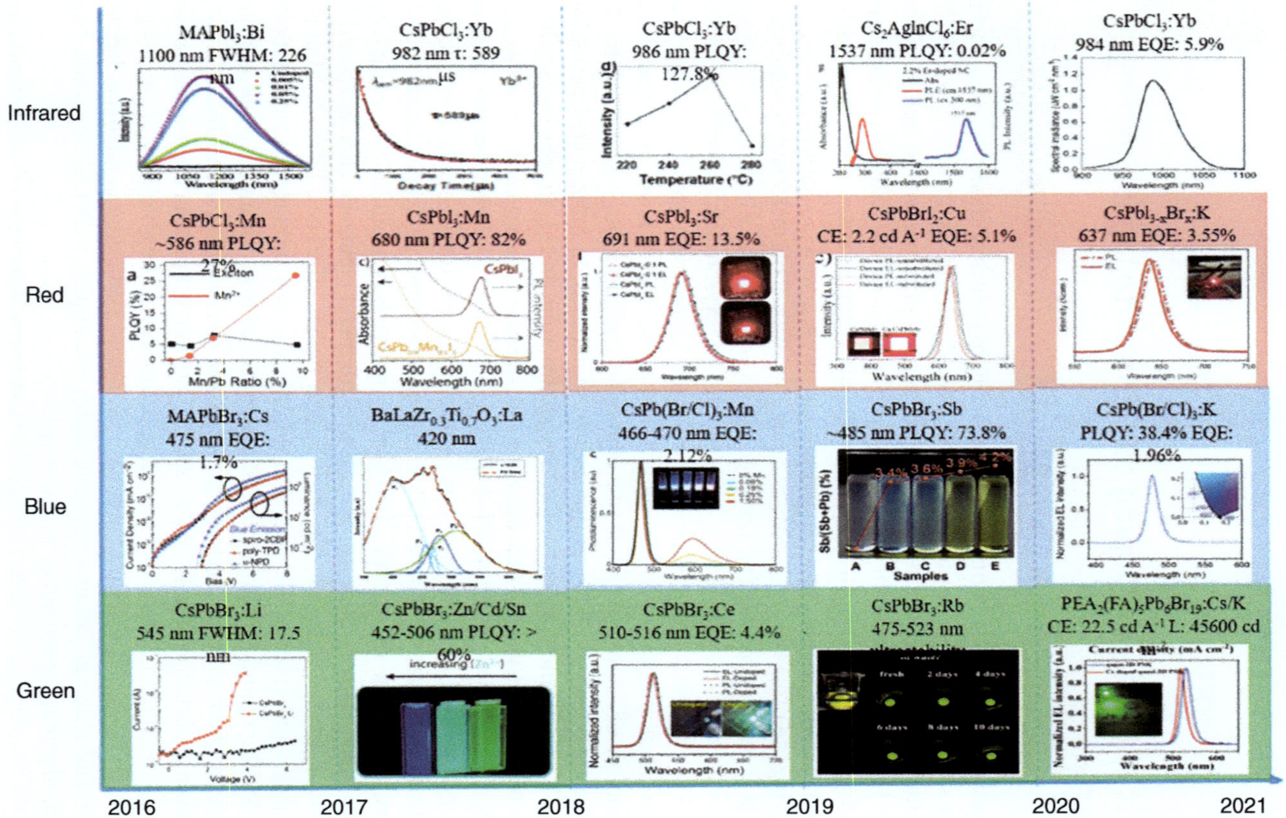

Figure 7.1 Recent research progress on metal ion doping in perovskite luminescent materials.

diodes across various colors, including green, blue, red, and infrared. Furthermore, we review the commonly used metal ion doping methods and discuss the existing challenges associated with metal ion-doped perovskite light-emitting diodes.

7.1 Metal Ion Doping Effect

Metal ion doping has proven to be an effective method to modify the fundamental properties of perovskite materials. The incorporation of metal ions into the perovskite lattice can alter its electronic and optical properties, which in turn can affect the performance of perovskite-based devices. By introducing dopants into the perovskite structure, researchers have been able to improve the stability, efficiency, and color purity of perovskite light-emitting diodes [16, 17]. Doping perovskite materials with suitable metal ions is a promising approach to stabilize the crystal structure, adjusting luminescence properties, and enhancing device performance [18, 19]. In this chapter, we will first introduce the differences in the energy band structures of perovskites in different crystal phases. In perovskite materials, maintaining a high degree of symmetry in the crystal structure requires that the ionic radii of A, B, and X, as well as the Goldschmidt tolerance factor t, satisfy certain equations.

$$\sqrt{2}(R_B + R_X)\,t = R_A + R_X \tag{7.1}$$

where R_A, R_B, and R_X represent the ionic radii of A, B, and X, respectively. The crystal structure can be maintained within a certain temperature range when the Goldschmidt tolerance factor t satisfies the equation, typically ranging between 0.8 and 1.1, in order to maintain a high degree of symmetry in the crystal structure [20]. Additionally, a range of 0.9 to 1 for t is particularly suitable for the formation of cubic structures (α-phase) (Figure 7.2a) [22]. Usually, smaller values of t result in the formation of tetragonal (β-phase) or orthorhombic (γ-phase) crystalline systems with lower symmetry (Figure 7.2b,c). However, larger values of t ($t > 1$) can cause the transformation of the three-dimensional perovskite structure into a two-dimensional structure, such as in the case of layered perovskites [23]. In contrast to the α-phase, the β- and γ-phases exhibit slight deformation, and the changes in properties such as bandgap are not significant. The crystal phases of perovskites, including the chalcocite phase (α, β, and γ phases), can undergo phase transitions at certain temperatures by deforming the bond angles. However, the B-X bond breakage limits the extent of such transformations [24]. The δ-phase is generally not favorable for improving the photoelectric properties as it cannot be transformed into the α-phase by distorting the B-X-B bond, as shown in Figure 7.2d [21].

For $MAPbI_3$, the transition temperatures from α-phase to β-phase and from β-phase to γ-phase are 330 K and 160 K, respectively [25]. Perovskite materials are direct bandgap semiconductors. Taking $CH_3NH_3PbI_3$ as an example, Figure 7.2e–h shows the theoretically calculated energy band structures of different perovskite crystal phases. The bandgaps of α-phase, β-phase, and γ-phase are 1.54 eV, 1.57 eV, and 1.46 eV, respectively. The electronic configurations of β-phase and γ-phase are relatively similar to that of α-phase, with only small distortions at the Pb—I—Pb

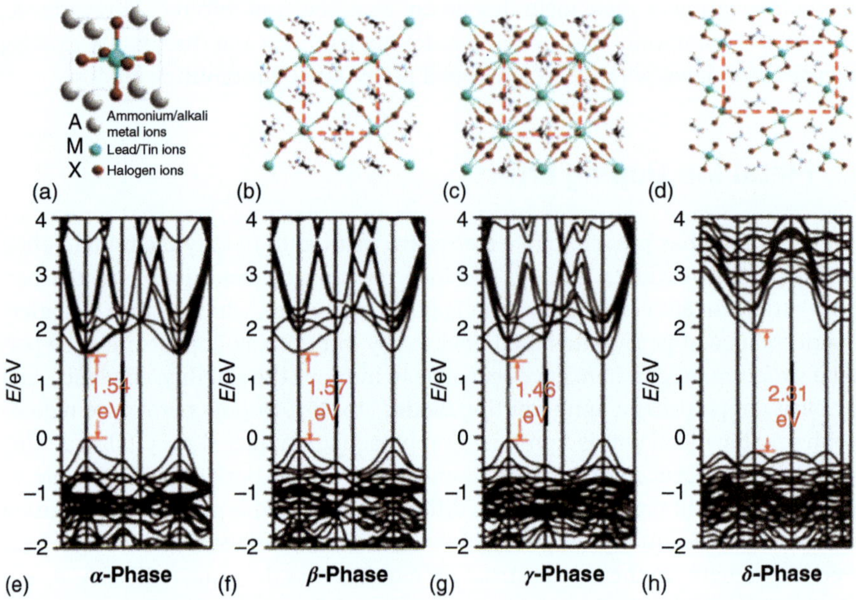

Figure 7.2 Different perovskite crystal structures and the energy band structures of $CH_3NH_3PbI_3$ (a, e) α-phase, (b, f) β-phase, (c, g) γ-phase, (d, h) δ-phase. Source: Adapted from Yin et al. [21].

bond angles. The δ-phase perovskites have a disrupted three-dimensional [7] skeleton structure, resulting in a bandgap increase of about 0.5 eV and an increase in the effective mass of holes [26]. This chapter will focus on the effects of A- and B-site metal ion doping in halogenated perovskite materials, specifically on their modulation of crystal structure, energy levels, and optoelectronic properties.

7.1.1 Effect of A-site Metal Ion Doping on Perovskite Materials

Equation (7.1) shows that a large radius for A is necessary to maintain t in the proper range; otherwise, the crystal structure will be distorted or the symmetry will be reduced [27]. Commonly used A-site cations include MA^+ (1.80 Å), FA^+ (1.90–2.20 Å), and Cs^+ (1.67 Å) (Figure 7.3a). Due to their appropriate sizes, $MAPbX_3$ and $FAPbX_3$ are ideal materials for obtaining cubic perovskite structures (Figure 7.3b) [30]. However, due to the organic nature of FA and MA components, composite perovskite materials face the issue of thermal instability [31]. Despite Cs^+ having a smaller ionic radius compared to FA^+ and MA^+, all-inorganic perovskite materials based on Cs are still a popular research topic. For A-site doping, introducing metal ions in a monovalent state and with larger sizes is preferred, as this helps stabilize the perovskite crystal structure [32, 33]. Alkali metal ions and silver ions are the main dopants used at the A-site in perovskite materials [34]. In the following sections, we will present several representative examples of A-site doping with these ions to illustrate the effects on crystal structure, bandgap, carrier dynamics, and luminescence properties [11, 29, 35].

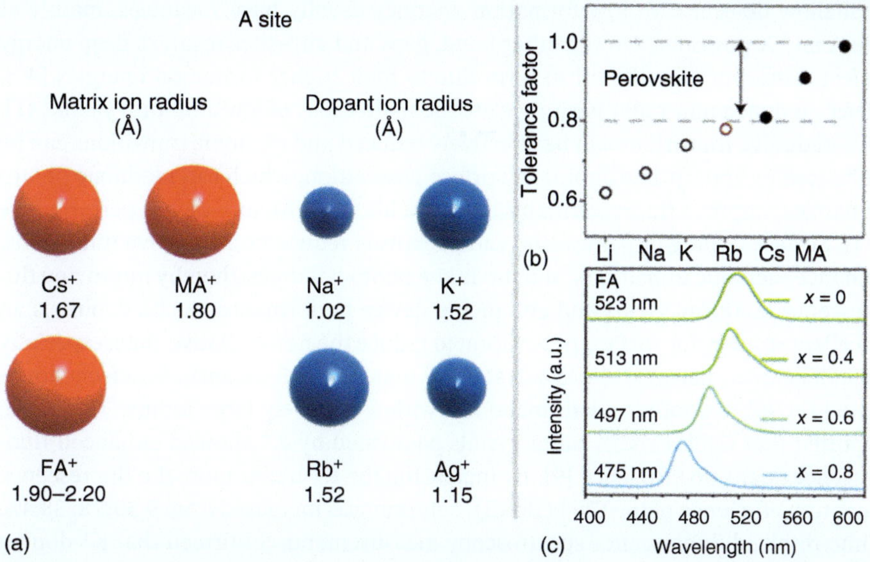

Figure 7.3 (a) Schematic diagram of matrix ions and doped metal ion species and sizes at the A-site; (b) tolerance factors of $APbI_3$-type perovskites. a = Li, Na, K, Rb, Cs, MA, and FA. Source: Adapted from Saliba et al. [28]; (c) PL emission spectra of $CsPbBr_3$:xRb ($x = 0$, 0.4, 0.6, 0.8) with different Rb doping contents in the matrix. Source: Zhangu et al. [29]/with permission of Elsevier.

The size of Rb^+ is smaller than that of Cs^+, which makes the crystal structure of perovskite materials using Rb^+ as the A-site cation theoretically less stable. In fact, Rb-based perovskite materials cannot be transformed from the δ phase to the α cubic phase through annealing like Cs-based perovskite materials [36]. Nevertheless, Saliba et al. experimentally demonstrated the effectiveness of Rb^+ as a dopant. They incorporated Rb^+ into the perovskite phase to form various A-site cationic compositions (RbCsMAFA) [28]. After adding Rb^+ and Cs^+ to the precursor solution, the films were able to crystallize in the photoactive perovskite phase. Compared to the undoped films, the MAFA-based films doped with Rb^+ showed improved phase stability, as evidenced by their crystallization in a homogeneous perovskite phase rather than an inhomogeneous non-calixarene phase. Furthermore, the introduction of alkali metal ions did not affect the valence band position [37], and the band energy orientation of both films was the same with respect to the Fermi energy level [38]. Zhang et al. conducted a study on the modulation of luminescence properties through Rb^+ doping. They synthesized $CsPbBr_3$:xRb ($x = 0$, 0.4, 0.6, 0.8) quantum dots and achieved tunable emission spectra ranging from 475 to 523 nm [29]. As the ratio of Rb/Cs increases, the lattice shrinks and the bandgap width increases, resulting in a significant blueshift of the fluorescence luminescence peak of $CsPbBr_3$:xRb ($x = 0$, 0.4, 0.6, 0.8) quantum dots (from 513 to 475 nm) (Figure 7.3c). It should be noted that while metal ion doping at the A-site can modulate the crystal structure and adjust the luminescence wavelength, relatively low doping concentrations typically do not alter the initial perovskite material's phase structure [39].

Shallow defects with low formation energies usually have vacancies, mainly at the A and X positions. On the other hand, gaps and anti-sites result in deep energy defect states but are difficult to form due to their higher formation energies [40]. Defect-induced non-radiative jumps are the main cause of low quantum yields [41]. Non-radiative transitions can be effectively reduced and excitonic transitions can be enhanced by appropriate doping or surface passivation, which can modulate photodynamics, improve fluorescence quantum yields, and enhance device performance [11]. Surface doping or passivation can effectively reduce non-radiative transitions, enhance excitonic transitions, and modulate photodynamics, thereby improving fluorescence quantum yields and enhancing device performance. A-site doping is an excellent strategy for surface passivation to reduce the non-radiative state, especially in halogenated perovskite nanocrystalline materials where surface defects dominate [42]. K^+ is another alkali metal ion with a relatively large radius. Yang et al. reported that $CsPb(Br/Cl)_3$ nanocrystals passivated by K^+ showed enhanced fluorescence quantum efficiency [9]. By increasing the K^+ molar ratio, the fluorescence quantum efficiency of the $CsPb(Br/Cl)_3$ solution was increased from 9.50% to 38.4%. Time-resolved fluorescence spectroscopy measurements confirmed that K^+ doping greatly reduced the non-radiative recombination rate, indicating that A-site metal ion doping can effectively improve the photoluminescence properties of the material and modulate carrier dynamics.

The radius of the A-site ion plays a crucial role in the formation and stability of the perovskite structure. If the dopant ion's radius is too small, the lattice will shrink, and vice versa, it will expand. Therefore, the size of the A-site dopant ion must be appropriate to ensure material stability. The bandgap and brightness are also directly related to the size of the dopant cation. Larger impurity ions will decrease the bandgap, while smaller ones, such as Rb^+ or K^+, will increase it. A wider bandgap will cause the fluorescence spectrum to show a redshift, while the opposite will show a blueshift. Furthermore, A-site doping with K^+ can be used to passivate surface defects and reduce defect trapping, thereby improving the fluorescence quantum efficiency and exciton lifetime of the perovskite material.

7.1.2 Effect of B-site Metal Ion Doping on Perovskite Materials

Unlike A-site doping, B-site doping is more complex because the B-site is located at the center of the $[BX]_6{}^{4-}$ octahedron, which has a greater impact on the phase stability and bandgap of the perovskite material [43]. The perovskite B-site can be doped with many ions, including main group metal ions (Sr^{2+}, Bi^{3+}, Ba^{2+}, Sb^{3+}, etc.) [12, 14, 44, 45], transition metal ions (Zn^{2+}, Mn^{2+}, Ni^{2+}, Cd^{2+}) [13, 46–48], and rare earth metal ions (Yb^{3+}, Ce^{3+}, Eu^{3+}, Er^{3+}) (Figure 7.4a) [15, 51–53]. For B-site doping, the octahedral factor μ is another semi-empirical geometric parameter in addition to the tolerance factor t. μ is calculated as follows:

$$\mu = R_B/R_X \tag{7.2}$$

For the stable $[BX]_6{}^{4-}$ octahedron, the value of μ should be in the range of 0.442–0.895 [54]. Pb^{2+} is an ideal choice for the B-position because it satisfies

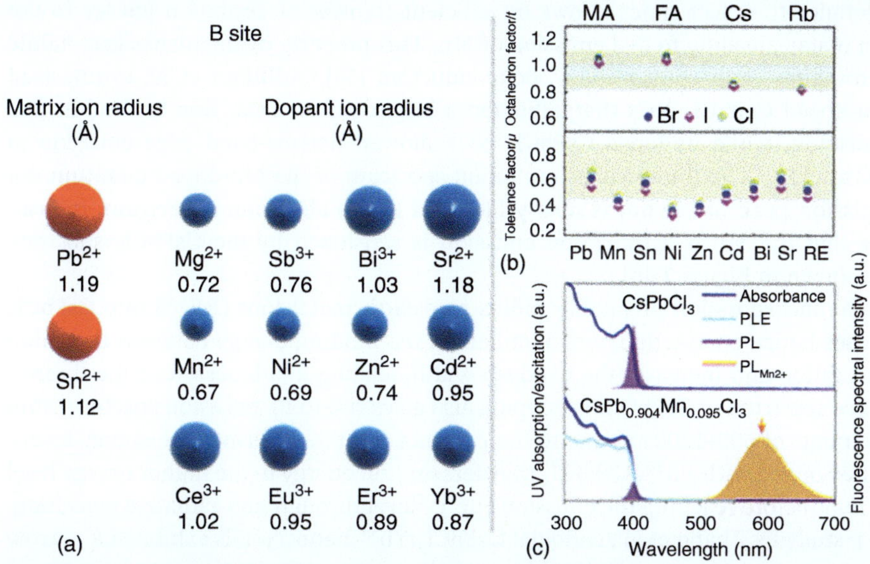

Figure 7.4 (a) Schematic diagram of matrix ion and doped metal ion species and sizes in the B-site; (b) tolerance factors of three-dimensional lead halide perovskite to A-site cations, X-site anions, and $[BX_6]^{4-}$ octahedral factors of B-site cations. Source: Xu et al. [49]/with permission of Elsevier; (c) optical properties of undoped and Mn-doped $CsPbCl_3$ nanocrystals [50].

not only the tolerance factor t but also the octahedral factor μ (Figure 7.4b) [43]. Although Pb-halide perovskite materials have been widely researched, their toxicity and potential leakage have prompted the development of Pb-free alternatives. However, the research on Pb-free perovskite materials is still in its early stages, and their efficiency and film formation ability are generally lower compared to their Pb-halide counterparts [55]. However, research on lead-free perovskite materials is still not as advanced as that on lead-based perovskites due to their low efficiency and poor film-forming ability. Doping small amounts of metal ions into the B-site of lead halide perovskites can significantly improve their morphology and optoelectronic properties without altering their basic properties [49]. This is because the B-site of lead halide perovskite does not affect the basic properties of the main perovskite but can greatly enhance its morphology and photovoltaic properties.

Sr^{2+} is an alkaline earth metal ion with a similar size and valence state as Pb^{2+}, making it an ideal dopant for lead halide perovskites. Doping with Sr^{2+} shortens the Pb—X bond and contracts the lattice due to Sr^{2+}'s slightly smaller radius than Pb^{2+}. Similar to A-site doping, B-site doping with smaller ions also results in a wider bandgap and a blueshift in the PL spectrum. Lu et al. observed a slight blueshift (about 3 nm) in both absorption and PL peaks with an increasing $SrCl_2$:PbI_2 ratio [12]. In addition, the increased formation energy of perovskite due to Sr^{2+} doping slightly improves its environmental tolerance, thus enhancing its stability. Doping with Mn^{2+} is a well-established technique for imparting novel optical and electrical properties to conventional semiconductors. In lead halide perovskites, the high tolerance to defect states allows for efficient transfer of excitation energy to the Mn d-state, leading to d–d emission of Mn. This property distinguishes lead halide perovskites from conventional semiconductors [56]. Adhikari et al. synthesized Mn-doped $CsPbCl_3$ NCs that exhibited a bright yellow color due to the Mn d–d transition, while undoped $CsPbCl_3$ NCs showed narrow-band edge emission at 402 nm [50]. The fluorescence excitation spectrum of the Mn-doped quantum dot emission peak at 586 nm is nearly identical to the absorption spectrum, indicating that the Mn d–d transition emission is sensitized by the $CsPbCl_3$ substrate (as shown in Figure 7.4c).

The incorporation of smaller radius rare earth metal ions (RE^{n+}) into $CsPbCl_3$ causes lattice contraction, which enhances the binding energy between the anion and cation and increases the bandgap width, leading to a blueshift in the fluorescence spectrum of $CsPbCl_3$. RE doping also gives rise to PL emission spectra within the range of 600–1200 nm due to the presence of more intermediate energy levels. When doped with Eu^{3+}, $CsPbCl_3$ transfers exciton energy to the higher energy level of Eu^{3+} before reaching the emission energy level through non-radiative relaxation. In a study by Zhang et al., colloidal $CsPbCl_3$:Yb^{3+} nanocrystals exhibited a narrow luminescence spectrum (full width at half maxima [FWHM] = 54 nm) with an emission peak at 984 nm. It is worth noting that the $CsPbCl_3$ host exciton leap may contribute to the ultrahigh fluorescence quantum efficiency of 127.8% observed in Yb^{3+}-doped near-infrared emission. This is likely due to a quantum splitting effect [15].

The effect of B-site ion doping on the lattice is similar to that of A-site doping. If the radius of the doping ion is smaller than Pb^{2+}, it causes lattice shrinkage. The $[BX]_6^{4-}$ octahedral factor has a significant effect on the halide perovskite, and the B-X-B angle plays a crucial role in the modulation of the valence and conduction bands. Therefore, B-site dopants can significantly modulate the bandgap. Materials doped with divalent metal ions (e.g. Mn^{2+}, Sr^{2+}) have a wider optical bandgap (absorption band blueshift) than undoped metal ions. However, there are exceptions such as Ni^{2+}, where Ni doping can change the lattice parameters leading to a spectral redshift. For trivalent rare earth metal ions, doping introduces intermediate energy levels in the nanocrystals, providing new radiative attenuation channels for excitons. Ions with phase equivalence and smaller radii (e.g. Zn^{2+}, Sr^{2+}) do not generate new radiative centers or defect states for the luminescence of perovskite substrates, contributing to improved luminescence properties. However, doping of Bi^{3+} into the main lattice may lead to a significant decrease in fluorescence intensity, which may be due to the formation of band defect states within the main bandgap. In conclusion, B-site doping has a complex effect due to the interference with both the size factor and electronic structure.

7.2 Metal Ion-Doped Materials and Devices

7.2.1 Near-infrared Optical Perovskite Materials

One of the main challenges for infrared or near-infrared light-emitting diodes based on organic compounds and colloidal quantum dots is low carrier mobility and intrinsically low luminescence efficiency due to carrier-defect complexation in the material. Doping with metal ions has been shown to be a promising strategy to modulate the luminescence of perovskite materials and enable near-infrared or infrared emission with high luminescence efficiency. Zhang et al. conducted a study on rare earth ion-doped colloidal nanocrystals of $CsPbCl_3:Yb^{3+}$, $CsPbCl_3:Er^{3+}$, and $CsPbCl_3:Yb^{3+}/Er^{3+}$. The $CsPbCl_3:Yb^{3+}$ nanocrystals demonstrated high fluorescence intensity near-infrared light emission at 986 nm, and the Yb^{3+}/Er^{3+} co-doping in $CsPbCl_3$ nanocrystals produced emission light at 1533 nm (Figure 7.5a). Notably, the fluorescence quantum efficiency of Yb^{3+}-doped $CsPbCl_3$ nanocrystals was 127.8%, the highest reported for perovskite materials with near-infrared or infrared light emission [15]. The Burstein-Moss effect in heavily doped n-type semiconductors can explain the slight blueshift in the absorption band and fluorescence spectra of Yb^{3+}-doped $CsPbCl_3$ nanocrystals due to the filling state of Yb^{3+} with electrons near the conduction band bottom, resulting in the broadening of some bandgaps. Inorganic perovskite nanocrystals doped with rare earth ions may have potential for use in infrared or near-infrared light-emitting diodes with high luminescence efficiency [59]. The substantial difference in fluorescence quantum efficiency between visible and near-infrared emission can be explained by the quantum cutting effect, and the incorporation of Yb^{3+} ions can introduce defect states in the middle of the bandgap of $CsPbCl_3$ nanocrystals. When excited by

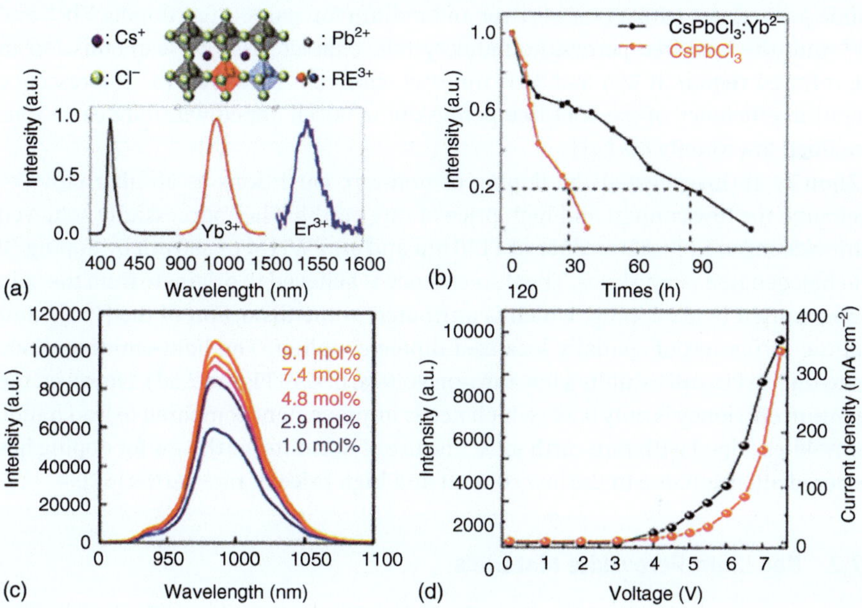

Figure 7.5 (a) Doping of Yb^{3+} or Er^{3+} ions into $CsPbCl_3$ nanocrystals extends the emission range to the IR region. Source: Zhang et al. [15]/Royal Society of Chemistry; (b) PL changes of $CsPbCl_3$ and $CsPbCl_3:Yb^{3+}$ nanocrystals after different times of 365 nm UV irradiation [15] (c) concentration dependence of Yb^{3+} on photoluminescence spectra (Yb^{3+} concentration defined as $Yb^{3+}/[Yb^{3+} + Pb^{2+}]$). Source: Ishii and Miyasaka [57]/John Wiley & Sons/CC BY 4.0; (d) current density and electroluminescence intensity versus driving voltage for $MAPbI_3$:Bi-based luminescent devices. Source: Zhou et al. [58]/American Chemical Society.

UV light, the $CsPbCl_3$ host produces excitons that combine through the bandgap, leading to light emission at 408 nm. The defect state induced by Yb^{3+} splits the exciton energy into two parts. One part is transferred to the first Yb^{3+} ion, and the exciton is then captured by the defect state via intermittent non-radiative relaxation. The remaining energy is released through exciton recombination between the electron and valence band hole of the defect state, exciting the second Yb^{3+} ion through an Auger non-radiative energy transfer. The excited Yb^{3+} ion transitions from the $2F_{5/2}$ energy level to the $2F_{7/2}$ energy level, resulting in photoemission at 986 nm. Moreover, the doping of Yb^{3+} significantly enhances the stability of the nanocrystals. The photoluminescence intensity of the undoped $CsPbCl_3$ NC solution decreased by 80% in approximately 27 hours, while the $CsPbCl_3$:Yb^{3+} NC solution maintained its fluorescence intensity for up to 85 hours (see Figure 7.5b).

Ishii et al. have also doped Yb^{3+} in $CsPbCl_3$ to enhance charge carrier mobility and balance charge injection in solid-state sensitizer films. In $CsPbCl_3$:Yb^{3+}, the intensity of NIR emission significantly increased with the increasing concentration of Yb^{3+} (as shown in Figure 7.5c). Light-emitting diodes based on doped $CsPbCl_3$ showed bright electroluminescence at 984 nm with an external quantum efficiency of up to 5.9%. However, due to the structural instability and toxicity associated with lead halide perovskites, researchers have started to focus on developing lead-free perovskite materials [57]. Lee et al. reported a double perovskite nanocrystal of $Cs_2AgInCl_6$, which has a crystal structure that replaces the two lead ions in lead halide perovskites with a pair of silver and indium ion pairs. After doping Yb^{3+} and Er^{3+} into the lead-free perovskite nanocrystals, characteristic f-f leap emission in the infrared region at 996 and 1537 nm was observed. However, the fluorescence quantum efficiency of the lead-free perovskite material doped with rare earth ions remained low at only 4% [53].

Zhou et al. investigated the doping of non-rare earth ions as an alternative to overcome the low content and high price of rare earths. They successfully achieved ultrawide infrared luminescence at 1140 nm and an FWHM of 380 nm by doping Bi into halogenated perovskites. The fluorescence is believed to originate from the spin center caused by Bi doping, which is attributed to the distortion of the $[PbI]_6$ unit and the formation of spatially localized dipole coupling. The light-emitting diode based on $MAPbI_3$:Bi exhibits a low turn-on voltage of 4 V (Figure 7.5d). However, the quantum efficiency is only 0.3%, which needs improvement compared to lead halide perovskites doped with rare earth ions. The use of non-rare earth ions for doping has garnered attention due to the low content and high price of rare earths [58].

7.2.2 Red Light Perovskite Materials

Protesescu et al. synthesized $Cs_{1-x}FA_xPbI_3$ nanocrystals by a facile colloidal synthesis method, combining the advantages of $CsPbI_3$ and $MAPbI_3$ to obtain red-emitting perovskite materials with higher stability. While $CsPbI_3$ NCs with red light emission tend to show a delayed phase transition to non-luminescence, wide-bandgap one-dimensional polycrystalline species, and $MAPbI_3$ has limited chemical durability, the $Cs_{1-x}FA_xPbI_3$ nanocrystals offer higher color purity and improved luminescence stability [60]. The doped NCs exhibited higher stability than the two individual materials with similar size and morphology. The $FA_{0.1}Cs_{0.9}PbI_3$ nanocrystals exhibited bright red luminescence with a wavelength of 690 nm and high quantum yields (>70%) in both colloidal state and thin films and were able to sustain their luminescence for months or longer. When used as the luminescent layer, the electroluminescence peak of the device was at 692 nm, with a maximum external quantum efficiency (EQE) of 0.12% and a maximum luminance of 4.3 cd m^{-2}. The focus of research on red-emitting perovskite materials is now on obtaining higher color purity and improving luminescence stability, as the efficiency of current materials and devices is already sufficiently high.

Cu^{2+} ions have a smaller ionic radius (0.72 Å) than Pb^{2+} ions (1.19 Å), which allows for partial substitution of Pb^{2+} with Cu^{2+} in the lattice structure of $CsPbBrI_2$ nanocrystals without disrupting the crystal structure. Zhang et al. have shown that Cu^{2+} doping can increase the formation energy and result in lattice contraction, leading to the stabilization of the cubic phase of the nanocrystals. Cu^{2+} can partially substitute Pb^{2+} in the lattice structure of $CsPbBrI_2$ nanocrystals due to its smaller ionic radius (0.72 Å) compared to Pb^{2+} (1.19 Å). The emission peaks of the nanocrystals were blueshifted from 630 to 621 nm with increasing Cu^{2+} doping (Figure 7.6a). The doping of Cu^{2+} resulted in $CsPbBrI_2$ nanocrystals with higher fluorescence quantum efficiency (94.8%) and improved stability. When used as the light-emitting layer in perovskite light-emitting diodes, $CsPbBrI_2$:Cu nanocrystals achieved a maximum EQE of 5.1% [61].

The ionic radius of Sr^{2+} (1.18 Å) is similar to that of Pb^{2+} (1.19 Å), allowing for the replacement of Pb^{2+} with Sr^{2+} in the perovskite lattice structure, resulting in a slight lattice shrinkage [12]. Sr^{2+} has a similar ionic radius to Pb^{2+} and can replace it in the perovskite lattice structure, causing slight lattice shrinkage. In the case of $CsPbI_3$ nanocrystals, Cl^- can passivate the surface defect state, converting the non-radiative defect state to the radiative state, and thus enhancing the photoluminescence quantum yield to 84%. The simultaneous doping of Sr^{2+} and surface Cl^- passivation can increase the emission lifetime and improve stability. Doping with Sr^{2+} also significantly increases the formation energy of α-$CsPbI_3$, reducing structural distortion and stabilizing the cubic phase at nanometer sizes. The maximum EQE of luminescent devices with doped $CsPbI_3$ nanocrystals as the luminescent layer is 13.5%, which is almost double the efficiency of devices based on undoped $CsPbI_3$ (Figure 7.6b), according to a study by Yao et al. [62]. Light-emitting devices using α-$CsPbI_3$ with 3.1% Sr^{2+} doping as the light-emitting layer were found to emit light around 678 nm at various driving voltages, corresponding to Commission Internationale d'Eclairage (CIE) color coordinates of (0.67, 0.26), which is highly suitable for red light displays.

Zou and colleagues have successfully prepared $CsPbX_3$:Mn perovskite materials with improved thermal stability and optical properties. The doped Mn^{2+} ions significantly enhance the thermal stability of the $CsPbX_3$, allowing it to remain stable even at high temperatures of up to 200 °C. The introduction of Mn^{2+} ions also leads to a broad emission band at 600 nm in the fluorescence spectrum of $CsPbCl_3$:Mn (5.8 mol%) quantum dots, which can be attributed to the $T^4{}_1 \rightarrow A^6{}_1$ leap radiation of

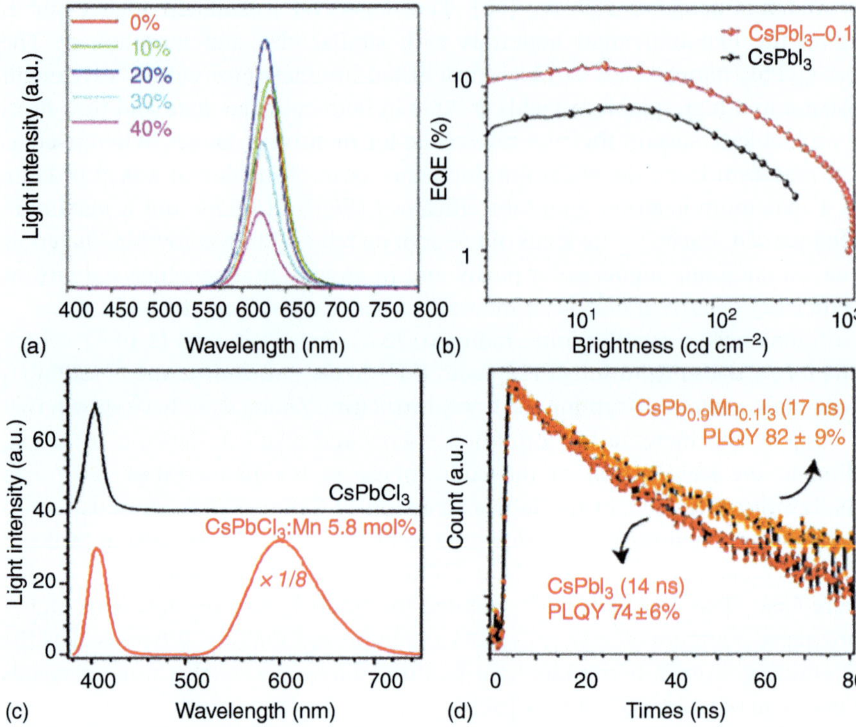

Figure 7.6 (a) PL spectra of pristine and doped $CsPbBrI_2$ nanocrystals with different Cu/Pb ratios (excitation wavelength 365 nm) [61]; (b) EQE versus luminance of light-emitting devices based on $CsPbI_3$ and $CsPbI_3$-0.1Sr nanocrystals [62]; (c) PL emission spectra of pure $CsPbCl_3$ and $CsPbCl_3$:Mn (5.8 mol%) quantum dots under UV excitation at 362 nm [63]; (d) Comparison of fluorescence lifetime and PLQY of $CsPbI_3$ and $CsPb_xMn_{1-x}I_3$ nanocrystals. Source: Adapted from Zhang et al. [13].

the Mn^{2+} ions (Figure 7.6c). In addition, the nanoscale α-$CsPbI_3$ maintains high fluorescence quantum efficiency (>80%) for up to two months, while solid films made from α-$CsPbI_3$ retain high fluorescence quantum efficiency (>40%) even when exposed to air for several weeks, indicating the stability of the cubic phase [63].

Akkerman et al. synthesized cubic-phase perovskite materials with a high Mn:Pb ratio (nearly 1 : 1), exhibiting an emission peak at 680 nm and excellent stability [13]. However, the alloying of Mn^{2+} into $CsPbI_3$ is limited due to the fast nucleation of the perovskite material. Other methods such as cation exchange, room temperature synthesis, or the insertion of MnI_2 into Cs_4PbI_6 nanocrystals can be considered to achieve higher Mn:Pb ratios [64]. The fluorescence quantum yields and fluorescence lifetimes were found to be increased in the alloyed samples compared to the original samples, as shown in Figure 7.6d.

7.2.3 Green Light Perovskite Materials

$CsPbBr_3$ nanocrystals, a commonly studied green luminescent material, are known for their low efficiency and instability. Despite its potential, $CsPbBr_3$ still suffers from low efficiency and instability [35]. To address this issue, Yang's group investigated the effect of Cs and K doping on the performance of quasi-dimensional perovskite light-emitting diodes. The addition of Cs reduces the efficiency roll-off at high current densities, leading to improved current efficiency. Furthermore, co-doping with K^+ (Cs/K co-doping) results in higher fluorescence intensity compared to Cs-doped quasi-dimensional perovskite films (Figure 7.7a). The green-light-emitting diodes based on polycationic perovskites achieved a maximum brightness of 45 600 cd m^{-2} with a current efficiency of 22.5 cd A^{-1}.

Ce^{3+} is an ideal dopant for $CsPbBr_3$ due to its similar ionic radius and higher conduction band energy level with bromine, which maintains the perovskite structure's integrity without introducing additional defect states. This allows Ce^{3+} to maintain the integrity of the perovskite structure without introducing other defect states [59]. The doping of Ce^{3+} in $CsPbBr_3$ nanocrystals can significantly enhance their fluorescence quantum efficiency, reaching up to 89%. Furthermore, the light-emitting devices incorporating the doped $CsPbBr_3$ as the light-emitting layer showed an increase in EQE from 1.6% to 4.4%, accompanied by an increase

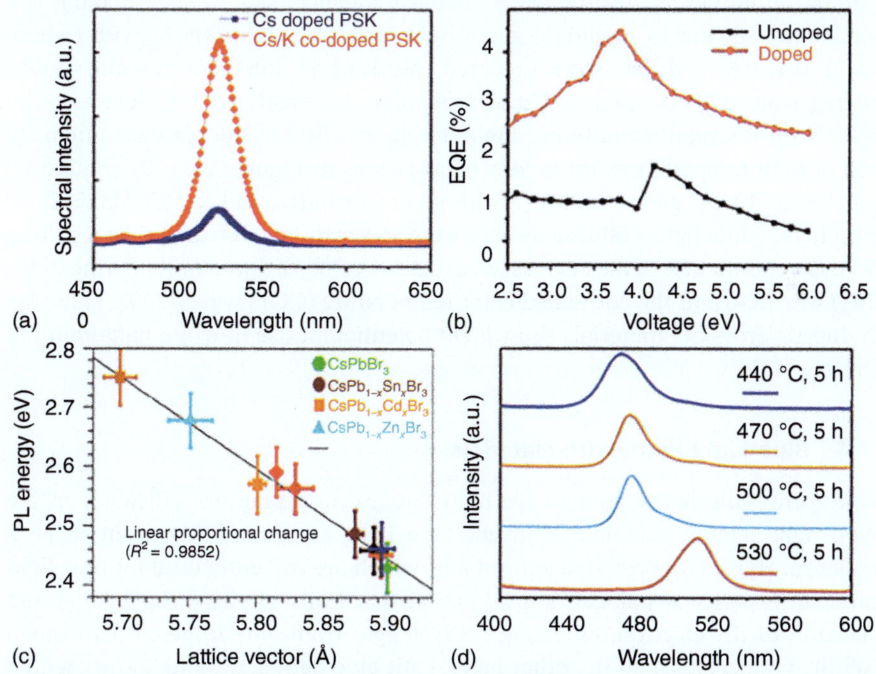

Figure 7.7 (a) Photoluminescence intensity of Cs-doped quasi-dimensional perovskite and Cs/K-doped quasi-dimensional perovskite with $PEA_2FA_2Pb_3Br_{10}$ for perovskite structure. Source: Yang et al. [35]/Royal Society of Chemistry; (b) EQE as a function of driving voltage for devices based on undoped $CsPbBr_3$ and Ce^{3+}-doped $CsPbBr_3$ nanocrystals. Source: Yao et al. [59]/American Chemical Society; (c) variation of PL energy with lattice vector in doped $CsPb_{1-x}M_xBr_3$ (M = Sn, Cd, and Zn) nanocrystals. Source: van der Stam et al. [65]/American Chemical Society; (d) fluorescence spectra of $CsPbBr_3$:0.8Rb glass substrates prepared at different heat treatment temperatures. Source: Zhang, et al. [29]/with permission of Elsevier.

in maximum current efficiency to 14.2 cd A^{-1} (Figure 7.7b). These improvements suggest that the introduction of Ce^{3+} suppresses non-radiative complexes and promotes the radiative recombination process.

In an attempt to replace Pb^{2+} with alternative metal ions, Stam et al. investigated the use of Sn^{2+}, Cd^{2+}, and Zn^{2+} in the synthesis of nanocrystals. They found that the size of the resulting nanocrystals was comparable to those synthesized using Pb^{2+}, although the lattice structure did undergo slight shrinkage upon doping with these metal ions [65]. Doping metal ions at the B-site of $CsPbBr_3$ nanocrystals can cause a blueshift in their spectrum while maintaining a high fluorescence quantum efficiency (>50%), broad absorption band, and narrow luminescence spectrum. Stam et al. found that the size of the nanocrystals remained essentially the same, except for a slight lattice shrinkage when Sn^{2+}, Cd^{2+}, or Zn^{2+} was doped. Interestingly, the blueshift of the spectrum was found to vary linearly with the lattice shrinkage (Figure 7.7c). Furthermore, all the divalent metal ion-doped samples showed a high fluorescence quantum efficiency of more than 60%.

Zhang's group successfully addressed the issue of poor stability by doping Rb^+ into $CsPbBr_3$ to obtain quantum dots with improved thermal stability. Rb^+ has a similar ionic radius and the same chemical valence state as Cs^+, making it a suitable dopant ion to modulate green light emission [29]. $CsPbBr_3$:xRb (where $x = 0$, 0.4, 0.6, and 0.8) were prepared, resulting in tunable emission spectra ranging from 475 to 523 nm. These substrates demonstrated excellent stability under hygrothermal conditions, maintaining consistent fluorescence intensity even at high temperatures up to 500 °C (as shown in Figure 7.7d). By combining blue InGaN chips, green $CsPbBr_3$:0.4Rb glass substrates, and red $CaAlSiN_3$:Eu^{2+} phosphors, white light-emitting devices were successfully fabricated. The resulting CIE color coordinates of the devices were (0.3275, 0.3303), the color rendering index (CRI) was 73.4, and the correlated color temperature (CCT) was 5732 K. Thus, the Rb-doped perovskite materials show great potential for use in white light-emitting diodes.

7.2.4 Blue-Light Perovskite Materials

While perovskite nanocrystals have high fluorescence quantum efficiency in the green light region, achieving efficient blue light emission has proven to be a challenge. Hybrid halogenated perovskites, which are currently used for blue light emission, experience halogen ion migration that leads to phase segregation and a redshift of the electroluminescence wavelength from blue to green. Ultrasmall $CsPbBr_3$ nanocrystals, on the other hand, emit blue light at around 460 nm with a narrow half-peak width and do not experience the phase separation issues of their Cl-Br counterparts. However, these nanocrystals tend to shift from blue to green light over time and have low fluorescence quantum efficiency. Thus, introducing specific metal ions to alter the bandgap, defect states, and exciton dynamics within or on the surface of the crystal is necessary to overcome these challenges.

Zhang and colleagues successfully improved the fluorescence quantum efficiency of blue light emission in $CsPbBr_3$ by doping Sb^{3+}. This approach reduced surface energy, improved lattice energy, and passivated defects, resulting in a fluorescence quantum efficiency of 73.8% with excellent spectral stability even at high temperatures [61]. The CIE coordinates of (0.14, 0.06) measured in accordance with the National Television Standards Committee (NTSC) color rendering standard closely match the standard blue coordinates (0.155, 0.070). In another study, Wang and colleagues investigated the effect of doping RbX precursors into $CsPbBr_3$ films during the film preparation [33]. Doping improved not only the blue light emission of the films but also greatly enhanced their uniformity. Light-emitting devices based on RbBr and RbCl-doped $CsPbBr_3$ displayed bright blue emission with electroluminescence peaks located at 492 nm and 468 nm, respectively, as shown in Figure 7.8a.

Recent studies have shown that doping perovskite with Mn can significantly improve its luminescence yield, despite introducing an attenuation pathway to the long-lived emitting state of the manganese ion. Hou and colleagues successfully enhanced blue luminescence using Mn doping, resulting in an on-state voltage reduction of approximately 1 V compared to the undoped device [20]. The device containing 0.25% Mn exhibited a maximum luminance of 389 cd m^{-2}, which was

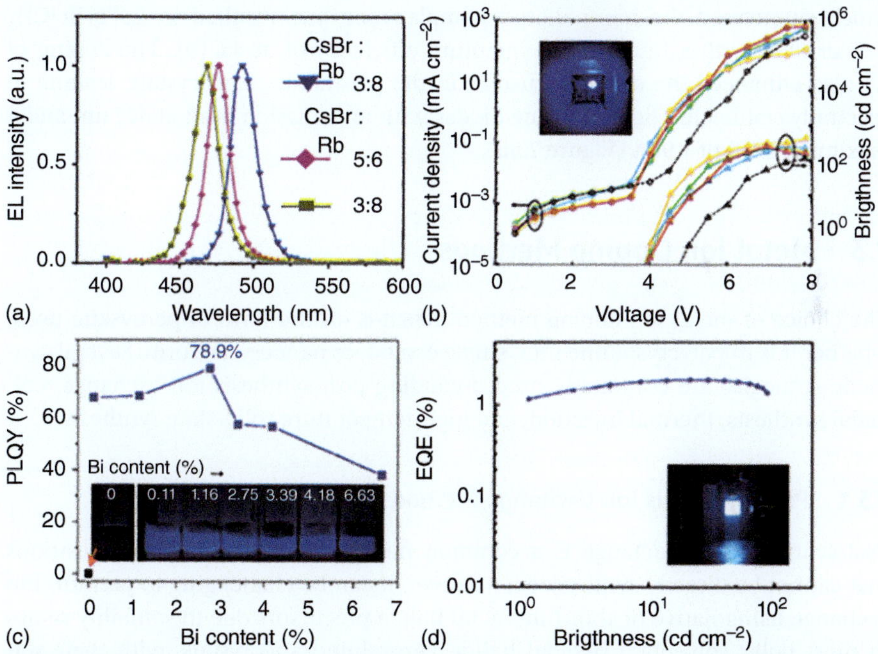

Figure 7.8 (a) Electroluminescence spectra of blue perovskite based on different RbX doping. Source: Wang et al. [10]/Royal Society of Chemistry; (b) *J-V-L* characteristic curves of the devices at different doping concentrations. The inset shows the images of the 0.19% Mn device. Source: Hou et al. [66]/Reproduced with permission from Elsevier; (c) room temperature PLQY of Cs_2SnCl_6:xBi with x at 0%, 0.11%, 1.16%, 2.75%, 3.39%, 4.18%, and 6.63%. The inset is a photograph of Cs_2SnCl_6:xBi under UV irradiation. Source: Tan et al. [14]/Reproduced from Springer Nature; (d) EQE curves of luminescent devices based on $CsPb(Br/Cl)_3$: 4.0% K. Source: Yang, et al. [9]/Reproduced from John Wiley & Sons.

three times brighter than the undoped device, as shown in Figure 7.8b. The Mn doping increased the photoluminescence yield and lifetime while reducing defect states, resulting in a perovskite light-emitting diode (LED) with a blue quantum efficiency of over 2%.

Blue-light halide lead perovskites possess excellent optoelectronic properties but raise concerns regarding stability and toxicity. To address this issue, Tan's group reported the use of Bi-doped lead-free inorganic perovskites Cs_2SnCl_6 as blue-light phosphorescence sources. Upon Bi doping, the initially non-emissive Cs_2SnCl_6 demonstrated efficient deep blue light emission at 455 nm with a Stokes shift of 106 nm and a fluorescence quantum efficiency of nearly 80%, as shown in Figure 7.8c. Cs_2SnCl_6:Bi also exhibited good hydrothermal stability due to its all-inorganic composition and the formation of a protective bismuth oxychloride (BiOCl) layer [29].

Other studies have focused on improving the optoelectronic properties of blue perovskite materials through surface engineering via alkali metal ion doping. For instance, Yang et al. have explored the surface engineering of alkali metal ion doping to improve the optoelectronic properties of blue perovskite materials. They introduced K^+ to occupy the position of the peripheral organic ligand, which suppressed the non-radiative complexation and resulted in $CsPb(Br/Cl)_3$ nanocrystals with a fluorescence quantum efficiency of 38.4% [9]. The doping of K^+ also enhanced the carrier transport in the perovskite nanocrystals, leading to spectrally stable blue-light-emitting diodes with an emission peak at 477 nm and a maximum EQE of 1.96% (Figure 7.8d).

7.3 Metal Ion Doping Methods

The choice of metal ion doping method depends on the form of perovskite used, whether it is in polycrystalline film, single crystal, or nanocrystal form. Several synthetic strategies are commonly used, including post-synthesis ion exchange, colloidal synthesis, thermal injection, and high-temperature solid-state synthesis.

7.3.1 Post-synthesis Ion Exchange Methods

Post-synthetic ion exchange is a common method used to access compositions that cannot be directly synthesized; however, it can be challenging to perform ion exchange using native or thin-film metal halide precursors due to solubility issues in most polar solvents. Colloidal halide perovskite nanocrystals, with their soft lattice, dynamic surface ligands, and high specific surface area, provide a suitable platform for ion exchange. Hazarika et al. proposed a cation exchange method to synthesize $Cs_{1-x}FA_xPbI_3$. In this method, $CsPbI_3$ and $FAPbI_3$ nanocrystals were dispersed in colloidal solutions of octane and mixed in different ratios to obtain the desired Cs/FA stoichiometric ratio [67]. The absorption spectra of each sample were measured before mixing, and the concentrations were adjusted to ensure that each solution had a similar optical density near the band edge. The mixture was

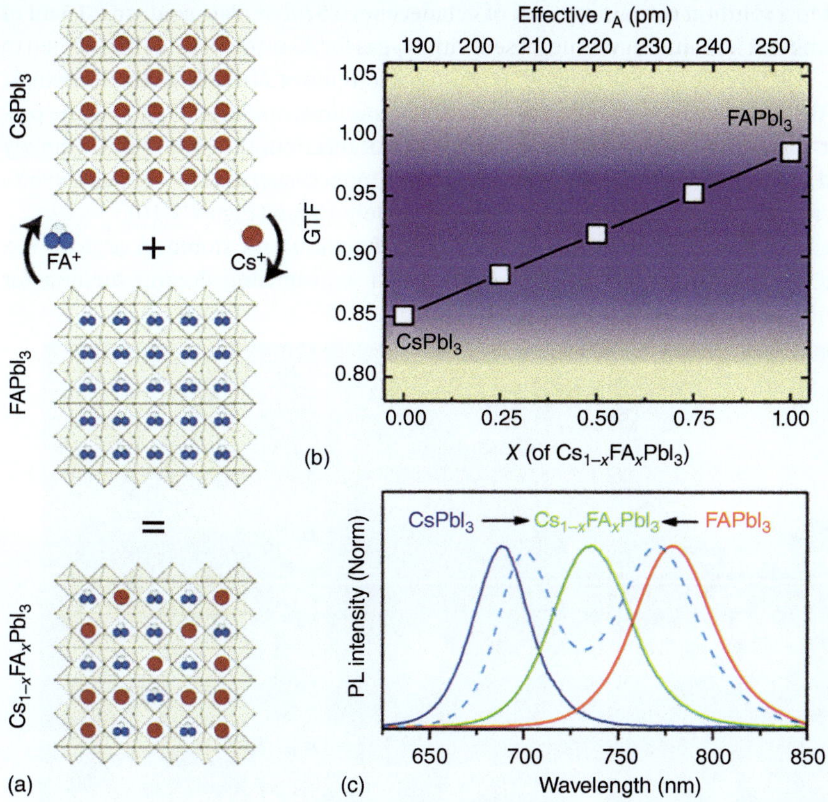

Figure 7.9 A-site ion exchange to form phase-stable $Cs_{1-x}FA_xPbI_3$ NCs. (a) Schematic diagram showing the cross-exchange of Cs^+ and FA^+ ions between $CsPbI_3$ and $FAPbI_3$ perovskite NCs. (b) Goldschmidt tolerance factor as a function of FA^+ ion concentration shows that all compositions are expected to be phase-stable (the purple region in the plot). The top axis shows the effective or average A-site radius. (c) PL emission spectra depict how $CsPbI_3$ and $FAPbI_3$ NCs convert into alloyed NCs; the dashed curve indicates the emission profile of one of the early stages of alloying. Source: Hazarika et al. [67]/American Chemical Society.

then reacted at room temperature for 48 hours to obtain the final product. These materials show bright and finely tunable emission in the red and near-infrared range between 650 and 800 nm. The activation energy for the miscibility between Cs^+ and FA^+ is measured (~0.65 eV) and is shown to be higher than reported for X-site exchange in lead halide perovskites (Figure 7.9).

7.3.2 Colloidal Synthesis Methods

Colloidal synthesis methods typically involve hydrolyzing inorganic salts or metal alcohol salts in an organic medium to obtain the target product, which is then dried or annealed using a colloidal dispersion system. Zou et al. used the colloidal synthesis method to synthesize $CsPbX_3$:Mn [63]. To synthesize Mn-doped $CsPbX_3$ via colloidal synthesis, a mixture of PbX_2 and MnX_2 (at the desired Mn/Pb molar ratio) was added to a solution containing 5 ml of octadecene, 0.5 ml of oleic acid, and 0.5 ml of oleylamine. The solution was degassed with N_2 gas for 20 minutes and then heated to 120 °C with continuous stirring for 60 minutes to remove water from the feedstock. Next, the mixture was heated to 150 °C with N_2 gas flow, and 0.4 ml of Cs-oleate precursor was rapidly injected. After 5–10 seconds of reaction, the solution was rapidly cooled to 0 °C using an ice water bath to obtain monodisperse Mn^{2+}-doped $CsPbX_3$ nanocrystals with an average size of approximately 10 nm (Figure 7.10).

The final products were collected via centrifugation and stored in a glove box under argon protection. Colloidal synthesis is a promising doping method for

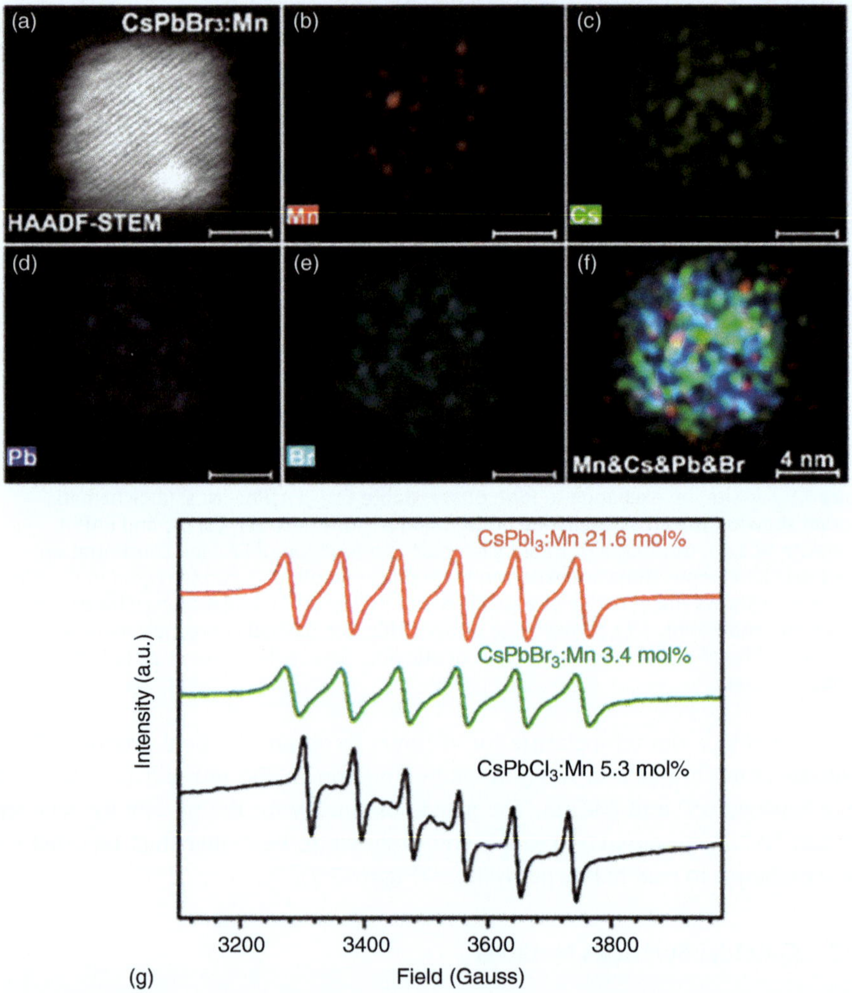

Figure 7.10 (a) HAADF-STEM image of a single randomly selected $CsPbBr_3$:Mn (3.4 mol%) QD and (b–f) its corresponding EDS element mapping images of Mn, Cs, Pb, and Br. (g) X-band EPR spectra for $CsPbX_3$:Mn QDs measured at room temperature, demonstrating the successful substitution of Pb^{2+} with Mn^{2+} in the lattices of $CsPbX_3$ QDs. Source: Shenghan Zou et al. [63]/Reproduced from American Chemical Society.

preparing materials with high purity and uniform particle size at low temperatures, which is beneficial for various applications.

7.3.3 The Thermal Injection Methods

The thermal injection method is a feasible and effective synthesis approach to doping ions into the lattice by high-temperature driving. Yb^{3+}-doped $CsPbCl_3$ nanocrystals were prepared by Zhang et al. using this method [15]. They employed the thermal injection method to synthesize Yb^{3+}-doped $CsPbCl_3$ nanocrystals. Initially, $YbCl_3$-$6H_2O$ and $Pb(OAc)_2$-$3H_2O$ were mixed with oleic acid, oleylamine, and octadecene in a flask. The reactants were degassed under vacuum at 120 °C for one hour until they were completely dissolved, and then the flask was switched to an N_2 atmosphere. The temperature was then rapidly increased to 260 °C and maintained for three minutes. Afterward, a preheated Cs-oleate solution was rapidly injected using a syringe, and the reaction was stopped by placing the flask in an ice bath to obtain the product. The Yb^{3+}-doped $CsPbCl_3$ NCs emitted strong 986 nm NIR light; the Yb^{3+}/Er^{3+} co-doped $CsPbCl_3$ NCs emitted at 1533 nm. The total photoluminescence quantum yield (PL QY) of the $CsPbCl_3$ NCs changed from 5.0% to 127.8% upon incorporating 2.0% Yb^{3+}, a factor of 25.6 times enhancement. The material's stability was tested under continuous ultraviolet (365 nm) irradiation. The doped $CsPbCl_3$ NCs exhibited better stability than the undoped ones. The PL intensity of the undoped $CsPbCl_3$ NCs dropped to 20% of the initial value in 27 hours, while the doped one took 85 hours (Figure 7.11).

7.3.4 High Temperature Solid-state Synthesis Methods

High-temperature solid-state synthesis is a simple method of synthesis, although it has low applicability and potential damage to the lattice. Ji's group employed

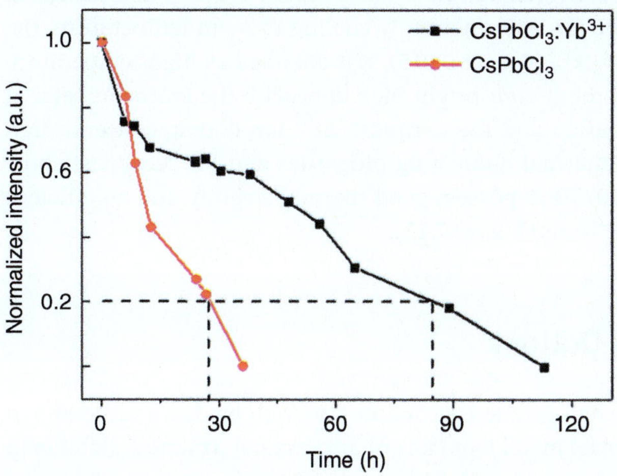

Figure 7.11 PL decrease of the $CsPbCl_3$ and $CsPbCl_3$:Yb^{3+} NCs under 365 nm UV light irradiation. Source: Zhang et al. [15]/Royal Society of Chemistry.

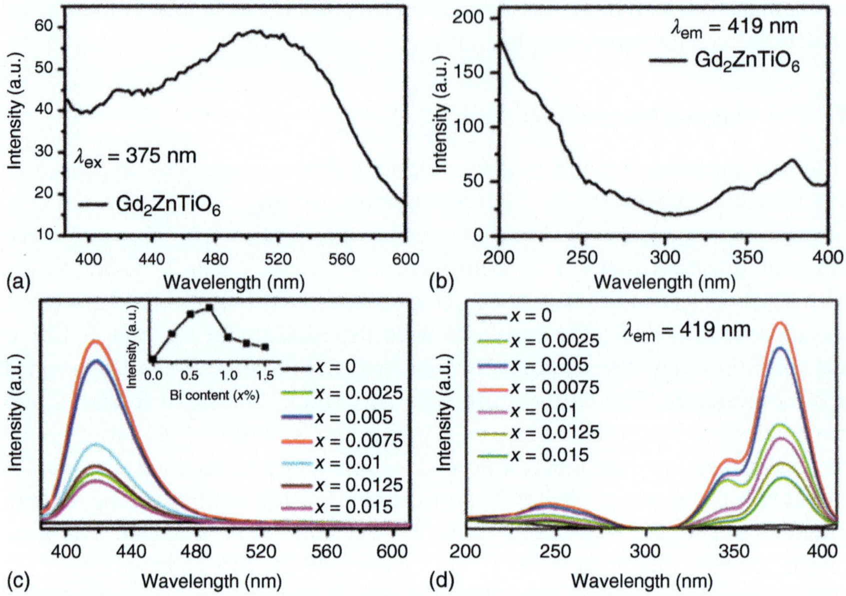

Figure 7.12 (a) The PL spectra of Gd_2ZnTiO_6 host lattice excited at 375 nm; (b) The PLE spectrum of Gd_2ZnTiO_6 phosphors monitoring at 419 nm; (c) The PL spectrum of $Gd_{2-x}ZnTiO_6$:$x$$Bi^{3+}$ ($0 \leq x \leq 0.015$) phosphors excited at 375 nm. Inset: the dependence of PL intensity on the content of Bi^{3+} ions. (d) The PLE spectrum of $Gd_{2-x}ZnTiO_6$:$x$$Bi^{3+}$ ($0 \leq x \leq 0.015$) phosphors monitored at 419 nm. Source: Ji et al. [68]/with permission of Elsevier.

this method to prepare $Gd_{2-x}ZnTiO_6$:xBi ($0 < x < 0.015$) powder [68]. The starting materials, Gd_2O_3, ZnO, TiO_2, and Bi_2O_3, were weighed in stoichiometric ratios and thoroughly mixed with a small amount of ethanol using an agate mortar. The resulting mixture was then transferred to a corundum crucible and heated at 1300 °C for four hours in the air. After naturally cooling to room temperature, the doped product, $Gd_{2-x}ZnTiO_6$:xBi ($0 < x < 0.015$), was obtained by high-temperature solid-state synthesis. All of them emit bright blue light with the emission peak at 419 nm under 375 nm excitation and the optimum Bi^{3+} ion doping concentration of 0.0075. In addition, the thermal quenching properties and PL decay test show that the $Gd_{1.9925}ZnTiO_6$:$0.0075Bi^{3+}$ possess good thermal stability and an efficient energy transfer between Bi^{3+} ions (Figure 7.12).

7.4 Conclusion and Outlook

To date, metal ion doping in halogenated perovskite materials has been successful in altering their properties. Doped metal ions can enhance crystal structure, stability in harsh conditions, and passivate surface defects, leading to improved optoelectronic properties in materials and devices. However, the field of doped halide perovskites is still in its early stages, and several issues need to be addressed.

Firstly, most A-site doping is limited to a few alkali metal ions, and exploring additional cations with suitable sizes is necessary for stabilizing the perovskite phase structure. Secondly, B-site doping is complicated, and further research is needed to establish the relationship between the dopant ion's electronic structure and carrier dynamics with the perovskite body. Thirdly, blue perovskite light-emitting diodes are still underperforming compared to green and red perovskite LEDs, indicating a need to explore their latent problems and solutions in theory and practice. Lastly, while significant progress has been made in developing perovskites, the exact crystallographic position occupied by doping ions like Mg^{2+} and the exact structural information still need investigation.

Despite these issues, doping and surface passivation remain effective strategies for improving perovskite properties and making them more practical for various applications. While many issues concerning metal ion doping require further study, it is believed that doping metal ions will play a crucial role in the development and application of the perovskite field.

In summary, doping metal ions into halogenated perovskite materials has proven to be an effective method to modulate the crystal structure, enhance stability, and improve optoelectronic properties. However, the field is still in its early stages, and there are several challenges that need to be addressed, such as exploring more cations for A-site doping, resolving controversies in B-site doping, improving the performance of blue perovskite LEDs, and investigating internal mechanistic issues. Despite these challenges, we believe that doping metal ions will continue to be a key strategy in advancing the perovskite field and its practical applications. Further research and development will undoubtedly shed more light on the potential of these materials.

References

1 Parobek, D., Roman, B.J., Dong, Y. et al. (2016). Exciton-to-dopant energy transfer in Mn-doped cesium lead halide perovskite nanocrystals. *Nano Letters* 16: 7376–7380.

2 Wang, Q., Lyu, M., Zhang, M. et al. (2015). Transition from the tetragonal to cubic phase of organohalide perovskite: the role of chlorine in crystal formation of $CH_3NH_3PbI_3$ on TiO_2 substrates. *The Journal of Physical Chemistry Letters* 6: 4379–4384.

3 Zheng, X., Wu, C., Jha, S.K. et al. (2016). Improved phase stability of formamidinium lead triiodide perovskite by strain relaxation. *ACS Energy Letters* 1: 1014–1020.

4 Li, N., Jia, Y., Guo, Y. et al. (2022). Ion migration in perovskite light-emitting diodes: mechanism, characterizations, and material and device engineering. *Advanced Materials* 34: e2108102.

5 Song, J., Li, J., Xu, J. et al. (2014). Superstable transparent conductive Cu@Cu_4Ni nanowire elastomer composites against oxidation, bending, stretching, and twisting for flexible and stretchable optoelectronics. *Nano Letters* 14: 6298–6305.

6 Mocatta, D., Cohen, G., Schattner, J. et al. (2011). Heavily doped semiconductor nanocrystal quantum dots. *Science* 332: 77–81.

7 Xu, L., Liu, G., Xiang, H. et al. (2022). Charge-carrier dynamics and regulation strategies in perovskite light-emitting diodes: from materials to devices. *Applied Physics Reviews* 9: 021308.

8 Akkerman, Q.A., Rainò, G., Kovalenko, M.V. et al. (2018). Genesis, challenges and opportunities for colloidal lead halide perovskite nanocrystals. *Nature Materials* 17: 394–405.

9 Yang, F., Chen, H., Zhang, R. et al. (2020). Efficient and spectrally stable blue perovskite light-emitting diodes based on potassium passivated nanocrystals. *Advanced Functional Materials* 30: 1908760.

10 Wang, H., Zhao, X., Zhang, B. et al. (2019). Blue perovskite light-emitting diodes based on RbX-doped polycrystalline $CsPbBr_3$ perovskite films. *Journal of Materials Chemistry C* 7: 5596–5603.

11 Lu, M., Zhang, X., Bai, X. et al. (2018). Spontaneous silver doping and surface passivation of $CsPbI_3$ perovskite active layer enable light-emitting devices with an external quantum efficiency of 11.2%. *ACS Energy Letters* 3: 1571–1577.

12 Lu, M., Zhang, X., Zhang, Y. et al. (2018). Simultaneous strontium doping and chlorine surface passivation improve luminescence intensity and stability of $CsPbI_3$ nanocrystals enabling efficient light-emitting devices. *Advanced Materials* 30: 1804691.

13 Akkerman, Q.A., Meggiolaro, D., Dang, Z. et al. (2017). Fluorescent alloy $CsPb_xMn_{1-x}I_3$ perovskite nanocrystals with high structural and optical stability. *ACS Energy Letters* 2: 2183–2186.

14 Tan, Z., Li, J., Zhang, C. et al. (2018). Highly efficient blue-emitting bi-doped Cs_2SnCl_6 perovskite variant: photoluminescence induced by impurity doping. *Advanced Functional Materials* 28: 1801131.

15 Zhang, X., Zhang, Y., Zhang, X. et al. (2018). Yb^{3+} and Yb^{3+}/Er^{3+} doping for near-infrared emission and improved stability of $CsPbCl_3$ nanocrystals. *Journal of Materials Chemistry C* 6: 10101–10105.

16 Song, J., Kulinich, S.A., Li, J. et al. (2015). A general one-pot strategy for the synthesis of high-performance transparent-conducting-oxide nanocrystal inks for all-solution-processed devices. *Angewandte Chemie International Edition* 127: 472–476.

17 Chen, W., Shao, H., Wu, X. et al. (2022). Highly stable and efficient Mn^{2+} doping zero-dimension $Cs_2Zn_xPb_{1-x}Cl_4$ alloyed nanorods toward white electroluminescent light-emitting diodes. *Journal of Physical Chemistry Letters* 13: 2379–2387.

18 Kubicki, D.J., Prochowicz, D., Hofstetter, A. et al. (2017). Phase segregation in Cs-, Rb-and K-doped mixed-cation $MA_xFA_{1-x}PbI_3$ hybrid perovskites from solid-state NMR. *Journal of the American Chemical Society* 139: 14173–14180.

19 Amgar, D., Binyamin, T., Uvarov, V. et al. (2018). Near ultra-violet to mid-visible band gap tuning of mixed cation $Rb_xCs_{1-x}PbX_3$ (X = Cl or Br) perovskite nanoparticles. *Nanoscale* 10: 6060–6068.

20 Travis, W., Glover, E.N.K., Bronstein, H. et al. (2016). On the application of the tolerance factor to inorganic and hybrid halide perovskites: a revised system. *Chemical Science* 7: 4548–4556.

21 Yin, W.J., Shi, T., and Yan, Y. (2014). Unique properties of halide perovskites as possible origins of the superior solar cell performance. *Advanced Materials* 26: 4653–4658.

22 Ju, M.G., Dai, J., Ma, L. et al. (2017). Lead-free mixed tin and germanium perovskites for photovoltaic application. *Journal of the American Chemical Society* 139: 8038–8043.

23 Shi, Z., Guo, J., Chen, Y. et al. (2017). Lead-free organic-inorganic hybrid perovskites for photovoltaic applications: recent advances and perspectives. *Advanced Materials* 29: 1605005.

24 Stoumpos, C.C., Malliakas, C.D., and Kanatzidis, M.G. (2013). Semiconducting tin and lead iodide perovskites with organic cations: phase transitions, high mobilities, and near-infrared photoluminescent properties. *Inorganic Chemistry* 52: 9019–9038.

25 Ball, J.M., Lee, M.M., Hey, A. et al. (2013). Low-temperature processed meso-superstructured to thin-film perovskite solar cells. *Energy & Environmental Science* 3: 18762.

26 Xiao, J. and Zhang, H.L. (2016). Recent progress in organic-inorganic hybrid perovskite materials for luminescence applications. *Acta Physico-Chimica Sinica* 32: 1894–1912.

27 Yantara, N., Bhaumik, S., Yan, F. et al. (2015). Inorganic halide perovskites for efficient light-emitting diodes. *The Journal of Physical Chemistry Letters* 6: 4360–4364.

28 Saliba, M., Matsui, T., Domanski, K. et al. (2016). Incorporation of rubidium cations into perovskite solar cells improves photovoltaic performance. *Science* 354: 206–209.

29 Zhang, H., Yuan, R., Jin, M. et al. (2020). Rb^+-doped $CsPbBr_3$ quantum dots with multi-color stabilized in borosilicate glass via crystallization. *Journal of the European Ceramic Society* 40: 94–102.

30 Filip, M.R., Eperon, G.E., Snaith, H.J. et al. (2014). Steric engineering of metal-halide perovskites with tunable optical band gaps. *Nature Communications* 5: 1–9.

31 Singh, S. and Kabra, D. (2018). Influence of solvent additive on the chemical and electronic environment of wide bandgap perovskite thin films. *Journal of Materials Chemistry C* 6: 12052–12061.

32 Li, Z., Yang, M., Park, J.S. et al. (2016). Stabilizing perovskite structures by tuning tolerance factor: formation of formamidinium and cesium lead iodide solid-state alloys. *Chemistry of Materials* 28: 284–292.

33 Chen, Z., Zhou, B., Yuan, J. et al. (2021). Cu^{2+}-doped $CsPbI_3$ nanocrystals with enhanced stability for light-emitting diodes. *Journal of Physical Chemistry Letters* 12: 3038–3045.

34 Li, Y.-h., Lou, Y.-H., Zhou, Y.-H. et al. (2023). Review on the promising roles of alkali metals toward highly efficient perovskite light-emitting diodes. *Journal of Materials Chemistry C* 11: 2011–2025.

35 Yang, G., Liu, X., Sun, Y. et al. (2020). Improved current efficiency of quasi-2D multi-cation perovskite light-emitting diodes: the effect of Cs and K. *Nanoscale* 12: 1571–1579.

36 Trots, D.M. and Myagkota, S.V. (2008). High-temperature structural evolution of caesium and rubidium triiodoplumbates. *Journal of Physics and Chemistry of Solids* 69: 2520–2526.

37 Yu, H., Chen, W., Fang, Z. et al. (2022). Alkalis-doping of mixed tin-lead perovskites for efficient near-infrared light-emitting diodes. *Science Bulletin (Beijing)* 67: 54–60.

38 Philippe, B., Saliba, M., Correa-Baena, J.P. et al. (2017). Chemical distribution of multiple cation (Rb^+, Cs^+, MA^+, and FA^+) perovskite materials by photoelectron spectroscopy. *Chemistry of Materials* 29: 3589–3596.

39 Hu, Y., Aygüler, M.F., Petrus, M.L. et al. (2017). Impact of rubidium and cesium cations on the moisture stability of multiple-cation mixed-halide perovskites. *ACS Energy Letters* 2: 2212–2218.

40 Wang, N., Cheng, L., Ge, R. et al. (2016). Perovskite light-emitting diodes based on solution-processed self-organized multiple quantum wells. *Nature Photonics* 10: 699–704.

41 Abdi-Jalebi, M., Andaji-Garmaroudi, Z., Cacovich, S. et al. (2018). Maximizing and stabilizing luminescence from halide perovskites with potassium passivation. *Nature* 555: 497–501.

42 Li, G., Rivarola, F.W.R., Davis, N.J.L.K. et al. (2016). Highly efficient perovskite nanocrystal light-emitting diodes enabled by a universal crosslinking method. *Advanced Materials* 28: 3528–3534.

43 Zhou, L., Liao, J.F., Huang, Z.G. et al. (2018). All-inorganic lead-free Cs_2PdX_6 ($X = Br$, I) perovskite nanocrystals with single unit cell thickness and high stability. *ACS Energy Letters* 3: 2613–2619.

44 Chan, S.H., Wu, M.C., Lee, K.M. et al. (2017). Enhancing perovskite solar cell performance and stability by doping barium in methylammonium lead halide. *Journal of Materials Chemistry A* 5: 18044–18052.

45 Zhang, X., Wang, H., Hu, Y. et al. (2019). Strong blue emission from Sb^{3+}-doped super small $CsPbBr_3$ nanocrystals. *The Journal of Physical Chemistry Letters* 10: 1750–1756.

46 Yong, Z.J., Guo, S.Q., Ma, J.P. et al. (2018). Doping-enhanced short-range order of perovskite nanocrystals for near-unity violet luminescence quantum yield. *Journal of the American Chemical Society* 140: 9942–9951.

47 Song, J., Fang, T., Li, J. et al. (2018). Organic-inorganic hybrid passivation enables perovskite QLEDs with an EQE of 16.48%. *Advanced Materials* 30: 1805409.

48 Swarnkar, A., Ravi, V.K., and Nag, A. (2017). Beyond colloidal cesium lead halide perovskite nanocrystals: analogous metal halides and doping. *ACS Energy Letters* 2: 1089–1098.

49 Xu, L., Yuan, S., Zeng, H. et al. (2019). A comprehensive review of doping in perovskite nanocrystals/quantum dots: evolution of structure, electronics, optics and light-emitting diodes. *Materials Today Nano* 6: 100036.

50 Das Adhikari, S., Dutta, S.K., Dutta, A. et al. (2017). Chemically tailoring the dopant emission in manganese-doped $CsPbCl_3$ perovskite nanocrystals. *Angewandte Chemie International Edition* 56: 8746–8750.

51 Zhou, D., Liu, D., Pan, G. et al. (2017). Cerium and ytterbium codoped halide perovskite quantum dots: a novel and efficient downconverter for improving the performance of silicon solar cells. *Advanced Materials* 29: 1704149.

52 Wang, L., Zhou, H., Hu, J. et al. (2019). A Eu^{3+}-Eu^{2+} ion redox shuttle imparts operational durability to Pb-I perovskite solar cells. *Science* 363: 265–270.

53 Lee, W., Hong, S., and Kim, S. (2019). Colloidal synthesis of lead-free silver-indium double-perovskite $Cs_2AgInCl_6$ nanocrystals and their doping with lanthanide ions. *The Journal of Physical Chemistry C* 123: 2665–2672.

54 Li, C., Lu, X., Ding, W. et al. (2008). Formability of ABX_3 (X = F, Cl, Br, I) halide perovskites. *Acta Crystallographica Section B: Structural Science* 64: 702–707.

55 Ogomi, Y., Morita, A., Tsukamoto, S. et al. (2014). $CH_3NH_3Sn_xPb_{1-x}I_3$ perovskite solar cells covering up to 1060 nm. *The Journal of Physical Chemistry Letters* 5: 1004–1011.

56 Brandt, R.E., Stevanović, V., Ginley, D.S. et al. (2015). Identifying defect-tolerant semiconductors with high minority-carrier lifetimes: beyond hybrid lead halide perovskites. *MRS Communications* 5: 265–275.

57 Ishii, A. and Miyasaka, T. (2020). Sensitized Yb^{3+} luminescence in $CsPbCl_3$ film for highly efficient near-infrared light-emitting diodes. *Advanced Science* 7: 1903142.

58 Zhou, Y., Yong, Z.J., Zhang, K.C. et al. (2016). Ultrabroad photoluminescence and electroluminescence at new wavelengths from doped organometal halide perovskites. *The Journal of Physical Chemistry Letters* 7: 2735–2741.

59 Yao, J., Ge, J., Han, B. et al. (2018). Ce^{3+}-doping to modulate photoluminescence kinetics for efficient $CsPbBr_3$ nanocrystals based light-emitting diodes. *Journal of the American Chemical Society* 140: 3626–3634.

60 Protesescu, L., Yakunin, S., Kumar, S. et al. (2017). Dismantling the "red wall" of colloidal perovskites: highly luminescent formamidinium and formamidinium-cesium lead iodide nanocrystals. *ACS Nano* 11: 3119–3134.

61 Zhang, J., Zhang, L., Cai, P. et al. (2019). Enhancing stability of red perovskite nanocrystals through copper substitution for efficient light-emitting diodes. *Nano Energy* 62: 434–441.

62 Yao, J.S., Ge, J., Wang, K.H. et al. (2019). Few-nanometer-sized α-$CsPbI_3$ quantum dots enabled by strontium substitution and iodide passivation for efficient red light emitting diodes. *Journal of the American Chemical Society* 141: 2069–2079.

63 Zou, S., Liu, Y., Li, J. et al. (2017). Stabilizing cesium lead halide perovskite lattice through Mn(II) substitution for air-stable light-emitting diodes. *Journal of the American Chemical Society* 139: 11443–11450.

64 Arunkumar, P., Gil, K.H., Won, S. et al. (2017). Colloidal organolead halide perovskite with a high Mn solubility limit: a step toward Pb-free luminescent quantum dots. *The Journal of Physical Chemistry Letters* 8: 4161–4166.

65 van der Stam, W., Geuchies, J.J., Altantzis, T. et al. (2017). Highly emissive divalent-ion-doped colloidal $CsPb_{1-x}M_xBr_3$ perovskite nanocrystals through cation exchange. *Journal of the American Chemical Society* 139: 4087–4097.

66 Hou, S., Gangishetty, M.K., Quan, Q. et al. (2018). Efficient blue and white perovskite light-emitting diodes via manganese doping. *Joule* 2: 2421–2433.

67 Hazarika, A., Zhao, Q., Gaulding, E.A. et al. (2018). Perovskite quantum dot photovoltaic materials beyond the reach of thin films: full-range tuning of A-site cation composition. *ACS Nano* 12: 10327–10337.

68 Ji, C., Huang, Z., Wen, J. et al. (2019). Blue-emitting bi-doped double perovskite Gd_2ZnTiO_6 phosphor with near-ultraviolet excitation for warm white light-emitting diodes. *Journal of Alloys and Compounds* 788: 1127–1136.

8

Non-lead Metal Halide Perovskite Materials

8.1 Development History of Non-lead Blue Perovskite Materials

Emerging organic-inorganic halide perovskite materials have garnered widespread attention in the academic community due to their excellent and unique optical and electrical properties. These materials exhibit outstanding optoelectronic performance, including tunable emission wavelength, narrow emission linewidth, high color purity, high charge carrier mobility, and high photoluminescence quantum yields (PLQYs). As a result, they have shown great potential for applications in the field of light-emitting diodes (LEDs), attracting significant interest from researchers worldwide. However, organic-inorganic halide perovskites also face a series of challenges, such as the presence of toxic lead elements in their structure and poor device stability, which have hindered their widespread adoption. The forefront of research aims to develop lead-free alternatives for perovskite devices while ensuring high efficiency and stability. In recent years, researchers have made significant breakthroughs by replacing lead ions with low-toxicity or nontoxic metal ions and forming materials with low-dimensional or zero-dimensional structures to effectively address the aforementioned issues. Lead-free metal halide emitters are increasingly moving toward environmentally friendly and highly efficient electroluminescence. In this chapter, we summarize the research progress on lead-free metal halide emitters and their applications in electrically driven LEDs.

In visible LEDs using perovskite materials, satisfactory efficiency values have been achieved for green and red LEDs. However, the performance of blue LEDs using this material is still relatively lagging [1, 2]. The main reasons why it is difficult to improve the performance of blue perovskite LEDs are the following: (1) The current strategy to achieve blue perovskite LEDs is the use of Cl/Br mixed halide perovskite material, but the migration of halogen ions under the action of voltage will lead to phase separation in the perovskite film, which makes the location of the luminescent peak shift [3]. (2) Quasi-two-dimensional perovskite materials can also be used to prepare blue LEDs, but due to quantum effects, the exciton binding energy in the quasi-two-dimensional system is too large, which will lead to rapid quenching of excitons in the octahedral framework, and the presence of the peripheral organic layer, the long-chain insulating amino group, will also

Perovskite Light Emitting Diodes: Materials and Devices, First Edition. Hong Meng.

inhibit the charge carrier transport, resulting in double- or triple-peak emission in the electroluminescence spectrum of the device [4]. (3) In addition to the above two impure emission peaks, the large bandgap of the blue perovskite material is difficult to match with the highest occupied molecular orbital (HOMO) energy level of existing commercial electron transport materials, thus increasing the charge injection barrier, which in turn leads to low device efficiency. Currently, it has been reported in the literature that the ion migration in the device can be suppressed by using a hybrid additive modification strategy to stabilize the electroluminescence spectrum and enhance the device's stability [3]. Secondly, quasi-two-dimensional perovskite materials can be used to prepare blue LEDs, but the exciton binding energy in the quasi-two-dimensional system is too large due to quantum effects, leading to the rapid quenching of excitons in the octahedral framework. Moreover, the presence of the peripheral organic layer with long-chain insulating amino groups also inhibits charge carrier transport, resulting in double- or triple-peak emission in the electroluminescence spectrum of the device [4]. Lastly, the large bandgap of the blue perovskite material is challenging to match with the HOMO energy level of existing commercial electron transport materials, resulting in an increased charge injection barrier and low device efficiency. Recent literature reports suggest that the ion migration in the device can be suppressed by using a hybrid additive modification strategy to stabilize the electroluminescence spectrum and enhance the device's stability. Despite these challenges, the development of efficient blue perovskite LEDs remains a priority in the field of optoelectronics.

However, the toxicity of lead in perovskite materials has drawn significant attention and has become an issue that cannot be ignored. The use of lead in perovskite materials can cause damage to the environment and human health, particularly to the brain development of children. Therefore, it is important to find alternatives to lead-based perovskite materials to mitigate these negative effects. The U.S. Environmental Protection Agency (U.S. EPA) has established strict limits for lead in air and water, with limits of 0.15 and 15 $\mu g\,l^{-1}$ of Pb^{2+}, respectively [5]. Therefore, there is a pressing need to develop lead-free perovskite materials. Many researchers have been working to replace lead with other more environmentally friendly elements, but this substitution often results in a significant decrease in the PLQY of the luminescent materials. As a result, the development of lead-free blue perovskite materials is still an active area of research.

Lead-free blue perovskite materials were first reported by Marcus L. Böhm in 2016. However, the synthesized $CsSnCl_3$ emitted a broad emission peak in the blue region, and the fluorescence quantum efficiency was not satisfactory [6]. Although many efforts have been made to improve the performance of lead-free blue perovskite materials, the results have been limited so far. Researchers have explored the use of bismuth, tin, antimony, copper, europium-based perovskite, and bichalcogenide materials as potential substitutes for lead. However, the highest fluorescence quantum efficiency achieved for each year from 2017 to 2023 is still relatively low, as shown in Figures 8.1 and 8.2.

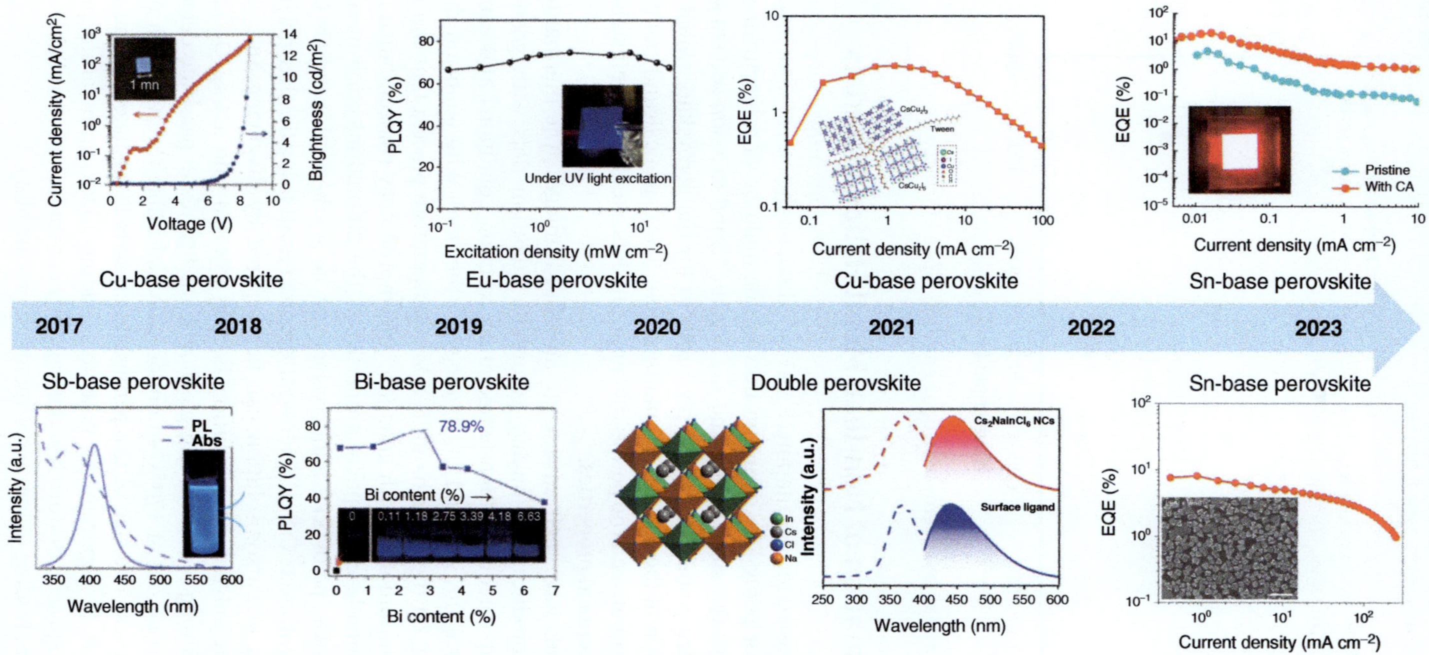

Figure 8.1 The development history of non-lead blue perovskite materials from 2016 to 2020.

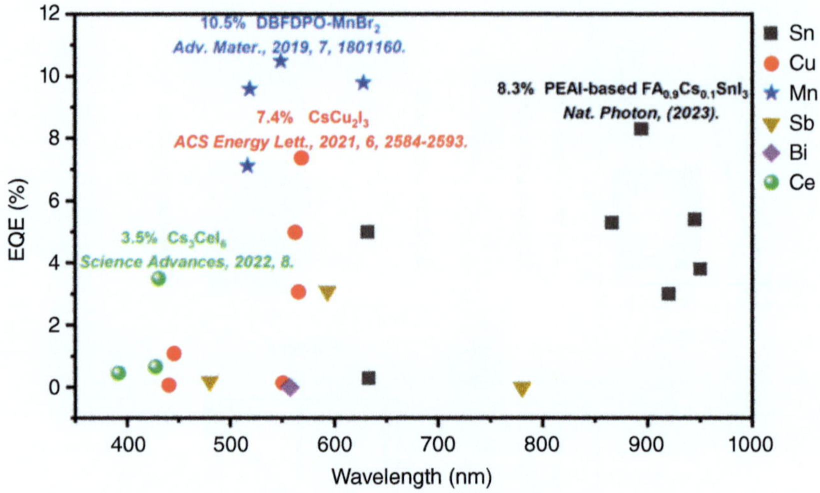

Figure 8.2 Non-lead blue perovskite light-emitting diode efficiency trends.

8.2 Preparation of Non-lead Metal Halide Materials

The ligand-assisted reprecipitation (LARP) method, on the other hand, involves the precipitation of perovskite quantum dots by the addition of a precursor solution containing both the metal cation and the organic ligand to a non-solvent solution. The non-solvent solution can be a poor solvent for the perovskite, which triggers the precipitation of the perovskite quantum dots. The organic ligand can provide a passivation layer around the perovskite quantum dots to enhance their stability and improve their optical properties.

Both methods have their advantages and disadvantages. The thermal injection method can produce high-quality perovskite quantum dots with good optical properties, but it requires high-temperature reaction conditions and can be difficult to control the size and morphology of the perovskite quantum dots. The LARP method, on the other hand, is a low-temperature process that can produce perovskite quantum dots with good size and morphology control, but it may not achieve the same level of optical properties as the thermal injection method. For example, $CsSnCl_3$ perovskite nanocrystals can be synthesized using Cs_2CO_3 and SnX_2 [6]. To synthesize $CsSnCl_3$ perovskite nanocrystals using the thermal injection method, a tin precursor solution can be prepared by dissolving SnX_2 in tri-*n*-octylphosphine solvent. This solution is then injected into a Cs_2CO_3 precursor solution containing oleic acid and oleylamine as ligands to passivate defects. The reaction is carried out at 170 °C to obtain colloidal stable $CsSnCl_3$ with desirable optical properties. The size of the perovskite crystals is regulated by the reaction temperature. It is important to note that adding lower-temperature precursor solutions to high-temperature reaction solvents can cause uncontrollable changes in temperature, leading to low experimental reproducibility.

The LARP method is a relatively mild synthetic method that utilizes the precipitation of perovskite nanocrystals under supersaturation conditions. For instance, to synthesize $Cs_3Bi_2Br_6$ quantum dots, a precursor solution can be prepared by dissolving a mixture of CsBr, $BiBr_3$, and oleylamine (ligand) in dimethyl sulfoxide (DMSO) (good solvent), followed by the slow dropwise addition of this solution into a poor solvent under stirring conditions [7]. The resulting supersaturated recrystallization causes the perovskite quantum dots to precipitate out. The choice of both the solvent and the poor solvent during the synthesis process can influence the fluorescence quantum efficiency of the quantum dots. Additionally, a series of halide quantum dots can be synthesized through anion exchange, whereby CsCl or CsI is added to the already prepared $Cs_3Sb_2Br_9$ quantum dot solution, and the anion-exchanged perovskite material can be obtained. Although the LARP method is relatively inexpensive and simple to perform at room temperature, the low-temperature reaction conditions often lead to a high density of structural defects and trap states, which may reduce the material's performance.

Encapsulating quantum dots can improve their stability, and one common method is to combine them with a polymer matrix. Tang Jiang et al. used the following method to prepare a $Cs_3Bi_2Br_9$ quantum dot/silica matrix: They first mixed tetraethyl orthosilicate, ethanol, and an ethanol solution of quantum dots, and then added deionized water while adjusting the pH with a trace amount of HBr mixture. After vigorous stirring, the solution turned colorless and clear. The solution was then allowed to naturally gel at room temperature for a few minutes in an airy environment after the cap was opened [7].

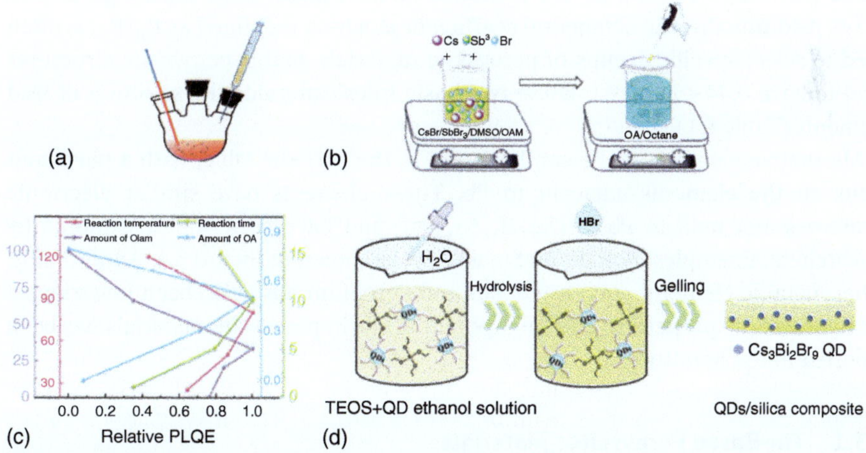

Figure 8.3 (a) Schematic diagram of the thermal injection method. Source: Wang et al. [8]/American Chemical Society; (b) schematic diagram of the ligand-assisted reprecipitation method. Source: Zhang et al. [9]/American Chemical Society; (c) relative PLQY of $Cs_3Bi_2Br_9$ quantum dots prepared under different conditions, where the PLQY is normalized for each test batch. Source: Leng et al. [7]/John Wiley & Sons; (d) schematic diagram of the encapsulation of perovskite quantum dots.

Apart from quantum dots, Hideo Hosono also reported the direct synthesis of zero-dimensional $Cs_3Cu_2I_5$ thin film materials. The synthesis involved preparing CsI-CuI solutions using a mixture of DMSO and DMF as solvents. The resulting solution was then stirred at 60 °C overnight, and the filtrate obtained after stirring was used to prepare thin films via spin-coating. Finally, the films were annealed at 60 °C in a glove box (Figure 8.3).

8.3 Types of Non-lead Metal Halide Materials

Perovskites can adapt to different crystal structures depending on the size of the A-atom radius and its interaction with the BX_6 octahedra, despite the fact that ABX_3 chalcocite compounds are normally cubic crystal systems. To express the stability and degree of deformation of the perovskite structure, Goldschmidt's tolerance factor (t) is frequently used. This factor is determined by the effective ionic radius of the ions in the perovskite material's structure [10].

$$t = \frac{R_A + R_X}{\sqrt{2}\,(R_B + R_X)}$$

where R_A, R_B, and R_X are the effective ionic radii of A, B, and X ions, respectively. The stability and degree of deformation of perovskite structures can be expressed using Goldschmidt's tolerance factor (t), which depends on the effective ionic radius of the ions in the perovskite material's structure. Perovskite materials with tolerance factor values less than 0.7 tend to have non-perovskite structures. For stable perovskite structures, the tolerance factor should be in the range of 0.75 to 1.0. Additionally, the octahedral coefficient μ, which is defined as R_B/R_X, is often used to select specific groups of perovskite materials. Stable perovskite structures tend to have $0.44 < \mu < 0.90$. These two basic rules can guide the selection of lead elements (Table 8.1).

Alternative elements to Pb can be found in the periodic table, with a particular focus on the elements adjacent to Pb. These elements have similar electronic structures and radii to Pb^{2+} ($[Xe]4f_{14}5d_{10}6s_2$) and have been widely explored by researchers. Examples include Bi, Sn, and Sb, as shown in Figure 8.4. Additionally, other metallic elements such as copper and europium have also been investigated for their unique properties, and their potential use in perovskite materials has been reported in the literature.

8.3.1 Tin-Based Perovskites Materials

Sn and Pb belong to the same group in the periodic table, and Sn has more similarities to Pb than other elements. Sn has a similar chemical valence and ionic radius to Pb, and there have been attempts to replace Pb with Sn as early as 2016. However, Cs_2SnCl_6 quantum dots obtained from these attempts exhibit a broad emission peak in the blue light band and a relatively low fluorescence quantum efficiency. The highest fluorescence quantum efficiency for this series of perovskite quantum

Table 8.1 Radii of common ions and ionophores in perovskite materials.

Ionic position	Ions	Ionic valence state	Number of allotments	Radius
A	Cs	+1	6–12, 16	1.52–1.83
	CH_3NH_3	+1	6, 8–12	2.17
	$NH_2(CH)NH_2$	+1		2.53
	Rb	+1		1.52–1.83
B	Pb	+2	4, 6–12	0.98–1.49
		+2	6	1.19
	Sn	+2	4, 6–12	0.96–1.46
		+2	6	1.18
		+4	4–8	0.55–0.81
		+4	6	0.69
	Bi	+3	3, 6, 8	0.96, 1.03, 1.17
	Sb	+3	4, 5, 6	0.76, 0.80, 0.76
	In	+3	4, 6, 8	0.62, 0.80, 0.92
	Na	+1	4–9, 12	0.99–1.39
		+1	6	1.02
	Cu	+1	2, 4, 6	0.46, 0.60, 0.77
		+2	4, 5, 6	0.57, 0.65, 0.73
	Ag	+1	2, 4–8	0.67–1.28
		+1	6	1.15
X	Cl	−1	6	1.81
	Br	−1	6	1.96
	I	−1	6	2.20

dots is only 0.14% [6]. It should be noted that the ease of oxidation of Sn^{2+} to Sn^{4+} limits the stability of tin-based perovskites. For this reason, some researchers have attempted to use Sn^{4+} in Cs_2SnI_6 to improve its stability, but the resulting perovskite material emits yellow light [8]. In 2018, Tang Jiang et al. successfully increased the fluorescence quantum efficiency of Cs_2SnCl_6 by doping it with Bi, reaching a value of 78.9%. Through experiments and calculations, it was found that Bi^{3+} was doped into the Cs_2SnCl_6 substrate at the Sn site, forming a $Bi_{Sn} + V_{Cl}$ defect complex that produced intense blue light. In a separate study, Chao Zhao et al. annealed Bi-doped Cs_2SnCl_6 at high temperatures to improve its stability, but this also reduced the fluorescence quantum efficiency to 50.8% [11]. Recently, a new Sn-based perovskite material, $(C_6N_2H_{16}Cl)_2SnCl_6$, was synthesized, and the resulting single crystals exhibited a broad blue emission peak attributed to self-trapped excitons. Moreover, the emission intensity remained stable for more than three months, indicating improved stability compared to other tin-based perovskites [12]. The newly

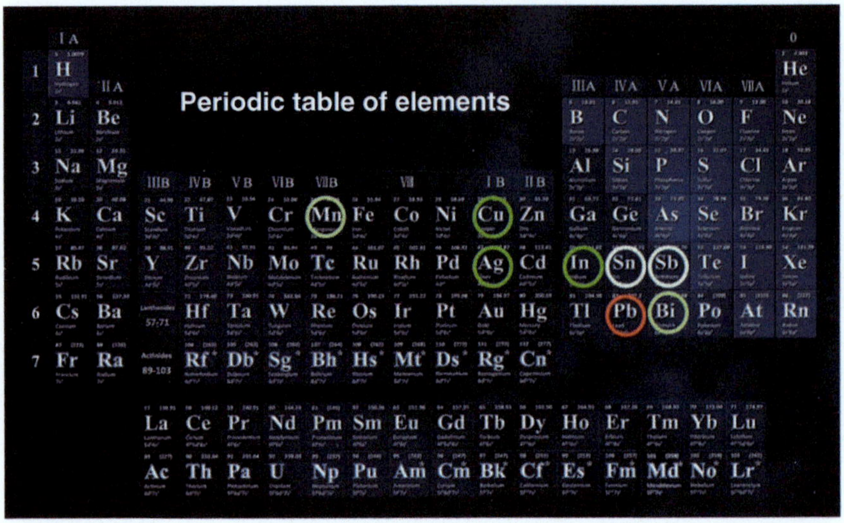

Figure 8.4 Elements often used as substitutes for Pb in the periodic table, with circles colored red–white–green indicating decreasing toxicity in that order.

synthesized $(C_6N_2H_{16}Cl)_2SnCl_6$ perovskites have shown promising characteristics, as their single crystals exhibit a stable blue emission originating from self-trapped excitons that remain unchanged for more than three months. Moreover, the perovskites exhibit good thermal stability, as even when the temperature was increased to 450 K, the luminescence intensity only decreased to around 50% (Figure 8.5).

8.3.2 Bismuth-Based Metal Halide Materials

Bismuth is a common element used to replace Pb in perovskite materials. Bi^{3+} is isoelectronic with Pb^{2+} and has similar properties. Additionally, Bi is less toxic and more environmentally friendly than Pb. In 2016, Tang Jiang et al. synthesized the first blue perovskite quantum dots, $MA_3Bi_2Br_9$ [13]. The group initially synthesized blue perovskite $MA_3Bi_2Br_9$ quantum dots with a fluorescence emission peak at 423 nm and a fluorescence quantum efficiency of 12% with a full width at half maxima (FWHM) of 62 nm. They then passivated the surface of the quantum dots using Cl^-, resulting in an increase in fluorescence quantum efficiency from 12% to 54.1% and a decrease in FWHM from 62 to 41 nm while maintaining constant emission peaks. This was achieved by taking advantage of the difference in crystallization rate and crystal structures of $MABiBr_3$ and $MABiCl_3$. They grew a thin shell of $MABiCl_3$ on the surface of $MABiBr_3$, suppressing the exchange of the shell with the inner halogen atoms [14]. The researchers also investigated the synthesis of non-lead blue perovskite materials using other organic cations. They successfully prepared quantum dots such as $FA_3Bi_2Br_9$, $(BA)_2BiBr_5$ (where BA = $C_4H_9NH^{3+}$), and $(C_8NH_{12})_4BiBr_7$-H_2O, with fluorescence quantum efficiencies of 52%, 1.26%, and 0.7%, respectively [15–17].

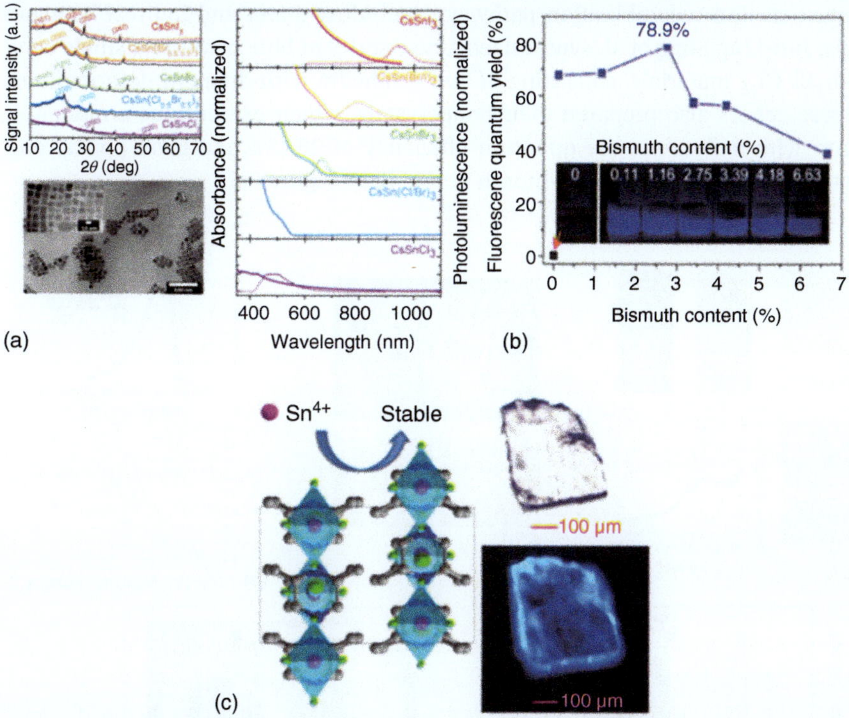

Figure 8.5 (a) XRD spectra, absorbance, and steady-state fluorescence spectra of $CsSnX_3$ perovskite nanocrystals, transmission electron microscope (TEM) images of $CsSnI_3$ nanocrystals [6]; (b) room temperature PLQY values of Cs_2SnCl_6:*x*Bi with different Bi doping values, inset is a photograph of Cs_2SnCl_6:*x*Bi under 365 nm UV irradiation, white values in the inset indicate *x* values [39]; (c) $(C_6N_2H_{16}Cl)_2SnCl_6$ crystal structure and optical images of single crystals under ambient light and UV light irradiation. Source: Song et al. [12]/ Reproduced with permission from American Chemical Society.

All-inorganic perovskites are known to have better stability compared to organic-inorganic hybrid perovskites. To explore this, Keli Han et al. used Cs^+ instead of MA^+ to synthesize all-inorganic halide perovskites. The resulting $Cs_3Bi_2Br_9$ quantum dots exhibited fluorescence quantum efficiencies of 0.2% (without ligand) and 4.5% (with oleic acid passivation), respectively, and demonstrated excellent air stability for over 30 days [18]. Subsequently, Tang et al. improved the synthesis process by using ethanol instead of isopropanol as the solvent while keeping the other steps essentially the same. This modification led to a significant increase in the fluorescence quantum efficiency to 19.4% [7]. Zhang Daoli et al. further optimized the synthesis conditions by utilizing a solvent system with relatively low polarity [19]. They optimized the synthesis conditions by using a relatively weakly polar solvent system to slow down the degradation of the particles caused by polar solvents in conventional synthesis conditions. The resulting $Cs_3Bi_2Br_9$ quantum dots exhibited blue double emission peaks at 414 and 433 nm with a fluorescence quantum efficiency of 29.6%. The double emission peaks may originate from different exciton recombination pathways, including direct and indirect bandgap jumps. Jun-Ling Song et al. synthesized a new series of blue perovskite single crystal $Rb_7Bi_3Cl_{16}$ materials, consisting of two octahedra with different deformations. The researchers also prepared their nanocrystals, which exhibit blue emission at 437 nm with a fluorescence quantum efficiency of 28.43% and good optical and water vapor stability over a one-month period [20] (Figure 8.6).

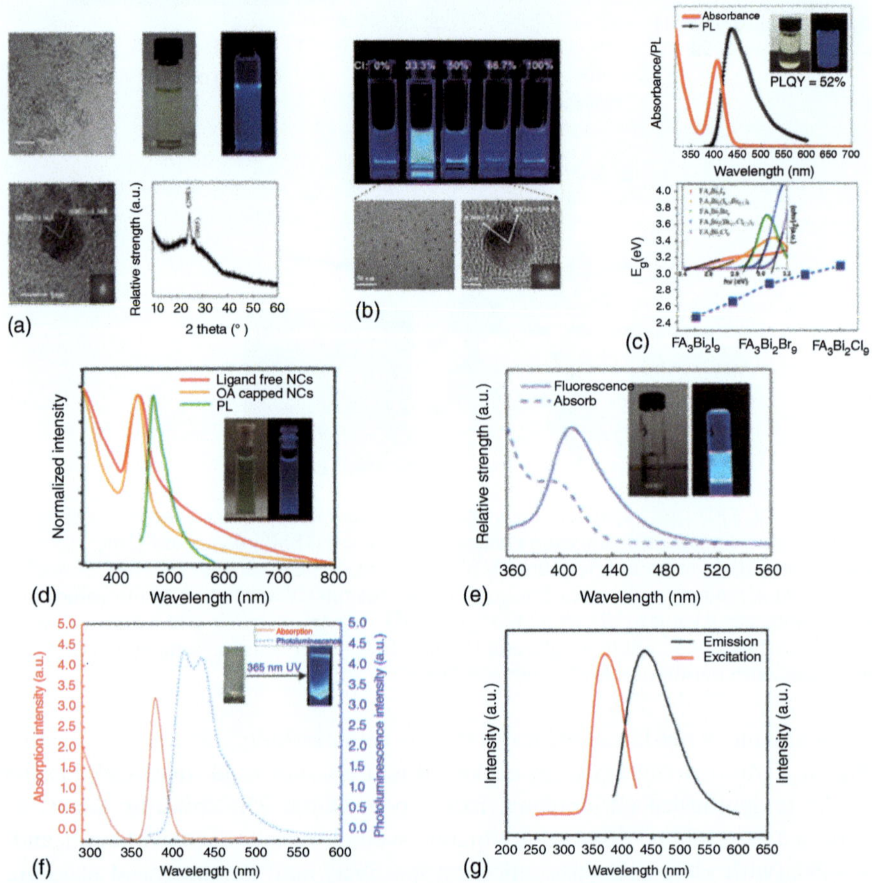

Figure 8.6 (a) Properties of $MA_3Bi_2Br_9$ quantum dots: TEM image with high resolution transmission electron microscope (HRTEM) image, typical optical image of $MA_3Bi_2Br_9$ quantum dot solution with solution under 365 nm UV light, XRD image. Source: Leng et al. [13]/Reproduced from John Wiley & Sons, Inc; (b) properties of Cl-$MA_3Bi_2Br_9$ QD: XRD image, photo of quantum dot solution under 325 nm UV light excitation and TEM image of quantum dots [14]; (c) normalized PL and UV-Vis absorption spectra of $FA_3Bi_2Br_9$ quantum dots and calculated bandgap results of $FA_3Bi_2Br_9$ quantum dots [15]; (d) steady-state fluorescence spectra of ligand-free $Cs_3Bi_2Br_9$ nanocrystals (red), oleic acid capped nanocrystals (orange) and $Cs_3Bi_2Br_9$ nanocrystals [18]; (e) absorption and PL spectra of $Cs_3Bi_2Br_9$ nanocrystals synthesized using ethanol poor solvents [14]; (f) Normalized absorption spectra and fluorescence spectra of $Cs_3Bi_2Br_9$ nanocrystals synthesized with weakly polar solvents [19]; (g) fluorescence spectra of $Rb_7Bi_3Cl_{16}$ nanocrystals. Source: Xie et al. [20]/Royal Society of Chemistry.

8.3.3 Antimony-Based Metal Halide Materials

Haisheng Song first synthesized $Cs_3Sb_2Br_9$ quantum dots using the LARP method. The luminescence peak of this material was observed at 410 nm, and it exhibited a high fluorescence quantum efficiency of 46% [9]. In their study, Haisheng Song and colleagues optimized the precursor concentration, solvent/ligand ratio, and reaction temperature to achieve a high fluorescence quantum efficiency of 46% for $Cs_3Sb_2Br_9$ quantum dots synthesized using LARP. Analysis of the perovskite single crystal structure revealed a two-dimensional layered $Cs_3Sb_2Br_9$ structure formed by removing Sb every three layers along the ⟨111⟩ direction. To investigate the main surface passivation mechanism, UV ozone and Ar^+ ion etching were used, and the results suggested that the higher fluorescence quantum efficiency was mainly generated by the material's abundant Br atomic effect, rather than ligand passivation. The relatively high exciton binding energy in the material (530 meV) explains the enhanced fluorescence intensity in $Cs_3Sb_2Br_9$. The air and light stability of the material were also tested for 35 days and 108 hours, respectively. In another study by Chongxin Shan and colleagues in 2019, $Cs_3Sb_2Br_9$ quantum dots were synthesized and found to have excellent photothermal and air stability. They further prepared $Cs_3Sb_2Br_9$ quantum dots with $Al/MoO_3/TCTA/Cs_3Sb_2Br_9/PEI/ZnO/ITO$ structures [21]. The LED based on $Cs_3Sb_2Br_9$ quantum dots shows emission spectra similar in shape and position to those of quantum dots, with an external quantum efficiency of 0.206%. The device also displays good stability, with the electroluminescence intensity decreasing to only 90% of the initial value after six hours of continuous operation. The stability and constant current density of the $Cs_3Sb_2Br_9$ LED can be explained by the high diffusion potential of V_{Sb+} in $Cs_3Sb_2Br_9$, which is much higher than that of $CsPbBr_3$ in the cubic crystal system. Theoretical calculations support this explanation (Figure 8.7).

8.3.4 Copper-Based Metal Halide Materials

Copper, a low-toxicity transition metal, has been used as a replacement for lead in non-lead perovskite materials by researchers. Since copper atomic orbitals have a small spatial dispersion, exciton diffusion can be limited. Therefore, researchers have focused on copper(I) halide compounds with a smaller valence band dispersion than that of $CsPbX_3$, consisting of hybrid Pb 6s and halogen p orbitals, which are more conducive to luminescence. $Cs_3Cu_2I_5$ quantum dots, thin films, and single crystals emitting blue light have fluorescence quantum efficiency of 29.5%, 62.5%, and 91.2%, respectively. Hosono's group successfully prepared the first electroluminescent diodes from $Cs_3Cu_2I_5$ films using $ITO/ZSO/Cs_3Cu_2I_5/NPD/MoO_x/Ag$, demonstrating good performance and stability [22, 23]. Tang Xiaosheng et al. improved the performance of the device by using ethyl acetate as an antisolvent to treat the perovskite films. Methyl acetate was added dropwise to the perovskite precursor solution during spin-coating to improve the nucleation growth of the $Cs_3Cu_2I_5$ films. This step effectively improved the roughness and crystal regularity of the films, resulting in an increased fluorescence quantum efficiency of 76%.

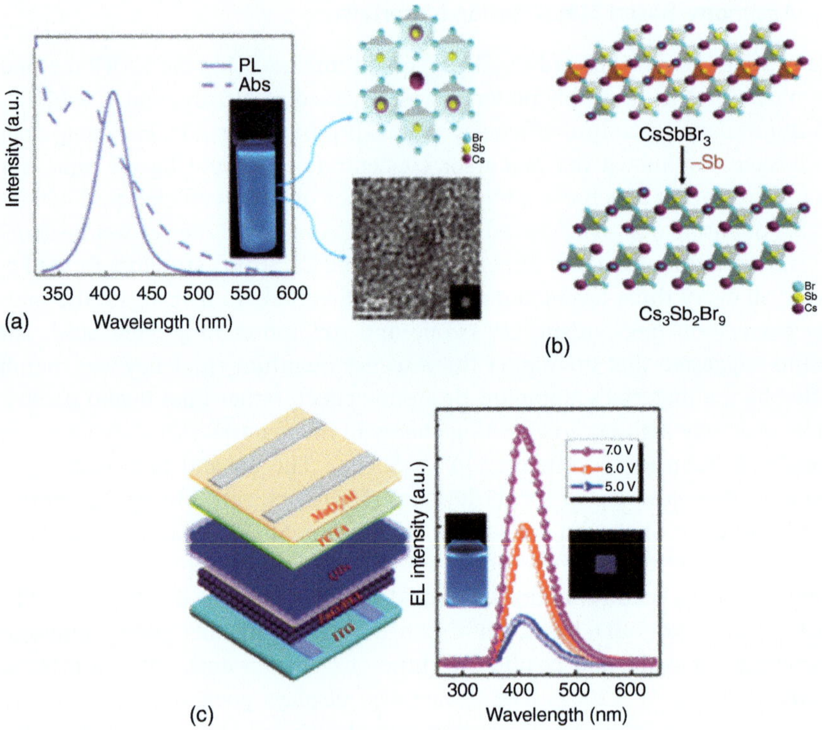

Figure 8.7 (a) Fluorescence and absorption spectra of $Cs_3Sb_2Br_9$ quantum dots under UV lamp illumination, optical images; (b) TEM images and crystal structure of $Cs_3Sb_2Br_9$ [9]; (c) device structure and luminescence spectra of $Cs_3Sb_2Br_9$ quantum dots. Source: Ma et al. [21]/Reproduced from American Chemical Society.

The film properties remained essentially unchanged after two months of storage in ambient air. However, the maximum brightness of the device was still only $\approx 10\,cd\,m^{-2}$ [24]. Wang Yongtian's group prepared $Cs_3Cu_2I_5$ crystalline material by solvent evaporation crystallization method [24]. However, the fluorescence quantum efficiency decreased from 89% to 60% after grinding the single crystal into powder. The group explored the application of this material in displays by using the precursor solution as a fluorescent ink, which showed stable fluorescence of the drawn graphics in the air for several months. Additionally, they investigated the use of laser writing technology in $Cs_3Cu_2I_5$/PVDF composite films to produce specific patterns. On the other hand, the luminescence of the divalent copper-based perovskite compound $Cs_2Cu_2X_4$ shows a different trend, with a red-shifted emission wavelength as the diameter of the halogen atom increases. The optical properties of these quantum dots can be tuned by adjusting the ratio of precursors to control the particle size, where the synthesized $Cs_2Cu_2Cl_4$ emits blue light [24, 25]. Zhanxin Wu and colleagues have developed a method for preparing perovskite films for use in electroluminescent devices, using dual-source thermal evaporation. Their work has resulted in an external quantum efficiency of approximately 0.1% and a maximum brightness of $70\,cd\,m^{-2}$ [26]. Bayrammurad Saparov's group recently reported the successful preparation of K_2CuCl_3 with a fluorescence quantum efficiency of 97%. However, they did not discuss the application of this material in devices [27]. In the same year, Shan Chongxin's group prepared $Cs_3Cu_2I_5$ nanocrystals via thermal injection and fabricated light-emitting devices using thin films with the device structure of Al/LiF/TPBi/$Cs_3Cu_2I_5$/NiO/ITO. They achieved dark blue emitted light with an exo-quantum efficiency of 1.12% and a device T_{50} exceeding 100 hours [28] (Figure 8.8).

8.3.5 Europium-Based Metal Halide Materials

To date, rare earth elements have rarely been used in metal halide perovskite blue-light materials. However, Peter Reiss et al. successfully introduced europium, which is octahedrally coordinated in the divalent state and has an ionic radius similar to that of Pb^{2+} in the hexa-coordinated configuration (117 pm/119 pm), into the perovskite structure [29]. The synthesis of europium-based perovskite materials using a hot injection method was reported by Peter Reiss et al. The materials have specific temperature requirements for synthesis; high temperatures (>130 °C) result in very low luminescence efficiency or partial oxidation of Eu^{2+}, while lower reaction temperatures (110 °C) lead to only partial precursor conversion, low reaction yields, and limited stability. The emitting source of this class of materials is Eu^{2+}, which has a $4f^7$-$4f^6 5d^1$ jump allowed by Laporte, and the energy cannot be adjusted by changing the particle size or halide nature. The corresponding perovskite $CsEuBr_3$ has a luminescence peak at 413 nm with a fluorescence quantum efficiency of 39% (the highest reported value for lead-free ABX_3 nanocrystals), excellent stability, and the narrowest FWHM (30 nm) of the current non-lead blue perovskite materials. These excellent properties make $CsEuBr_3$ nanocrystals promising for use in blue-light devices as light-emitting materials. It is also worth exploring the application of other rare earth elements (e.g. Sm^{2+}) in perovskite LEDs (Figure 8.9).

8.3.6 Bimetallic Cationic Halide Perovskites Materials

Reducing the dimensionality of perovskite structures, as reported in the literature, typically results in a reduced energy band spread and lower effective carrier mass and mobility, which can have a negative impact on the material's electrical properties [13]. Electroluminescent device preparation is often limited due to reduced energy band spread, lower effective carrier mass, and mobility resulting from the reduced dimensionality of perovskite structures reported in the literature. However, Keli Han et al. attempted to overcome this limitation by synthesizing $ABB'X_6$-type double perovskite with a monovalent Ag^+ and a trivalent Bi^{3+} instead of two Pb^{2+}. By using a monovalent B cation and a trivalent B′ cation, a three-dimensional cubic perovskite structure was formed. The fluorescence quantum efficiency of $Cs_2AgBiCl_6$ was found to be 6.7%. Interestingly, the introduction of oleic acid to passivate surface defects resulted in a 100-fold increase in the fluorescence quantum efficiency of this class of materials, making them

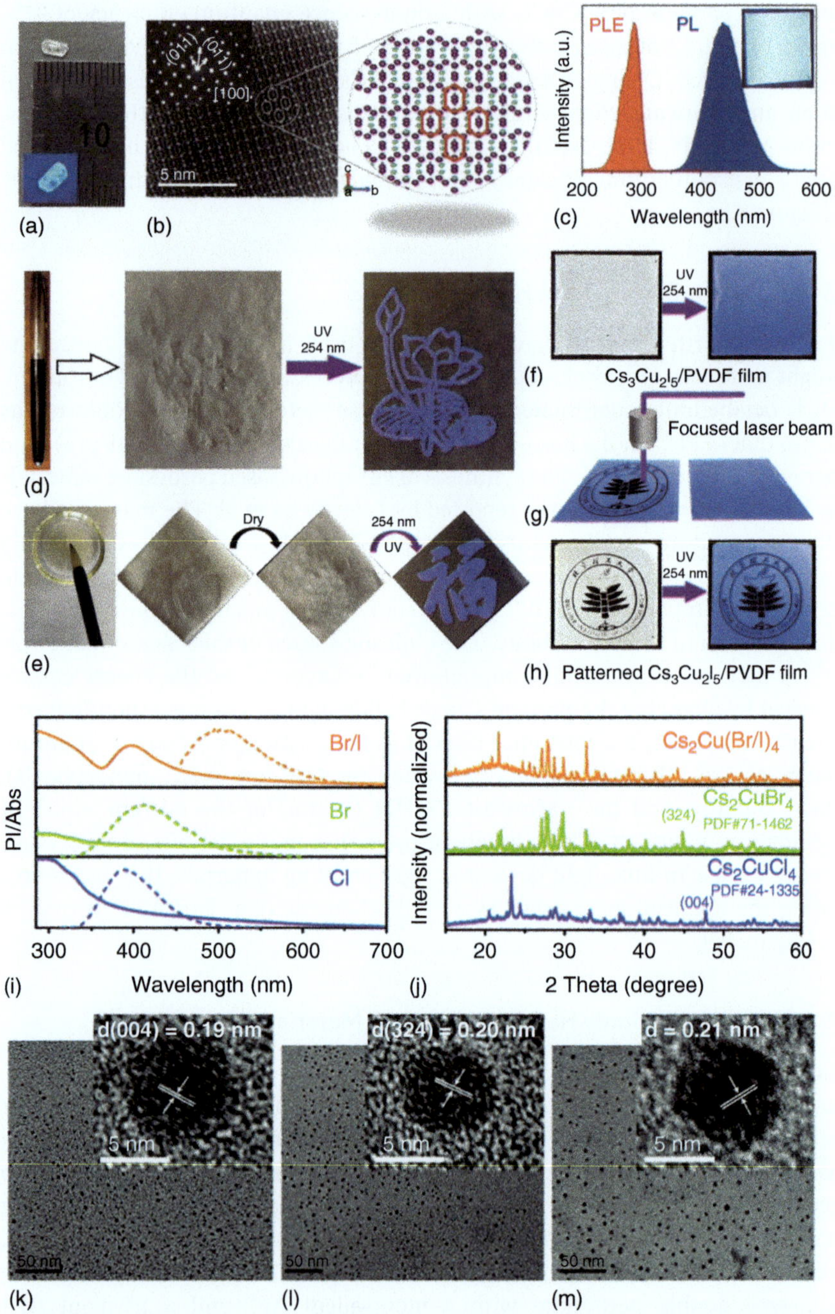
(011)
(01̄1)
[100]
5 nm
PLE
PL
Intensity (a.u.)
200 300 400 500 600
Wavelength (nm)
(a)
(b)
(c)
UV
254 nm
(d)
Dry
254 nm
UV
(e)
(f) Cs3Cu2I5/PVDF film
Focused laser beam
(g)
(h) Patterned Cs3Cu2I5/PVDF film
Br/I
Br
Cl
Pl/Abs
300 400 500 600 700
(i) Wavelength (nm)
Cs2Cu(Br/I)4
Cs2CuBr4
(324) PDF#71-1462
Cs2CuCl4
PDF#24-1335
(004)
Intensity (normalized)
20 30 40 50 60
(j) 2 Theta (degree)
d(004) = 0.19 nm
d(324) = 0.20 nm
d = 0.21 nm
5 nm
50 nm
(k)
(l)
(m)

Figure 8.8 (a) Photographs of $Cs_3Cu_2I_5$ single crystals under photoexcitation (245 nm light) and (b) TEM images and electron diffraction patterns of selected regions of $Cs_3Cu_2I_5$ single crystals (right: projected view of $Cs_3Cu_2I_5$ (100); Cs, Cu, and I are represented in green, blue, and purple, respectively), (c) photoluminescence (PL) and photoluminescence excitation (PLE) spectra of $Cs_3Cu_2I_5$ films prepared on glass substrates (excited by 245 nm monochromatic light; inset shows their photographs) [22]. (d) Photograph of a pen with $Cs_3Cu_2I_5$ precursor solution as an ink inside, colorless lotus picture on the weighing paper under daylight and corresponding blue emissive picture under UV lamp. (e) Ink for the brush calligraphy writing, wet Chinese character for "blessing" when it was still wet, and after drying naturally, and the corresponding blue emissive picture of the latter taken under UV lamp. (f) Photograph off the $Cs_3Cu_2I_5$/PVDF composite film on the glass substrate, and its blue emission picture under UV lamp. (g) Schematic illustration of a direct laser writing technology. (h) Photograph of a patterned $Cs_3Cu_2I_5$/PVDF film with a BIT logo under daylight and under UV lamp. Source: Adapted from Zhang et al. [24]; (c) (i) steady-state absorption and PL spectra, (j) XRD spectra and (k–m) TEM images of Cs_2CuCl_4, Cs_2CuBr_4 and $Cs_2Cu(Br/Cl)_4$; inserts: the corresponding HRTEM images. Source: Yang et al. [25]/Royal Society of Chemistry.

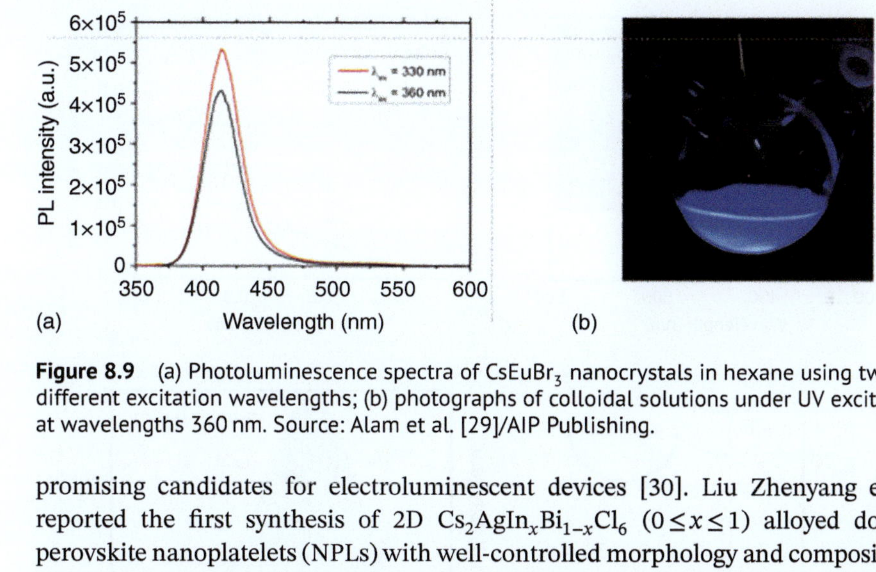

Figure 8.9 (a) Photoluminescence spectra of $CsEuBr_3$ nanocrystals in hexane using two different excitation wavelengths; (b) photographs of colloidal solutions under UV excitation at wavelengths 360 nm. Source: Alam et al. [29]/AIP Publishing.

promising candidates for electroluminescent devices [30]. Liu Zhenyang et al. reported the first synthesis of 2D $Cs_2AgIn_xBi_{1-x}Cl_6$ ($0 \leq x \leq 1$) alloyed double perovskite nanoplatelets (NPLs) with well-controlled morphology and composition. The obtained NPLs show unique optical properties with the highest PLQY of 40.1% [31]. Shilin Jin et al. successfully doped Ln^{3+} and Bi^{3+} ions into $Cs_2AgInCl_6$ DP microcrystals using a modified coprecipitation method. Bi^{3+} doping reduced the excitation energy required for self-trapped exciton (STE) recombination to the UV region (330–400 nm) and significantly increased the PLQY from around 1% to approximately 25% by modifying the band-edge structure of $Cs_2AgInCl_6$ DP. Importantly, co-doping Ln^{3+} further raised the PLQY to 43–56% by combining Ln^{3+} emissions with broadband STE luminescence [32]. Lou Yongbing et al. reported the synthesis of highly stable bis-$CaTiO_3$ $Cs_2AgBiCl_6$ quantum dots using ethyl acetate as a poor solvent. The prepared quantum dots exhibited excellent stability even after 90 days and 500 hours in air and had a high fluorescence quantum efficiency of 31.33% [33]. The material synthesized by Zou et al. has demonstrated exceptional stability in an air environment for up to six months, which surpasses the stability of any other reported non-lead blue light material. Theoretical calculations suggest that the valence band top and conduction band bottom are primarily derived from the antibonding orbitals of Ag 4d-Br 5p and the bonding orbitals of Bi 6p-Br 4p,

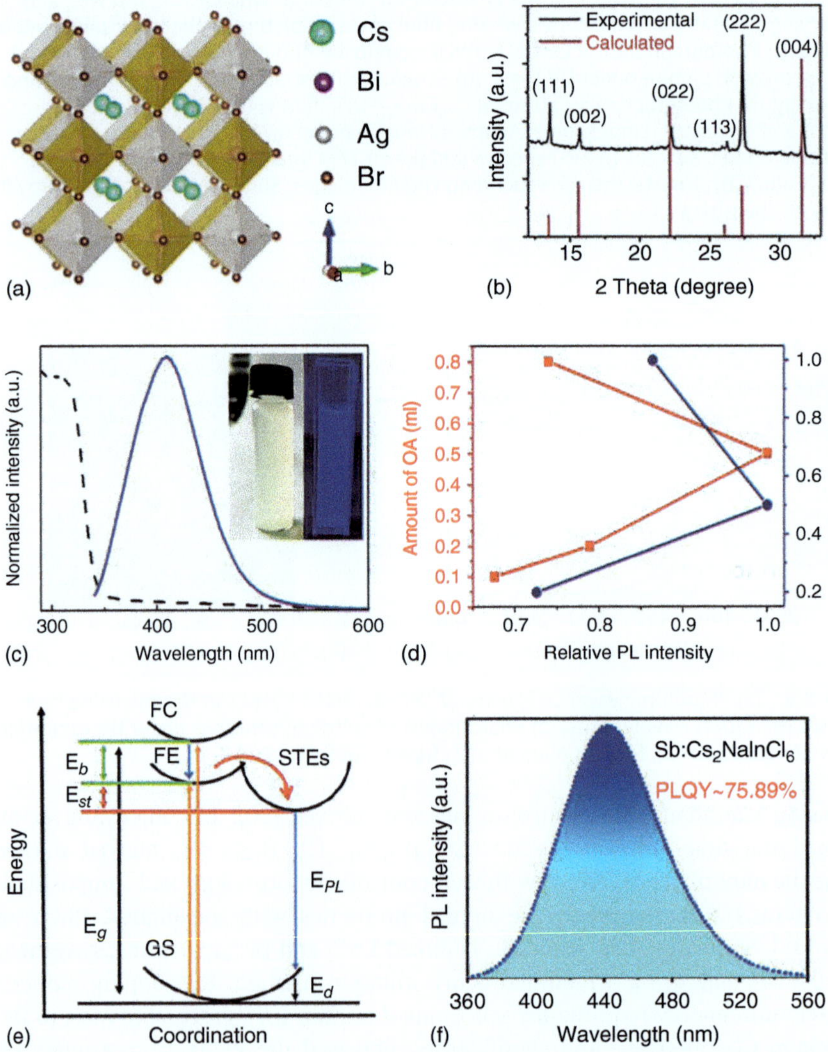

Figure 8.10 (a) Crystal structures of $Cs_2AgBiBr_6$; (b) XRD patterns from experiment and theoretical calculation [30]; (c) steady-state absorption and fluorescence spectra. Insets: images of quantum dot solutions in normal environment and under 325 nm UV lamp irradiation, and (d) relative fluorescence intensities of perovskite colloidal solutions under different conditions, where for each experiment, the PL intensity was normalized to the batch that yielded the highest intensity value. Source: Lv et al. [33]/Royal Society of Chemistry; (e) schematic of the energy levels of Sb^{3+}:$Cs_2NaInCl_6$ with (f) PL spectra [34].

respectively, which confine the wave functions of carriers to the [$AgBr_6$] and [$BiBr_6$] octahedra. This accounts for the low carrier mobility and large effective mass of carriers in biperovskite materials, highlighting that the electronic size has a more significant effect on the optoelectronic properties of halide perovskites than the structural size. Zou's group has synthesized Sb-doped $Cs_2NaInCl_6$ double perovskite materials, which have achieved a fluorescence quantum efficiency of 65.5% and a peak emission wavelength of 479 nm. These results demonstrate the potential of Sb-doped $Cs_2NaInCl_6$ as a promising candidate for blue-light-emitting devices [34]. The composite perovskite emits blue light by self-trapped excitons, and Zou et al. synthesized Sb^{3+}-doped $Cs_2NaInCl_6$ double perovskite materials. The fluorescence quantum efficiency of the $Cs_2NaInCl_6$:Sb^{3+} material was as high as 75.89% at a doping concentration of 10%. The material also showed excellent photostability, with the luminescence intensity decreasing by only 10% after 1000 hours of 365 nm UV irradiation. The diffraction peaks of X-ray diffraction (XRD) patterns indicated the high purity of the prepared quantum dots. X-ray photoelectron spectroscopy and energy scattering spectrometry results demonstrated that the doping of Sb^{3+} occupied In^{3+} lattice sites, forming $Cs_2NaIn_{1-x}Sb_xCl_6$. Therefore, the Sb^{3+} doping did not introduce the possibility of vacancies and defects that reduce the fluorescence quantum efficiency (Figure 8.10).

8.4 Methods for Optimizing the Fluorescence Quantum Efficiency of Non-lead Metal Halide Materials

8.4.1 Surface Passivation

Surface passivation is a proven technique to enhance the fluorescence quantum efficiency of perovskite materials. Perovskite crystals typically form at relatively low temperatures, resulting in various defect states in the material structure due to ionic bonding. Additionally, perovskite crystals have soft lattices that are susceptible to surface deformation, accelerating the formation of defects on the nanoparticle surface. Furthermore, electron–phonon coupling and ion migration can also generate defect states with energy levels ranging from 100 to 400 meV below the perovskite material's bandgap, thereby trapping charge carriers and limiting luminescence [36].

Surface passivation using ligands is an effective way to improve the fluorescence quantum efficiency of perovskite materials. Perovskite crystals are usually formed at relatively low temperatures, leading to various defect states in the ionic bonds of the material structure. Additionally, perovskite crystals have a soft lattice structure that is prone to surface deformation, accelerating the formation of defects on the nanoparticle surface. Defect states with energy levels 100–400 meV below the bandgap of perovskite materials can also be formed by electron–phonon coupling and ion migration, capturing charge carriers and limiting luminescence. The addition of surfactants, such as oleic acid, during the synthesis process can increase the fluorescence quantum efficiency of $Cs_3Bi_2Br_9$ quantum dots from 0.2% to 4.5% [18]. Keli Han et al. reported a significant increase in fluorescence intensity by a factor of 100 for partially passivated nanocrystals using oleic acid. This resulted in an improved fluorescence quantum efficiency of 6.7% [30]. On the other hand, these insulating organic ligands can severely hinder charge injection and transport in the crystal films [36]. The effects of passivation introduced during the synthesis process, as shown in Figure 8.4, have also proven to be effective. $Cs_3Bi_2Br_9$ quantum dots with Br-rich surfaces are beneficial for the optical properties of europium, as the wider bandgap on the surface can introduce an internal electric field and prevent excited electrons from being trapped by surface defects. The quasi-quantum well energy band structure facilitates effective radiative recombination, resulting in enhanced fluorescence quantum efficiency. Tang Jiang et al. designed Cl^- anions to act as passivating ligands that cover the surface of $MA_3Bi_2Br_9$ quantum dots. The incompatible crystal structure between $MA_3Bi_2Br_9$ and $MA_3Bi_2Cl_9$ effectively suppresses surface defects and halogen atom shuttling to enhance fluorescence quantum efficiency. Additionally, Tang Jiang et al. observed that water can react with the material surface and passivate the $Cs_3Bi_2Br_9$ colloidal quantum dots due to the self-passivation of the reaction product BiOBr. However, insulating organic ligands may inhibit charge injection and transport in crystal films [7].

8.4.2 Selection of Solvents and Undesirable Solvents

The choice of solvents and poor solvents plays a critical role in the synthesis and optical properties of perovskite materials. Tang Jiang et al. demonstrated that by replacing the unfavorable solvent isopropanol with ethanol during the synthesis of $Cs_3Bi_2Br_9$ quantum dots, the fluorescence quantum efficiency was significantly improved from 4.5% to 19.4% [7]. Dao-Li Zhang et al. successfully slowed down the degradation of crystal particles by using a relatively weak polar solvent in their experiment [19]. In his experiment, He synthesized $Cs_3Bi_2Br_9$ quantum dots with blue double emission peaks at 414 and 433 nm and achieved a high fluorescence quantum efficiency of 29.6%. He also compared the effect of different solvents on the synthesis of quantum dots using the conventional LARP method. In this method, DMSO acted as the 'good' solvent, toluene acted as the 'bad' solvent, and oleylamine and oleic acid were used as auxiliary ligands. The fluorescence quantum efficiency of the synthesized quantum dots was only 4.3%. Additionally, the yield of the conventional method was very low due to the solubilizing and destructive effects of the polar solvent. This highlights the superiority of using a less polar solvent in the synthesis process.

8.4.3 Doping

Doping is a useful technique to adjust the physical properties of perovskite materials. Bi^{3+} ions, with similar ionic radius and electronic configuration as Pb^{2+} ions, are often used as the luminescence center in halogenated perovskites. Zhou et al. used Bi-doping to prepare halogenated lead perovskites with ultrawide photoluminescence and electroluminescence [35]. Doping has been proven to be an effective method for modulating the physical properties of perovskite materials. Bi^{3+}, with a similar ionic radius and electronic configuration to Pb^{2+}, is commonly used for doping and is expected to be the luminescence center of halogenated perovskites.

While there is limited literature on doping in lead-free blue perovskite quantum dots, the highest fluorescence quantum efficiency achieved to date is 78.9% by doping Cs_2SnCl_6 with Bi. Although Sn elements have been previously attempted by many researchers for the preparation of perovskite luminescent materials, the quantum dots obtained were both unstable and exhibited broad emission in the blue region [6, 8]. Tang Jiang et al. successfully synthesized high-quality perovskite materials with excellent luminescence by doping bismuth into tin-based perovskite materials. They attributed the luminescence enhancement effect to the formation of $Bi_{Sn} + V_{Cl}$ defect complexes, which produce intense blue light. However, previous attempts to use Sn elements for perovskite luminescent materials resulted in unstable and broad blue-emitting quantum dots. Similarly, Binglock Zou et al. synthesized $Cs_2NaInCl_6:Sb^{3+}$ double perovskite material and achieved the highest fluorescence quantum efficiency of 75.89% after optimizing the doping concentration. X-ray diffraction spectra by Qiu et al. indicate the high purity of the prepared quantum dots, while the X-ray photoelectron spectroscopy and energy dispersive X-ray spectroscopy results also demonstrate that the doped Sb^{3+} occupies In^{3+} lattice sites and forms $Cs_2NaIn_{1-x}Sb_xCl_6$. The appropriate Sb doping concentration can break the cosmic ring ban leap rule and effectively modulate the electronic density of states, thus significantly improving the fluorescence quantum efficiency without introducing a large number of vacancies and defects (Figure 8.11).

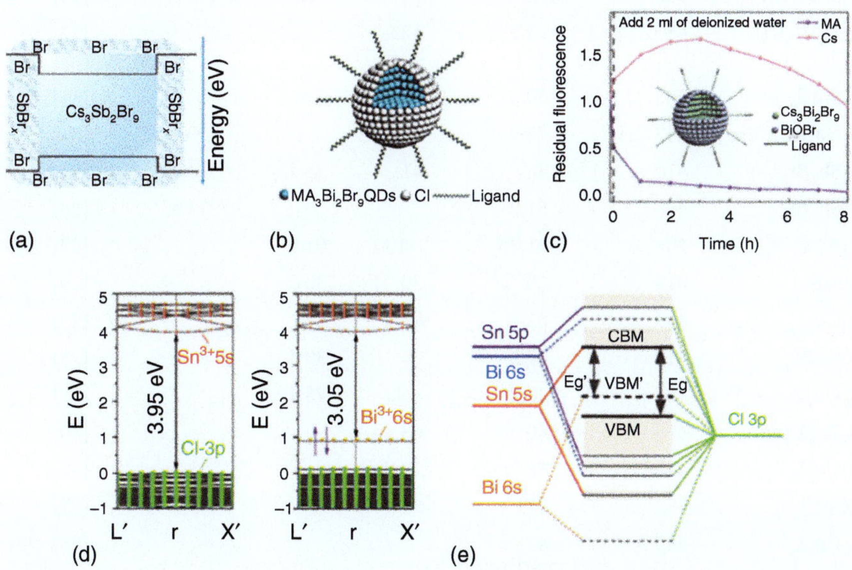

Figure 8.11 (a) Proposed band structure diagram of $Cs_3Sb_2Br_9$ IPQDs [9]; (b) passivation model of Cl-passivated $MA_3Bi_2Br_9$ QDs [14]; (c) stability of $Cs_3Bi_2Br_9$ QDs after adding DI water (2 mL DI water was added to 2 mL $Cs_3Bi_2Br_9$ QD solutions, and then 2 mL mixed solution was taken out to make characterization). Insets: Scheme of $Cs_3Bi_2Br_9$ QDs passivated by BiOBr after adding water [7]; (d) calculated energy band structures of pure Cs_2SnCl_6 and Bi-doped Cs_2SnCl_6; (e) Schematic energy diagram for Bi-doped Cs_2SnCl_6. The antibonding states between Bi 6s orbital and Cl 2p orbital form the VBM'. Transition between VBM' and CBM explains the observed blue emission [39].

Table 8.2 Properties of non-lead blue perovskite materials reported in the literature.

ABX_3	Luminous peak (nm)	Fluorescence quantum yield (%)	Half-peak width (nm)	Air stability	Light stability	Literature sources
$Cs_3Bi_2Br_9$	468	4.5	40		400 min	[18]
$Cs_3Bi_2Br_9$	410	19.4	48		78 h	[7]
$Cs_3Bi_2Br_9$	414 & 433	29.6	50	21 d		[19]
$Cs_3Bi_2Br_9$	414	22	38			[37]
$Rb_7Bi_3Cl_{16}$	437	28.43	93		12 h	[20]
$FA_3Bi_2Br_9$	437	52	65	30 d		[15]
$MA_3Bi_2Br_9$	423	12	62		25 h	[13]
Cl-$MA_3Bi_2Br_9$	422	54.1	41	26 d	12 h	[14]
$(BA)_2BiBr_5$	415	1.26	57			[16]
$(C_8NH_{12})_4BiBr_7$-H_2O	450	0.7	~50			[17]
$Cs_2AgSb_{1-x}Bi_xCl_6$	452		~60			[38]
$Cs_2AgBiCl_6$	395	6.7	68	90 d	500 h	[30]
Cs_2SnCl_6:Bi	455	78.9	66	96 h		[39]
$(C_6N_2H_{16}Cl)_2SnCl_6$	450	8.1	125			[12]
Cs_2SnCl_6:Bi	483	50.8	80	3 m		[11]
$Cs_2NaInCl_6$:Sb	442	75.9	66		1000 h	[21]
$Cs_3Sb_2Br_9$	410	46	41	35 d	108 h	[9]
$Cs_3Sb_2Br_9$	408	51.2			72 h	[34]
$Cs_2AgSbCl_6$	409	31.33	~100	6 m		[33]
$CsEuBr_3$	413	39	30	30 d		[29]
$Cs_3Cu_2I_5$	445	62.1	~70	2 m		[22]
Cs_2CuBr_4	410	37.5		30 d	83 h	[25]
$Cs_3Cu_2I_5$	445	29.2	~90	45 d		[23]
Cs_2ZrCl_6:xBi	456	50	63			[40]
$Cs_3Cu_2I_5$	440	58	70			[26]
K_2CuCl_3	392	97	54			[27]
$Cs_3Cu_2I_5$	445	87				[28]
$Cs_3Cu_2I_5$	440	96.6			300 h	[41]
Cs_3CeBr_6	421	91		100 h		[42]

Where h is the hour; d is the day; and m is the month.

8.5 Conclusion and Outlook

This chapter focuses on the development of non-lead blue perovskite materials and the methods used to optimize their fluorescence quantum efficiency. Currently, there are two main methods for synthesizing non-lead blue perovskite titanites: thermal injection and LARP. Many elements have been investigated as possible replacements for Pb, with Bi-doped Sn-based halide perovskite quantum dots showing promising results. Methods for improving the fluorescence quantum efficiency of non-Pb blue perovskite materials include optimizing solvents and antisolvents, surface passivation, and doping. However, more recent methods such as posttreatment and ligand engineering are still under development.

The performance of current electro-luminescent devices based on non-lead materials is still unsatisfactory due to poor electrical properties resulting from low fluorescence quantum efficiency or a large bandgap. The half-peak width of non-lead metal halide blue-emitting materials is typically 40–60 nm, which is much wider than that of lead halide perovskite materials and unfavorable for display applications. Although rare-earth Eu-based perovskites have reported the narrowest half-peak width, there is still much room for improvement in the synthesis and optimization of this material (Table 8.2).

Despite the many advances in perovskite research, the specific design principles for optimal material performance remain unknown. Further exploration of these principles can advance our understanding and research in this field. A major breakthrough for perovskite LEDs would be the realization of non-lead blue perovskite materials with both the excellent performance of lead-containing perovskites and nontoxic properties.

References

1 Tian, Y., Zhou, C., Worku, M. et al. (2018). Highly efficient spectrally stable red perovskite light-emitting diodes. *Advanced Materials* 30: 1707093.

2 Lin, K., Xing, J., Quan, L.N. et al. (2018). Perovskite light-emitting diodes with external quantum efficiency exceeding 20 per cent. *Nature* 562: 245–248.

3 Sadhanala, A., Ahmad, S., Zhao, B. et al. (2015). Blue-green color tunable solution processable organolead chloride-bromide mixed halide perovskites for optoelectronic applications. *Nano Letters* 15: 6095–6101.

4 Xing, J., Zhao, Y., Askerka, M. et al. (2018). Color-stable highly luminescent sky-blue perovskite light-emitting diodes. *Nature Communications* 9: 3541.

5 Hailegnaw, B., Kirmayer, S., Edri, E. et al. (2015). Rain on methylammonium lead iodide based perovskites: possible environmental effects of perovskite solar cells. *The Journal of Physical Chemistry Letters* 6: 1543–1547.

6 Jellicoe, T.C., Richter, J.M., Glass, H.F. et al. (2016). Synthesis and optical properties of lead-free cesium tin halide perovskite nanocrystals. *Journal of the American Chemical Society* 138: 2941–2944.

7 Leng, M., Yang, Y., Zeng, K. et al. (2018). All-inorganic bismuth-based perovskite quantum dots with bright blue photoluminescence and excellent stability. *Advanced Functional Materials* 28: 1704446.

8 Wang, A., Yan, X., Zhang, M. et al. (2016). Controlled synthesis of lead-free and stable perovskite derivative Cs_2SnI_6 nanocrystals via a facile hot-injection process. *Chemistry of Materials* 28: 8132–8140.

9 Zhang, J., Yang, Y., Deng, H. et al. (2017). High quantum yield blue emission from lead-free inorganic antimony halide perovskite colloidal quantum dots. *ACS Nano* 11: 9294–9302.

10 Chatterjee, S. and Pal, A.J. (2018). Influence of metal substitution on hybrid halide perovskites: towards lead-free perovskite solar cells. *Journal of Materials Chemistry A* 6: 3793–3823.

11 Yan, A., Li, K., Zhou, Y. et al. (2020). Tuning the optical properties of Cs_2SnCl_6:Bi and Cs_2SnCl_6:Sb lead-free perovskites via post-annealing for white LEDs. *Journal of Alloys and Compounds* 822: 153528.

12 Song, G., Li, M., Yang, Y. et al. (2020). Lead-free tin(IV)-based organic-inorganic metal halide hybrids with excellent stability and blue-broadband emission. *The Journal of Physical Chemistry Letters* 11: 1808–1813.

13 Leng, M., Chen, Z., Yang, Y. et al. (2016). Lead-free, blue emitting bismuth halide perovskite quantum dots. *Angewandte Chemie International Edition* 55: 15012–15016.

14 Leng, M., Yang, Y., Chen, Z. et al. (2018). Surface passivation of bismuth-based perovskite variant quantum dots to achieve efficient blue emission. *Nano Letters* 18: 6076–6083.

15 Shen, Y., Yin, J., Cai, B. et al. (2020). Lead-free, stable, high-efficiency (52%) blue luminescent $FA_3Bi_2Br_9$ perovskite quantum dots. *Nanoscale Horizons* 5: 580–585.

16 Wang, J., Li, J., Wang, Y. et al. (2019). A new organic-inorganic bismuth halide crystal structure and quantum dot bearing long-chain alkylammonium cations. *Organic Electronics* 70: 155–161.

17 Zhang, R., Mao, X., Yang, Y. et al. (2019). Air-stable, lead-free zero-dimensional mixed bismuth-antimony perovskite single crystals with ultra-broadband emission. *Angewandte Chemie International Edition* 58: 2725–2729.

18 Yang, B., Chen, J., Hong, F. et al. (2017). Lead-free, air-stable all-inorganic cesium bismuth halide perovskite nanocrystals. *Angewandte Chemie International Edition* 56: 12471–12475.

19 Gao, M., Zhang, C., Lian, L. et al. (2019). Controlled synthesis and photostability of blue emitting $Cs_3Bi_2Br_9$ perovskite nanocrystals by employing weak polar solvents at room temperature. *Journal of Materials Chemistry C* 7: 3688–3695.

20 Xie, J.L., Huang, Z.Q., Wang, B. et al. (2019). New lead-free perovskite $Rb_7Bi_3Cl_{16}$ nanocrystals with blue luminescence and excellent moisture-stability. *Nanoscale* 11: 6719–6726.

21 Ma, Z., Shi, Z., Yang, D. et al. (2019). Electrically-driven violet light-emitting devices based on highly stable lead-free perovskite $Cs_3Sb_2Br_9$ quantum dots. *ACS Energy Letters* 5: 385–394.

22 Jun, T., Sim, K., Iimura, S. et al. (2018). Lead-free highly efficient blue-emitting $Cs_3Cu_2I_5$ with 0D electronic structure. *Advanced Materials* 30: 1804547.

23 Luo, Z., Li, Q., Zhang, L. et al. (2020). 0D $Cs_3Cu_2X_5$ (X = I, Br, and Cl) nanocrystals: colloidal syntheses and optical properties. *Small* 16: 1905226.

24 Zhang, F., Zhao, Z., Chen, B. et al. (2020). Strongly emissive lead-free 0D $Cs_3Cu_2I_5$ perovskites synthesized by a room temperature solvent evaporation crystallization for down-conversion light-emitting devices and fluorescent inks. *Advanced Optical Materials* 8: 1901723.

25 Yang, P., Liu, G., Liu, B. et al. (2018). All-inorganic Cs_2CuX_4 (X = Cl, Br, and Br/I) perovskite quantum dots with blue-green luminescence. *Chemical Communications* 54: 11638–11641.

26 Liu, X., Yu, Y., Yuan, F. et al. (2020). Vacuum dual-source thermal-deposited lead-free $Cs_3Cu_2I_5$ films with high photoluminescence quantum yield for deep-blue light-emitting diodes. *ACS Applied Materials and Interfaces* 12: 52967–52975.

27 Creason, T.D., McWhorter, T.M., Bell, Z. et al. (2020). K_2CuX_3 (X = Cl, Br): all-inorganic lead-free blue emitters with near-unity photoluminescence quantum yield. *Chemistry of Materials* 32: 6197–6205.

28 Wang, L., Shi, Z., Ma, Z. et al. (2020). Colloidal synthesis of ternary copper halide nanocrystals for high-efficiency deep-blue light-emitting diodes with a half-lifetime above 100 h. *Nano Letters* 20: 3568–3576.

29 Alam, F., Wegner, K.D., Pouget, S. et al. (2019). Eu^{2+}: a suitable substituent for Pb^{2+} in $CsPbX_3$ perovskite nanocrystals? *The Journal of Chemical Physics* 151: 231101.

30 Yang, B., Chen, J., Yang, S. et al. (2018). Lead-free silver-bismuth halide double perovskite nanocrystals. *Angewandte Chemie International Edition* 57: 5359–5363.

31 Liu, Z., Sun, Y., Cai, T. et al. (2023). Two-dimensional $Cs_2AgIn_xBi_{1-x}Cl_6$ alloyed double perovskite nanoplatelets for solution-processed light-emitting diodes. *Advanced Materials* 35: e2211235.

32 Jin, S., Li, R., Huang, H. et al. (2022). Compact ultrabroadband light-emitting diodes based on lanthanide-doped lead-free double perovskites. *Light: Science & Applications* 11: 52.

33 Lv, K., Qi, S., Liu, G. et al. (2019). Lead-free silver-antimony halide double perovskite quantum dots with superior blue photoluminescence. *Chemical Communications* 55: 14741–14744.

34 Zeng, R., Zhang, L., Xue, Y. et al. (2020). Highly efficient blue emission from self-trapped excitons in stable Sb^{3+}-doped $Cs_2NaInCl_6$ double perovskites. *The Journal of Physical Chemistry Letters* 11: 2053–2061.

35 Zhou, Y., Yong, Z.J., Zhang, K.C. et al. (2016). Ultrabroad photoluminescence and electroluminescence at new wavelengths from doped organometal halide perovskites. *Journal of Physical Chemistry Letters* 7: 2735–2741.

36 Kim, Y.H., Kim, J.S., and Lee, T.W. (2019). Strategies to improve luminescence efficiency of metal-halide perovskites and light-emitting diodes. *Advanced Materials* 31: 1804595.

37 Lou, Y., Fang, M., Chen, J. et al. (2018). Formation of highly luminescent cesium bismuth halide perovskite quantum dots tuned by anion exchange. *Chemical Communications* 54: 3779–3782.

38 Kshirsagar, A.S. and Nag, A. (2019). Synthesis and optical properties of colloidal $Cs_2AgSb_{1-x}BixCl_6$ double perovskite nanocrystals. *The Journal of Chemical Physics* 151: 161101.

39 Tan, Z., Li, J., Zhang, C. et al. (2018). Highly efficient blue-emitting bi-doped Cs_2SnCl_6 perovskite variant: photoluminescence induced by impurity doping. *Advanced Functional Materials* 28.

40 Xiong, G., Yuan, L., Jin, Y. et al. (2020). Aliovalent doping and surface grafting enable efficient and stable lead-free blue-emitting perovskite derivative. *Advanced Optical Materials* 8: 2000779.

41 Gao, F., Zhu, X., Feng, Q. et al. (2022). Deep-blue emissive $Cs_3Cu_2I_5$ perovskites nanocrystals with 96.6% quantum yield via InI_3-assisted synthesis for light-emitting device and fluorescent ink applications. *Nano Energy* 98: 107270.

42 Wang, L., Guo, Q., Duan, J. et al. (2021). Exploration of nontoxic Cs_3CeBr_6 for violet light-emitting diodes. *ACS Energy Letters* 6: 4245–4254.

9

Perovskite White Light-emitting Materials and Devices

9.1 Background of WPeLED

White light-emitting diodes (WLEDs) have advantages such as high brightness, low energy consumption, and a long lifespan, making them a substitute for traditional incandescent bulbs. The development of efficient and cost-effective white light-emitting materials and light-emitting diode (LED) devices has been a goal in the field of lighting technology [1].

Organic light-emitting diodes (OLEDs) have high light utilization efficiency and can be compatible with flexible substrates, resulting in lighter, thinner, and more transparent characteristics. However, OLED production suffers from small-scale manufacturing, low material utilization (<10%) during the evaporation process, low yield, complex organic material synthesis processes, and expensive production equipment, all of which contribute to high costs.

Quantum dot light-emitting diodes (QLEDs) based on Cd and InP have become the mainstream technology in commercial QLED applications. They exhibit an external quantum efficiency (EQE) exceeding 20% and offer an extended device lifetime of over 1 million hours for red, green, and blue devices. Compared to OLED technology, the main advantages of QLED technology lie in the solution processing of large-area panels and the extended device lifespan due to the excellent durability of inorganic materials. However, QLED technology also faces inherent drawbacks such as the toxicity of cadmium and the scarcity of indium resources. QLED technology is still in its early stages with low reliability, low efficiency, an unstable component lifespan, and other limiting factors.

Micro/mini-LEDs are self-emissive and minimize traditional inorganic LED arrays, allowing each pixel to be individually addressed and hit. However, the high cost of micro-LEDs has hindered their large-scale commercialization. In conclusion, many lighting technologies have emerged, demonstrating significant potential to replace traditional lighting methods in the future. However, there are still challenges that need to be addressed. OLED and QLED manufacturing costs are high, and their lifespans often fall short compared to traditional LEDs. The high cost of micro/mini-LEDs also makes them unsuitable for lighting applications [2].

Perovskite materials offer advantages such as low cost, high efficiency, and tunable full-visible spectrum emissions, showing great potential in the fields of lighting

Perovskite Light Emitting Diodes: Materials and Devices, First Edition. Hong Meng.

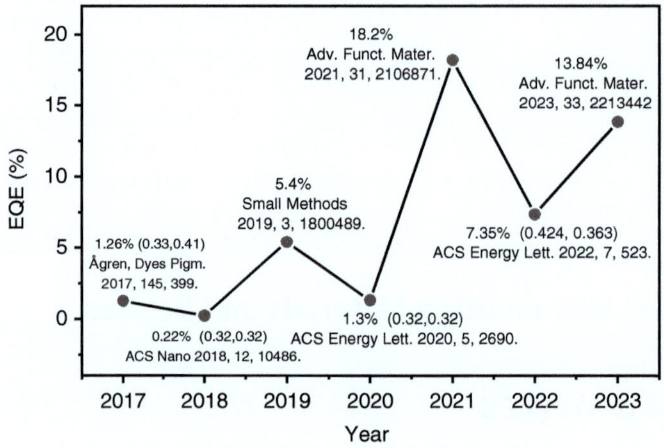

Figure 9.1 The development in EQEs of white WPeLEDs.

and displays. While the luminescence efficiency of perovskite materials and white perovskite light-emitting diodes (WPeLEDs) devices has rapidly improved in recent years, achieving efficient and stable WPeLEDs remains a significant challenge. Over the past 50 years, many research groups have made significant progress, producing various materials and devices with excellent optoelectronic properties.

Figure 9.1 illustrates the decade-long evolution of external quantum efficiencies (EQEs) of white perovskite light-emitting diodes (WPeLEDs), demonstrating marked advancements in efficiency over this period. The roadmap began with the synthesis and stabilization of white-emitting perovskite materials, a pivotal first step toward efficient white PeLEDs. In 2017, the technology was nascent, with EQEs for white PeLEDs hovering around a modest 1.26% and CIE coordinates at (0.33, 0.41). The subsequent goal set out in this period was to engineer materials with a broad emission spectrum (EM) spanning the entire visible light range. This necessitated a fine-tuning of the material's composition – adjusting the types of cations, anions, and halide components – as well as its structural parameters, such as dimensionality and crystalline structure.

Achieving stable white light emission under operational conditions emerged as a key research focus during these years. As we traverse the timeline, we observe a steady ascent in EQE values, culminating in 5.4% by 2019. This uptick reflects the successful development and implementation of perovskite LEDs capable of efficient white light production. Strategic considerations, including multi-layer device design and the integration of perovskites with other semiconductors, contributed significantly to these advancements. Further, enhancements in charge injection and transport layers in device architecture bolstered device performance. The ascending trend seen in the EQE values, particularly the impressive 18.2% achieved in 2021, reflects the relentless efforts in technology advancements and successful strategies for optimizing the devices. A core focus of this roadmap is to improve device performance by targeting high luminous efficiency, excellent color rendition, and a stable, tunable color temperature. Achieving these targets inevitably requires iterative material optimization and device design cycles. Despite the substantial progress charted, the persistent research endeavors, as depicted by the most recent data points, indicate a sustained commitment to push the efficiency frontier even further. This roadmap, therefore, presents an ambitious blueprint set to guide research and development in the realm of white-emitting perovskites. By delineating clear goals and potential challenges, it seeks to align the efforts of researchers and expedite technological advancements in this dynamic field.

There are typically five methods for preparing white-light PeLEDs, including the down-conversion method, full electroluminescent PeLEDs, single white-light perovskite materials or self-trapped excitons (STEs), POC white-light PeLEDs, and others. The down-conversion method is widely applied for realizing WPeLEDs, where perovskite is used as a phosphor. However, the lifespan of phosphor, complex device structures, and the intensity balance between different light-emitting centers in the device are significant obstacles to the development of micro, integrated, wearable electronic products, and other cutting-edge technologies. Full electroluminescent WLEDs are a mainstream technology, simplifying device structures, especially by coupling perovskites with low-cost and high photoluminescent quantum yield (PLQY), making WPeLEDs a candidate technology. However, these WPeLEDs with mixed emissive layer (EML) [3] or stacked EML [4] structures face problems of energy distribution imbalance and ion migration [5–7]. Recently, some single white light perovskite materials or some WPeLEDs utilizing STEs or metal ion doping have been reported. However, their luminescent efficiency is still in the initial stage and far from practical application. Meanwhile, the introduction of organic luminescent molecules to form perovskite–organic coupled white-light-emitting diodes (POC-WLEDs) has become an important method for realizing white light emission. In this system, the luminescent function of organic molecules and their inherent charge injection capabilities are key to achieving coupled emission. The connection state of organic molecules with perovskite crystals and their direct impact on perovskite crystals are the core issues of these POC-WLEDs [2]. In addition to this, many innovative methods have been used to prepare white-light PeLEDs.

9.2 Down-conversion Method

Chen's team successfully optimized the SiO_2–B_2O_3–ZnO–CaF_2 glass network, perovskite component concentration, and glass crystallization conditions, preparing high-quality $CsPbX_3$@ glass. The PLQYs of the green $CsPbBr_3$@ glass and red $CsPbBr_{1.5}I_{1.5}$@ glass are as high as 93.8% and 78.1%, respectively. The as-prepared LED devices yield extremely bright green, red, and white luminescence at an operating bias of 3 V (Figure 9.2g–i). The backlight source was assembled by embedding 36 $CsPbX_3$@glass-based WLEDs on both sides of the light guide plate (LGP) (Figure 9.2a–d). The main role of the LGP is to direct the speckled white light emitted by the lead to the planar white light. The photoluminescence (PL) spectrum recorded by the LPG shows blue/green/red narrow-band emission (Figure 9.2e),

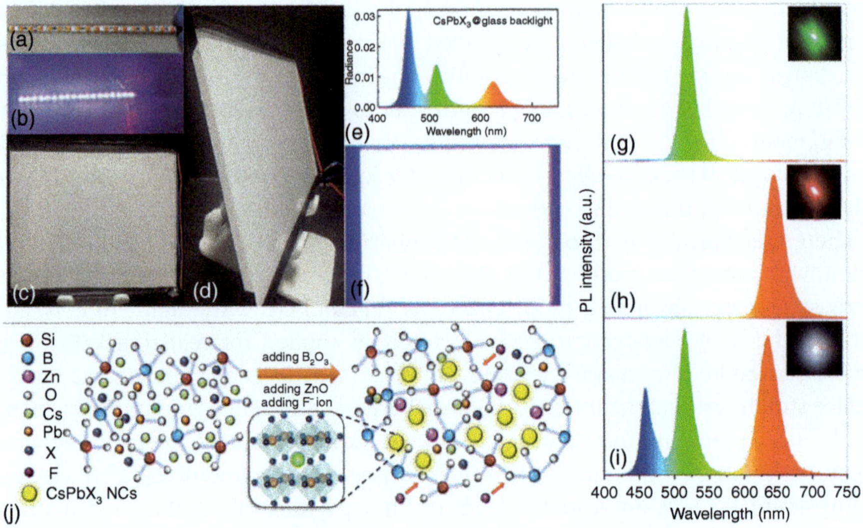

Figure 9.2 Photographs of (a and b) $CsPbX_3$@glass-based WLED array and (c and d) LGP coupled with two WLED arrays. (e) PL spectrum and (f) luminescent photograph of LGP under an operating voltage of 12 V. (g–i) PL spectra for the constructed $CsPbX_3$@glass converted green LED, red LED, and white LED. (j) Schematic illustration of $SiO_2-B_2O_3-ZnO-CaF_2$ glass network structure and in situ growth of $CsPbX_3$ PeQDs. Source: Chen et al. [8]/Reproduced with permission from John Wiley & Sons, Inc.

and the high brightness and uniform planar white light can be clearly observed at a driving voltage/current of 12 V/2 A (Figure. 9.2f). They designed and prepared a high-brightness WLED with an EQE of 13.84% and a brightness of 500 000 cd m^{-2} and applied it to backlight modules, achieving a wide color gamut display with a brightness of 127% of NTSC [8]. Figure 9.2j shows the schematic structure of the SiO_2–B_2O_3–ZnO–CaF_2 glass network and the in situ growth of $CsPbX_3$ PeQDs.

In 2019, Wang's group published a paper on the synthesis of high-emissivity two-dimensional halogenated tin perovskites $(OCTAm)_2SnX_4$ (X = Br, I, or a mixture thereof) using an aqueous acid-based method [9]. The synthetic method

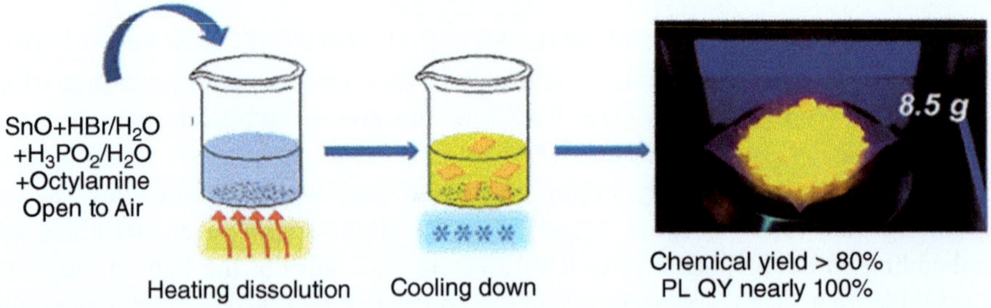

Figure 9.3 $(OCTAm)_2SnBr_4$ schematic diagram of a simple aqueous acid synthesis method under air. Source: Wang et al. [9]/Reproduced from Royal Society of Chemistry/CC BY-NC 3.0.

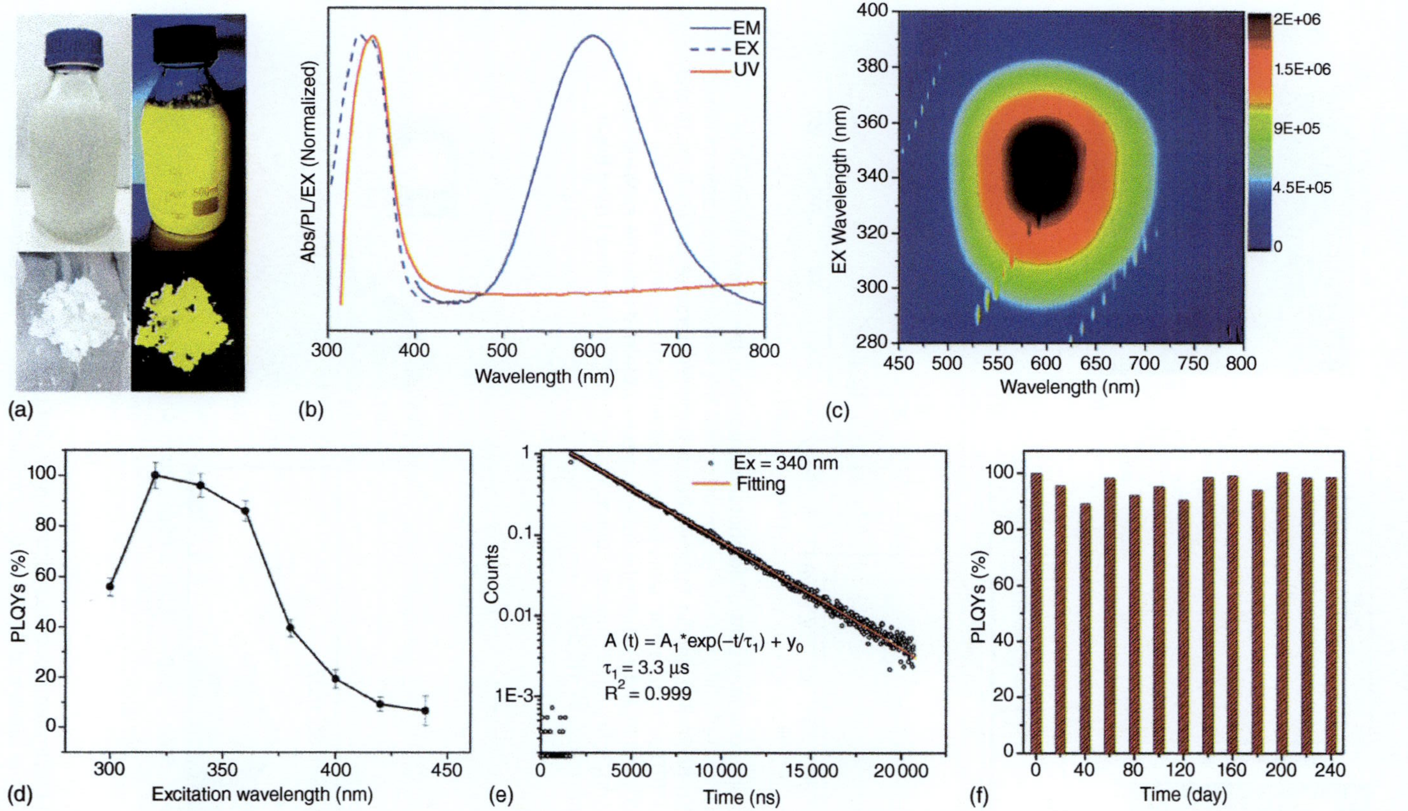

Figure 9.4 Plots of the optical properties of two-dimensional tin perovskites bromide: (a) 0.5 L of a two-dimensional $(OCTAm)_2SnBr_4$ colloidal solution in room light (left) and UV (right); (b) UV–vis absorption spectrum, fluorescence excitation spectrum (EX), and fluorescence emission spectrum (EM) of the sample shown in (c) three-dimensional excitation–emission matrix (EEM) fluorescence spectrum; (d) plots of absolute PLQYs at different excitation wavelengths; (e) light attenuation and fitted curves at 600 nm for a time-resolved excitation wavelength of 340 nm; and (f) two-dimensional stability study of $(OCTAm)_2SnBr_4$ (monitored by PLQY), plotted with exposure time (0–240 days in normal air humidity). Source: Wang et al. [9]/Reproduced from Royal Society of Chemistry/CC BY-NC 3.0.

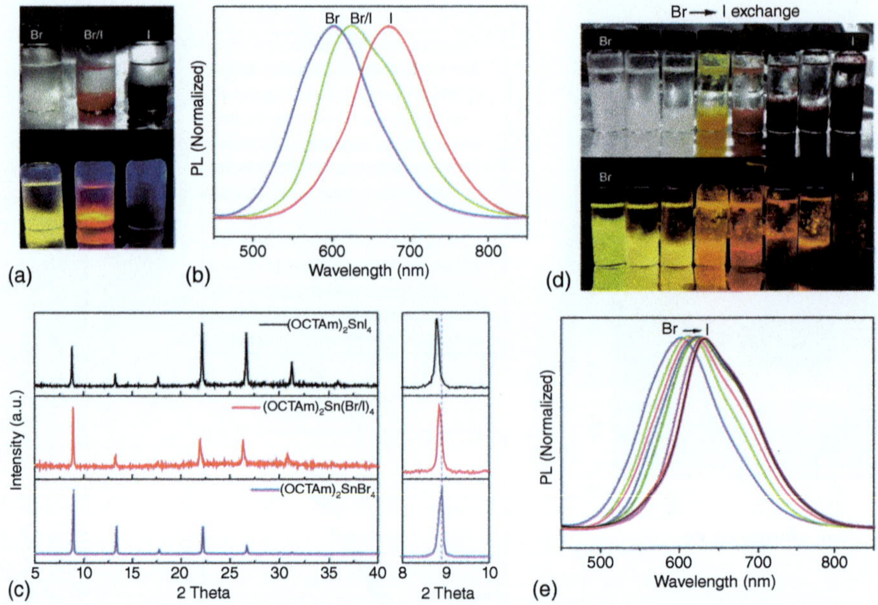

Figure 9.5 Anion-modulated controlled photoluminescence: (a–c) direct synthesis of two-dimensional phase halide perovskite precursors; (a) left to right, photographs of mixed perovskites of Sn-Br, Sn-Br/I, and Sn-I perovskites at room temperature and UV light. (b and c) Spectra and PXRD patterns of the same perovskite sample; (c) right panel is a fine PXRD pattern at 8–10°; and (d and e) photographs and spectra obtained from the exchange of synthetic $(OCTAm)_2SnBr_4$ halide anions. Source: Wang et al. [9]/Reproduced from Royal Society of Chemistry/CC BY-NC 3.0.

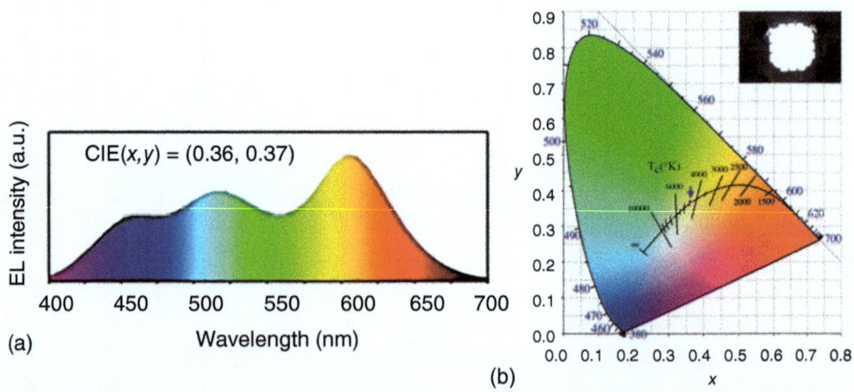

Figure 9.6 (a) Electroluminescence spectrum of Cs_2SnCl_6:2.75% Bi luminescent material, glowing warm white diode and (b) CIE color coordinate diagram (inset shows device operating conditions). Source: Tan et al. [10]/Reproduced from John Wiley & Sons, Inc.

resulted in a high absolute PLQY, approaching 100% in the solid state, with a luminescence wavelength centered at 600 nm and a bandwidth of 136 nm. Additionally, the material exhibited a large Stokes shift of 250 nm, a long lifetime ($\tau = 3.3$ ms), and zero overlap between the absorption and emission spectra. These excellent optical properties make this new yellow phosphor a suitable candidate for the preparation of UV-excited WLEDs (Figures 9.3–9.5).

In 2018, Tan's group proposed a blue phosphor with bismuth (Bi)-doped lead-free inorganic perovskite Cs_2SnCl_6 [10]. After doping, Cs_2SnCl_6, which was initially nonluminescent, exhibits highly efficient deep blue luminescence at 455 nm with a photoluminescence quantum efficiency (PLQE) close to 80%. This phosphor was then combined with commercially available yellow phosphors, such as $Ba_2Sr_2SiO_4:Eu^{2+}$ and $GaAlSiN_3:Eu^{2+}$, to prepare white light devices, as depicted in Figure 9.6.

9.3 Full Electroluminescent PeLEDs

Achieving full-color emission of the three primary colors in the luminescent layer is a prerequisite for obtaining a composite white-light perovskite. However, differing luminescence decay rates among the layers can lead to energy losses and alterations in the overall emission color over time. Currently, most white light is achieved through multilayer structures, utilizing highly efficient yellow devices synthesized with blue devices. Although there have been many breakthrough advancements in single PeLEDs such as green, red, and near-infrared light, with over 20% high EQE achieved, progress in blue and yellow PeLEDs has been far from satisfactory, with highly disappointing efficiency.

The development of highly efficient blue and yellow light-emitting devices is a key step in manufacturing white light-emitting devices, which have enormous commercial potential in the lighting industry. Although it is not possible to synthesize blue light directly, yellow light can be obtained by combining red and green light. Therefore, the focus of research is primarily on synthesizing highly efficient blue light-emitting devices. However, if high-efficiency yellow light-emitting devices can be realized in one step, their efficiency would still be significantly higher than the method of producing yellow light using red and green light. Therefore, there are ample opportunities for further research into the development of high-performance blue and yellow light-emitting perovskite diodes.

9.3.1 Yellow Perovskite Light-emitting Diodes

9.3.1.1 Zero-dimensional Sn-doped Halide Perovskites [11]

A study by Zhou's group reported on zero-dimensional tin-doped halide perovskite nanocrystals (NCs) with the molecular formula $(C_4N_2H_{14}Br)_4SnBr_xI_{6-x}$ (as depicted in Figure 9.7). The excited state undergoes structural reorganization, resulting in a strong Stokes-shifted broadband yellow emission peak at 582 nm (Figure 9.8), a half-peak width of 126 nm, and a high PLQE of 85% at room temperature. When combined with a commercial Eu^{2+}-doped barium magnesium aluminate blue

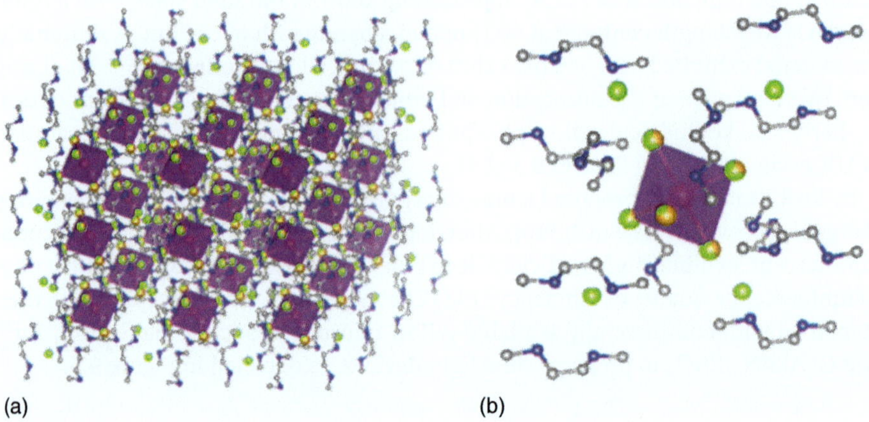

Figure 9.7 (a) Single crystal structure of Sn mixed halide perovskite and (b) a Sn mixed halide octahedron completely surrounded by organic ligands. Red spheres: tin atoms; green spheres: bromine atoms; orange spheres: iodine atoms; blue spheres: nitrogen atoms; gray spheres: carbon atoms; hydrogen atoms are hidden for clarity. Source: Zhou, et al. [11]/American Chemical Society.

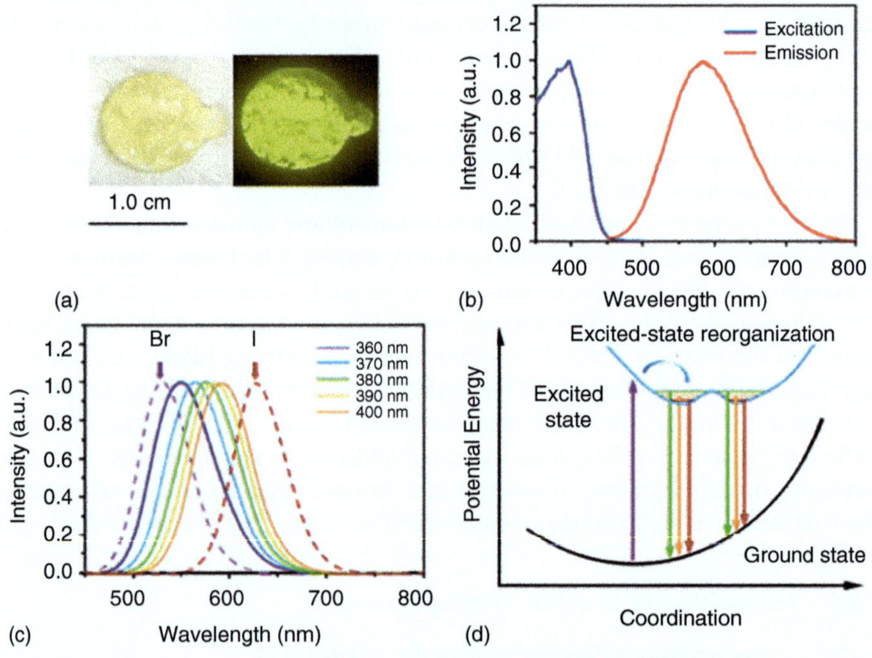

Figure 9.8 (a) Sn mixed halide perovskite crystals irradiated with ambient light and UV lamp (365 nm); (b) excitation (blue line) and emission (red line) spectra of Sn mixed halide perovskite crystals at room temperature; (c) $(C_4N_2H_{14}Br)_4SnBr_xI_{6-x}$ ($x = 3$) emission spectra at different wavelengths (360–400 nm) for excitation, $(C_4N_2H_{14}Br)_4SnBr_xI_{6-x}$ (dashed line, purple) and $(C_4N_2H_{14}Br)_4SnBr_xI_{6-x}$ (dashed line, red) at 77 K); and (d) Schematic space of potential energy curves of mixed halide perovskite crystals. Source: Zhou et al. [11]/Reproduced with permission from American Chemical Society.

Table 9.1 CIE coordinates, CCTs, and CRIs for UV-excited LEDs with different weight ratios of phosphor.

Blue:Yellow	CIE (*x*, *y*)	CCT/K	CRI
1:0	(0.15, 0.09)	N/A	N/A
4:1	(0.32, 0.32)	6160	84
2:1	(0.36, 0.38)	4600	85
1:1	(0.41, 0.44)	3740	78
0:1	(0.46, 0.49)	3200	67

phosphor ($BaMgAl_{10}O_{17}$:Eu^{2+}), the yellow emitter is used for white light devices with a color rendering index (CRI) of up to 85. This work is advantageous in that the yellow phosphors do not contain rare earth elements, leading to a significant reduction in cost, and there is no loss of CRI in the WLEDs tested with UV lamp excitation. Although this work does not directly produce a highly efficient yellow device, it provides a new idea for the future of WLEDs, which is crucial for the future of commercial lighting (Table 9.1).

9.3.1.2 2D $(C_{18}H_{35}NH_3)_2SnBr_4$ Perovskite

In 2018, Zhang's group et al. proposed a lead-free, two-dimensional Ruddlesden–Popper-type $(C_{18}H_{35}NH_3)_2SnBr_4$ perovskite [12]. The UV photoelectron spectroscopy of this two-dimensional material shows that its HOMO and LUMO levels are −5.4 and −3.4 eV, respectively, with a bandgap of 2 eV (Figure 9.9a,b). This material has strong self-trapped emission and broad emission characteristics, with a PLQY of 88% in colloidal suspensions and 68% in thin films. The insulating characteristics of organic oleylamine cations prevent the formation of electronic bands between the $[SnBr_6]^{4-}$ octahedral layers, resulting in orange emission with a 140 nm FWHM Stokes-shifted at 620 nm under 365 nm excitation. Orange LEDs based on $(OAm)_2SnBr_4$ achieved a maximum brightness of 350 cd m^{-2} and an EQE of 0.1% (Figure 9.9c,d). The spectra of the devices were in good agreement with the fluorescence spectra of the films, indicating that it is the $(C_{18}H_{35}NH_3)_2SnBr_4$ material itself that is subject to electrogenic excitation (Figure 9.9e). The fluorescence spectra of the films remained stable at different temperatures, indicating good temperature stability (Figure 9.9f). These findings represent a significant breakthrough in the development of bright and environmentally friendly perovskite lead-free LEDs suitable for practical applications in solid-state lighting.

9.3.1.3 Colloidal Undoped and Double-doped $Cs_2AgInCl_6$ Nanocrystals

In 2019, Liu's group reported the synthesis of undoped and double-doped $Cs_2AgInCl_6$ NCs through a simple thermal injection process, using nontoxic and readily available precursors. The resulting NCs exhibited broadband orange emission peaks at 580 nm, with fluorescence quantum efficiencies as high as 11.4%. This work presents a promising approach to improving the optical properties of

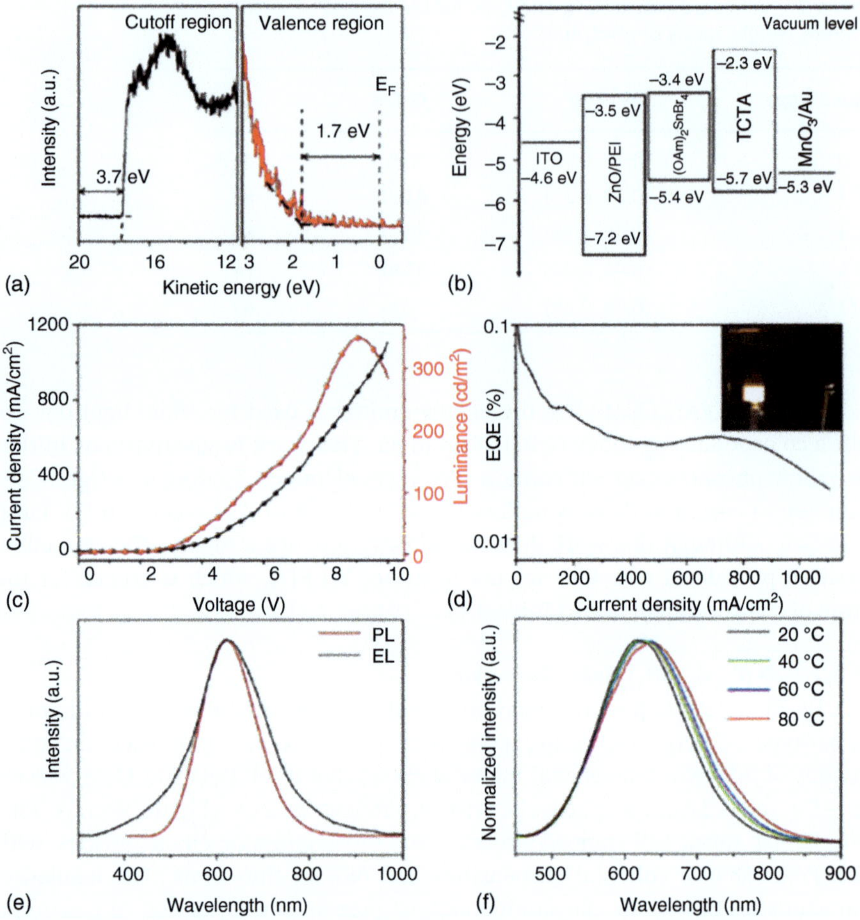

Figure 9.9 (a) UPS spectra of $(OAm)_2SnBr_4$ perovskite films on ITO glass substrates, E_F: Fermi energy level; (b) energy band diagram of the inverted device structure based on $(OAm)\ Sn_2Br_4$ perovskite; (c) current density (black) and brightness (red) of the current-emitting device versus driving voltage; (d) EQE of the light-emitting diode versus current density (inset: photo of the device in operation) (inset: photo of the device in operation); (e) fluorescence spectrum (red) of $(OAm)_2SnBr_4$ fluorescence spectrum (red) and electroluminescence spectrum (black) of perovskite in solution; and (f) gage normalized PL spectrum $(OAm)_2SnBr_4$ 2D perovskite film measured over a temperature range of 20–80 °C. Source: Zhang et al. [12]/Reproduced with permission from American Chemical Society.

new lead-free perovskite NCs [13]. Regrettably, this work did not include any experiments on device preparation to provide further characterization of the properties (Figures 9.10 and 9.11).

9.3.1.4 Introducing Separated Emitting Centers

In 2022, Zhang's group reported that pure hexagonal-phase 3D single crystal $CsCdCl_3$ was synthesized through a simple solvothermal method, exhibiting weak green emission under ultraviolet excitation with a PLQY of 4.8%. Structural

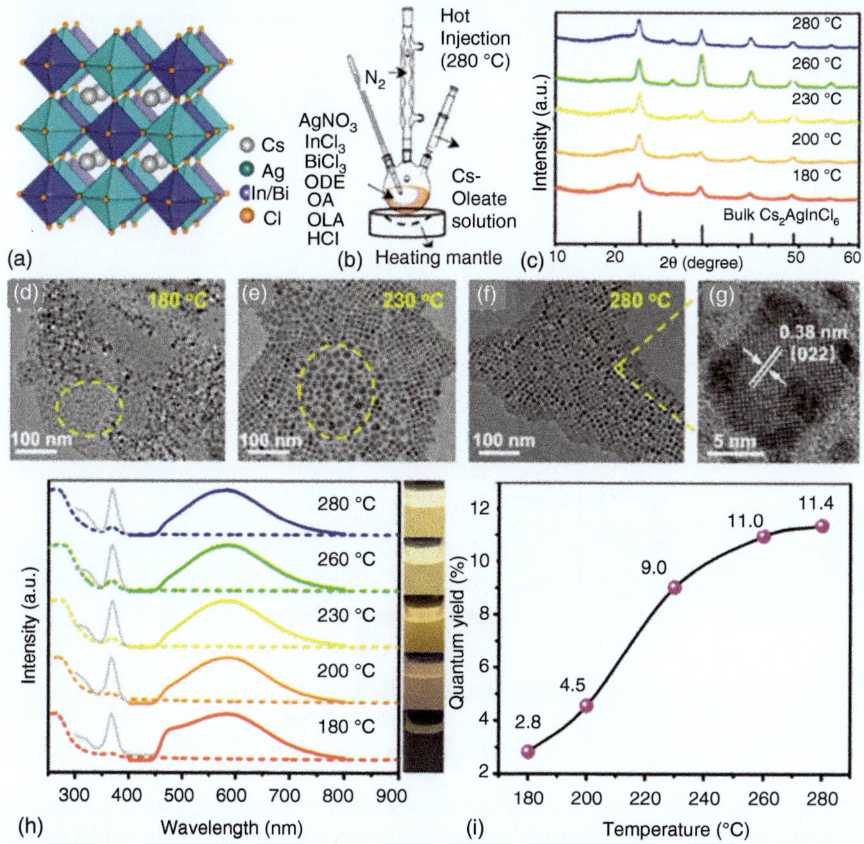

Figure 9.10 (a) Nanocrystal structure model of $Cs_2AgInCl_6$:Bi and (b) synthesis illustration; (c) XRD spectra of Bi-doped $Cs_2AgInCl_6$ nanocrystals at different synthesis temperatures (180–280 °C); TEM images of synthesized nanocrystals at (d) 180 °C, (e) 230 °C, (f) 280 °C, (g) 280 °C HRTEM images of the crystals; (h) UV–vis absorption (dashed line), photoluminescence (368 nm excitation), and photoluminescence double-doped $Cs_2AgInCl_6$ nanocrystals excitation (580 nm emission) spectra, right panel shows the dispersion of hexane at different synthesis temperatures (180–280 °C); inset shows the sample irradiated with UV lamp at 365 °C; and (i) different synthesis temperature variations in photoluminescence quantum yields. Source: Liu et al. [13]/Reproduced with permission from American Chemical Society.

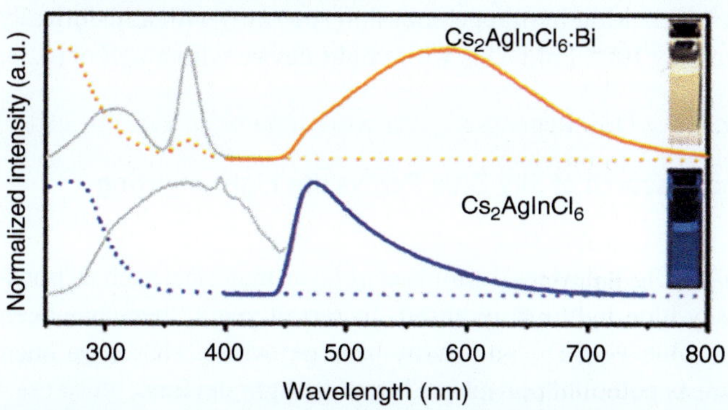

Figure 9.11 Plot of Bi-doped and undoped double perovskite luminescence. Source: Liu et al. [13]/Reproduced with permission from American Chemical Society.

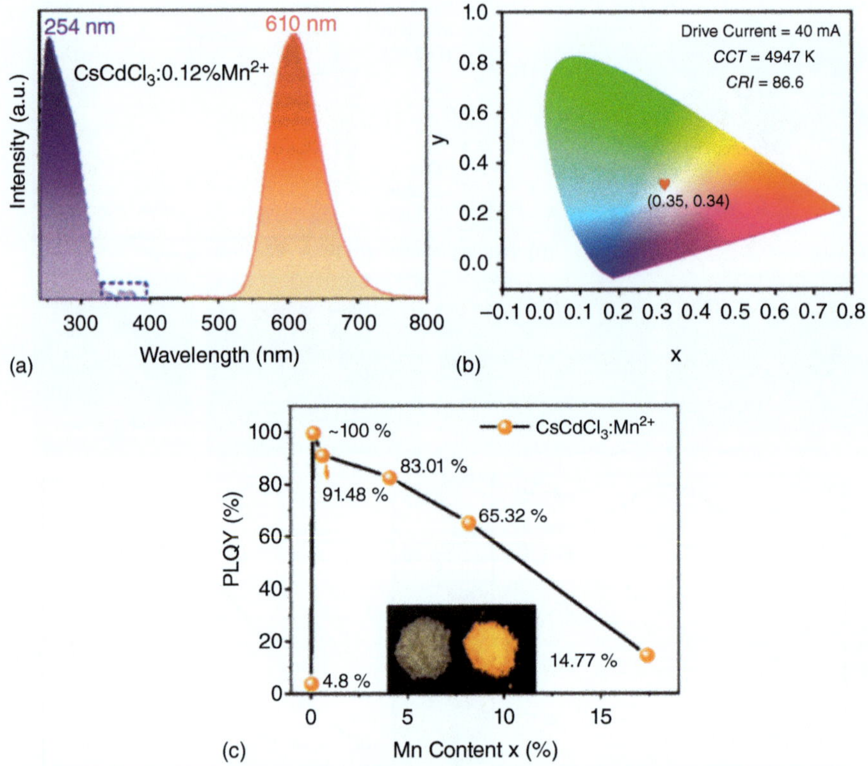

Figure 9.12 (a) Room-temperature PL and PLE spectra of $CsCdCl_3$:0.12%Mn^{2+}; (b) CIE chromaticity diagram of the fabricated white LED device; and (c) PLQY of the samples versus different dopant contents for $CsCdCl_3$:Mn^{2+}. Source: Zhang et al. [14]/Reproduced with permission from American Chemical Society.

analysis revealed that $CsCdCl_3$ is composed of alternating $[CdCl_6]^{4-}$ octahedra and $[Cd_2Cl_9]^{5-}$ dimers, providing possibilities for tuning the material's optoelectronic properties through compositional engineering. In this case, photoactive Mn^{2+} and Sn^{2+} ions, which are highly dependent on the coordination environment, were introduced to enhance the luminescence performance. For $CsCdCl_3$:Mn^{2+}, the unstable STE emission in $CsCdCl_4$ was suppressed by the efficient d–d orbital transition ($^4T_1$–$^6A_1$) of Mn^{2+}, resulting in orange emission (610 nm) with a historically high PLQY of approximately 100% [14]. This white light device with a CIE of (0.35, 0.34) (Figure 9.12).

9.3.2 Progress in the Research of Sky-Blue Perovskite Light-emitting Diodes

Generally, to prepare white light devices, luminescent layer materials such as phosphors for yellow and sky-blue light are required. In recent years, there has been continuous research on blue PeLEDs, and many blue perovskite NCs have been reported in the literature as potential phosphors for white light devices.

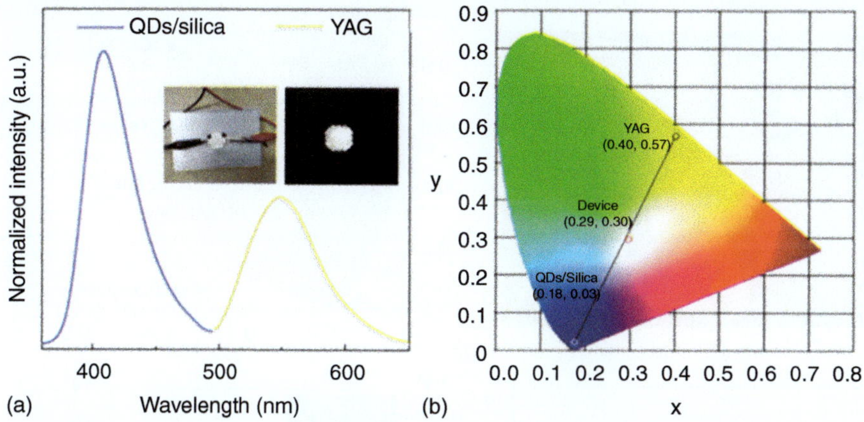

Figure 9.13 White light-emitting devices: (a) blue-emitting $Cs_3Bi_2Br_9$ QDs and yellow-emitting rare earth YAG light-emitting diodes EL spectra. Inset: device image of YAG with perovskite quantum dot/SiO_2 composite (switch) and (b) CIE color coordinates corresponding to $Cs_3Bi_2Br_9$ quantum dots, YAG, and white light-emitting diodes. Source: Leng et al. [15]/Reproduced with permission from John Wiley & Sons, Inc.

In 2018, Leng's group reported a new blue LED material based on $Cs_3Bi_2Br_9$ perovskite quantum dots. The material had a high fluorescence quantum efficiency (19.4%) and good stability and was synthesized in a simple and environmentally friendly way using ethanol as a solvent. The emission wavelength of the material was 410 nm, and the luminescence could be adjusted from 393 to 545 nm by mixing the precursors. Furthermore, $Cs_3Bi_2Br_9$ quantum dots were combined with $Y_3Al_5O_{12}$ phosphor and silica as the light-emitting layer, resulting in the production of white light, as shown in Figure 9.13 [15].

Jun's group published a work in 2018 on the preparation of $Cs_3Cu_2I_5$ perovskite quantum dots from non-lead solutions [16]. The perovskite quantum dot emits blue light at approximately 445 nm with fluorescence quantum efficiencies of 90% and 60% for single crystals and thin films, respectively. By doping this perovskite quantum dot with a commercial yellow phosphor, a white light device with CIE coordinates of (0.33, 0.33) was obtained, as shown in Figure 9.14.

In 2019, Ji's group presented a study on the synthesis of Bi-doped double perovskite (DP) Gd_2ZnTiO_6 phosphor [17]. The $Gd_{1.9925}ZnTiO_6$:$0.0075Bi^{3+}$ phosphor, with an irregular crystal shape, smooth surface, and clear edges, showed bright blue emission with a peak at 419 nm upon excitation with 375 nm light. The optimum doping concentration for Bi^{3+} ions was found to be 0.0075. The material exhibited good thermal stability, as confirmed by thermal quenching and photoluminescence experiments. When combined with a commercial yellow phosphor, $Gd_{1.9925}ZnTiO_6$:$0.0075Bi^{3+}$ produced white light with a color coordinate CIE of (0.4485, 0.4222) and a CRI of 89.4, indicating its promising application in white light devices (Figure 9.15).

Liu's research group has successfully synthesized a type of environmentally friendly Mn^{2+}-doped $Cs_3Cu_2Br_5$ perovskite-like derivative material using a one-step

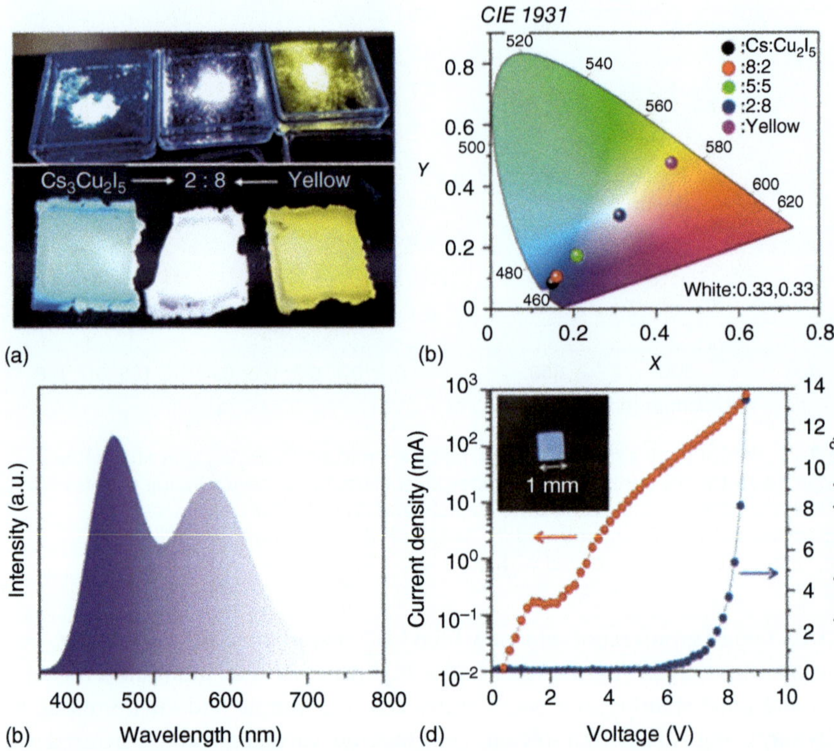

Figure 9.14 (a) UV photograph of the yellow phosphor $Cs_3Cu_2I_5$ and its mixture (from left to right: $Cs_3Cu_2I_5$, 2 : 8 mixture, yellow phosphor); light (top) and film prepared by kneading the powder into a polydimethylsiloxane matrix (bottom); (b) PL spectrum of $Cs_3Cu_2I_5$ mixture and yellow phosphor (mixture ratio = 2 : 8); (c) PL in powder mixture CIE coordinates; and (d) current density–brightness–voltage characteristics plot (blue LED using $Cs_3Cu_2I_5$ film as the light-emitting layer). Source: Jun et al. [16]/Reproduced with permission from John Wiley & Sons, Inc.

solid-phase method. This material can be used in LEDs with a high CRI and a low blue light band. Theoretical and experimental results show that Mn^{2+} can be incorporated into $Cs_3Cu_2Br_5$ perovskite-like derivative by replacing Cu^+ ions. Simultaneously, the dopant of Mn^{2+} could effectively modulate the optical properties of $Cs_3Cu_2Br_5$ perovskite-like derivatives, leading to broadening the EM (400–620 nm). Thanks to the high exciton absorption of Mn^{2+}, excitons in the excited state have been successfully introduced into the d-state orbital of Mn^{2+}. The blue emission peak at 460 nm is reduced by a maximum of 80%. Compared with white LEDs fabricated from $Cs_3Cu_2Br_5$ and 17.5% Mn^{2+}-doped $Cs_3Cu_2Br_5$ perovskite-like materials, the CRI of the LED produced by 17.5% Mn-doped $Cs_3Cu_2Br_5$ is as high as 94.5, and more importantly, the blue band (400–480 nm) decreases by 71.45%. These research results suggest that the introduction of Mn^{2+}

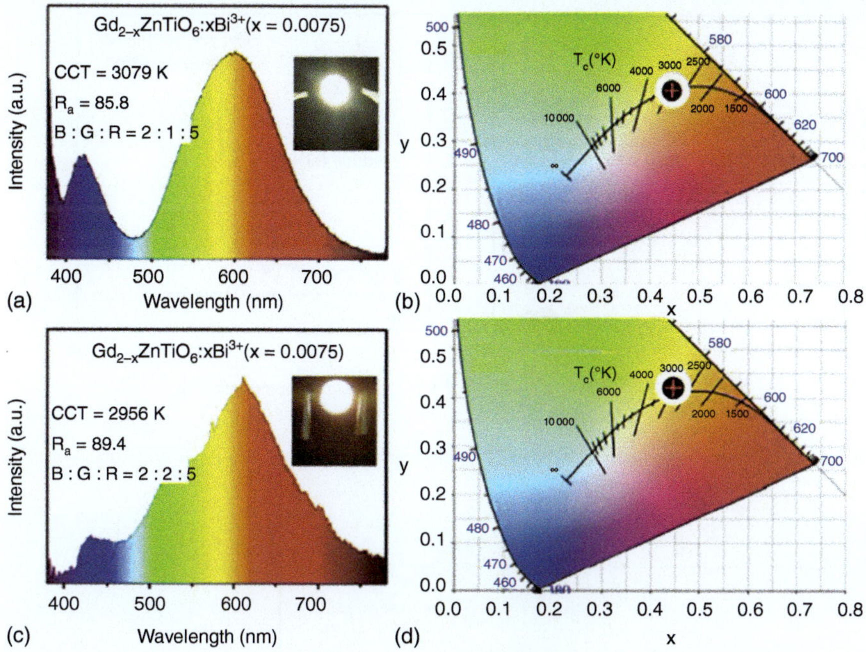

Figure 9.15 (a) EL spectrum of WLED-1 with mass ratio $Gd_{1.9925}ZnTiO_6$:$0.0075Bi^{3+}$ at 370 nm excitation [(Ba, Sr) $_2SiO_4$:Eu^{2+} (λ_{em} = 525 nm): (Ca, Sr)$AlSiN_3$:Eu^{2+} = 2 : 1 : 5]; (b) EL spectrum of WLED-2 with mass ratio $Gd_{1.9925}ZnTiO_6$:$0.0075Bi^{3+}$ at 370 nm near UV chip excitation EL spectra of WLED-2 with Bi^{3+} [(Ba, Sr)$_2SiO_4$:Eu^{2+} (λ_{em} = 525 nm): (Ca, Sr)$AlSiN_3$:Eu^{2+} = 2 : 2 : 5]; and (c and d) CIE chromaticity diagrams of WLED-1 and WLED-2. Source: Ji et al. [17]/Reproduced from Elsevier.

can broaden the EM and reduce the blue light emission band, offering a new promising strategy for fabricating high-performance, blue-light-friendly white LEDs [18] (Figure 9.16).

In 2023, Li's group reported that Sb^{3+}-doped Cs_2ZrCl_6 crystals were synthesized using a co-precipitation method, where Sb^{3+} ions partially replaced Zr^{4+} ions. Cs_2ZrCl_6:$x$$Sb^{3+}$ powder exhibited blue and orange–red emissions under 254 and 365 nm light excitation, respectively, attributed to the $[ZrCl_6]^{2-}$ octahedra and $[SbCl_6]^{3-}$ octahedra. The Cs_2ZrCl_6:$x$$Sb^{3+}$ (x = 0.1) crystals showed a high PLQY of 52.5%. Based on experimental and computational results, the luminescence mechanism of Cs_2ZrCl_6:$x$$Sb^{3+}$ crystals was proposed. On one hand, the broad blue emission with a large Stokes shift was attributed to STEs generated by the excitation of $[ZrCl_6]^{2-}$ octahedra at 260 nm. On the other hand, the emission from $[SbCl_6]^{3-}$ octahedra involved two transitions: $^1P_1 \rightarrow {}^1S_0$ (490 nm) and $^3P_1 \rightarrow {}^1S_0$ (625 nm) [19]. This light device with a CIE of (0.203, 0.233) under 254 nm light excitation (Figure 9.17).

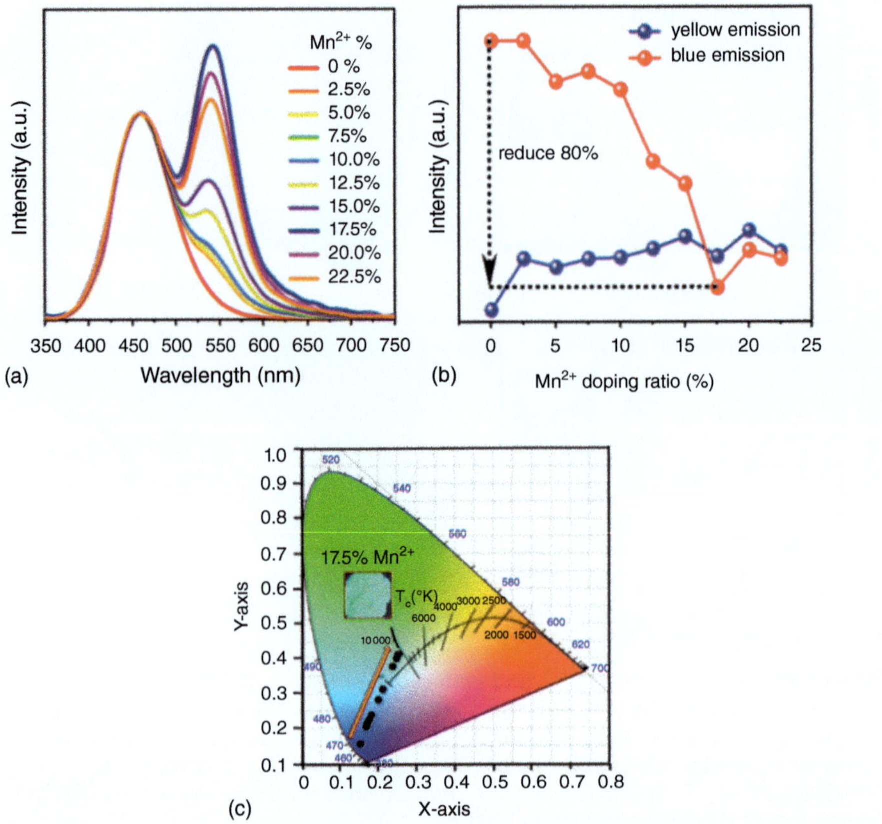

Figure 9.16 (a) Emission spectrums of $Cs_3Cu_2Br_5$ with different ratios of Mn^{2+}; (b) blue and yellow PL intensities under different ratios of Mn^{2+} ions; and (c) CIE coordination of PL from the perovskite-like derivatives of $Cs_3Cu_2Br_5$ with different doping ratios of Mn^{2+}. Source: Liu et al. [18]/John Wiley & Sons/CC BY 4.0.

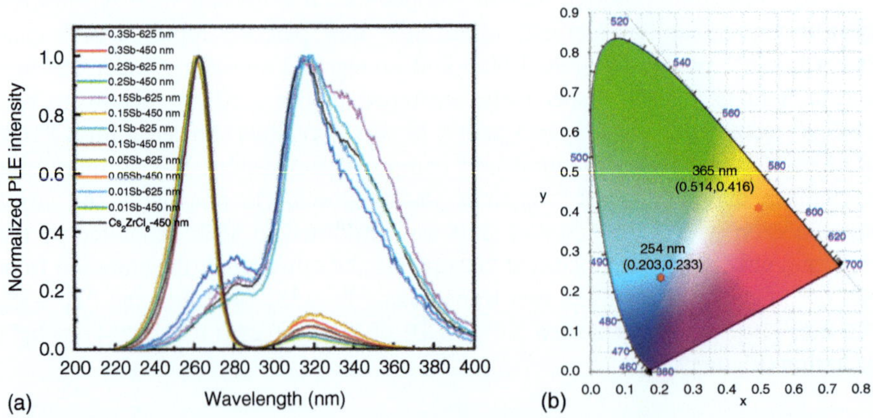

Figure 9.17 (a) Excitation spectra of the Cs_2ZrCl_6:$x$$Sb^{3+}$ (0–0.3) crystals under 460 and 625 nm emissions and (b) corresponding CIE color coordinates of Cs_2ZrCl_6:$x$$Sb^{3+}$ (0.1) crystals at 254 and 365 nm. Source: Li et al. [19]/American Chemical Society.

9.4 Single White Light Perovskite Materials and Self-trapped Excitons

A single white light perovskite is a material that is capable of emitting broad-spectrum white light (ranging from 400 to 700 nm) on its own, without requiring any additional components [20]. Different from ordinary free excitons (FE), researchers have explored the use of STE in materials with large electron–phonon coupling and soft lattices, such as alkali metal halides and organic molecular crystals. Upon excitation, electrons and holes cause rapid deformation of the lattice into a metastable state, leading to self-localization of the electron and hole and hence the term STE. Changes in the excited state crystal structure give rise to both FE and STE, and a shift in the excitonic energy level and ground state conformation occurs (as illustrated in Figure 9.18). Compared to band-edge emission or free exciton emission, STE emission has two notable features: a larger Stokes shift and a broader EM, making it an ideal choice for single-substrate white light illumination.

9.4.1 Single White Light Perovskite Materials

9.4.1.1 (110) Perovskite with Corrugated Inorganic Layers

When excited by UV radiation, the majority of lead halide perovskites with a flat (001) inorganic layer produce narrow free exciton photoluminescence [21, 22]. The perovskite (N-MEDA)$PbBr_4$ with a corrugated inorganic layer on (110) plane emits broadband white light around 560 nm when excited by UV radiation, allowing for continuous emission from 400 to 700 nm (as shown in Figure 9.20) [23]. A shoulder peak at around 420 nm can also be observed, which is attributed to the contribution of FE [24]. The light absorption spectra of this (110) perovskite and other (001)

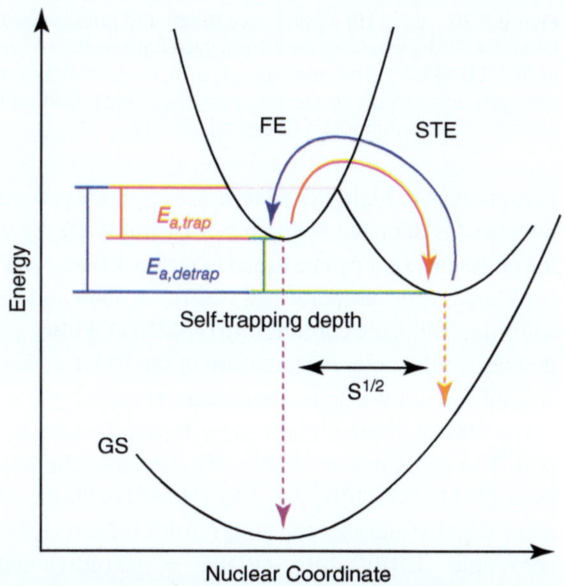

Figure 9.18 Schematic diagram of the self-induced exciton energy band. Source: Smit et al. [21]/Royal Society of Chemistry/CC BY 3.0.

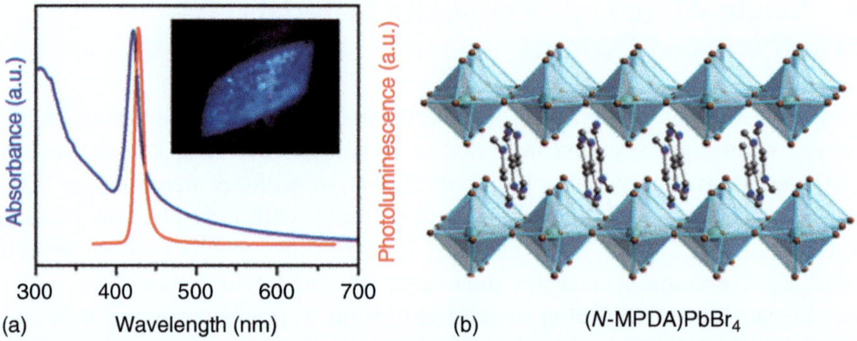

Figure 9.19 (a) (001) Absorbance (blue) and photoluminescence spectra (PL, red) of (N-MPDA)$PbBr_4$ powder emitting blue light under UV irradiation and (b) crystal structure of (N-MPDA)$PbBr_4$. The turquoise, brown, blue, and gray spheres represent lead, bromine, nitrogen, and carbon atoms, respectively. Source: Dohner et al. [23]/Reproduced with permission from American Chemical Society.

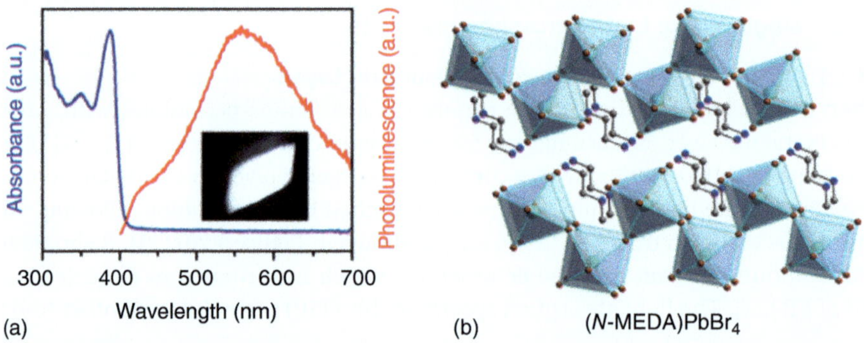

Figure 9.20 (a) (110) Absorbance (blue) and photoluminescence spectra (PL, red) of (N-MEDA)$PbBr_4$ powders emitting white light under UV irradiation and (b) crystal structure of (N-MEDA)$PbBr_4$. The turquoise, brown, blue, and gray spheres represent lead, bromine, nitrogen, and carbon atoms, respectively. Source: Dohner et al. [23]/Reproduced with permission from American Chemical Society.

perovskites are highly similar; however, their emission spectra show significant differences (as depicted in Figures 9.19 and 9.20). By varying the halide composition, the emission color can be tuned to produce both "warm" and "cold" white light, with correlated color temperatures (CCTs) of 4669 and 6502 K, respectively. Introducing additional chlorine content into (N-MEDA)$PbBr_{4-x}Cl_x$ results in a blue shift and an increase in the color temperature of the EM. Furthermore, a shoulder peak appears around 420 nm due to the contribution of FE.

The CRI of commercial indoor-lighting white light sources needs to be over 80 [25]. The substitution of chloride enhances the CRI of the phosphor to 85, rising from (N-MEDA)$PbBr_{2.8}Cl_{1.2}$ to (N-MEDA)$PbBr_4$. Additionally, chloride replacement slightly elevates the PLQY from 0.5% to 1.5%. The (110) white luminescent perovskite (EDBE)$PbBr_4$ (EDBE = 2,2′(ethylenedioxy)bis(ethylammonium); as shown in Figure 9.21), demonstrates a higher fluorescence quantum efficiency (9%)

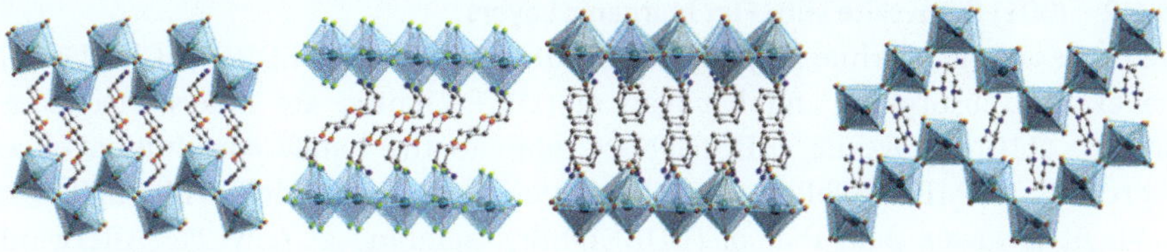

(a) (EDBE)PbBr$_4$ (b) (EDBE)PbCl$_4$ (c) (CyA)$_2$PbBr$_4$ (d) α-(DMEN)PbBr$_4$

Figure 9.21 Crystal structures of (a) (EDBE)PbBr$_4$ (EDBE = 2,2′-(ethylenedioxy)bis (ethylammonium); (b) (EDBE)PbCl$_4$; (c) (CyA)$_2$PbBr$_4$ (CyA = cyclohexylamine); and (d) α-(DMEN)PbBr$_4$ (DMEN = N,N-dimethylethane-1,2-diamine). Source: Adapted from Dohner et al. [26].

under warm white light with a CRI of 84 [26]. The broad photoluminescence of (EDBE)PbBr$_4$ (Figure 9.21a) was not compromised by chloride substitution, while a small amount of iodine doping caused the broad component of the luminescence to burst, resulting in a narrower green emission in line with (EDBE)PbBr$_4$. Before the observation of white emission from (NMEDA)PbBr$_4$ and (EDBE)PbBr$_4$, (API)PbBr$_4$ (API = 1-[3-aminopropyl]imidazole), a (110) perovskite, was reported to exhibit a deep yellow color emission in the region of 400–650 nm [27]. Here, it was initially believed that the broadband luminescence in (API)PbBr4 was due to a π*–π jump in the organic API$^+$ chromophore. However, since the emission spectra of (API)PbBr$_4$ and (110) perovskite (which contain simple cations such as N-MEDA) are similar, this mechanism is now considered unlikely [23]. A study has reported that the broad white light emission from (110) Pb-Cl perovskite [28], templated by the API$^+$ cation, is similar to that of (110)Pb-Br and (001)Pb-Cl perovskites [26]. While α-(DMEN)PbBr$_4$ (DMEN = N,N-dimethylethane-1,2-diammonium) is not strictly a perovskite, it has larger ripple repetition units than (110) perovskite (Figure 9.21d) and has been found to emit cool white light at room temperature with a CRI of 73 (Table 9.2) [26].

Table 9.2 PLQE, CRI, and CIE values for white light perovskites.

Grouping	PLQY (%)	CRI	CIE (*x*, *y*)
(N-MEDA)PbBr$_4$	0.5	82	(0.36, 0.41)
(EDBE)PbBr$_4$	9	84	(0.39, 0.42)
(EDBE)PbCl$_4$	2	81	(0.33, 0.39)
α-(DMEN)$_2$PbBr$_4$	/	73	(0.28, 0.36)
(AEA)PbBr$_4$	/	87	(0.29, 0.34)
(PEA)$_2$PbCl$_4$	<1	84	(0.37, 0.42)
(EA)$_4$Pb$_3$Cl$_{10}$	/	66	(0.27, 0.39)
(API)PbCl$_4$	<1	93	(0.36, 0.37)
(MPenDA)PbBr$_4$	3.4	91	(0.24, 0.23)
(CyBMA)PbBr$_4$	1.5	/	(0.23, 0.29)

9.4.1.2 (001) Perovskite with Flat Inorganic Layers

The first example of white light-emitting (001) perovskite is (EDBE)$PbCl_4$ [25, 26]. The exciton absorption and bandgap of (DMEN)$PbBr_4$ are blue-shifted compared to (110) perovskite (EDBE)PbBr4, akin to the halide substitution effect observed in (N-MEDA)$PbBr_{4-x}Cl_x$. This material can emit "cool" white light, but its PLQE is lower than that of (EDBE)$PbBr_4$, standing at only 2%. Afterward, Boukheddaden's group synthesized the first white-emitting (001) Pb-Br perovskite $(C_6H_{11}NH_3)_2PbBr_4$ [29] with optical properties very similar to those of the two (EDBE)PbX_4 (X = Cl, Br) perovskites. They also observed a broad luminescence burst in $(C_6H_{11}NH_3)_2PbBr_{4-x}I_x$ with increasing iodine substitution [30]. Although broad spectral emission was observed at low temperatures in $(C_6H_{11}NH_3)_2PbI_4$ [31]. The highly distorted (001) perovskite (AEA)$PbBr_4$ (AEA = $^+H_3NC_6H_4[CH_2]_2NH_3^+$) emits white room-temperature light with a relatively high CRI of 87. Further results for white light-emitting Pb–X (X = Br, Cl) perovskites reported by Mathews et al. are $(C_6H_5[CH_2]_2NH_3)_2PbCl_4$ and $(H_3NCH_2C_6H_{10}CH_2NH_3)PbBr_4PbBr_4$ [32, 33]. The Kanatzidis found that n = 3 perovskites $(C_2H_5NH_3)_4Pb_3Cl_{10-x}Br_x$ emitted white light and that the Cl : Br ratio also modulated the emission color [34]. Although only a few (001) perovskites are white emitters at room temperature, Karunadasa et al. found that broad photoluminescence is in fact common to the entire (001) Pb-Br perovskite series, although their appearance is highly temperature-dependent [21].

9.4.2 Self-trapped Excitons

In 2021, a study conducted by Zeng et al. reported a WLED based on solution-processed heterophase halide perovskites. The device structure can be seen in Figure 9.22. Unlike GaN white LEDs, this approach features only a single broadband emission layer and does not involve phosphors. Its LED achieved a peak luminance of 12 200 cd m^{-2} at a bias of 6.6 V and a maximum EQE of 6.5% at a current density of 8.3 mA cm^{-2}. In situ and ex situ characterizations reveal that the mechanism for efficient electroluminescence includes charge injection into the α-phase of $CsPbI_3$, charge transfer from α to δ, and radiative recombination balanced between α and δ. They anticipate that advancements in fabrication techniques and mechanistic understanding will further enhance the efficiency and brightness of the devices in the future. Unfortunately, this work does not include any device fabrication experiments to further characterize the device's properties [35].

Ma's group developed a new dual-component strategy by assembling two broadband luminescent materials with STEs together to achieve a high CRI and stable WLEDs. Based on first-principles thermodynamic calculations, a copper-based ternary halide composite material, $CsCu_2I_3@Cs_3Cu_2I_5$, was synthesized through a simple one-step solution method. This composite material possesses ideal white light emission and cold/warm white light tuning capabilities, and it shows excellent stability against heat, UV light, and oxygen/humidity in the environment. A series of cold/warm tunable WLEDs exhibited a maximum brightness of 145 cd m^{-2}, an EQE of 0.15%, and a CRI of up to 91.6. The device shows excellent operational stability in continuous current mode, with a half-life of 238.5 min. These results

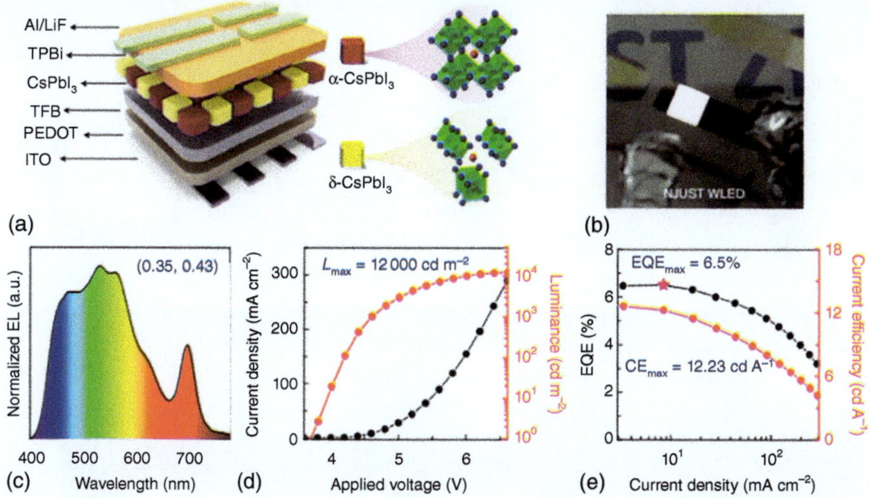

Figure 9.22 Demonstration of typical Pe-WLEDs with an α/δ-$CsPbI_3$ hetero-phase as a single emissive layer. (a) Pe-WLED structure, in which the α–δ hetero-phase synergistic effect plays a crucial role. (b) Photograph of the WLED under operation, showing uniform, bright, and warm white emission. (c) Typical electroluminescence spectra, covering the whole visible band very well. (d) Current density–voltage (J–V) curve and luminance–voltage (L–V) curve of the PeWLEDs. (e) External quantum efficiency and current efficiency of the WLEDs, exhibiting efficiencies of 6.5% and 12 200 cd m^{-2}, respectively. Source: Chen et al. [35]/Reproduced with permission from Springer Nature.

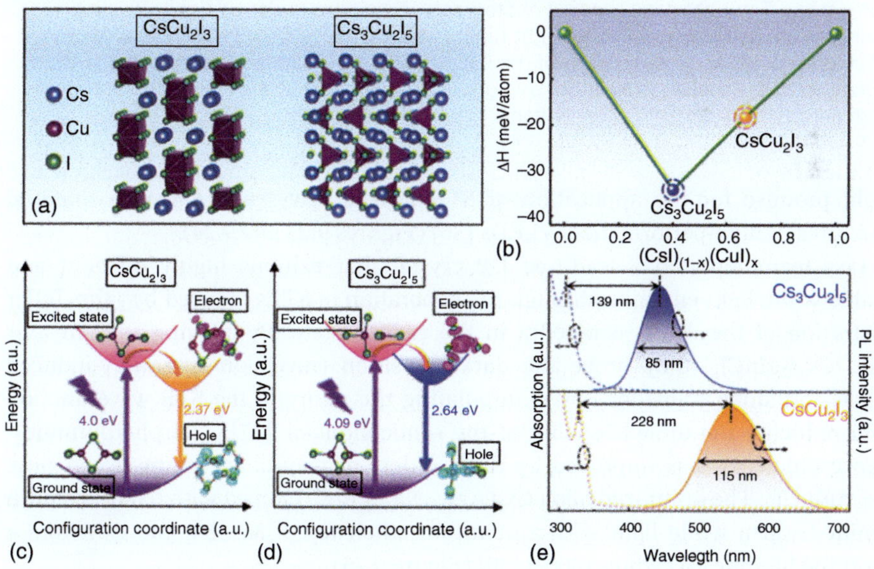

Figure 9.23 First-principles calculations of thermodynamic stability and light emission for copper-based ternary halides. (a) Crystal structures of $CsCu_2I_3$ and $Cs_3Cu_2I_5$ (blue, Cs; violet, Cu; green, I; violet polyhedrons, $[Cu_2I_3]^-$ tetrahedron, and $[Cu_2I_5]^{3-}$ tetrahedron). (b) The phase stability of copper-based ternary halides with respect to decomposition into binary compounds CsI and CuI. Both $CsCu_2I_3$ and $Cs_3Cu_2I_5$ being on the convex hull indicate their thermodynamically stable condition. (c and d) The configuration coordinate diagram of STE process of $CsCu_2I_3$ and $Cs_3Cu_2I_5$, respectively. (e) The absorption and PL spectra of the $CsCu_2I_3$ and $Cs_3Cu_2I_5$ films were measured at room temperature. Source: Ma et al. [36]/John Wiley & Sons.

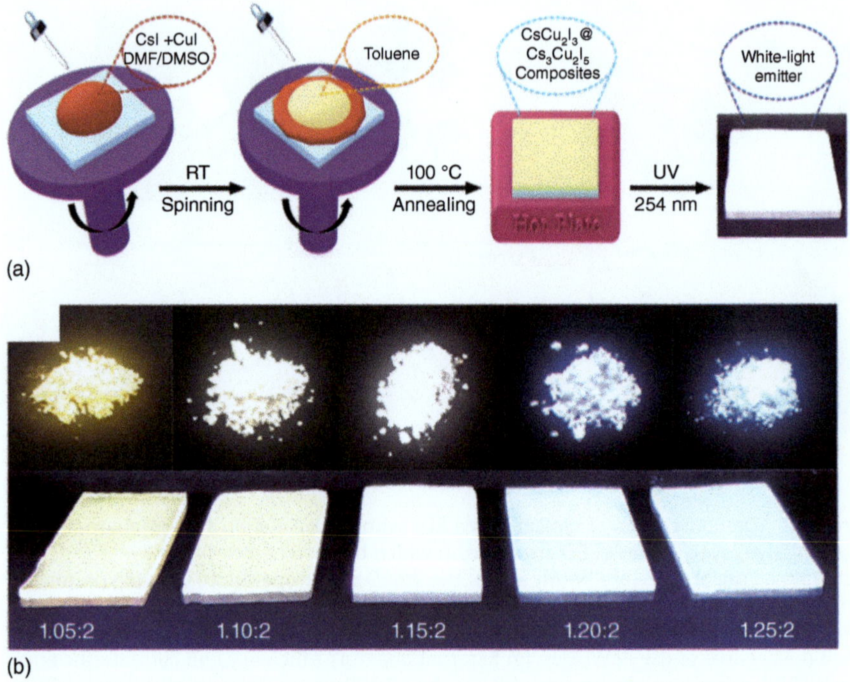

Figure 9.24 (a) Schematic diagram of the preparation process of $CsCu_2I_3@Cs_3Cu_2I_5$ composites. The right pane shows a photograph of the films under UV light (254 nm) illumination. (b) Cold/warm white-light tuning of the $CsCu_2I_3@Cs_3Cu_2I_5$ composites. Photographs of powders and films of the $CsCu_2I_3@Cs_3Cu_2I_5$ composites prepared with different CsI/CuI molar ratios under UV light irradiation. Source: Ma et al. [36]/Reproduced with permission from John Wiley & Sons.

hold promise for the application of STE-derived broadband emission material mixtures in high-performance WLEDs [36] (Figures 9.23 and 9.24).

Lu's team reported a lead-free DP crystal that exhibits highly efficient and stable white light emission through the generation of STEs induced by Jahn-Teller distortion of the $AgCl_6$ octahedra in the excited state. By alloying sodium ions into $Cs_2AgInCl_6$, they broke the dark transition (inversion symmetry-induced parity-forbidden transition) by manipulating the parity of the STE wavefunction and reducing the dimensionality of the semiconductor [37]. The photoluminescence efficiency was improved by three orders of magnitude compared to pure $Cs_2AgInCl_6$. The optimized alloy $Cs_2(Ag_{0.60}Na_{0.40})InCl_6$ doped with 0.04% bismuth emitted warm white light with a quantum efficiency of $86 \pm 5\%$ and exhibited a working lifetime exceeding 1000 h [38] (Figure 9.25).

In 2022, Li's group synthesized Al^{3+}-$Cs_2AgBiCl_6$. By controlling the doping level of Al^{3+} ions, Al^{3+}-$Cs_2AgBiCl_6$ exhibited bright STE-red emission. The PLQY

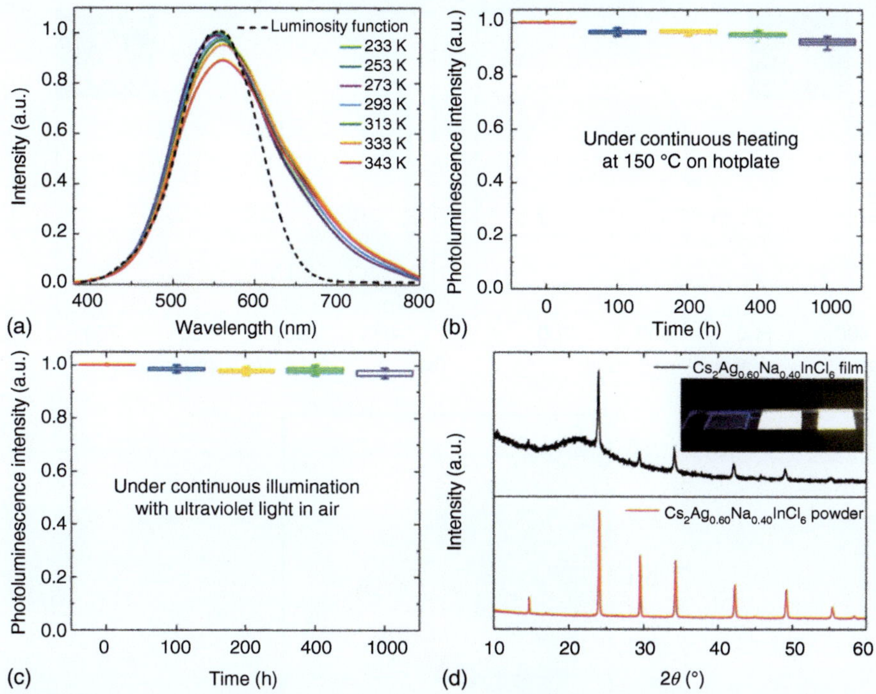

Figure 9.25 (a) Luminosity function (dashed line) and photoluminescence spectra (solid lines) of $Cs_2Ag_{0.60}Na_{0.40}InCl_6$ measured at different temperatures from 233 to 343 K. (b) Photoluminescence stability of $Cs_2Ag_{0.60}Na_{0.40}InCl_6$ against continuous heating at 150 °C on a hotplate, measured after cooling to room temperature. (c) Operational stability of $Cs_2Ag_{0.60}Na_{0.40}InCl_6$ down-conversion devices, measured in the air without any encapsulation. The box plot shows the results for five different samples measured separately, with the box edges representing quartiles, the band inside the box showing the median, and the ends of the whiskers representing the minimum and maximum of the data. (d) XRD patterns of a $Cs_2Ag_{0.60}Na_{0.40}InCl_6$ film (black line) and powder (red line). The inset shows a 300-nm-thick quartz substrate and 500-nm-thick $Cs_2Ag_{0.60}Na_{0.40}InCl_6$ films under 254-nm ultraviolet illumination. Source: Luo et al. [38]/Reproduced with permission from Springer Nature.

significantly increased from 0.98% to 17.2%, surpassing the highest reported value (16%) for lead-free DP red-emitting materials. The PL enhancement in lead-free DP NCs was achieved by doping Al^{3+} ions, facilitating an indirect bandgap to direct bandgap transition, supported by density functional theory (DFT) calculations. Moreover, Al^{3+}-$Cs_2AgBiCl_6$ demonstrated exceptional stability, maintaining its original structure for over 7 years. Finally, WLEDs were fabricated; the PL spectrum of the WLED (the inset is a photo of the device operating at 3.0 V) and the CIE coordinates of the perovskite NC-based LED are shown in Figure 9.26 [39].

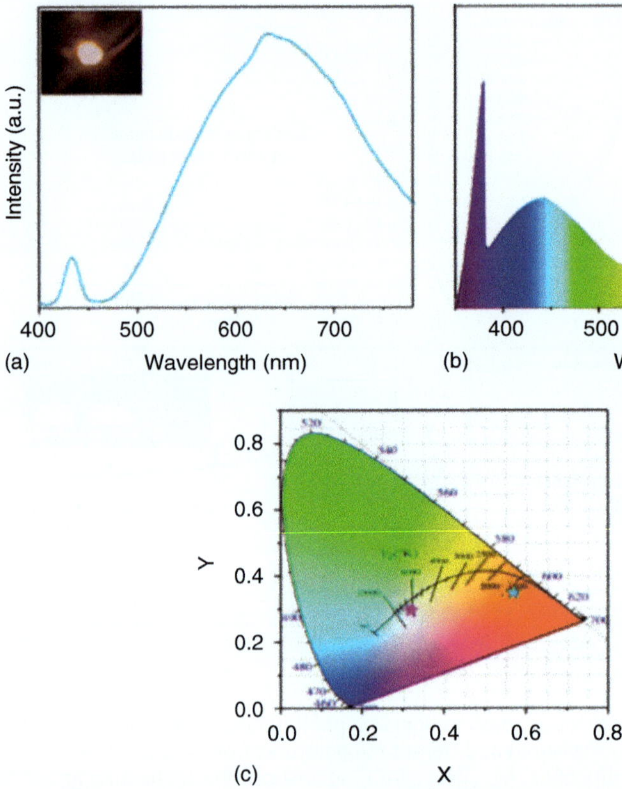

Figure 9.26 (a) The PL spectra of LED based on Al^{3+}-$Cs_2AgBiCl_6$ NCs excited by a UV-chip. Inset is the photograph of the device operated at 3.0 V; (b) the PL spectra of the WLED. Inset is the photograph of the device operated at 3.0 V; and (c) CIE coordinates of LED based on perovskite NCs. Source: Li et al. [39]/Reproduced with permission from Elsevier.

9.5 Perovskite–Organic Coupling White PeLEDs

In order to solve the problems of energy distribution imbalance and ion migration [5–7] faced by WPeLEDs with hybrid EMLs [3] or stacked EML structures [4], introducing organic luminescent molecules becomes a considerable approach to achieve white light emission by forming POC-WLEDs.

In this system, the emissive properties of organic molecules and their inherent charge injection capabilities are crucial for achieving coupling emission (Figure 9.27a). The connection state between organic molecules and perovskite crystals and their direct influence on perovskite crystals are the core issues in these POC-WLEDs. As shown in Figure 9.27b, the organic molecules can bind to perovskite crystals as surface ligands, interlayer intercalation, or even as A-site cations. As illustrated in Figure 9.27c, depending on the form of emissive centers in perovskite and organic molecules, coupling emission can be divided into the following three types: (i) hybrid coupling emission system, (ii) stacking coupling emission system, and (iii) inter-bond emission coupling. In the hybrid coupling

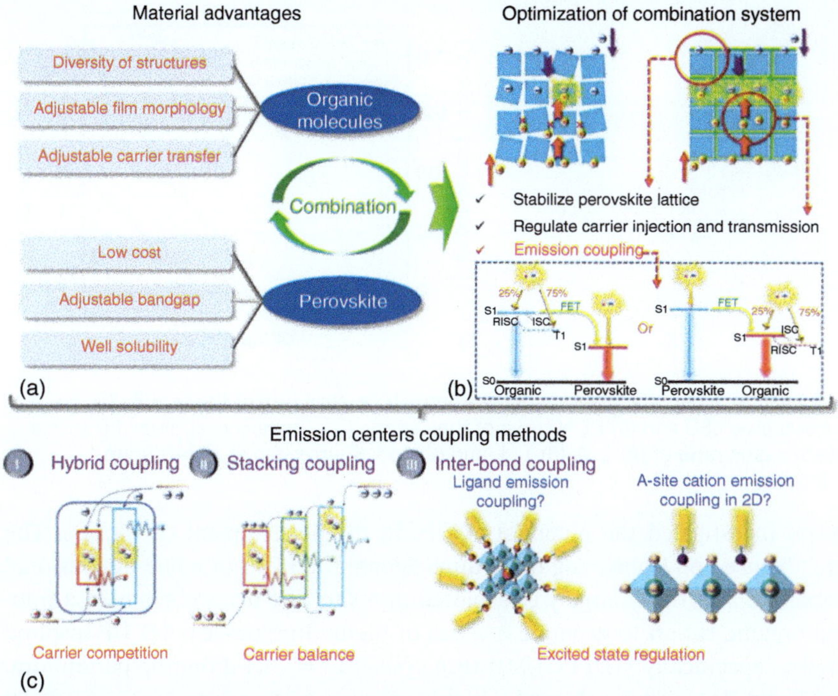

Figure 9.27 Schematic illustration of ideas in this article. (a) Summary of advantages of organic and perovskite materials; (b) improvement of material properties by perovskite–organic combination system; and (c) construction methods of coupling emission systems. Source: Chen et al. [2]/John Wiley & Sons.

emission system, organic emissive molecules are mixed with perovskite materials to form an EML where charge carriers compete between emission centers. In the stacked coupling emission system, organic molecules and perovskite are combined through stacking, and charge carriers undergo exciton recombination for coupling emission over a wide range of layers. In the intermolecular emission coupling system, organic molecules and perovskite are chemically bonded. Organic molecules can participate in lattice construction while-emitting light and can regulate the electrical properties of the perovskite–organic system through their charge transfer capabilities [2].

In 2017, Yao's group reported on blue perovskite NCs and presented a composite NC of $CsPbBr_xCl_{3-x}$, which was utilized as a luminescent layer to prepare blue light-emitting devices. The device structure can be seen in Figure 9.28 [40].

They developed a composite NC of $CsPbBr_xCl_{3-x}$, which was used as a luminescent layer to prepare blue light-emitting devices, as shown in Figure 9.29. However, the obtained blue light devices had unsatisfactory brightness (0.18 cd A^{-1}) and EQE (0.07%). To improve the color purity of the resulting white light, in a later study, an orange polymeric material, MEH:PPV (poly[2-methoxy-5-(2-ethylhexyloxy)-1,4-phenylvinyl]), was wrapped on $CsPbBr_xCl_{3-x}$ NCs, which resulted in a white light device with a CIE of (0.33, 0.34).

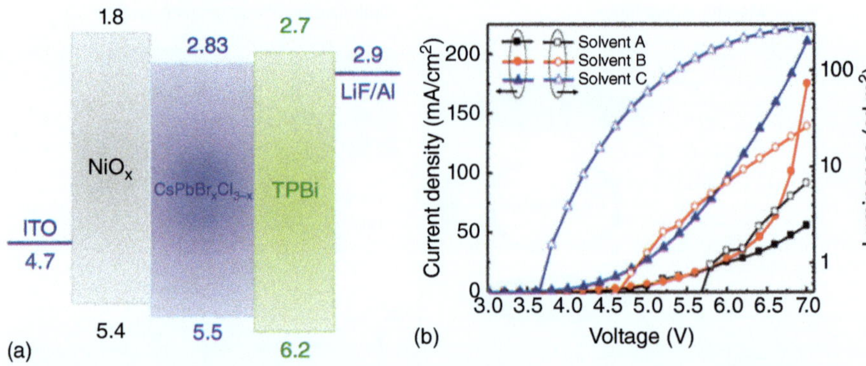

Figure 9.28 (a) Schematic diagram of the energy level structure of the $CsPbBr_xCl_{3-x}$ nanocrystal blue LED and (b) J-L-V curve of the $CsPbBr_xCl_{3-x}$ nanocrystal blue LED with a toluene : hexane ratio of (A) 1 : 0, (B) 1 : 1, and (C) 1 : 3. Source: Yao et al. [40]/John Wiley & Sons.

A team investigated the potential of DPs in the development of WLEDs. The research focused on translating three-dimensional (3D) DPs into one-dimensional (1D) analogs, a territory largely unexplored thus far. The researchers used a fluorinated organic cation to generate a series of highly luminescent 1D DP-inspired materials, specifically $(DFPD)_2MIInBr6$ (DFPD = 4,4-difluoropiperidinium, MI = K^+ and Rb^+). They achieved a highly efficient warm-white PLQY of 92% by doping 0.3% Sb^{3+} in $(DFPD)_2KInBr_6$. The inset of Figure 9.29b presents a typical photograph of the prepared thin film, which exhibits very stable and uniform warm white-light emission under 365 nm UV light. Moreover, the team fabricated single-component warm-WLEDs with $(DFPD)_2KInBr_6$:Sb, achieving a luminance of 300 cd m^{-2}, one of the best-performing outcomes among lead-free metal halides WLEDs reported to date. This study significantly broadened the scope of In-based metal halides from 3D to 1D, showcasing superior optical performances and broad application prospects. However, it is important to note that this work did not include any experiments on device preparation or further characterization, leaving room for future research in this domain [41].

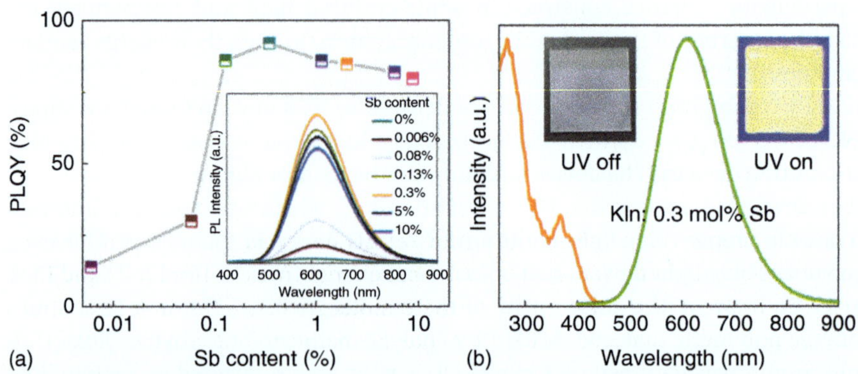

Figure 9.29 (a) Dependence of PLQY on intentionally doped Sb^{3+} concentration. Inset shows the corresponding PL spectra. (b) UV/vis absorption, PL spectrum. Source: Bai et al. [41]/John Wiley & Sons.

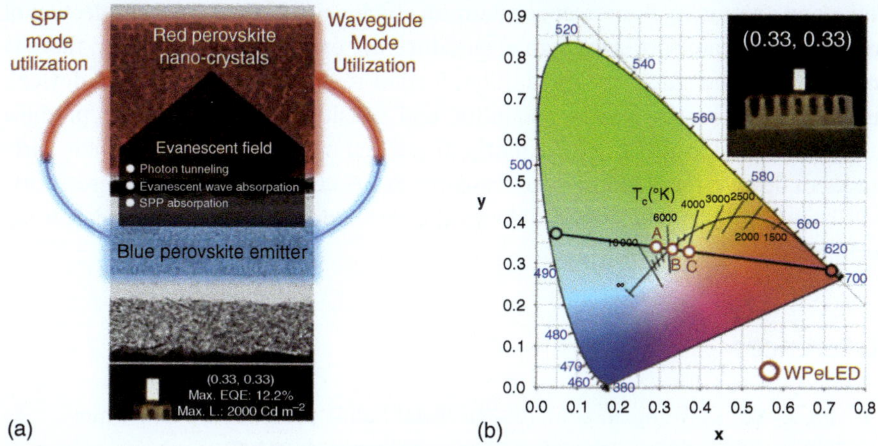

Figure 9.30 (a) Schematic diagram of the device design. (b) CIE coordinates of WPeLEDs and Points A (0.29, 0.34), B (0.33, 0.33), and C (0.37, 0.32) represent three devices with various thicknesses of red PeNC layer. The inset is a photo of a WPeLED (point B). Source: Chen et al. [42]/Reproduced with permission from Elsevier.

9.6 Others

In a paper published in Joule in 2021, Chen and colleagues reported an innovative and efficient method to construct high-performance white perovskite light emitting device with significantly improved light extraction efficiency (LEE). The conventional LEE in the metal halide-PeLED field is typically less than 20%, and achieving white emission has been a major challenge. The authors addressed these issues by coupling a blue PeLED with a layer of red perovskite NC down-converter through a carefully designed multilayer semitransparent electrode (LiF/Al/Ag/LiF). This configuration enabled the extraction of the trapped waveguide mode and surface plasmon polariton mode in a blue PeLED, converting them into red emission, which resulted in over a 50% improvement in LEE. Simultaneously, the complementary EM of blue photons and down-converting red photons contributed to a white PeLED with a high EQE and luminance of more than 12% and approximately 2000 cd m^{-2}, respectively. These results represent significant advancements in the field [42] (Figure 9.30).

9.7 Conclusion and Outlook

In conclusion, the low efficiency of white PeLED devices can be attributed to the balance of charge carrier transport and overall charge carrier injection in the EML, which can be improved through appropriate molecular design and device fabrication. Suitable molecular design can regulate the bandgap and exciton generation, resulting in narrower emission peaks, lower difficulty of exciton injection, and higher exciton binding efficiency. Meanwhile, appropriate device processes can improve exciton transport and binding, avoiding the excessive generation of defect states and enhancing fluorescence quantum efficiency and EQE. Proper selection of injection and transport layers can also reduce the exciton transfer barrier, thereby improving device efficiency. In addition, further research is needed on the device's lifespan, ambient temperature stability, and sensitivity to humidity to produce efficient perovskite WLEDs. Currently, these technologies are still in their early stages of development, and further research is needed in the future. These efforts will bring significant breakthroughs in the development of WLEDs and lighting technology.

References

1 Yao, J., Xu, L., Wang, S. et al. (2022). Metal halide perovskites-based white light-emitting diodes. *Journal of Physics: Photonics* 4: 042001.

2 Chen, X., Xiang, H., Wang, R. et al. (2023). Perovskite-organic coupling WLED: progress and perspective. *Advanced Functional Materials* 2304750.

3 Yu, H., Wang, H., Pozina, G. et al. (2020). Single-emissive-layer all-perovskite white light-emitting diodes employing segregated mixed halide perovskite crystals. *Chemical Sicence* 11: 11338–11343.

4 Wang, C., Xue, D., Shen, X. et al. (2019). White light-emitting devices based on ZnCdS/ZnS and perovskite nanocrystal heterojunction. *Nanotechnology* 30: 465201.

5 Draguta, S., Sharia, O., Yoon, S.J. et al. (2017). Rationalizing the light-induced phase separation of mixed halide organic-inorganic perovskites. *Nature Communications* 8: 200.

6 Motti, S.G., Meggiolaro, D., Martani, S. et al. (2019). Defect activity in lead halide perovskites. *Advanced Materials* 31: 1901183.

7 Hoke, E.T., Slotcavage, D.J., Dohner, E.R. et al. (2015). Reversible photo-induced trap formation in mixed-halide hybrid perovskites for photovoltaics. *Chemical Sicence* 6: 613–617.

8 Chen, S., Lin, J., Zheng, S. et al. (2023). Efficient and stable perovskite white light-emitting diodes for backlit display. *Advanced Functional Materials* 33: 2213442.

9 Wang, A., Guo, Y., Zhou, Z. et al. (2019). Aqueous acid-based synthesis of lead-free tin halide perovskites with near-unity photoluminescence quantum efficiency. *Chemical Sicence* 10: 4573–4579.

10 Tan, Z., Li, J., Zhang, C. et al. (2018). Highly efficient blue-emitting Bi-doped Cs_2SnCl_6 perovskite variant: photoluminescence induced by impurity doping. *Advanced Functional Materials* 28: 1801131.

11 Zhou, C., Tian, Y., Yuan, Z. et al. (2017). Highly efficient broadband yellow phosphor based on zero-dimensional tin mixed-halide perovskite. *ACS Applied Materials & Interfaces* 9: 44579–44583.

12 Zhang, X., Wang, C., Zhang, Y. et al. (2018). Bright orange electroluminescence from lead-free two-dimensional perovskites. *ACS Energy Letters* 4: 242–248.

13 Liu, Y., Jing, Y., Zhao, J. et al. (2019). Design optimization of lead-free perovskite $Cs_2AgInCl_6$:Bi nanocrystals with 11.4% photoluminescence quantum yield. *Chemistry of Materials* 31: 3333–3339.

14 Zhang, Y., Zhou, L., Li, D. et al. (2022). Realizing efficient emission in three-dimensional $CsCdCl_3$ single crystals by introducing separated emitting centers. *Inorganic Chemistry* 61: 17902–17910.

15 Leng, M., Yang, Y., Zeng, K. et al. (2018). All-inorganic bismuth-based perovskite quantum dots with bright blue photoluminescence and excellent stability. *Advanced Functional Materials* 28: 1704446.

16 Jun, T., Sim, K., Iimura, S. et al. (2018). Lead-free highly efficient blue-emitting $Cs_3Cu_2I_5$ with 0D electronic structure. *Advanced Materials* 30: 1804547.

17 Ji, C., Huang, Z., Wen, J. et al. (2019). Blue-emitting Bi-doped double perovskite Gd_2ZnTiO_6 phosphor with near-ultraviolet excitation for warm white light-emitting diodes. *Journal of Alloys and Compounds* 788: 1127–1136.

18 Liu, Y., Zheng, X., Ji, Z. et al. (2022). Blue light hazard optimization for white light-emitting diode of Mn^{2+}-activated 0D $Cs_3Cu_2Br_5$ perovskite materials. *Advanced Materials Interfaces* 10.

19 Li, Z., Li, Q., Cao, M. et al. (2023). Multimodal luminescent low-dimension Cs_2ZrCl_6:xSb^{3+} crystals for white light-emitting diodes and information encryption. *Langmuir* 39: 3792–3799.

20 Smith, M.D., Connor, B.A., and Karunadasa, H.I. (2019). Tuning the luminescence of layered halide perovskites. *Chemical Reviews* 119: 3104–3139.

21 Smith, M.D., Jaffe, A., Dohner, E.R. et al. (2017). Structural origins of broadband emission from layered Pb-Br hybrid perovskites. *Chemical Science* 8: 4497–4504.

22 Tanaka, K., Takahashi, T., Kondo, T. et al. (2005). Electronic and excitonic structures of inorganic-organic perovskite-type quantum-well crystal $(C_4H_9NH_3)_2PbBr_4$. *Japanese Journal of Applied Physics* 44: 5923–5932.

23 Dohner, E.R., Hoke, E.T., and Karunadasa, H.I. (2014). Self-assembly of broadband white-light emitters. *Journal of the American Chemical Society* 136: 1718–1721.

24 Hu, T., Smith, M.D., Dohner, E.R. et al. (2016). Mechanism for broadband white-light emission from two-dimensional (110) hybrid perovskites. *The Journal of Physical Chemistry Letters* 7: 2258–2263.

25 Bera, D., Qian, L., and Holloway, P.H. (2016). *Luminescent Materials and Applications*, 41–74. John Wiley & Sons, Ltd.

26 Dohner, E.R., Jaffe, A., Bradshaw, L.R. et al. (2014). Intrinsic white-light emission from layered hybrid perovskites. *Journal of the American Chemical Society* 136: 13154–13157.

27 Li, Y.Y., Lin, C.K., Zheng, G.L. et al. (2006). Novel ⟨110⟩-oriented organic-inorganic perovskite compound stabilized by *N*-(3-aminopropyl)imidazole with improved optical properties. *Chemical Materials* 18: 3463–3469.

28 Wu, Z., Ji, C., Sun, Z. et al. (2018). Broadband white-light emission with a high color rendering index in a two-dimensional organic-inorganic hybrid perovskite. *Journal of Materials Chemistry C* 6: 1171–1175.

29 Yangui, A., Garrot, D., Lauret, J.S. et al. (2015). Optical investigation of broadband white-light emission in self-assembled organic-inorganic perovskite $(C_6H_{11}NH_3)_2PbBr_4$. *The Journal of Physical Chemistry C* 119: 23638–23647.

30 Yangui, A., Pillet, S., Lusson, A. et al. (2017). Control of the white-light emission in the mixed two-dimensional hybrid perovskites $(C_6H_{11}NH_3)_2[PbBr_{4-x}I_x]$. *Journal of Alloys and Compounds* 699: 1122–1133.

31 Yangui, A., Pillet, S., Mlayah, A. et al. (2015). Structural phase transition causing anomalous photoluminescence behavior in perovskite $(C_6H_{11}NH_3)_2[PbI_4]$. *The Journal of Chemical Physics* 143: 224201.

32 Thirumal, K., Chong, W.K., Xie, W. et al. (2017). Morphology-independent stable white-light emission from self-assembled two-dimensional perovskites driven by strong exciton-phonon coupling to the organic framework. *Chemistry of Materials* 29: 3947–3953.

33 Neogi, I., Bruno, A., Bahulayan, D. et al. (2017). Broadband-emitting 2D hybrid organic-inorganic perovskite based on cyclohexane-bis(methylammonium) cation. *ChemSusChem* 10: 3765–3772.

34 Mao, L., Wu, Y., Stoumpos, C.C. et al. (2017). Tunable white-light emission in single-cation-templated three-layered 2D perovskites $(CH_3CH_2NH_3)_4Pb_3Br_{10-x}Cl_x$. *Journal of the American Chemical Society* 139: 11956–11963.

35 Chen, J., Wang, J., Xu, X. et al. (2020). Efficient and bright white light-emitting diodes based on single-layer heterophase halide perovskites. *Nature Photonics* 15: 238–244.

36 Ma, Z., Shi, Z., Yang, D. et al. (2021). High color-rendering index and stable white light-emitting diodes by assembling two broadband emissive self-trapped excitons. *Advanced Materials* 33: 2001367.

37 Xiao, Z., Meng, W., Wang, J. et al. (2017). Searching for promising new perovskite-based photovoltaic absorbers: the importance of electronic dimensionality. *Materials Horizons* 4: 206–216.

38 Luo, J., Wang, X., Li, S. et al. (2018). Efficient and stable emission of warm-white light from lead-free halide double perovskites. *Nature* 563: 541–545.

39 Li, L., Shao, H., Wu, X. et al. (2022). Aluminum-doped lead-free double perovskite $Cs_2AgBiCl_6$ nanocrystals with ultrahigh stability towards white light emitting diodes. *Materials Research Bulletin* 147.

40 Yao, E.P., Yang, Z., Meng, L. et al. (2017). High-brightness blue and white LEDs based on inorganic perovskite nanocrystals and their composites. *Advanced Materials* 29: 1606859.

41 Bai, T., Wang, X., Wang, Z. et al. (2023). Highly luminescent one-dimensional organic-inorganic hybrid double-perovskite-inspired materials for single-component warm white-light-emitting diodes. *Angewandte Chemie International Edition* 62: e202213240.

42 Chen, Z., Li, Z., Chen, Z. et al. (2021). Utilization of trapped optical modes for white perovskite light-emitting diodes with efficiency over 12%. *Joule* 5: 456–466.

10

Electron and Hole Transport Materials

10.1 Background of Charge Transport Materials

Perovskite materials have attracted widespread attention from researchers in the field of luminescent semiconductors over the past decade due to their excellent properties such as tunable bandgap, high luminescent quantum efficiency, and narrow band excitation. However, perovskite luminescent materials have performance deficiencies such as inevitable defect formation, vulnerable film morphology, instability of ion migration and phase segregation, as well as inferior light outcoupling. These defects greatly hinder the commercialization process of perovskite light-emitting diodes (PeLEDs). To solve these intrinsic defects of perovskite luminescent materials, many recent reports have focused on charge transport materials and charge injection layers (CILs) situated between the electrodes and the luminescent layers. Charge transport materials are a class of semiconductor materials that can achieve the directional, ordered, and controlled migration of carriers under the influence of an electric field when a carrier, electron, or hole is injected, resulting in charge transport. Charge transport materials are fundamentally essential for achieving efficient and balanced charge transport, which is the core performance requirement of these types of devices. Thus, introducing a charge transport layer (CTL) into the device has become an obligatory option.

As an emerging light-emitting semiconductor device that competes with organic light-emitting diodes (OLEDs), PeLEDs continue to use some traditional charge transport materials that have performed excellently in OLEDs, such as those widely used in the electron transport layer (ETL) of OLEDs. In recent times, several traditional electron transport materials (ETMs) like 2,2′,2″-(1,3,5-Benzinetriyl)-tris (1-phenyl-1-*H*-benzimidazole) (TPBi), 4,6-Bis(3,5-di-3-pyridylphenyl)-2-methyl-pyrimidine (B3PYMPM), bathophenanthroline (BPhen), 1,3,5-Tris(3-pyridyl-3-phenyl)benzene (TmPyPB), Tris(2,4,6-trimethyl-3-(pyridin-3-yl)phenyl)borane (3TPYMB), and 2,4,6-tris[3-(diphenylphosphinyl)phenyl]-1,3,5-triazine (PO-T2T), which were initially utilized in OLEDs, have found their way into PeLEDs. However, their thermal stability still poses a challenge. Inorganic oxide metals like TiO_2, ZnO, and SnO_2 have also emerged as promising candidates for electron injection layers, which exhibit good thermal stability. Nevertheless, further improvement is still needed to boost the electron injection characteristics and increase the bonding

Perovskite Light Emitting Diodes: Materials and Devices, First Edition. Hong Meng.

(interface contact) between the ETL and the perovskite film. In terms of HTLs (hole transport layers), the traditional PEDOT:PSS (poly(3,4-ethylenedioxythiophene): poly(styrenesulfonate)) transport material has received extensive research in the organic optoelectronics field due to its high conductivity and transparency in the visible light range. Moreover, this material is currently commercialized, which provides significant advantages for research centered around it in terms of cost control and research cycle. However, PEDOT:PSS has many intrinsic defects, in the field of PeLEDs, including serious light quenching at the interface with perovskites, significant barriers when injecting charges between the HTL and the perovskite layer, poor crystal growth properties of the perovskite film on top of PEDOT:PSS, and corrosion of the indium tin oxide (ITO) electrode due to the material's strong acidity. Therefore, the development of new hole transport materials or careful tuning of existing materials is an essential requirement for developing high-performance PeLEDs. The use of appropriate hole transport materials is crucial in promoting energy level alignment between the electrode and the light-emitting layer, leading to improved hole transport efficiency. This, in turn, facilitates exciton recombination in the light-emitting layer, resulting in reduced device operating voltage and increased efficiency. Therefore, the development of efficient hole transport materials is a critical aspect of PeLEDs research, and ongoing efforts are focused on identifying new materials with enhanced performance and stability [1].

This chapter will discuss recent developments in the optimization of electron and hole transport materials in PeLEDs and provide insights into future directions in the design of these materials.

10.1.1 Charge Transport of Metal Halide Perovskite Materials

The charge–carrier mobilities in metal halide perovskites (MHPs) are suitably compared with those of typical inorganic ionic semiconductors, like gallium arsenide (GaAs). For lead-iodide perovskites such as $MAPbI_3$ and $FAPbI_3$ – presently used in the highest-efficiency solar cells – the electron and hole mobility values are in the range of fundamentally restricted to approximately $200\,cm^2/(V\,s)$. These values fall short of those for GaAs, underscoring the relative ionicity of the metal-halide bond. A survey of the current literature reports suggests electron mobility in the range of $10–66\,cm^2V^{-1}s^{-1}$ and hole mobility in the range of $10–30\,cm^2V^{-1}s^{-1}$ for thin films of lead halide perovskites such as $MAPbI_3$. Table 10.1 provides a literature overview of the charge carrier mobility values of $MAPbI_3$ at room temperature. The limiting factors for charge–carrier mobility in MHPs can be neatly categorized into intrinsic and extrinsic effects [17]. Intrinsic effects originate from charge–carrier interactions with the underlying lattice. Extrinsic effects, however, are constrained by material imperfections, such as grain size, grain boundaries, energetic disorder, or impurities, which can be mitigated through appropriate material processing techniques. It should be noted that for any particular stoichiometry of MHPs, reported charge–carrier mobility values cover a considerable range. The charge–carrier mobilities in MHPs can be greatly influenced by quasi-two-dimensional perovskite

Table 10.1 Charge–carrier mobility values of $MAPbI_3$ at room temperature.

Architecture	Fabric. route	Measurement	Electron, hole, sum	Mobility $cm^2(V s)^{-1}$	Refs.
Film	Sol.	THzC	Σ	35	[2]
Film	Sol.	THzC	Σ	20	[3]
Film	Sol.	MWC	Σ	30	[4]
Film	Sol.	MWC	Σ	29	[5]
Film	Sol.	MWC	Σ	71	[6]
Film	Sol.	PLQ	e	1.4	[7]
Film	Sol.	PLQ	h	0.9	[7]
Film	Sol.	PLQ	e	0.7	[8]
Film	Sol.	PLQ	h	0.4	[8]
Film	Sol.	micros	e/h	2–3	[9]
Meso	Sol.	THzC	Σ	8	[10]
Meso	Sol.	MWC	Σ	9	[4]
SingleC	Sol.	SCLC	h	164	[11]
SingleC	Sol.	SCLC	h	67	[12]
SingleC	Sol.	SCLC	h	2.5	[13]
SingleC	Sol.	THzC	Σ	600	[14]
SingleC	Sol.	MWC	Σ	115	[15]
SingleC	Sol.	TOF	h	24	[11]
SingleC	Sol.	Hall	h	105	[11]
PolyC	Vap.	Hall	e	66	[16]

structures. These structures incorporate layers of nonconducting organic cations that are too large to fit within the perovskite structure. In lead perovskites, smaller halides are associated with lower charge carrier mobilities. Conversely, substituting a significantly lighter metal, like tin, results in higher mobility values. Changes to the A-cation in MHPs have little impact on the intrinsic charge–carrier mobilities.

Charge mobilities were normally investigated by using various techniques including time-resolved laser spectroscopies [transient absorption (TA), time-resolved photoluminescence (TRPL), time-resolved terahertz spectroscopy (TRTS), and time-resolved microwave conductivity (TRMC)]; electrical transient techniques [charge extraction by linearly increasing voltage (CELIV) and time-of-flight (TOF)]; and steady-state methods [field-effect transistor (FET), Hall effect, and space charge limited current (SCLC)]. Notably, the carrier mobility of perovskite samples either in thin films or within devices obtained using different techniques can vary by up to 7–8 orders of magnitude. Relevant data differences and trends are shown in Figure 10.1.

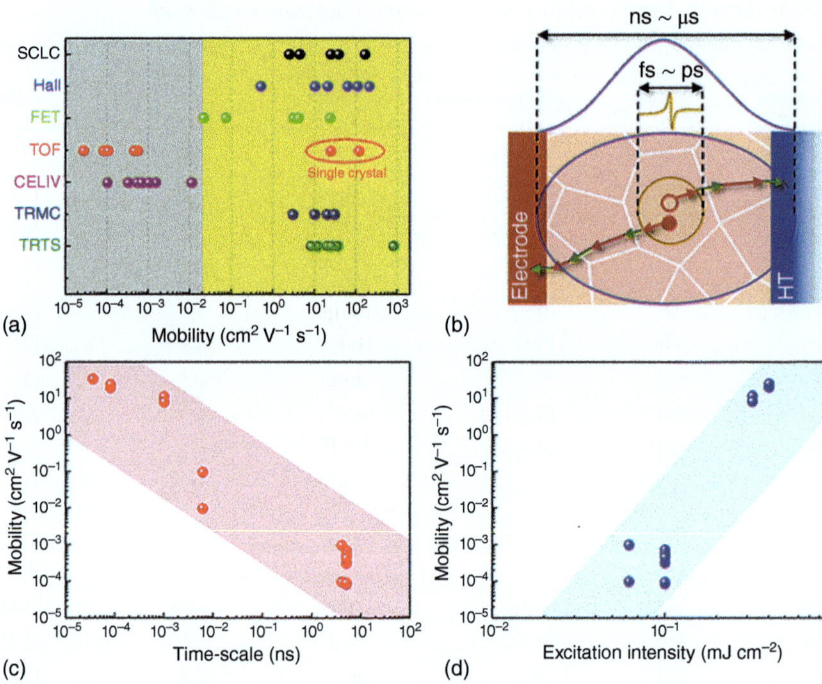

Figure 10.1 (a) Charge–carrier mobilities of perovskite films and single crystals measured by various techniques, (b) schematic illustrating the main detection regime of those techniques with different time windows, (c) dependence of the probing time-scale, and (d) light excitation intensities on the charge mobilities obtained using different techniques for $MAPbI_xCl_{3-x}$ films. Source: Peng et al. [18]/Royal Society of Chemistry/CC BY 3.0.

10.1.2 Charge Transport Materials in PeLED

In a conventional PeLED architecture, as illustrated in Figure 10.2a, the perovskite emissive layer (EML) is strategically situated between an electron and a hole transport/injection layer. Under a forward bias condition, charge carriers are introduced from the injection layers into the perovskite EML, subsequently generating electron–hole pairs. These pairs undergo radiative recombination, which culminates in light emission.

The charge-transport characteristics of the perovskite materials along with the optimization of the interfaces and naturally, selecting or tailoring the ETLs and HTLs, have been one of the main sources of PeLED performance. All these materials' charge–carrier mobilities are thus necessary to maximize the efficiency of the PeLED devices. Figure 10.2b shows the control of the interfacial energetics is essential to minimize efficiency losses.

Matching the carrier mobility from ETL transport and from the HTL is very important to achieve high-performance PeLEDs. Figure 10.3b lists the carrier mobilities of the commonly used ETL and HTL materials. Therefore, it is necessary to balance the carrier mobility between the ETL and HTL. This strategy aims to avoid carrier loss due to ineffective carrier compound formation in the light-emitting layer, which can result in decreased device efficiency.

In the following parts, we will discuss the charge transport/injection materials of electrons and holes in PeLEDs.

10.2 Electron Transport Materials in PeLEDs

In PeLED devices, the ETL primarily serves three functions. First, it effectively transports electrons from the cathode to the perovskite light-emitting layer. This is mainly achieved through the energy level alignment of the materials. Specifically, the lowest unoccupied molecular orbital (LUMO) energy level (or the conduction band minimum, CBB) of the ETM should align with the work function of the cathode. At the same time, the LUMO energy level (or CBB) of the ETM should be slightly below the CBB of the perovskite light-emitting material, thereby reducing the electron injection barrier and improving injection efficiency. The second function is to prevent holes in the perovskite layer from injecting the ETL and non-radiatively recombining with the electrons inside. For this, the highest occupied molecular orbital (HOMO) energy level (or valence band top, VBT) of the ETM should be slightly below the VBT of the perovskite light-emitting material, effectively blocking hole injection. It's important to note that the difference between the HOMO (or VBT) of the ETM and the VBT of the perovskite light-emitting material should not be too large. Otherwise, it could result in an excessive reverse injection barrier, causing an accumulation of holes at the interface between the two interlayers, leading to a substantial negative voltage. This would undeniably increase the device's driving voltage, negatively affecting energy efficiency and device stability. Lastly, the introduction of the ETL prevents the perovskite light-emitting layer from directly contacting the cathode, which to some extent safeguards the perovskite layer from degradation, enhancing the stability of the device.

10.2.1 Inorganic Oxides Electron Transport Materials

Metal oxides have been widely used in all-inorganic PeLEDs due to their better fitting and temperature stability. Among them, ZnO nanoparticles (NPs) or quantum dots are particularly favored for their suitable function, good valence bonding, and easy low-temperature synthesis [21–23]. However, defects formed by oxygen adsorption can lead to the decomposition of perovskite films and the degradation of electron transport properties, which can be problematic [24, 25]. Defects resulting from oxygen adsorption can break down perovskite films and adversely affect their electron transport properties. Additionally, controlling the agglomeration of NPs or quantum dots is a major challenge in achieving a smooth, uniform ETL [26]. In a study by Heyong Wang et al., it was found that by combining the organic polymer polyethylenimine ethoxyl (PEIE) with ZnO nanocrystals (NCs) and in situ formation of perovskite NC composite films in an electron transport matrix, they

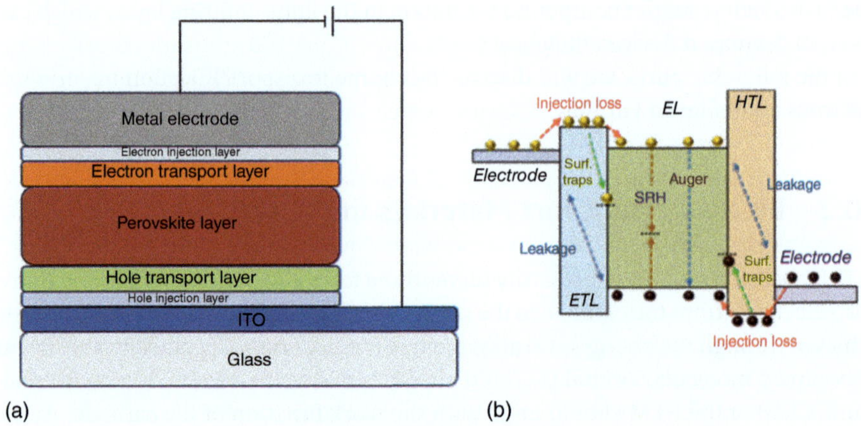

Figure 10.2 (a) A typical architecture of a perovskite LED. The possible active layers constituting a PeLED are labeled on the left, and their contributions to various processes are listed on the right. (b) Schematic representation of the processes responsible for the electron charge injection losses in PeLED. Source: Fakharuddin et al. [19]/Springer Nature.

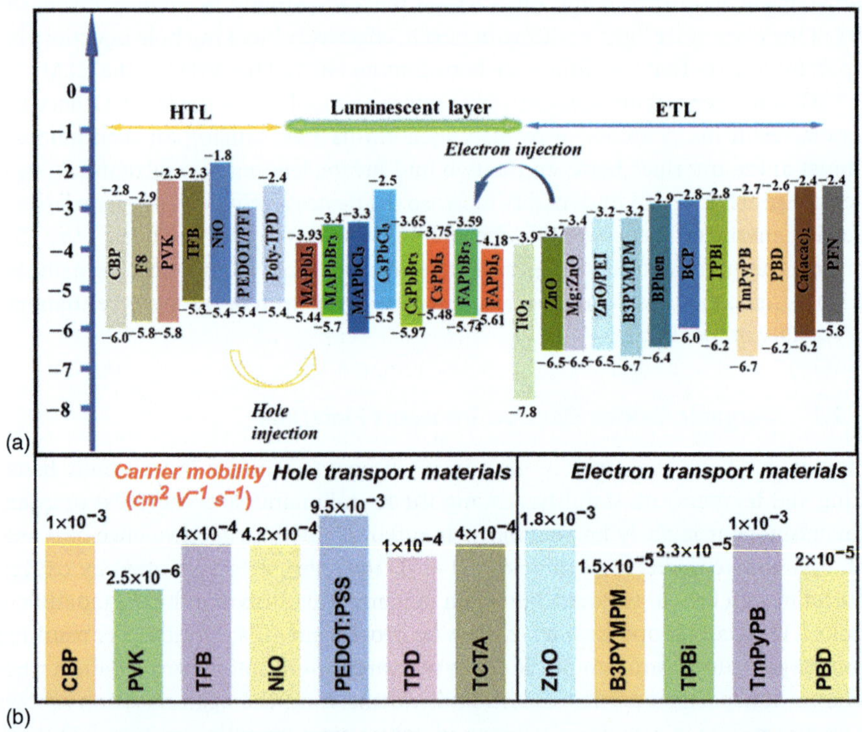

Figure 10.3 Basic properties of the carrier transport materials. (a) Energy level. (b) Carrier mobility. Source: Liu et al. [20]/John Wiley & Sons.

were able to control the nucleophilic process of perovskite, resulting in a maximum external quantum efficiency (EQE) of 17.3% and a T_{50} of 100 hours for PeLEDs [27]. In particular, the study found that the decay mechanism of PeLEDs differs at high voltages compared to low voltages, which offers a new design approach for achieving efficient and stable PeLEDs.

To enhance the solution-processability of materials for PeLEDs, it is necessary to avoid particle agglomeration during the preparation process. Yun Cheol Kim et al. successfully addressed this issue by modifying ZnO quantum dots with oleic acid (OA) for surface treatment. This modification prevented the agglomeration of the quantum dots, allowing for improved solution-processability and spin-coating properties [28]. The use of OA-modified ZnO quantum dots has successfully addressed the issue of particle agglomeration. The modified ZnO quantum dot ETL achieves an root mean square (RMS) roughness of 1.62 nm while maintaining a transmittance of 88.7% in the visible range (380–750 nm), indicating that reducing the agglomeration of ZnO quantum dots can lead to a smooth and highly transmissive ETL. The study found that the use of OA-modified ZnO quantum dots resulted in a maximum luminescence of 11 095.1 cd m^{-2} at 6 V and a maximum current efficiency of 6.79 cd A^{-1} at 4 V. These results demonstrate that OA modification is an effective method for optimizing the ETL to produce smooth and highly transparent films.

Besides, Fan Cao et al. reported a core–shell ZnO/ZnS NP as a solution-processable PeLED ETL [29]. The deposition of a ZnS shell can attenuate the trap states of the ZnO core by anchoring sulfur to the vacancies of oxygen and concurrently removing involved hydroxyl groups, which favors the suppression of interface trap-assisted non-radiative recombination and the deprotonation reaction between the perovskite layer and ZnO. This strategy enables the modified ZnO NPs to display a reduced energy barrier between the perovskite luminescent layer and the ETL compared to pure ZnO control group and improved charge injection balance in PeLEDs. Ultimately, the ETL with this core–shell nano-encapsulation structure allows the corresponding PeLEDs to display a peak luminance of 32 400 cd m^{-2}, 10.3% EQE, and a 20-fold life extension compared to devices using pure ZnO as the ETL.

In Heyong Wang's study, they found that SnO_2 had better chemical compatibility with perovskite materials than ZnO, and they used a solution-processed SnO_2 layer as the ETL in their PeLEDs. They also used an n-i-p (inverted) device structure, where the ETL is sandwiched between the perovskite light-emitting layer and the HTL, to improve device performance. Their study showed that using a SnO_2 ETL can lead to a higher maximum EQE and a longer device lifetime compared to using a ZnO ETL [30]. And SnO_2 has demonstrated excellent characteristics such as good transparency, film-forming morphology, and a suitable energy level, leading to the achievement of an EQE_{max} of 7.9%. These features make SnO_2 a promising candidate for three-dimensional and lower-dimensional PeLEDs. However, modified SnO_2 still presents a non-negligible photoluminescence (PL) quenching, which significantly hinders device efficiency and poses a challenge to further optimization.

10.2.2 Inorganically Doped Electron Transport Materials

Khan Qasim et al. reported that the energy band structure of ZnO can be tuned by Ca doping to create a cascade of conduction energy level from the cathode to the perovskite, resulting in the formation of Ca-doped ZnO (CZO) [21]. Khan Qasim et al. demonstrated that Ca doping of ZnO can match better with the energy level of the perovskite emitter layer/can realize a better energy level alignment with the adjacent perovskite layer, resulting in a CZO ETL with enhanced conductivity and electron mobility. This allows for controlled electron injection at sub-bandgap start-up voltages, achieving high performance in terms of current efficiency and EQE for red, yellow, and green light emissions. The power efficiency of these devices can reach up to 36 lm W^{-1} for red, 28 lm W^{-1} for yellow, and 30 lm W^{-1} for green PeLEDs. Additionally, the two-step thermal injection technique enables the use of porous perovskite layers, achieving high luminance for red, yellow, and green light. This approach offers potential for future development of PeLEDs with tunable ETLs and a combined solution method.

The exploration of HTLs continues unabated. In 2020, Han's team exhaustively investigated various physical phenomena in PeLED devices from the mechanism of electron transport and thoroughly introduced, summarized, and classified a large number of existing materials, which is conducive to the later research [31].

10.2.3 Organic Monolayer Electron Transport Materials

Sudhir Kumar et al. have developed a new ETL using 9,10-bis(*N*-benzimidazole) anthracene (BBIA), which can offer enhanced electron mobility and thermal stability [32]. The BBIA-based device showed an electron mobility of $4.17 \times 10^{-4}\ \text{cm}^2\ \text{V}^{-1}\ \text{S}^{-1}$ at an electric field strength of 105 V cm^{-1} and a T_d (the thermal decomposition temperature at which the mass drops to 95% of the initial value) of up to 325 °C. Compared to devices using TPBi (2,2′,2″-(1,3,5-benzinetriyl)-tris(1-phenyl-1-H-benzimidazole)) as the ETL, the BBIA-based devices exhibited a twofold performance enhancement with an increase in current efficiency from 6.25 to 12.2 cd A^{-1} and an increase in EQE from 1.51% to 2.96%. The appropriate LUMO energy level of BBIA (3.0 eV) helped in achieving carrier transport balance in the light-emitting layer. The BBAI-based device also showed a low turn-on voltage of 2.35 V, indicating an increased electron injection capability, and a small efficiency roll-off of $8 \pm 1\%$ at a luminous brightness of 2000 cd m^{-2}. Additionally, the BBIA complex played a role in enhancing device lifetime, with LT_{50} estimated at 37 hours at a luminescence of 100 cd m^{-2}, whereas the TPBi-based device had a lifetime T_{50} of only 0.13 hours at a low current density of 20 mA cm^{-2}. The use of anthracene-based complexes in PeLEDs could open up new opportunities for perovskite research.

10.2.4 Organic Multilayer Electron Transport Materials

In their research, Khan Qasim et al. have proposed a novel strategy to improve the performance of TPBi by incorporating the conventional organic ETL Alq_3

(tris(8-hydroxyquinoline) aluminum) to form a multilayer electron transport structure TPBi/Alq_3/TPBi. This combination enhances carrier equilibrium and binding, leading to improved device performance [21]. This combination also effectively promotes carrier balancing and binding, leading to a significant increase in performance. When applied to green PeLEDs, a maximum EQE of 1.43%, a current efficiency of 4.69 cd A^{-1}, and a power efficiency of 1.84 lm W^{-1} can be achieved, which is almost 200% higher than that of PeLEDs using a single TPBi as the ETM. In addition to the high electron mobility, the EQE is increased by a multiple of 16 compared to devices using TPBi alone as the electron injection material.

Khan Qasim and colleagues considered this design at three levels. When selecting the appropriate HTL level, enhancing the nature of the ETL alone does not ensure the availability of enough holes, which can limit exciton formation. To address this, the researchers used PVK (poly(9-vinylcarbazole)) as the hole transport material because it possesses the suitable HOMO energy level (5.6 eV) [33, 34]. In terms of electron mobility, materials such as Alq_3, BPhen (4,7diphenyl-1,10-phenanthroline), and TPBi are commonly used as ETLs in green PeLEDs. However, they differ in their electron mobility and carrier-binding capabilities, as indicated by their HOMO values, and are therefore chosen based on the specific experimental requirements. The thickness of the ETL was fixed at 35 nm to avoid lots of degradation caused by exciton quenching at the electrode surface, as thinner ETLs (less than 25 nm) are more susceptible to this. This approach presents a new opportunity for optimizing the performance of all-inorganic PeLEDs.

In addition, Song's group used TPBi and PO-T2T (2,4,6-tris[3-(diphenylphosphinyl)phenyl]-1,3,5-triazine) as a bilayer ETL to regulate the electron transport ability and thus achieve carrier injection equilibrium and efficient compounding [35]. The resulting device exhibited saturated green luminescence with a low turn-on voltage (2 V) and high EQE (21.63%). Therefore, the regulation of the charge balance is also particularly important for the construction of high-performance PeLEDs.

10.2.5 Doped Organic Electron Transport Materials

The instability of the interface between the ETL and the perovskite layer is a crucial factor contributing to the instability of PeLEDs. During operation, the electric field induces ion migration, leading to the formation of defects and damage to the perovskite NCs. This process can even occur at room temperature, ultimately resulting in decay of the perovskite diode and loss of the intrinsic region [36]. Additionally, similar to conventional p-i-n junction LEDs, p-i-n NCs can be used to produce high-quality light. As the bias voltage is increased, the p- and n-doped regions move toward the intrinsic region, causing a shift in the intrinsic region. However, in the case of PeLEDs, the instability of the interface between the ETL and the perovskite layer, as well as between the electron injection layer and the cathode electrode, can lead to defects and damage to the perovskite NCs, resulting in the decay and loss of the intrinsic region [37].

Taejun Kim's group has attempted to enhance the stability of perovskite LEDs by using PFN (poly[(9,9-bis(3′-(*N*,*N*-dimethylamino)propyl)-2,7-fluorene)-alt-2,7-(9,9-dioctylfluorene)]) and PEI (polyethylenimine) as dopants, along with TPBi as the ETL for the devices [38]. The dopants PEI or PFN were utilized to achieve this goal. The high surface functionality of the TPBi doped with PEI or PFN allows for balanced charge transport into the perovskite layer, leading to a maximum luminance of up to 14 358 and 11 465 cd m^{-2}, respectively, in PeLEDs. Additionally, the PeLEDs with the doped ETL exhibit a 3.6-fold increase in T_{50} at 4 V compared to the undoped TPBi control group. It is worth noting that only a doping level of 5% is effective in increasing the luminance and reducing the turn-on voltage when comparing different doping concentrations.

10.2.6 Organic–Inorganic Hybrid Electron Transport Materials

Won-Chul Choi and his teammates have developed a multilayer structure by combining PEO (polyethylene oxide) with a metal oxide ETL. This innovative approach leads to a significant fivefold increase in luminescence compared to the conventional single-layer structure [39]. When PEO is not involved, the device exhibits minimal luminescence. However, when a spacer is used, the PeLEDs show a maximum current efficiency of 0.8 cd A^{-1} at a luminance of 1300 cd m^{-2}. By optimizing the thickness of the ETL, a maximum current efficiency of 7.7 cd A^{-1} can be achieved at a luminous brightness of 11 300 cd m^{-2}. It is observed that reducing the thickness of the ETL can improve carrier balance during operation, but the use of the thinnest ETL results in a slight decay in current density. Additionally, adjusting the molar concentration of the metal precursor solution from 0.7 to 0.3 M increases the luminous brightness from 350 to 1300 cd m^{-2} and the current efficiency from 0.6 to 7.7 cd A^{-1}.

10.3 Hole Transport Materials in PeLEDs

The study of hole transport materials in OLEDs has been extensive, and since OLEDs and PeLEDs share certain similarities, the fundamental issues encountered by the HTL in PeLEDs can be inferred from the knowledge acquired in the OLEDs field. These fundamental issues include the following:

(1) Appropriate HOMO energy level and LUMO energy level
The hole transport process mainly involves the HTL, which, under the application of an electric field, transfers an electron from its own HOMO orbital to the electrode, and then takes away an electron from the HOMO orbital of the light-emitting layer, thus creating a hole transfer from the electrode to the light-emitting layer. Therefore, the HOMO energy level of the HTL should be selected carefully to avoid producing too large an energy barrier with the electrode and the light-emitting layer, which can cause a significant transmission barrier. Moreover, the LUMO energy level of the HTL should not be lower than the LUMO energy level of the light-emitting layer, as this can allow electrons

injected into the LUMO orbital of the light-emitting layer to escape to the LUMO orbital of the HTL, resulting in exciton loss. Furthermore, suitable energy alignment between CTL and perovskite layer can not only increase the charge injection capability but also better balance the injection of holes and electrons, which facilitates to improve many properties of devices. Figure 10.3a shows the energy levels of the general hole and ETMs.

(2) Appropriate hole transfer rate and carrier mobility
In OLEDs, the hole transport materials have been extensively researched. Since OLEDs and PeLEDs share some similarities, the core issues faced by the HTL in PeLEDs can be inferred from the experience of OLEDs. One of the core issues is that the hole transport rate is much higher than the electron transport rate, which requires the hole transport material to reduce the transport rate to some extent. Besides, for PeLEDs, the most common hole-transport materials present much slower carrier mobility than ETMs.

(3) Appropriate absorption spectral range
In general, the absorption spectra of the HTL need to be tested to avoid any overlap between its absorption band and the emission band of the light-emitting layer. An overlap could affect the luminous brightness and efficiency of the device.

(4) Good thermal stability (high T_g) and chemical stability, long life
A high level of thermal stability is essential to prevent defects caused by material degradation and extend the device's lifespan. This is typically measured by the glass transition temperature (T_g). As the device may operate in various environments, the chemical stability and longevity of the HTL are equally critical factors to consider.

(5) Sufficiently high triplet state energy levels (E_T)
Many excitons exhibit excited triplet state properties, and hole transport materials need to have a triplet state energy level (E_T) that is high enough to prevent exciton loss caused by exciton escape from the light-emitting layer.

(6) Influence of the device processing process
The material requirements for each layer of the device depend on the specific device processing technique used. If the material does not meet the processing requirements, it can result in nonuniform film formation and the formation of various defect states. These defect states can decrease the efficiency of exciton transfer, cause unnecessary exciton losses, and ultimately reduce the overall device efficiency. Therefore, it is important to select a hole transport material that is suitable for the specific device processing technique being used.

10.4 Progress in the Study of Hole Transport Layers and Hole Injection Layers in Perovskite Light Emitting Diodes

First of all, from a mechanistic point of view, an in-depth understanding of the electron and hole transfer mechanisms inside the working device has a guiding role in the study of device performance. For example, in 2022, Tang's team investigated the effect of the HTL on the charge transfer and recombination processes in PeLEDs by impedance spectroscopy [40]. The results show that a reasonable interfacial energy level arrangement can improve radiative recombination by reducing the leakage current and carrier transport resistance. Shockley–Red–Hall recombination and Auger recombination extend the lifetime of carrier transport in the working device, which is determined by the electroluminescence spectrum. This work provides a unique and reliable approach to explore the properties of charge transfer and highlights the importance of interfaces to facilitate charge transfer as well as the importance of interfaces to improve the performance of PeLEDs.

Secondly, a large number of studies have focused on the material selection and structural design of the HTL, which directly determine the device's performance, and the various structures of HTLs are elaborated below.

10.4.1 PVK-Doped TPD, TCTA

The luminescence efficiency of PeLEDs is highly dependent on the device morphology and crystalline quality, which places high demands on the device processing. Therefore, the hole transport material often needs to be matched with the device preparation process; otherwise, the exciton complex efficiency will be reduced due to the presence of defects. The most common processing method is generally solution spin coating, which puts clear limitations on the solubility of the hole transport material in commonly used organic solvents. In 2017, Meng et al. reported a paper on the solution processing of $MAPbBr_3$ PeLEDs hole transport materials [41]. In their study, the researchers employed polyvinylcarbazole (PVK, a polymeric hole transport material) as a substrate. To enhance its performance, they doped PVK with two novel hole-transport materials: *N*,*N*′-bis(3-methylphenyl)-*N*,*N*′-bis(phenyl)benzidine (TPD) and 4,4′,4″-tris(carbazol-9-yl)triphenylamine (TCTA). The study compared the performance of the devices under different doping cases. Figure 10.4 illustrates the device structures, while Table 10.2 and Figure 10.5 summarize their device properties.

The researchers aimed to challenge Spiro-OMeTAD, the most commonly used small molecule in solution processing methods for HTLs. However, the synthesis of

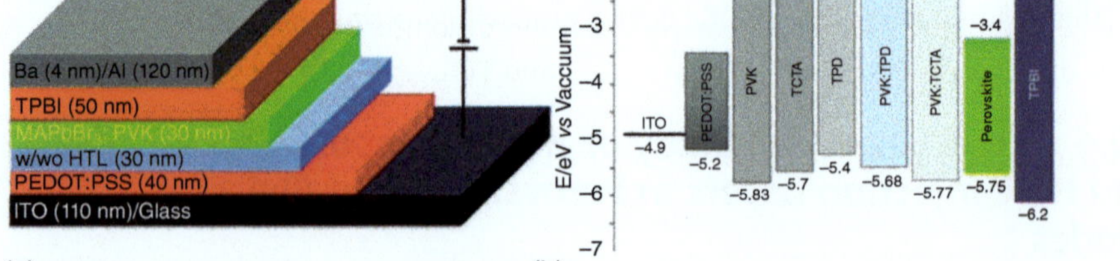

Figure 10.4 (a) Device structure and (b) energy level diagram of a PeLED with different hole transport layers [41].

Table 10.2 Performance of devices with different hole injection layers [41].

HTLs	Von (V)	CE_{max} (cd A^{-1})	EQE_{max} (%)	L_{max} (cd m^{-2})
PVK	3.0	4.18	1.21	613
PVK:TCTA	2.9	5.44	1.57	1615
PVK:TPD	2.9	6.50	1.88	1427

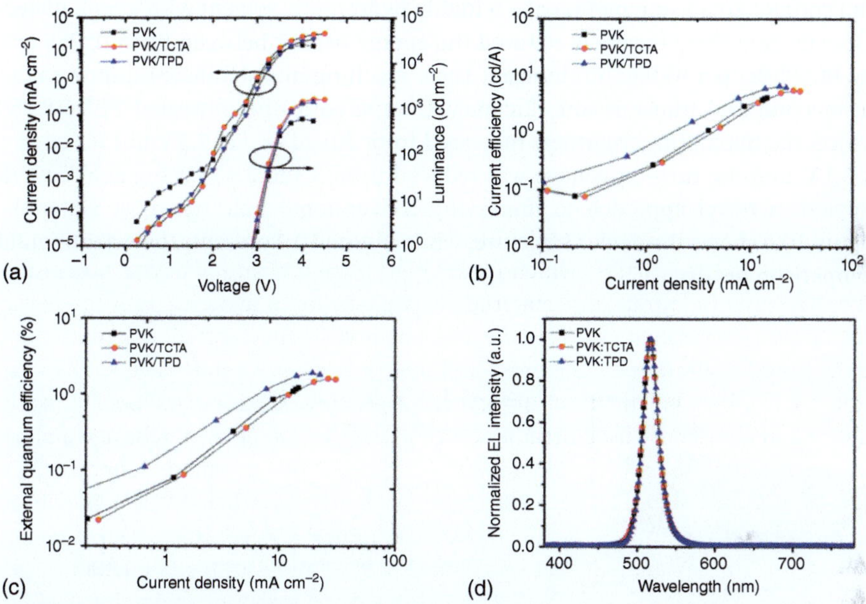

Figure 10.5 (a) Current density and brightness versus voltage (*J–L–V*). (b) Current efficiency versus current density (*CE–J*). (c) External quantum efficiency versus current density (*EQE–J*) characteristics. (d) Normalized EL spectrum (10 mA cm^{-2}) of perovskite light-emitting diodes based on different HTLs with hybrid films of PVK:$MAPbBr_3$ (w/w, 2 : 20) after crystal modified (MABr+TPBI) as luminescent layer [41].

Spiro-OMeTAD is complex and expensive. Therefore, the researchers used TPD and TCTA, which are commercially available and simpler to obtain. However, TPD and TCTA alone are less soluble in organic solvents than Spiro-OMeTAD, which can result in morphological defects during direct solution preparation. To overcome this, the researchers doped PVK with TPD and TCTA, which substantially improved their solubility and led to good results. This breakthrough paves the way for the preparation of many HTLs that are not suitable for spin-coating processes.

10.4.2 PEDOT:PSS After Methanol Treatment

PEDOT:PSS is a commercially popular hole injection material known for its good solubility and high hole injection efficiency. It is well suited for the preparation of optoelectronic devices through solution processing. However, when used as a hole's transport material, PEDOT:PSS generates large energy barriers with the light-emitting layer material due to its low HOMO energy level. This reduces the hole transport efficiency. To use PEDOT:PSS as an integrated hole injection and transport layer, two reported strategies can be used. One is to dope other hole transport materials, such as PVK, to improve its performance. Another method was reported by Wang's group in 2017, which involves treating PEDOT:PSS with methanol to achieve better results [42]. After coating the PEDOT:PSS film using conventional methods, the researchers dissolved the hydrophilic and insulating PSS film components using methanol – a highly hydrophilic solvent with a high dielectric constant. This effectively reduced the energy barrier between PEDOT:PSS and the $MAPbBr_3$ perovskite luminescent layer, resulting in a significant improvement in injection and transmission efficiency. Compared to the untreated PEDOT:PSS device, the maximum brightness increased from 201 cd m^{-2} at 7.5 V to 1565 cd m^{-2} at 5.1 V, and the turn-on voltage was reduced from 3.3 to 2.4 V. Wang et al.'s work provides a novel approach to improving conventional hole transport materials. Figure 10.6 shows the device structure, while Figure 10.7 presents the experimental comparison results.

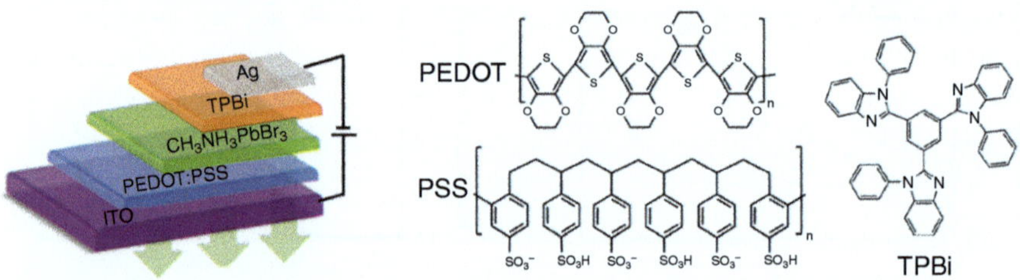

Figure 10.6 Schematic diagram of the device structure and the PEDOT:PSS, TPBi structure formula [42].

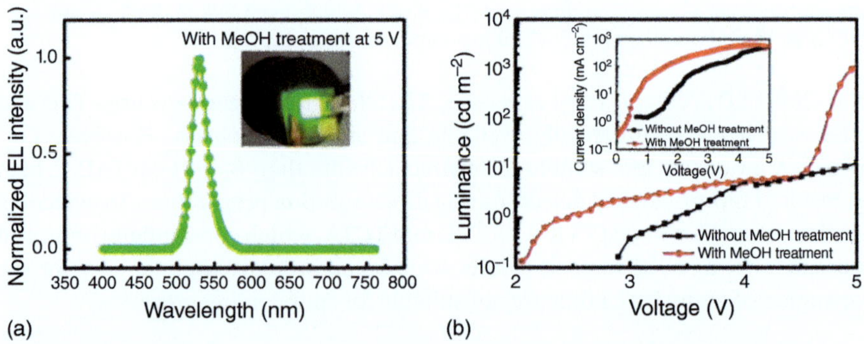

Figure 10.7 (a) Normalized EL spectrum and device photograph (inset); (b) luminance-voltage curve. Inset: current density versus voltage curves. Source: Wang et al. [42]/ Reproduced from AIP Publishing LLC.

10.4.3 TB(MA) Instead of PEDOT:PSS

Recently, Xu's group synthesized and developed a new neutral polyelectrolyte material TB(MA), which was used to replace the conventional acidic PEDOT:PSS as the HTL and successfully prepared highly efficient quasi-two-dimensional blue chalcogenide LEDs [43]. TB(MA) layer has good hydrophilicity and good wettability with polar chalcogenide precursor solutions. This ensures the growth of chalcogenide crystals in solution and the successful preparation of complete and flat high-quality chalcogenide films. In addition, the MA^+ functional group in TB(MA) can effectively passivate the halogen vacancy defects on the surface of the chalcogenide films, thus reducing the concentration of surface defects in the chalcogenide films and promoting charge transport. The fluorescence quantum efficiency (PLQY) of the quasi-dimensional chalcogenide film based on the TB(MA) HTL reached 60.0%, and finally, the authors successfully prepared a high-efficiency blue quasi-dimensional chalcogenide LED based on neutral TB(MA) as the HTL, which improved the electroluminescence EQE from 7.8% to 13.5% compared with the conventional acidic PEDOT:PSS, with the operating stability significantly improved. This is the highest efficiency of quasi-dimensional blue chalcogenide LEDs reported so far. In addition, the authors also successfully prepared a flexible quasi-2D blue chalcogenide LED based on (polyethylene naphthalate) film-Indium tin oxide coated (PEN-ITO) flexible substrate, and the EQE of the device also reached 8.3%. This work will greatly facilitate the commercialization of quasi-2D blue chalcogenide LEDs, and the device structure and related experimental results are shown in Figure 10.8.

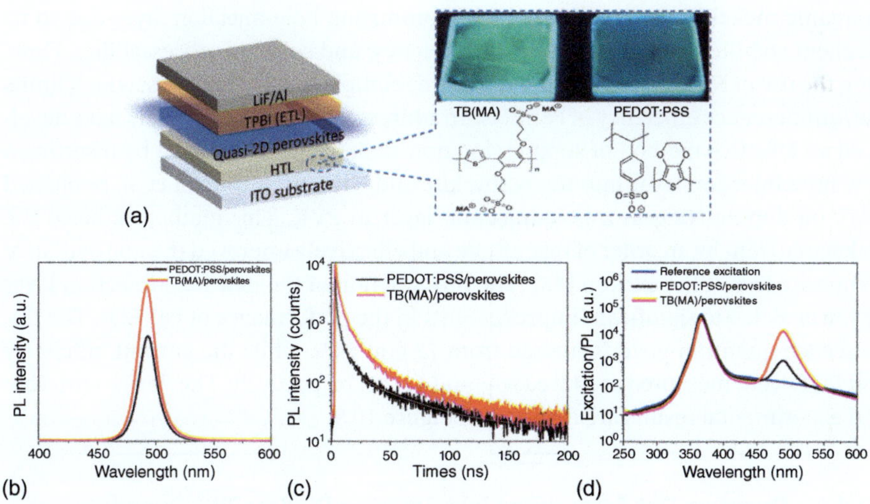

Figure 10.8 (a) Device structure and material structure; (b) photoluminescence (PL); (c) time-resolved photoluminescence (TRPL); (d) photoluminescence quantum yield (PLQY) of PEDOT:PSS and TB(MA)-based quasi-2D perovskite films. Source: Liu et al. [43]/ Reproduced with permission from John Wiley & Sons, Inc.

10.4.4 PSS-Doped Na

In 2023, Mathews' group showed that the efficiency of blue PeLEDs can be improved by adding PSS-Na to the PEDOT:PSS HTL [44]. By doing so, a PSS-rich layer is formed at the HTL/perovskite interface, minimizing the direct contact between PEDOT and perovskite and ultimately inhibiting exciton quenching. And the addition of PSS-Na also deepens the ionization potential of the HTL, thereby lowering the injection barrier into the perovskite. Despite the lowered injection barrier, the detected cavity current is lower due to the presence of the PSS-rich surface, which may increase the balance between electron and hole injection into the emitter layer. In addition, the addition of PSS-Na increases the probability of forming higher m domains in the RP perovskite that act as the main emitter centers. During the merging process, these effects lead to an increase in PeLED efficiency at an optimal concentration of 6% PSS-Na addition. At this concentration, the best blue PeLED device achieves an EQE of 4% at 480 nm and 6.36% at 496 nm, approximately twice that of the device using the original PEDOT:PSS HTLs. In addition, the 6% PSS-Na addition improves the device lifetime by more than four times, from 40 to 175 seconds, probably due to the reduced Joule heating. Thus, in the pursuit of high-efficiency blue PeLEDs, the main findings reported in this work provide a feasible route to modify PEDOT:PSS HTLs, which could complement other methods targeting the interface and emitting layers.

10.4.5 PVK-Doped NiO_x

Inorganic nickel oxide (NiO_x) is a highly promising hole injection layer due to its excellent stability, high hole transport efficiency, and solution processability. However, the use of NiO_x can lead to radiation-free complexation of carriers, which limits the luminescence efficiency of PeLEDs. To address this issue, researchers have developed an effective method of suppressing non-radiative complexation by inserting a new hole-injection layer into the perovskite and NiO_x. In 2019, Liu et al. published work on doping NiO_x as a hole-injection layer in PVK. This method reduced the leakage current by an order of magnitude and effectively improved the non-radiative composite loss of the exciton [45]. The introduction of the new hole-injection layer by Liu et al. led to significant improvements in the performance of PeLEDs. The fluorescence quantum yield increased from 23% to 54%, while the current efficiency and EQE were measured at 34.2 cd A^{-1} and 11.2%, respectively. The device structure and experimental results are depicted in Figure 10.9.

10.4.6 Quantum Dot Perovskite Light Emitting Diodes: PVK-Doped PTAA

In order to improve device performance, different solution-treated high-temperature superconducting materials such as PVK, poly[bis(4-butylphenyl)-bis(phenyl) biphenyl] (poly TPD), and PVK doped with small molecules have been investigated

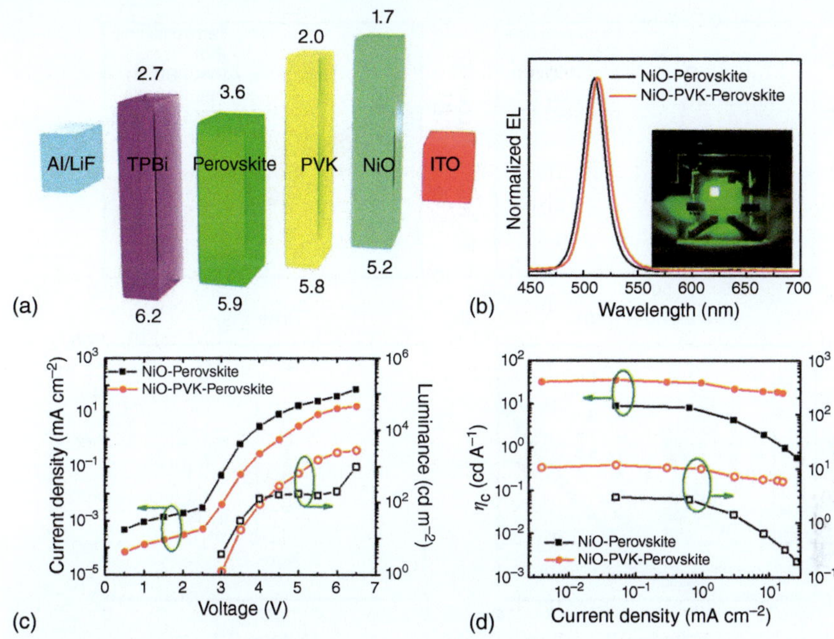

Figure 10.9 (a) Device structure and energy band diagram; (b) EL spectrum. Inset is a photograph of the NiO/PVK device under operating conditions; (c) *J*–*V* and brightness-voltage curves; (d) EQE and η of the device$_c$ plots. Source: Liu et al. [45]/Reproduced from AIP Publishing LLC/CC BY 4.0.

due to the typically large potential barrier between the hole injection layer and the quantum dot (>1 eV). However, intrinsic PVK films exhibit relatively low hole mobility, and the low HOMO energy level of poly TPD limits its use as a hole transport material. Meanwhile, the current performance of small-molecule-doped PVK is promising. To address these challenges, Lin's group developed a method for doping PTAA (poly[bis(4-phenyl)(2,4,6-trimethylphenyl)amine]) with PVK as a hole injection layer for fluorescent colloidal $MAPbBr_3$ quantum dot luminescent layer in hexane solution. The details of this approach are presented in their research work [46]. Due to the potential barrier between the hole injection layer and the quantum dot, the performance of optoelectronic devices is often limited. To address this issue, different high-temperature superconducting materials have been used, such as PVK, poly TPD, and small-molecule-doped PVK. However, the intrinsic PVK films have low hole mobility, and poly TPD has a low HOMO energy level. Small-molecule-doped PVK has shown promising results in improving device performance. The maximum brightness after doping was 7352 cd m^{-2} at 9 V, with a maximum current efficiency of 11.1 cd A^{-1} and turn-on voltages as low as 3.2 V.

10.4.7 PVK Blended with PBD

Besides, for the sake of overcoming the charge balance issue, Yu et al. developed a polymer/small-molecule binary-blend HTL with PVK and PBD[poly(9-vinyl-

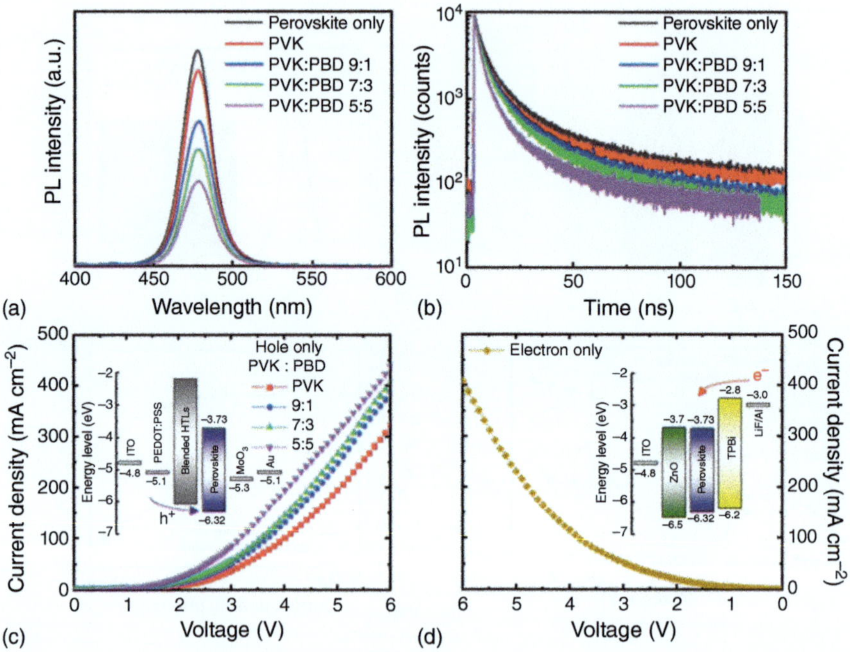

Figure 10.10 (a) Photoluminescence spectrum; (b) photoluminescence decay curves of perovskite and perovskite with various HTL films; (c) *J–V* characteristics of hole-only device; (d) *J–V* characteristics of electron-only device.

carbazole) (PVK) and 2-(4-biphenyl)-5-(4-tert-butylphenyl)-1,3,4-oxadiazole] [47]. The binary-blend HTL allows a systematic modulation of the hole injection barrier, enabling the charge transport between electrons and holes to be balanced for effective recombination within the perovskite emitting layer. The hybrid HTL provides a path for hole injection, reduces energy barriers, and effectively forms an energy ladder from the electrode to the perovskite emitter. By taking advantage of the achieved balanced charge transport and efficient recombination of electrons and holes in the perovskite layer, an optimal device based on a PVK:PBD weight ratio of 9 : 1 shows the maximum energy ladder, with a maximum luminance of 167 cd m^{-2} and a maximum EQE of 5.30%, and the peak emission is 478 nm. Relative results are shown in Figure 10.10.

10.4.8 Double HTLs with PVK and TFB

In 2021, Wang et al. reported a simple method to fabricate green light-emitting PeLEDs with double HTLs, which can greatly facilitate hole injection and charge balance [48]. To solve the solvent erosion problem, they employed 1,4-dioxane as a solvent for PVK and cast it onto the surface of poly[(9,9-dioctylfluorene-2,7-diyl)-*co*-(4,40-(*N*-(*p*-butylphenyl))-diphenylamine] (TFB). Also, the orthogonal solvent 1,4-dioxane was used to improve the surface smoothness of the perovskite film on

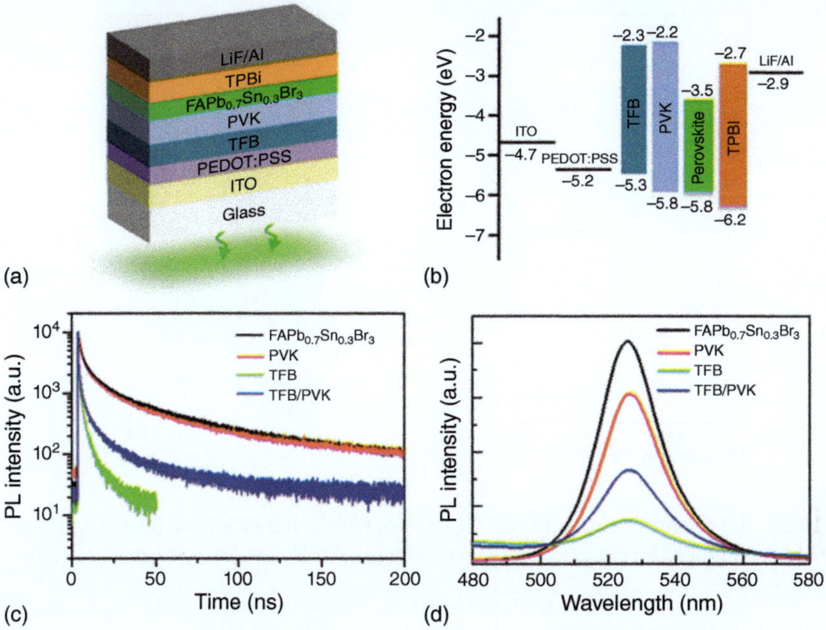

Figure 10.11 (a) Device configuration; (b) the energy level diagram for each layer in the PLED structure; (c) TRPL decay curves; (d) PL intensities of the $FAPb_{0.7}Sn_{0.3}Br_3$ films cast on top of glass, PVK, TFB, and TFB/PVK, respectively. Source: Li et al. [49]/John Wiley & Sons.

the bis-HTL. The PeLEDs with TFB/PVK dual HTLs showed maximum current efficiency (CE) and EQE of 53.5 cd A^{-1} and 12.9%, respectively. The EQE of this device is 1.7 and 3 times higher than that of the single HTL device with PVK and TFB, respectively. This significant improvement is mainly attributed to the cascaded energy arrangement of the dual HTLs, which prevents charging in the perovskite NCs. The device structure and relative results are shown in Figure 10.11. Besides, Jing Li et al. significantly improved the morphology of the perovskite layer doped with divalent strontium ions and simultaneously reduced trap states by adding a small amount of TFB to the traditional PVK HTL [49]. Ultimately, the blue LED based on this HTL achieved an EQE of 2.24% at a wavelength of 470 nm and has a stable electroluminescence spectrum under a forward bias as high as 8 V.

10.4.9 Polyfluorenylbenzene Anion-Conjugated Polyelectrolytes with Counter Ions

The presence of defect states in perovskite crystals is a significant challenge in the development of efficient PeLEDs. Defect states can result in the loss of excitons through radiation-free combinations, by making it crucial to reduce their severity. To address this issue, researchers have explored various approaches, such as

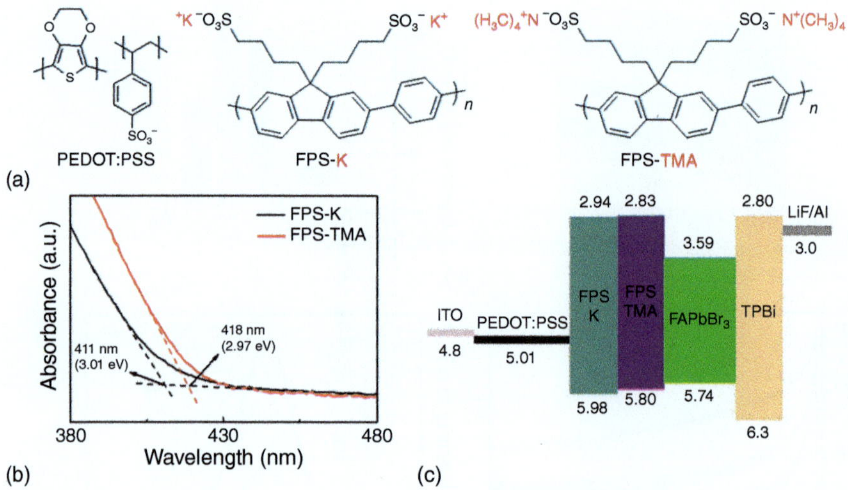

Figure 10.12 (a) Chemical structures of PEDOT:PSS, FPS-K, and FPS-TMA. (b) Absorption spectra of FPS-K and FPS-TMA. (c) Energy level diagram of PeLEDs. Source: Lee et al. [50]/ John Wiley & Sons.

reducing grain size and selecting quasi-two-dimensional perovskite crystals, as well as surface passivation to reduce the loss of defect states to excitons. Recent studies have shown that functional groups such as amino (—NH_2), hydroxyl (—OH), phosphine oxide (P=O), carbonyl (C=O), carboxyl (—COOH), and ammonium bromide can passivate defects. However, the addition of these groups may also cause an insulating effect, reducing carrier transport efficiency and increasing turn-on voltage. In 2019, Lee et al. reported a promising approach using different counterions [K^+ and tetramethylammonium (TMA^+)] in polyfluorenyl anionic conjugated polyelectrolytes (CPEs) as hole transport materials for PeLEDs [50]. The device structure is shown in Figure 10.12.

The researchers conducted a thorough investigation of the impact of perovskite crystal growth and passivated CPEs on interfacial defects. They accomplished this by combining multifunctional anionic polyfluorenylbenzene-coated potassium (K^+) or tetramethylammonium (TMA^+) electrolytes with quasi-dimensional $FAPbBr_3$ crystals as the luminescent layer. The TMA^+–containing CPEs proved to be more effective than K-containing CPEs in suppressing interfacial defect generation and enhancing the properties of PL and electroluminescence. The electroluminescent devices underwent aging tests to further optimize them. The results indicate a significant reduction in the surface defects of the CPE and perovskite layers, resulting in a maximum brightness of 14 800 cd m^{-2}, a maximum current efficiency of 43.6 cd A^{-1}, and a maximum EQE of 10.2%. CPE was used as the hole injection layer, and TMA^+ was used as the counterion. Figure 10.13 displays the luminescence of the UV lamp irradiated at 375 nm.

Figure 10.13 Emission of light from perovskite films based on different hole injection layers under 365 nm UV light irradiation [50].

10.5 Conclusion and Outlook

After reviewing the literature, it is clear that the choice of charge injection and transport layer plays a critical role in enhancing the performance of PeLEDs. While organic materials can improve device efficiency, they are often susceptible to degradation at moderate temperatures and humidity. On the other hand, inorganic materials offer better stability, but the control of charge injection and carrier balance remains a challenge. To optimize the transport characteristics of the ETL and improve luminescence efficiency, a multilayer composite of organic small molecules or polymers and inorganic metal oxides can be a promising solution. Additionally, doping metal particles into the metal oxide layer may also offer a viable way to regulate the ETL and ensure carrier transport balance. Further research in this area is necessary to fully exploit the potential of PeLEDs.

Currently, the main focus of research on PeLEDs is to optimize the performance of the light-emitting layer and improve the materials used. However, it is essential to further explore the injection and transport layers to improve the device's efficiency. This section summarizes the core issues related to hole transport materials for PeLEDs.

Firstly, the processing method is the most significant influencing factor for the HTL. As most reported materials are conducive to solution processing, it is essential to ensure that the hole transport material matches the corresponding process. Otherwise, there may be issues with morphological and crystallographic mismatches, too many defect states, and reduced efficiency. A potential future research direction is doping processable hole-transporting materials with good optoelectronic properties using PVK as a substrate to enhance the injection efficiency of the HTL.

Secondly, inorganic hole transport materials have several limitations, including radiation loss and excessive leakage currents. Inorganic materials also have poor energy level tunability compared to organic hole transport materials, and high-temperature processing makes them more expensive. Organic–inorganic hole transport material doping can improve performance and processing effectiveness. Therefore, inorganic hole transport materials will mainly be used as performance enhancers in conjunction with organic materials.

Organic hole transport materials are the future choice for the HTL of PeLEDs, but they have limited molecular designability compared to OLEDs. Computational simulations can help with molecular screening, but relevant research is scarce, making it a potential future hot topic. Combining the hole injection layer with the HTL and optimizing PEDOT:PSS and material doping may also be promising research options.

Lastly, the thermal and chemical stability of the materials used in PeLEDs are critical factors affecting the device's performance. However, there is currently little research work in this area. Therefore, improving the stability of all materials is an important research direction.

References

1 Mahmood, K., Sarwar, S., and Mehran, M.T. (2017). Current status of electron transport layers in perovskite solar cells: materials and properties. *RSC Advances* 7: 17044–17062.

2 Milot, R.L., Eperon, G.E., Snaith, H.J. et al. (2015). Temperature-dependent charge-carrier dynamics in $CH_3NH_3PbI_3$ perovskite thin films. *Advanced Functional Materials* 25: 6218–6227.

3 Ponseca, C.S. Jr., Savenije, T.J., Abdellah, M. et al. (2014). Organometal halide perovskite solar cell materials rationalized: ultrafast charge generation, high and microsecond-long balanced mobilities, and slow recombination. *Journal of the American Chemical Society* 136: 5189–5192.

4 Hutter, E.M., Eperon, G.E., Stranks, S.D. et al. (2015). Charge carriers in planar and meso-structured organic-inorganic perovskites: mobilities, lifetimes, and concentrations of trap states. *The Journal of Physical Chemistry Letters* 6: 3082–3090.

5 Reid, O.G., Yang, M., Kopidakis, N. et al. (2016). Grain-size-limited mobility in methylammonium lead iodide perovskite thin films. *ACS Energy Letters* 1: 561–565.

6 Kim, D.H., Park, J., Li, Z. et al. (2017). 300% enhancement of carrier mobility in uniaxial-oriented perovskite films formed by topotactic-oriented attachment. *Advanced Materials* 29: 1606831.

7 Xing, G., Mathews, N., Sun, S. et al. (2013). Long-range balanced electron and hole-transport lengths in organic-inorganic $CH_3NH3PbI_3$. *Science* 342: 344–347.

8 Stranks, S.D., Eperon, G.E., Grancini, G. et al. (2013). Electron-hole diffusion lengths exceeding 1 micrometer in an organometal trihalide perovskite absorber. *Science* 342: 344–347.

9 Guo, Z., Manser, J.S., Wan, Y. et al. (2015). Spatial and temporal imaging of long-range charge transport in perovskite thin films by ultrafast microscopy. *Nature Communications* 6: 7471.

10 Wehrenfennig, C., Eperon, G.E., Johnston, M.B. et al. (2014). High charge carrier mobilities and lifetimes in organolead trihalide perovskites. *Advanced Materials* 26: 1584–1589.

11 Qingfeng Dong, Y.F., Shao, Y., Mulligan, P. et al. (2015). Electron-hole diffusion lengths >175 mm in solution-grown $CH_3NH_3PbI_3$ single crystals. *Science* 347: 967–970.

12 Saidaminov, M.I., Abdelhady, A.L., Murali, B. et al. (2015). High-quality bulk hybrid perovskite single crystals within minutes by inverse temperature crystallization. *Nature Communications* 6: 7586.

13 Shi, D., Adinolfi, V., Comin, R. et al. (2015). Low trap-state density and long carrier diffusion in organolead trihalide perovskite single crystals. *Science* 347: 519–522.

14 Valverde-Chávez, D.A., Ponseca, C.S., Stoumpos, C.C. et al. (2015). Intrinsic femtosecond charge generation dynamics in single crystal $CH_3NH_3PbI_3$. *Energy & Environmental Science* 8: 3700–3707.

15 Semonin, O.E., Elbaz, G.A., Straus, D.B. et al. (2016). Limits of carrier diffusion in n-type and p-type $CH_3NH_3PbI_3$ perovskite single crystals. *Journal of Physical Chemistry Letters* 7: 3510–3518.

16 Stoumpos, C.C., Malliakas, C.D., and Kanatzidis, M.G. (2013). Semiconducting tin and lead iodide perovskites with organic cations: phase transitions, high mobilities, and near-infrared photoluminescent properties. *Inorganic Chemistry* 52: 9019–9038.

17 Herz, L.M. (2017). Charge-carrier mobilities in metal halide perovskites: fundamental mechanisms and limits. *ACS Energy Letters* 2: 1539–1548.

18 Peng, J., Chen, Y., Zheng, K. et al. (2017). Insights into charge carrier dynamics in organo-metal halide perovskites: from neat films to solar cells. *Chemical Society Reviews* 46: 5714–5729.

19 Fakharuddin, A., Gangishetty, M.K., Abdi-Jalebi, M. et al. (2022). Perovskite light-emitting diodes. *Nature Electronics* 5: 203–216.

20 Liu, A., Bi, C., Guo, R. et al. (2021). Electroluminescence principle and performance improvement of metal halide perovskite light-emitting diodes. *Advanced Optical Materials* 9: 2002167.

21 Qasim, K., Wang, B., Zhang, Y. et al. (2017). Solution-processed extremely efficient multicolor perovskite light-emitting diodes utilizing doped electron transport layer. *Advanced Functional Materials* 27: 1606874.

22 Baek, S.D., Biswas, P., Kim, J.W. et al. (2016). Low-temperature facile synthesis of Sb-doped p-type ZnO nanodisks and its application in homojunction light-emitting diode. *ACS Applied Materials & Interfaces* 8: 13018–13026.

23 Liu, J., Wu, J., Shao, S. et al. (2014). Printable highly conductive conjugated polymer sensitized ZnO NCs as cathode interfacial layer for efficient polymer solar cells. *ACS Applied Materials & Interfaces* 6: 8237–8245.

24 Cheng, Y., Yang, Q.D., Xiao, J. et al. (2015). Decomposition of organometal halide perovskite films on zinc oxide nanoparticles. *ACS Applied Materials & Interfaces* 7: 19986–19993.

25 Zhao, X., Shen, H., Zhang, Y. et al. (2016). Aluminum-doped zinc oxide as highly stable electron collection layer for perovskite solar cells. *ACS Applied Materials & Interfaces* 8: 7826–7833.

26 Chen, S., Manders, J.R., Tsang, S.W. et al. (2012). Metal oxides for interface engineering in polymer solar cells. *Journal of Materials Chemistry* 22: 24202–24212.

27 Wang, H., Kosasih, F.U., Yu, H. et al. (2020). Perovskite-molecule composite thin films for efficient and stable light-emitting diodes. *Nature Communications* 11: 891.

28 Kim, Y.C., Baek, S.D., and Myoung, J.M. (2019). Enhanced brightness of methylammonium lead tribromide perovskite microcrystal-based green light-emitting diodes by adding hydrophilic polyvinylpyrrolidone with oleic acid-modified ZnO quantum dot electron transporting layer. *Journal of Alloys and Compounds* 786: 11–17.

29 Cao, F., Wu, Q., Li, W. et al. (2023). Core/shell ZnO/ZnS nanoparticle electron transport layers enable efficient all-solution-processed perovskite light-emitting diodes. *Small* 19: 2207260.

30 Wang, H., Yu, H., Xu, W. et al. (2018). Efficient perovskite light-emitting diodes based on a solution-processed tin dioxide electron transport layer. *Journal of Materials Chemistry C* 6: 6996–7002.

31 Jeong, J.E., Park, J.H., Jang, C.H. et al. (2020). Multifunctional charge transporting materials for perovskite light-emitting diodes. *Advanced Materials* 32: e2002176.

32 Kumar, S., Marcato, T., Vasylevskyi, S.I. et al. (2019). Efficient perovskite nanocrystal light-emitting diodes using a benzimidazole-substituted anthracene derivative as the electron transport material. *Journal of Materials Chemistry C* 7: 8938–8945.

33 Huang, H., Lin, H., Kershaw, S.V. et al. (2016). Polyhedral oligomeric silsesquioxane enhances the brightness of perovskite nanocrystal-based green light-emitting devices. *The Journal of Physical Chemistry Letters* 7: 4398–4404.

34 Mashford, B.S., Stevenson, M., Popovic, Z. et al. (2013). High-efficiency quantum-dot light-emitting devices with enhanced charge injection. *Nature Photonics* 7: 407–412.

35 Fang, T., Wang, T., Li, X. et al. (2021). Perovskite QLED with an external quantum efficiency of over 21% by modulating electronic transport. *Science Bulletin(Beijing)* 66: 36–43.

36 Berhe, T.A., Su, W.N., Chen, C.H. et al. (2016). Organometal halide perovskite solar cells: degradation and stability. *Energy & Environmental Science* 9: 323–356.

37 Shan, X., Li, J., Chen, M. et al. (2017). Junction propagation in organometal halide perovskite-polymer composite thin films. *The Journal of Physical Chemistry Letters* 8: 2412–2419.

38 Kim, T., Kim, J.H., Triambulo, R.E. et al. (2019). Improving the stability of organic-inorganic hybrid perovskite light-emitting diodes using doped electron transport materials. *Physica Status Solidi A: Applications and Material Science* 216: 1900426.

39 Won-Chul, C., Kanwat, A., Moyen, E. et al. (2019). Role of interlayers in inorganic perovskite thin films LEDs. *SID Symposium Digest of Technical Papers* 50: 859–860.

40 Han, Y., Shen, Y., Shen, K. et al. (2022). Unraveling the hole-transport-layer-manipulated carrier transfer dynamics in perovskite light-emitting diodes. *Journal of Physical Chemistry Letters* 13: 10455–10463.

41 Meng, F., Zhang, C., Chen, D. et al. (2017). Combined optimization of emission layer morphology and hole-transport layer for enhanced performance of perovskite light-emitting diodes. *Journal of Materials Chemistry C* 5: 6169–6175.

42 Wang, Z., Li, Z., Zhou, D. et al. (2017). Low turn-on voltage perovskite light-emitting diodes with methanol treated PEDOT:PSS as hole transport layer. *Applied Physics Letters* 111: 233304.

43 Liu, Y., Zhang, L., Chen, S. et al. (2021). Water-soluble conjugated polyelectrolyte hole transporting layer for efficient sky-blue perovskite light-emitting diodes. *Small* 17: e2101477.

44 Chua, H.M., Yantara, N., Tay, Y.B. et al. (2023). Influence of ionic additives in the PEDOT:PSS hole transport layers for efficient blue perovskite light emitting diodes. *ACS Applied Materials & Interfaces* 15: 14614–14623.

45 Liu, Y., Wu, T., Liu, Y. et al. (2019). Suppression of non-radiative recombination toward high efficiency perovskite light-emitting diodes. *APL Materials* 7: 021102.

46 Lin, X., Wu, X., Zheng, J. et al. (2019). Enhanced performance of green perovskite quantum dots light-emitting diode based on co-doped polymers as hole transport layer. *IEEE Electron Device Letters* 40: 1479–1482.

47 Yu, Z., Jeong, W.H., Kang, K. et al. (2022). A polymer/small-molecule binary-blend hole transport layer for enhancing charge balance in blue perovskite light emitting diodes. *Journal of Materials Chemistry A* 10: 13928–13935.

48 Wang, W., Wu, Z., Ye, T. et al. (2021). High-performance perovskite light-emitting diodes based on double hole transport layers. *Journal of Materials Chemistry C* 9: 2115–2122.

49 Li, J., Zhang, D., Gao, Y. et al. (2022). Hole transport layer modification for highly efficient divalent ion-doped pure blue perovskite light-emitting diodes. *Advanced Optical Materials* 11: 2201883.

50 Lee, S., Jang, C.H., Thanh Luan, N. et al. (2019). Conjugated polyelectrolytes as multifunctional passivating and hole-transporting layers for efficient perovskite light-emitting diodes. *Advanced Materials* 31: 1900067.

11

Stability of Perovskite Light-emitting Diodes

In recent times, organic metal halide perovskite light-emitting diodes (PeLEDs) have taken center stage in materials science, primarily due to their impressive increases in device efficiency. Metal halide perovskites uniquely blend outstanding electronic and optical properties, such as remarkable defect tolerance and high photoluminescence efficiency. Additionally, these materials offer the advantages of lower cost, scalable production, and solution processing, adding to their appeal in the realm of cutting-edge materials science.

Perovskites, inherently characterized by tunable composition and size, have been skillfully used to create vibrant and efficient light-emitting diodes (LEDs). Despite the dramatic increase in the external quantum efficiency of PeLEDs – an impressive jump from just 1% to over 30% within a few years – the operational stability of these devices has yet to reach a satisfactory level. This lack of resilience is largely due to the adverse effects of environmental factors and inherent operational degradation.

Moreover, perovskite materials are vulnerable to ion migration, a major factor in the degradation process intrinsic to PeLEDs. Simultaneously, achieving both operational and spectral stability in these innovative devices poses a substantial challenge that remains to be effectively overcome.

This chapter scrutinizes the intrinsic instability embedded in perovskite materials, shedding light on the principal factors leading to their degradation when employed in LEDs. It also delineates approaches to enhance the stability of these materials, emphasizing the significance of incorporating inorganic charge-transporting layers, fine-tuning charge balance, and judicious thermal management to reinforce the operational durability of perovskite LEDs. Lastly, it summarizes the current challenges in this sphere and, drawing upon the path of recent research advancements, provides insights to further refine and amplify device stability.

11.1 Sources of Instability in Metal Halide Perovskites and Perovskite Light-emitting Diodes

The root of instability in PeLEDs stems from a combination of the inherent characteristics of metal halide perovskites and the mutual interactions of external conditions that affect these materials, particularly during the operation of PeLEDs.

Perovskite Light Emitting Diodes: Materials and Devices, First Edition. Hong Meng.

The pursuit of effective strategies to prevent the emergence of defects and associated trap states is critically important for boosting both the efficiency and longevity of PeLEDs. In this section, our goal is to illuminate the twofold factors – internal and external – that instigate instability in PeLEDs. Concurrently, we explore the intricate interplay and relationships among these diverse factors that ultimately determine the device's overall stability. We also offer a comprehensive overview of the primary catalysts of PeLED instability, considering a variety of factors such as thermal flux [1], atmospheric moisture [2], and photonic irradiation [3], which are ubiquitously recognized as the predominant causes of perovskite material degradation. It is worthy of note that such ion migration induced by electric fields is deleterious to the operational stability of these devices, adding a further dimension of complexity to the issue of PeLED stability.

11.1.1 Intrinsic Instability of PeLEDs

Metal halide perovskites encompass a broad category of semiconductor materials distinguished by their unique ABX_3 perovskite crystallographic lattice structures, as depicted in Figure 11.1. This formula indicates a setup where the A-site is filled by a cationic component like Cs^+, $CH_3NH_3{}^+$ (often referred to as methylammonium or MA), or $CH(NH_2)_2{}^+$ (formamidinium or FA). Conversely, the B-site is commonly occupied by Pb^{2+}, though instances of Sn^{2+} or a combination of monovalent/trivalent cations can also be observed. Lastly, the X-site hosts halide anions such as Cl^-, Br^-, or I^-.

The fundamental building blocks of these structures, the $[BX_6]$ octahedra, are stabilized through ionic interactions and further connect with the organic A-site cations via hydrogen bonds. These bonding arrangements contribute significantly to the low formation energy, flexibility under stress, and softness in the lattice structure of perovskites.

However, perovskites' notable ionic nature is a double-edged sword that also leads to their inherent instability. This is primarily because highly ionized perovskites are

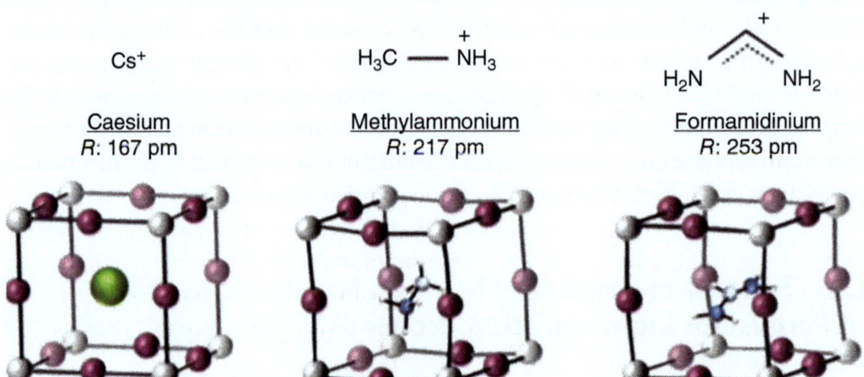

Figure 11.1 Three common A-site cations in metal halide perovskites, with their effective ionic radii *R*. Source: Kong et al. [4]/John Wiley & Sons.

prone to creating a variety of defect types. These include Schottky defects (revealed as vacancies in both cations and anions), Frenkel defects (which denote interstitial sites), and misalignments in cation/anion. External factors can trigger or further intensify these defects, making perovskites predisposed toward decomposition.

11.1.2 Extrinsic Instability of PeLEDs

As previously discussed, several external factors, such as temperature, humidity, ultraviolet (UV) light, and bias voltage during PeLED operation, can affect metal halide perovskites. In this section, we compile data on the external instability of perovskites under these influences, placing special emphasis on their effect on the light emission properties of both perovskites and PeLEDs.

During continuous operation at low bias voltage, PeLEDs' device temperature may escalate to between 30 and 60 °C, even exceeding 100 °C at high voltages. This is significantly higher than the operating temperature of perovskite solar cells [5]. At these elevated temperatures, the semiconductor properties of perovskites may degrade. For example, the excitons within perovskite grains may dissociate into free carriers, which hinders the radiative recombination process in PeLEDs. When comparing lead halide perovskites with Cs^+, MA^+, or FA^+ A^- cations, the all-inorganic perovskites ($CsPbX_3$) exhibit better thermal resistance than their organic–inorganic (hybrid) counterparts ($MAPbX_3$ and $FAPbX_3$). This is mainly because organic cations are more prone to volatility. Notably, MA-based perovskites display the least thermal stability due to their lower enthalpy and decreased activation energy (E_a) involved in their decomposition reactions compared to those with FA cations.

Perovskites display a susceptibility to air due to their exposure to both moisture and oxygen, which significantly impacts the performance of perovskite devices. It has been established that humidity in the surrounding environment is the principal culprit for perovskite degradation, as evidenced by perovskite films kept under a consistent flow of dry air that exhibited no alterations in the absorption spectrum for a period exceeding 2 weeks [6]. The initiation of the degradation sequence is heavily contingent on the level of relative humidity. Conversely, an optimal level of humidity has been observed to boost the crystallinity of perovskites and neutralize the nonradiative recombination centers due to the hydrogen bonds established between the hydroxyl groups of water molecules and uncoordinated ions on the perovskite film surface. Thus, this is deemed advantageous during the fabrication of PeLEDs. The degradation of perovskites instigated by humidity is primarily confined to surfaces and interfaces, which often leads to the emergence of non-perovskite phases beginning at the grain boundaries and subsequently advancing into the grain interiors. The incorporation of hydrophobic protective molecules equipped with defect-sealing functional groups can impede the penetration of water molecules through the surface defects, thereby providing effective protection for perovskites against degradation in humid air [7].

Photocatalyzed breakdown of perovskites serves as an additional reason for their deterioration. Specifically, UV light can incite the separation of organic cations in perovskites, a process that is expedited by the presence of oxygen in the air.

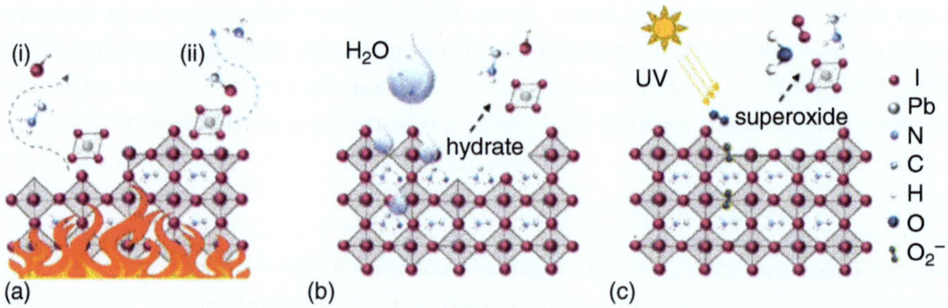

Figure 11.2 Schematic representations of perovskite crystal degradation under the effects of heat (a), humidity (b), and UV light and oxygen exposure (c). Source: Kong et al. [4]/John Wiley & Sons.

Halide vacancies also hold a significant position in ion migration when exposed to light. Light-induced ion rearrangement can mend iodine Frenkel defects due to the decreased activation energy E_a during light stimulation, culminating in an escalation of PL intensity over time under exposure to light [8] (Figure 11.2).

Given that PeLEDs function under significantly higher bias voltages relative to perovskite solar cells, the electric field emerges as another vital element, potentially leading to perovskite deterioration and triggering an intense EL roll-off. Beyond impairing the performance of PeLEDs, ion migration also negatively affects the EL color purity when mixed-halide perovskites are used as emitters. Another mechanism through which ion migration can detrimentally affect PeLED performance stems from the interplay between dispersed ions and adjacent functional layers (either ETL and/or HTL), which may induce irreversible and lasting degradation of device efficiency (Figure 11.3).

11.2 Analysis of the Current Stability of Perovskite Light-emitting Diodes

At present, the EQE of high-performance PeLEDs has exceeded 30%, which marks a major breakthrough in device efficiency in the research of chalcogenide LEDs and fully demonstrates the great potential of PeLEDs in the field of full-color display and solid-state lighting applications. However, the stability problem is still one of the important influencing factors limiting PeLEDs to large-scale industrialization [9–11]. Table 11.1 shows the published EQE, lifetime data, and corresponding test conditions of some high-performance, high-brightness PeLEDs, from which it can be seen that even for PeLEDs with EQE higher than 20%, their device lifetime is still very low, and the green system has the longest lifetime under T_{50} conditions with an initial brightness of 1000 cd m^{-2}, while the green system has the longest lifetime of 105 h at an initial brightness of 1000 cd m^{-2} (efficiency drops to half of the initial value) and the shortest lifetime of 10 min at 4.4 V bias.

Currently, the longest T_{50} lifetime of an organic diode stands at 1 000 000 h (at an initial brightness of 1000 cd m^{-2}), while that of a quantum dot (QD) diode is

22 600 000 h (at an initial brightness of 100 cd m^{-2}). This underscores the significant stability issues that PeLEDs still face and highlights the pressing need for further research in this area. The operational stability of contemporary PeLEDs is still much inferior to that of organic LEDs (OLEDs) (T_{50} > 9 500 000 h at 1000 cd m^{-2}) [18] and QDs-based LEDs (QLEDs) (T_{50} = 125 000 000 h at 100 cd m^{-2}) [24]. We analyze the operational lifetimes of the reported PeLEDs, which are summarized in Table 11.1 (the abbreviations used in the table are as follows: PCs refer to polycrystals, QDs stand for quantum dots, quasi-2D represents quasi-2D, and T_{50} denotes the time it takes for the device's luminance to decay to half of its initial value).

11.3 Factors Affecting Efficiency Roll-off

When a device is subjected to high current densities, its external quantum efficiency often decreases, resulting in a phenomenon called efficiency roll-off. This effect is commonly used to evaluate the stability of light-emitting devices, as a smaller efficiency roll-off at high current densities typically indicates better device stability [25]. Efficiency roll-off can also serve as a metric for determining the critical current density at which a device's external quantum efficiency drops from its maximum value

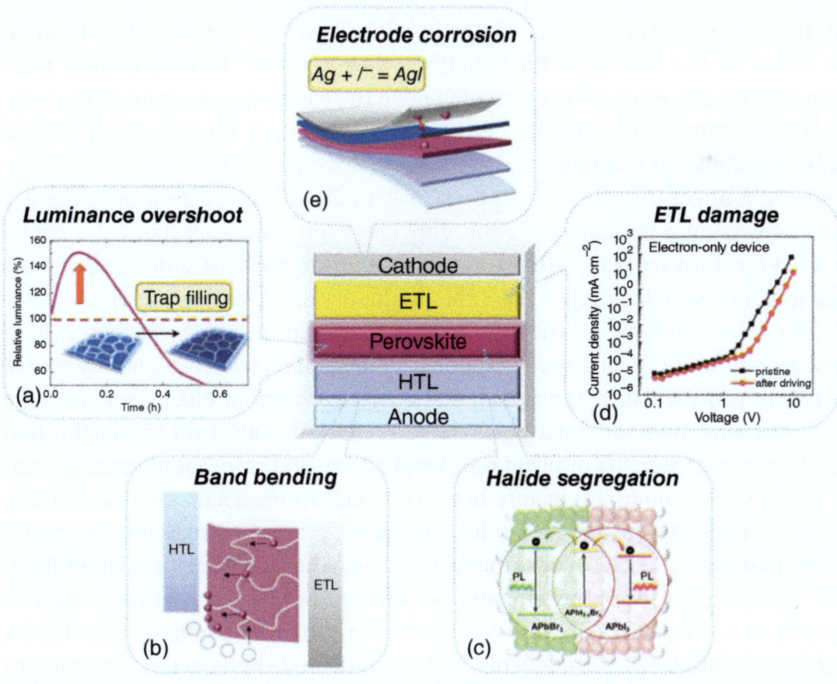

Figure 11.3 Consequences of the electric field-induced ion migration for PeLEDs, including luminance overshoot (a), band bending (b), halide segregation (c), ETL damage (d), and electrode corrosion (e). The *blue hollow circles and pink balls* stand for injected holes and halide ions, respectively. Source: Kong et al. [4]/John Wiley & Sons.

Table 11.1 Survey of the reported PeLEDs showing their emission characteristics and stability (lifetime).

Perovskite	EQE_{max} (%)	EL_{peak} (nm)	T_{50}	T_{50} testing conditions	Year [Ref.]
$FAI:PbI_2$:ODEA PCs	21.6	800	20 h	25 mA cm^{-2}	2017 [12]
$FAPbI_3$-5AVA PCs	20.7	803	20 h	100 mA cm^{-2}	2018 [13]
NMAI:FAI:PbI_2:Poly-HEMA PCs	20.1	~795	46 h	0.1 mA cm^{-2}	2018 [14]
$CsPbBr_{0.6}I_{2.4}$QDs	21.3	653	5 min	100 cd m^{-2}	2018 [15]
$CsPbBr_3$/MABr PCs	20.3	525	105 h	100 cd m^{-2}	2018 [16]
Quasi-2D $CsPbCl_{0.9}Br_{2.1}$	5.7	480	10 min	4.4 V	2019 [17]
$FAPbI_3$-ODEA	21.6	800	20 h	25 mA m^{-2}	2019 [18]
$(FA_{0.9}Cs_{0.1})PbI_3$	19.6	780	7 h	20 mA m^{-2}	2019 [19]
$FAPbI_3$	20.2	799	20 h	57 mA m^{-2}	2020 [20]
$FA_{0.33}Cs_{0.67}Pb(I_{0.7}Br_{0.3})_3$	20.9	694	14 h	2.5 mA m^{-2}	2020 [21]
$CsPbBr_3$	21.6	520	180 h	6 mA cm^{-2}	2021 [22]
$CsPbBr_xCl_{3-x}$	>2.5	471	1.1 h	1 mA m^{-2}	2022 [23]

to half of that value. The relative change in critical current density is proportional to the extent of the efficiency roll-off effect, with a greater relative change indicating a smaller efficiency roll-off. To maximize the electroluminescence efficiency of perovskite emitters, one ideally needs to simultaneously increase the radiative recombination rate and suppress the non-radiative recombination rate [26].

Factors that affect the efficiency roll-off of PeLEDs at high current densities are currently receiving significant attention. Efficiency roll-off is an issue that must be considered for most types of LEDs, and while its origin remains somewhat controversial, it is now widely accepted that the oscillatory combination mechanism is a significant factor influencing the efficiency roll-off that occurs in devices at high current densities. The oscillatory combination mechanism involves a non-radiative combination process called the Osher–Stark effect, where an electron and a hole combine in a semiconductor and transfer energy to another electron or hole through collisions, causing them to jump and resulting in a non-radiative compound. In principle, the Auger recombination rate is proportional to materials' exciton binding energy (E_b). Thus, Auger recombination can be suppressed by reducing the corresponding materials' E_b [27]. This process can cause the sudden extinction of light, thereby reducing the luminous efficiency of the device [28–30]. The non-radiative compound caused by the oscillatory combination mechanism can have a significant impact on the efficiency roll-off. Zou et al. investigated the effect of intermittent recombination on the efficiency roll-off by adjusting the trap width of the multiple quantum well structure (2D/3D) through changes in the perovskite precursor solution's component ratio. Their findings suggest that a larger quantum well width can effectively mitigate the efficiency roll-off process. Some reports indicate that a significant roll-off in EQE occurs at higher current densities ($J \approx 100$ mA cm^{-2}), with peak EQEs observed at current densities $J \approx 10$–30 mA cm^{-2}. This suggests that there is significant competition between trap-induced non-radiative and radiative complexes in these devices, with high EQEs at low current densities and high external quantum efficiency at low current densities, with low-efficiency roll-off at high current densities not yet being achieved [31, 32].

Similar to the improvement in PeLED performance, the stability of these devices is often related to the presence of defects in perovskites. The most common kind of defects in perovskites are halide vacancies, which result in trap states located within their bandgap; their presence not only leads to decreased PLQY [33] and broadened emission bandwidths [34] of redshifted emission peaks [35], but also provides air-sensitive sites where degradation of perovskite lattice in atmosphere may be triggered [36]. The related trap states with inhomogeneous energy level distributions [19] may cause a carrier imbalance along with the increase in driving current for PeLEDs, and thus the undesired roll-off of the EQE (decrease of its value at high current densities) (Figure 11.4).

Joule heating is a phenomenon in which heat is generated when an electric current passes through a conductor, such as a PeLED. This heating effect can cause changes in the device's electrical and optical properties, leading to a decrease in efficiency. Unbalanced charge injection, on the other hand, refers to the uneven distribution of charges within the device, which can also lead to a decrease in efficiency. Kim et al. used pulsed current drives to minimize the Joule heating effect and found that the efficiency roll-off was mainly caused by unbalanced charge injection. They proposed that optimizing the device structure and interface could help reduce the unbalanced charge injection and improve the device's efficiency roll-off [37].

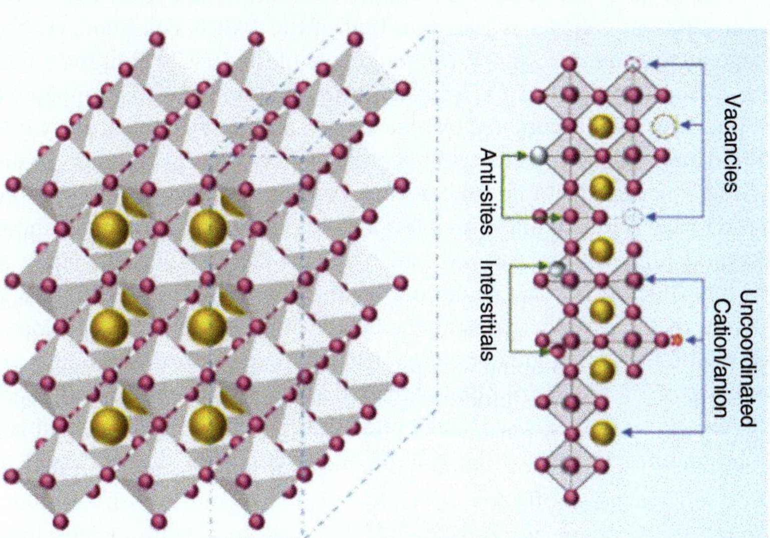

Figure 11.4 Illustration of the common cubic crystal structure and possible defects of metal halide perovskites. Source: Kong et al. [4]/John Wiley & Sons.

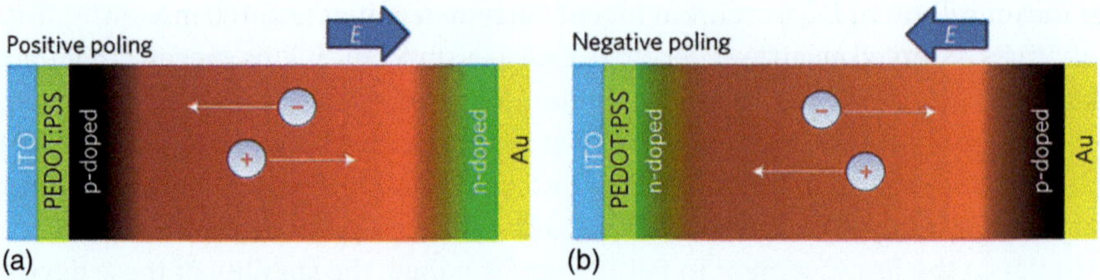

Figure 11.5 Schematics of ion drift in perovskite during positive and negative poling, respectively, showing that accumulated ions in the perovskite near the electrodes induced p- and n-doping. Source: Xiao et al. [38] Copyright © 2014, Springer Nature Limited.

The switchable photovoltaic scene is shown in Figure 11.5, taking vertical structure devices as an example. Positively charged ions or vacancies move to the Au side during the positive polarization process and accumulate there, resulting in n doping in the Au side perovskite. Similarly, the remaining negative space charge layer can be p-doped with a perovskite layer close to PEDOT: PSS, forming a p-i-n homogeneous structure. Reverse bias can flip the p-i-n structure into n-i-p by forcing ions or ion vacancies to drift in the opposite direction.

For certain 3D perovskite nanocrystals, degradation of the material at high current densities and disruption of the film morphology can play a significant role in the efficiency roll-off process.

Figure 11.6 showcases the variations in the morphology of perovskite films under different bias pressures. Specifically, the addition of 10 μL of thiocyanate, referred to as 10SCN, induced changes in the perovskite precursors, which were captured in the image. It can be observed that for the 30SCN condition, the overall roughness of the films at different voltages is less than that of the 10SCN condition. As the applied voltage increased from 3 to 7 V, the roughness of the 30SCN films increased directly from 2.2 to 3.1 nm, and at 7 V, a few large areas without coverage appeared. The current image at 3 V is dominated by blue-white and yellow-green areas, while at 5 V, small red areas begin to increase, indicating effective charge injection in the nanocrystal and consistent with optimal device performance. However, at 7 V, the majority of the image is blue-white with a few large red spots, indicating enhanced charge leakage at high voltage and reduced radiation combination efficiency at high current densities. At 3 V, the whole film is at a low current level, but at 5 V, the total current of the film increases as the current in the red region increases, with the red region indicating greater charge leakage than the rest of the film. At an applied voltage of 7 V, the distribution of the red and blue-white regions changes significantly, creating large exposed areas and localized high current densities, which may be another reason for the roll-off in the efficiency of the PeLED at high current densities. Overall, the morphology of the perovskite nanocrystal film deteriorates as the applied voltage increases, possibly leading to enhanced charge leakage [39].

11.4 Strategies for Dealing with Efficiency Roll-off

11.4.1 Perovskite Structure Modulation

Tilting of $[PbX_6]^{4-}$ octahedra induces destabilization in the perovskite structure [29]. To overcome the efficiency roll-off of PeLEDs at high current densities, researchers have developed a new type of perovskite structure called the multi-quantum well structure. Wei Huang's team has demonstrated that using a mixture of formamidine (FA) and cesium (Cs) cations to create multi-quantum well PeLEDs can further suppress the efficiency roll-off at high current densities. The use of Cs as a partial replacement for FA allows for the formation of wider quantum wells, effectively reducing the efficiency roll-off by suppressing intermittent complexation. The effectiveness of this approach has been confirmed by the multi-quantum well perovskite diode's external quantum efficiency of 7.8% at 125 mA cm^{-2} and 6.6% at 500 mA cm^{-2}, with only a slight decrease in efficiency at high current densities. The multi-quantum well structure suppresses ion migration, leading to high EQE and good stability of PeLEDs. This structure promotes carrier confinement and increases the probability of radiative recombination, similar to III–V LEDs. Using perovskite multiple quantum well structures significantly improves the operational stability of light-emitting devices, allowing for reliable assessment of their physical properties. The presence of defects in perovskites makes them sensitive to external stimuli, thus producing perovskite crystals with low trap density, high formation energy, and appropriate core/shell structure in the case of perovskite nanocrystals is extremely important for improving their stability [4].

The researchers found that at high current densities, the photoluminescence quantum efficiency (PLQE) decreases in the same trend as the external quantum efficiency, suggesting that photoluminescence bursts are the main cause of the efficiency roll-off, likely caused by the oscillatory combination. Another way to tune the width of a perovskite quantum well is to use a 2D/3D hybrid perovskite, which has been shown to work well for several hours at high current densities of 1 A cm^{-2}. Extensive spectroscopic measurements of optimized perovskite films and devices have shown that performance improvements come from three sources: (i) stronger beneficial constraints on carriers in the 3D region of the luminescent layer using optimized 2D/3D hybrid perovskites; (ii) effective control of the spatial electron/hole distribution in the perovskite semiconductor; and (iii) balanced carrier injection in the optimized device [40].

Compared with the organic carrier transport layers widely used in PeLEDs at present, the inorganic carrier transport layers represented by ZnO and NiO_x would prolong the working lifetime of PeLEDs [41]. Three strategies are important to suppress ion migration to improve the lifetime: (i) suppressing bulk ion migration

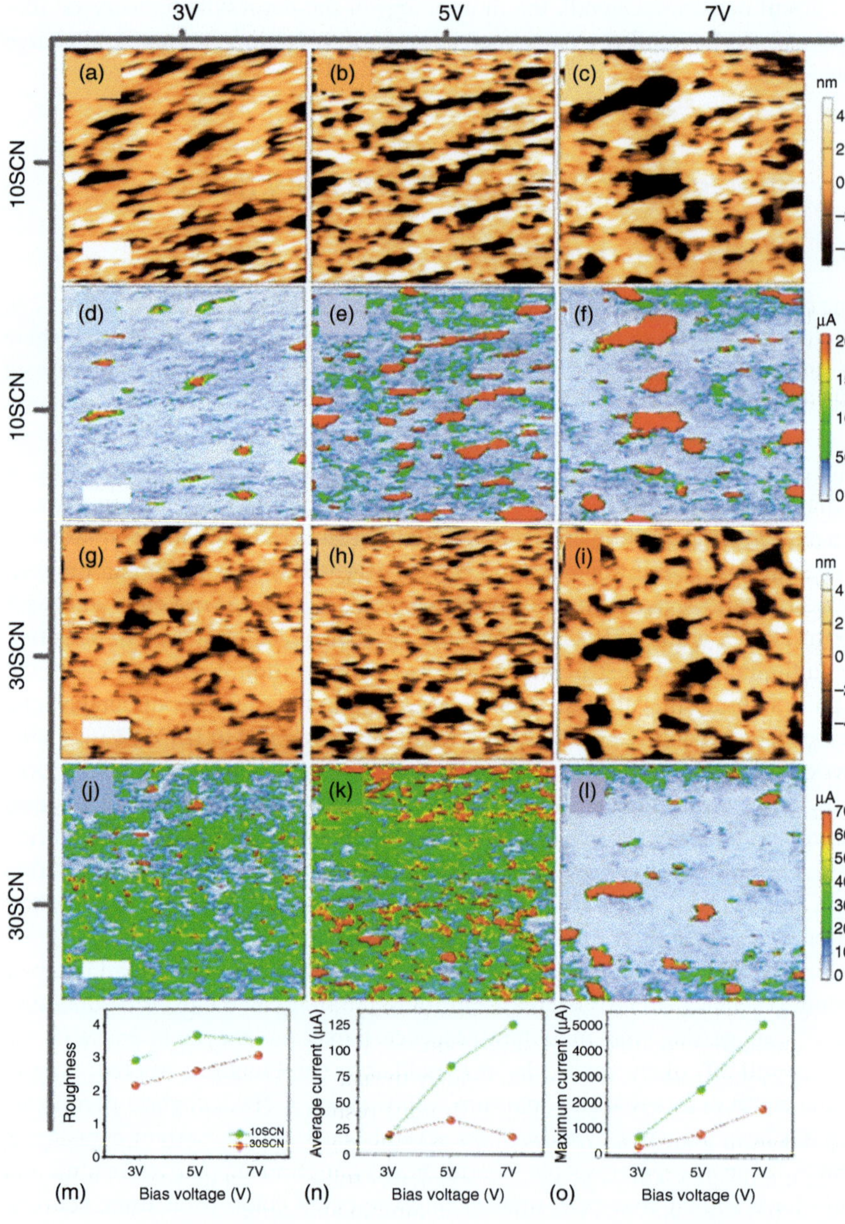

Figure 11.6 (a-c) Topography and (d-f) current mapping images of 10SCN NC film under 3 V, 5 V and 7 V; (g-i) topography and (j-l) current mapping images of 30SCN NC film under 3 V, 5 V and 7 V; comparison of (m) roughness, (n) average current, and (o) maximum current of 10SCN and 30SCN NC films under voltages. The scale bars are 200 nm [39].

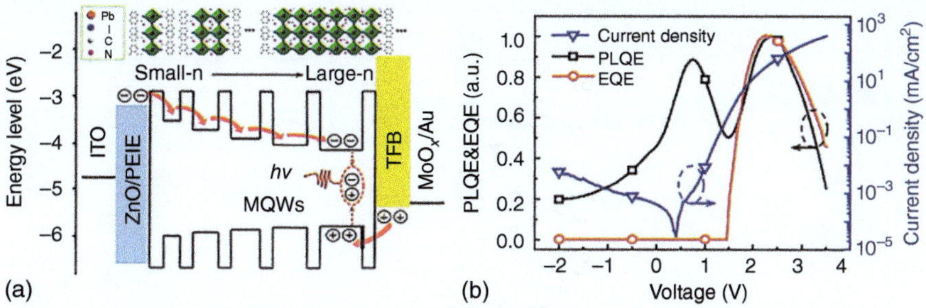

Figure 11.7 (a) shows the device structure of a perovskite light-emitting diode based on a multi-quantum well structure and (b) represents the current density (blue triangles), normalized PLQE (black squares), and EQE (red circles) versus drive voltage. Source: Adapted from Song et al. [46].

through compositional engineering (A-site, B-site, and quasi-2D perovskite); (ii) suppressing grain boundary ion migration by defect passivation; and (iii) suppressing interlayer ion migration by optimal device structure [42]. It is possible to achieve the ultra-high stability of perovskite QDs by using polymer as a macromolecular ligand [43]. According to the reports about metal-doped PQDs, it can be found that metal doping is also a powerful strategy to optimize both the PL performance and stability of PQDs [44].

One way to modify the structure of perovskites is to use ligand engineering to passivate the perovskite surface and improve its stability. Thiocyanate (SCN), a quasi-halide, has been shown to be effective in this regard. By increasing the grain size of perovskite films and reducing defects on the perovskite surface, SCN can enhance the device performance of PeLEDs, particularly in terms of stability and hysteresis suppression. Another promising approach is the use of a core-shell structure, which has been shown to improve device performance and reduce efficiency roll-off. There are also other strategies for regulating perovskite structures. For example, Antonella Giuri et al. have developed a simple method for fine-tuning the size of perovskite grains using a starch biopolymer as a template to prepare $FAPbI_3$ three-dimensional perovskite grains [45]. Wang et al. investigated the origin of the efficiency roll-off, by simultaneously measuring EL and photoluminescence (PL) on a working LED device. This technique results in size regulation, which has been shown to maintain an increase in current density along with an increase in radiation efficiency and suppress the efficiency roll-off of the device. The findings indicate that this approach can not only maintain an increased current density, but also enhance radiation efficiency and mitigate efficiency roll-off in the device [46] (Figure 11.7).

11.4.2 Hole Injection Layer Modulation

Charge-transport layers, which are generally divided into ETLs and HTLs, play an important function in facilitating charge injection. These functional layers can be built up from organic or inorganic constituents. In PeLEDs, suppressing

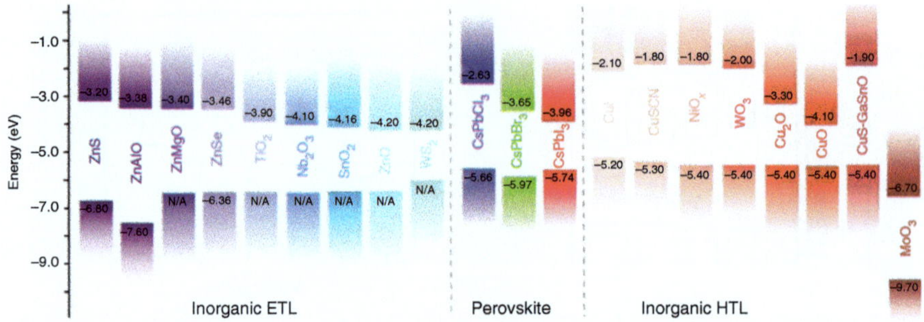

Figure 11.8 Positions of energy levels for commonly used inorganic ETLs, CsPbX3 perovskites emission layers, and inorganic HTLs. Source: Kong et al. [4]/John Wiley & Sons.

the ion migration within the perovskites and preventing ions from diffusion into charge-transport layers are the key points in enhancing the device's stability (Figure 11.8).

The device structure commonly used for PeLEDs is known to be anode (transparent substrate)/hole injection layer/perovskite light-emitting layer/electron injection layer (EIL)/cathode. However, unbalanced carrier injection into the device can lead to significant efficiency roll-off, making the regulation of the hole injection layer critical. To address this, HIL modulation has been developed as a technique to improve PeLED performance by optimizing the properties of the HIL layer. By adjusting the HIL layer's properties, it is possible to increase hole injection efficiency, reduce charge recombination, and improve overall device performance.

There are different ways to modulate the HIL layer in PeLEDs. One common approach is to use a dopant material to modify the electronic properties of the HIL layer. For example, doping the HIL layer with PEI can increase the injection efficiency of holes and reduce charge recombination, leading to improved device performance. Another approach to modulating the HIL layer is to use a material with a higher work function than the standard material used for the HIL layer. For instance, using molybdenum oxide (MoO_x) as the HIL material can improve hole injection and device performance compared to the standard material, PEDOT:PSS.

Additionally, Some commonly used charge-transport layers, such as acidic PEDOT:PSS, can etch the ITO electrodes, causing the release of In and Sn that may subsequently diffuse into the EML [47]. A hybrid HIL made of PEDOT:PSS and MoO_x has been shown to enhance hole injection and reduce charge recombination, thus improving the performance of PeLEDs.

Richard H. Friend's team investigated the application of various conjugated polymeric electrolytes (CPEs) as a hole injection layer and found that PCPDT-K can effectively transfer holes, block electron transfer from the perovskite to the underlying ITO layer, and reduce luminescence bursts at the perovskite/PCPDT-K interface [48]. Optimizing PeLEDs using PCPDT-K showed a ~4-fold enhancement in EQE compared to using PEDOT:PSS as the hole injection layer, with an EQE value of 5.66% and excellent device stability. This provides an important direction for tackling the efficiency roll-off problem by tuning the hole injection layer.

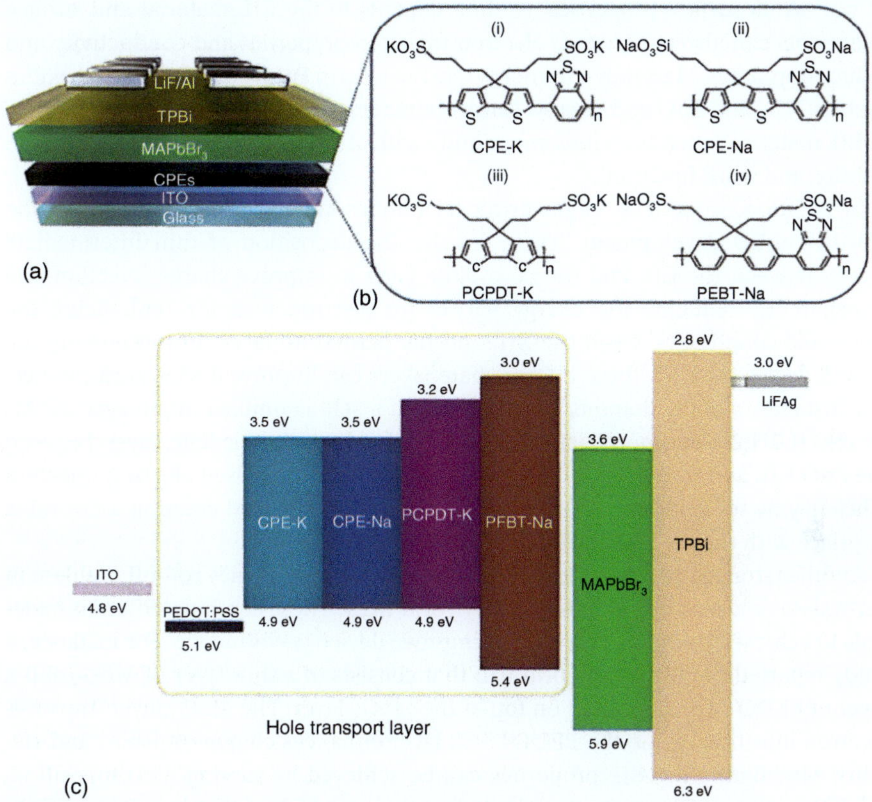

Figure 11.9 Schematic diagram of a perovskite light-emitting diode: (a) device architecture of a perovskite light-emitting diode; (b) chemical structure of conjugated polymeric electrolytes (CPEs); and (c) energy band diagram of various CPEs for a perovskite light-emitting diode. Source: Adapted from Friend et al. [49].

Moreover, the performance of PeLEDs can be further improved by using highly polar alcohols to treat PEDOT:PSS films prior to annealing treatment. Highly polar alcohols introduce a shielding effect between the positively charged PEDOT and the negatively charged PSS, which removes some of the excess PSS insulators from the PEDOT:PSS during the spinning and coating process, improving the injection of cavities from the PEDOT:PSS layer to the perovskite light-emitting layer. The PEDOT:PSS film is also smoother after treatment with highly polar alcohols, resulting in smaller perovskite particles and better perovskite coverage, which reduces the internal defects of the perovskite film and improves the device's efficiency roll-off problem (Figure 11.9).

11.4.3 Electron Injection Layer Modulation

To combat the efficiency roll-off in perovskite diodes, modulating the EIL is an effective strategy. This involves tailoring the properties of the EIL to optimize electron injection and transport. The composition and thickness of the EIL can be varied, as well as its surface properties. Adding dopants to the EIL material and surface treatments can also enhance its electron injection properties and conductivity and reduce trap states. Materials commonly used as EILs in PeLEDs include metal oxides such as TiO_2 and ZnO and organic semiconductors such as PEDOT:PSS. The choice of EIL material depends on its compatibility with other device components, electron affinity, and work function.

In addition, interfacial engineering techniques can be used to modulate the EIL in PeLED development. These involve the deposition of thin intermediate layers between the EIL and the perovskite layer to improve charge injection and transport. By reducing the energy barrier for electron injection, enhancing the electronic coupling between the EIL and the perovskite layer, and improving the overall device stability, these intermediate layers can improve device performance. For instance, a study demonstrated the use of a self-assembled monolayer (SAM) of 1H,1H,2H,2H-perfluorodecanethiol (PFDT) as an intermediate layer between the ZnO EIL and the perovskite layer. The PFDT SAM increased electron injection efficiency by reducing the work function of the ZnO EIL and creating a favorable interface with the perovskite layer.

Another strategy to modulate the EIL and tackle the efficiency roll-off problem in perovskite diodes is through the use of hybrid EILs. These combine different materials to achieve synergistic effects that improve device performance. For instance, a study reported the use of a hybrid EIL that consists of a thin layer of MoO_3 and a layer of PEDOT:PSS deposited on top of the MoO_3 layer. The MoO_3 layer improves electron injection, while the PEDOT:PSS layer enhances charge transport and stability. Modulating the EIL properties can be achieved by varying its composition, thickness, and surface properties, including the use of interfacial engineering techniques. Common materials used as EILs in PeLEDs include metal oxides such as TiO_2 and ZnO, as well as organic semiconductors like PEDOT:PSS. The selection of an appropriate EIL material depends on factors such as its compatibility with other device components, its electron affinity, and its work function.

Azhar Fakharuddin et al. designed a PeLED with a tuned EIL to reduce the efficiency roll-off of the device. The device used Bphen as the EIL and a small molecule Alq_3 to modify the electron injection and transport processes (Figure 11.10) [50]. The cross-section of the PeLED layers was observed under the scanning electron microscope, with the edges covered to ensure accurate external quantum efficiency calculation. The effect of Alq3 doping on the external quantum efficiency of the PeLEDs was studied, and the EQE_{max} of undoped Bphen decreased by 72% at a voltage of 6.5 V, while the PeLEDs doped with 5% and 10% Bphen behaved differently, with a decrease of no more than 10% in EQE_{max} and voltage values up to 7 V. The external quantum efficiency of the Bphen-doped devices was statistically analyzed, showing that the average external quantum efficiency of the Bphen-doped devices was significantly higher, and the external quantum efficiency of the 5% and 10% Bphen-doped devices was more stable. Furthermore, controlling ZnO nanoparticles as the ETL can prepare devices with low-efficiency roll-off due to its excellent optical bandgap and stability to transport electrons efficiently. However, preparing highly homogeneous nano-ZnO layers on the substrate remains a challenge, and this is an area for further

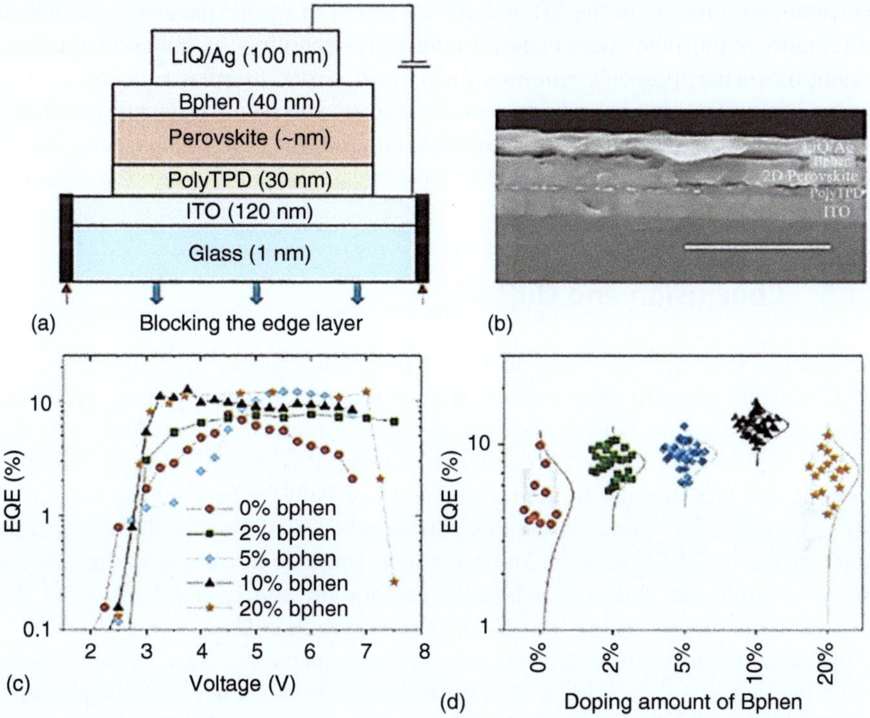

Figure 11.10 (a) Schematic diagram of the individual functional layers of the perovskite light-emitting diode and their thicknesses. (b) SEM cross-section of the device with a scale bar of 500 nm. (c) Shows the effect of different levels of Bphen-doped electron injection layers on the EQE of the perovskite light-emitting diode. (d) EQE distribution of devices with more than 150 different levels of Bphen-doped electron injection layers. Source: Fakharuddin et al. [50]/Reproduced with permission from John Wiley & Sons, Inc.

investigation [51]. These reports highlighting the modulation of the EIL underscore the significance of proper space charge carrier distribution in perovskite luminescent layers. By achieving a balance in carrier injection and transport, it is possible to prevent excessive carrier leakage through the device at high current densities, which leads to a reduction in efficiency roll-off [39, 52, 53].

Additionally, other effective approaches to counteract efficiency roll-off in PeLEDs include phase engineering, interface engineering, and metal electrode optimization. These strategies aim to reduce luminescence quenching caused by intermittent recombination, minimize Joule heating within the device, balance carrier injection and transport, and optimize film morphology [54, 55].

Additionally, alternative approaches to EIL modulation in PeLEDs include the use of different electrode materials like carbon nanotubes, graphene, and metal grids, as well as employing interface engineering techniques such as chemical doping and surface functionalization. These methods aim to optimize carrier injection/transport, reduce luminescence quenching, and minimize Joule heating to counteract the efficiency roll-off of PeLEDs.

Overall, modulation of the EIL is a critical aspect of PeLED development, which can significantly affect device performance and stability. To optimize the EIL, a comprehensive approach is needed that considers the interplay between various device components and the impact of the EIL on the device's overall efficiency. Ongoing research in this field aims to further enhance the performance and stability of PeLEDs and unlock their potential for practical applications in lighting, displays, and other optoelectronic devices.

11.5 Conclusion and Outlook

The issue of efficiency roll-off constitutes a paramount hurdle to surmount should PeLEDs aspire to secure their place in large-scale industrial deployment. Presently, the investigation into PeLEDs remains somewhat inchoate, as evidenced by the paucity of published scholarly research addressing this particular issue. Notwithstanding the fact that the luminous efficiency of PeLEDs may eclipse a value of 20%, a tangible discrepancy in stability persists when juxtaposed with their organic counterparts or QDLEDs. Such an observation could be attributed to the existing deficit in profound and comprehensive exploration and comprehension of the internal dynamics governing the efficiency roll-off in PeLEDs.

To contemplate future enhancements and advancements, three primary facets merit focused attention with a view to tackling the efficiency roll-off quandary intrinsic to PeLEDs. Firstly, the utilization of all-inorganic material systems, known for their resilience in enduring higher voltage and showcasing superior stability, might provide an effective remedy. Secondly, implementing strategies that stymie film decomposition and ion migration could significantly bolster efficiency, given the critical role that the quality of the film occupies. Lastly, meticulously calibrating the injection and transmission processes of space charge carriers within the device holds promise for rectifying this issue.

Hence, the efficiency roll-off predicament in PeLEDs necessitates deeper, more comprehensive scholarly investigation. Achieving a resolution to this cardinal challenge would unlock the considerable potential of PeLEDs, enabling them to assume a pivotal and irreplaceable role within the fields of full-color display technology and solid-state illumination.

References

1 Zhidkov, I.S., Boukhvalov, D.W., Akbulatov, A.F. et al. (2021). XPS spectra as a tool for studying photochemical and thermal degradation in $APbX_3$ hybrid halide perovskites. *Nano Energy* 79: 105421.

2 Ho, K., Wei, M., Sargent, E.H. et al. (2021). Grain transformation and degradation mechanism of formamidinium and cesium lead iodide perovskite under humidity and light. *ACS Energy Letters* 6: 934–940.

3 Chen, Z., Brocks, G., Tao, S. et al. (2021). Unified theory for light-induced halide segregation in mixed halide perovskites. *Nature Communications* 12: 2687.

4 Kong, L., Zhang, X., Zhang, C. et al. (2022). Stability of perovskite light-emitting diodes: existing issues and mitigation strategies related to both material and device aspects. *Advanced Materials* 34: 2205217.
5 Conings, B., Drijkoningen, J., Gauquelin, N. et al. (2015). Intrinsic thermal instability of methylammonium lead trihalide perovskite. *Advanced Energy Materials* 5: 1500477.
6 Yang, J., Siempelkamp, B.D., Liu, D. et al. (2015). Investigation of $CH_3NH_3PbI_3$ degradation rates and mechanisms in controlled humidity environments using in situ techniques. *ACS Nano* 9: 1955–1963.
7 Yang, J., Liu, C., Cai, C. et al. (2019). High-performance perovskite solar cells with excellent humidity and thermo-stability via fluorinated perylenediimide. *Advanced Energy Materials* 9: 1900198.
8 Xing, J., Wang, Q., Dong, Q. et al. (2016). Ultrafast ion migration in hybrid perovskite polycrystalline thin films under light and suppression in single crystals. *Physical Chemistry Chemical Physics* 18: 30484–30490.
9 Jeong, B., Han, H., Choi, Y.J. et al. (2018). All-inorganic $CsPbI_3$ perovskite phase-stabilized by poly(ethylene oxide) for red-light-emitting diodes. *Advanced Functional Materials* 28: 1706401.
10 Yuan, F., Xi, J., Dong, H. et al. (2018). All-inorganic hetero-structured cesium tin halide perovskite light-emitting diodes with current density over 900 A cm^{-2} and its amplified spontaneous emission behaviors. *Physica Status Solidi-Rapid Research Letters* 12: 1800090.
11 Chiba, T., Hayashi, Y., Ebe, H. et al. (2018). Anion-exchange red perovskite quantum dots with ammonium iodine salts for highly efficient light-emitting devices. *Nature Photonics* 12: 681–687.
12 Xu, B., Wang, W., Zhang, X. et al. (2017). Bright and efficient light-emitting diodes based on MA/Cs double cation perovskite nanocrystals. *Journal of Materials Chemistry C* 5: 6123–6128.
13 Zou, Y., Ban, M., Yang, Y. et al. (2018). Boosting perovskite light-emitting diode performance via tailoring interfacial contact. *ACS Applied Materials & Interfaces* 10: 24320–24326.
14 Clasen Hames, B., Sánchez Sánchez, R., Fakharuddin, A. et al. (2018). A comparative study of light-emitting diodes based on all-inorganic perovskite nanoparticles ($CsPbBr_3$) synthesized at room temperature and by a hot-injection method. *ChemPlusChem* 83: 294–299.
15 Ye, S., Sun, J.-Y., Han, Y.-H. et al. (2018). Confining Mn^{2+}-doped lead halide perovskite in zeolite-Y as ultrastable orange-red phosphor composites for white light-emitting diodes. *ACS Applied Materials & Interfaces* 10: 24656–24664.
16 Li, D., Liu, D., Wang, M. et al. (2018). Cyanopyridine based bipolar host materials for phosphorescent light-emitting diodes with low efficiency roll-off: importance of charge balance. *Dyes and Pigments* 159: 230–237.
17 Leyden, M.R., Terakawa, S., Matsushima, T. et al. (2019). Distributed feedback lasers and light-emitting diodes using 1-naphthylmethylamnonium low-dimensional perovskite. *ACS Photonics* 6: 460–466.

18 Xu, W., Hu, Q., Bai, S. et al. (2019). Rational molecular passivation for high-performance perovskite light-emitting diodes. *Nature Photonics* 13: 418–424.

19 Chen, B., Rudd, P.N., Yang, S. et al. (2019). Imperfections and their passivation in halide perovskite solar cells. *Chemical Society Reviews* 48: 3842–3867.

20 Zhao, X. and Tan, Z.-K. (2020). Large-area near-infrared perovskite light-emitting diodes. *Nature Photonics* 14: 215–218.

21 Fang, Z., Chen, W., Shi, Y. et al. (2020). Dual passivation of perovskite defects for light-emitting diodes with external quantum efficiency exceeding 20%. *Advanced Functional Materials* 30: 1909754.

22 Wang, T., Fang, T., Li, X. et al. (2021). Controllable transient photocurrent in photodetectors based on perovskite nanocrystals via doping and interfacial engineering. *The Journal of Physical Chemistry C* 125: 5475–5484.

23 Yuan, S., Zheng, X., Shen, W.-S. et al. (2022). Overcoming degradation pathways to achieve stable blue perovskite light-emitting diodes. *ACS Energy Letters* 7: 1348–1354.

24 Lee, T., Kim, B.J., Lee, H. et al. (2022). Bright and stable quantum dot light-emitting diodes. *Advanced Materials* 34: 2106276.

25 Yan, F., Xing, J., Xing, G. et al. (2018). Highly efficient visible colloidal lead-halide perovskite nanocrystal light-emitting diodes. *Nano Letters* 18: 3157–3164.

26 Pacchioni, G. (2021). Highly efficient perovskite LEDs. *Nature Reviews Materials* 6: 108–108.

27 Jiang, Y., Cui, M., Li, S. et al. (2021). Reducing the impact of Auger recombination in quasi-2D perovskite light-emitting diodes. *Nature Communications* 12: 336.

28 Yu, J.C., Kim, D.B., Baek, G. et al. (2015). High-performance planar perovskite optoelectronic devices: a morphological and interfacial control by polar solvent treatment. *Advanced Materials* 27: 3492–3500.

29 Yang, G., Liu, X., Sun, Y. et al. (2020). Improved current efficiency of quasi-2D multi-cation perovskite light-emitting diodes: the effect of Cs and K. *Nanoscale* 12: 1571–1579.

30 Yang, D., Zou, Y., Li, P. et al. (2018). Large-scale synthesis of ultrathin cesium lead bromide perovskite nanoplates with precisely tunable dimensions and their application in blue light-emitting diodes. *Nano Energy* 47: 235–242.

31 Cho, Y.J. and Lee, J.Y. (2011). Low driving voltage high quantum efficiency high power efficiency and little efficiency roll off in red, green, and deep-blue phosphorescent organic light-emitting diodes using a high-triplet-energy hole transport material. *Advanced Materials* 23: 4568–4572.

32 Chen, M., Shan, X., Geske, T. et al. (2017). Manipulating ion migration for highly stable light-emitting diodes with single-crystalline organometal halide perovskite microplatelets. *ACS Nano* 11: 6312–6318.

33 Quan, L.N., García de Arquer, F.P., Sabatini, R.P. et al. (2018). Perovskites for light emission. *Advanced Materials* 30: 1801996.

34 Yuan, J., Zhang, L., Bi, C. et al. (2018). Surface trap states passivation for high-performance inorganic perovskite solar cells. *Solar RRL* 2: 1800188.

35 de Quilettes, D.W., Vorpahl, S.M., Stranks, S.D. et al. (2015). Impact of microstructure on local carrier lifetime in perovskite solar cells. *Science* 348: 683–686.

36 Saidaminov, M.I., Kim, J., Jain, A. et al. (2018). Suppression of atomic vacancies via incorporation of isovalent small ions to increase the stability of halide perovskite solar cells in ambient air. *Nature Energy* 3: 648–654.

37 Cao, X., Yang, Y., and Guo, H. (2008). On the origin of efficiency roll-off in InGaN-based light-emitting diodes. *Journal of Applied Physics* 104: 093108.

38 Xiao, Z., Yuan, Y., Shao, Y. et al. (2015). Giant switchable photovoltaic effect in organometal trihalide perovskite devices. *Nature Materials* 14: 193–198.

39 Chen, F., Boopathi, K.M., Imran, M. et al. (2020). Thiocyanate-treated perovskite-nanocrystal-based Light-emitting diodes with insight in efficiency roll-off. *Materials* 13: 367.

40 Yang, M., Wang, N., Zhang, S. et al. (2018). Reduced efficiency roll-off and enhanced stability in perovskite light-emitting diodes with multiple quantum wells. *The Journal of Physical Chemistry Letters* 9: 2038–2042.

41 Xie, M. and Tian, J. (2022). Operational stability issues and challenges in metal halide perovskite light-emitting diodes. *The Journal of Physical Chemistry Letters* 13: 1962–1971.

42 Dong, Q., Lei, L., Mendes, J. et al. (2020). Operational stability of perovskite light emitting diodes. *Journal of Physics: Materials* 3: 012002.

43 Wen, Z., Xie, F., and Choy, W.C. (2022). Stability of electroluminescent perovskite quantum dots light-emitting diode. *Nano Select* 3: 505–530.

44 He, H., Mei, S., Wen, Z. et al. (2022). Recent advances in blue perovskite quantum dots for light-emitting diodes. *Small* 18: 2103527.

45 Song, M., Huang, J., Bai, F. et al. (2016). A series of pure orange-yellow iridium complexes with low efficiency roll-off: a computational study. *Chemical Research in Chinese Universities* 32: 451–454.

46 Zou, W., Li, R., Zhang, S. et al. (2018). Minimising efficiency roll-off in high-brightness perovskite light-emitting diodes. *Nature Communications* 9: 608.

47 Yang, D., Zhao, B., Yang, T. et al. (2022). Toward stable and efficient perovskite light-emitting diodes. *Advanced Functional Materials* 32: 2109495.

48 Friend, R.H., Gymer, R., Holmes, A. et al. (1999). Electroluminescence in conjugated polymers. *Nature* 397: 121–128.

49 Lee, B.R., Yu, J.C., Park, J.H. et al. (2018). Conjugated polyelectrolytes as efficient hole transport layers in perovskite light-emitting diodes. *ACS Nano* 12: 5826–5833.

50 Fakharuddin, A., Qiu, W., Croes, G. et al. (2019). Reduced efficiency roll-off and improved stability of mixed 2D/3D perovskite light emitting diodes by balancing charge injection. *Advanced Functional Materials* 29: 1904101.

51 Yu, Y., Ma, L., Feng, Z. et al. (2019). Strategy for achieving efficient electroluminescence with reduced efficiency roll-off: enhancement of hot excitons spin mixing and restriction of internal conversion by twisted structure regulation using an anthracene derivative. *Journal of Materials Chemistry C* 7: 5604–5614.

52 Song, M.X., Wang, G.F., Wang, J. et al. (2015). Theoretical study on a series of iridium complexes with low efficiency roll-off property. *Spectrochimica Acta Part A: Molecular and Biomolecular Spectroscopy* 134: 406–412.

53 Wang, H., Zhang, X., Wu, Q. et al. (2019). Trifluoroacetate induced small-grained $CsPbBr_3$ perovskite films result in efficient and stable light-emitting devices. *Nature Communications* 10: 665.

54 Wu, W.K., Wang, C.M., Chan, M.C. et al. (2018). Tuning the color temperature of white-light-emitting electrochemical cells by laser-scanning perovskite-nanocrystal color conversion layers. *ChemPlusChem* 83: 239–245.

55 Giuri, A., Yuan, Z., Miao, Y. et al. (2018). Ultra-bright near-infrared perovskite light-emitting diodes with reduced efficiency roll-off. *Scientific Reports* 8: 15496.

12

Perovskite Materials for Laser Applications

12.1 Physics Principles of Laser

Laser is the abbreviation for "light amplification by stimulated emission of radiation," which represents light from stimulated atomic radiation. Laser was first discovered at Bell Laboratories by Schawlow and Townes in 1958 [1]. The principles of lasers originated from Einstein in 1916, when electrons in atoms absorb energy and transition from low to high energy levels, and then fall back from high to low energy levels, the released energy is released in the form of photons. Three conditions are required to generate a laser: a laser gain medium with an appropriate energy level structure (can achieve particle number inversion) that can amplify stimulated radiation light; an energy source (excitation energy) that enables the activated gain medium to emit laser; and an optical resonant cavity that is used for optical oscillation. The laser emission emerged when the induced (excited) photons achieved light amplification by stimulated emission. The principles of laser are shown in Figure 12.1. In general, atoms can be in the ground state E_1 for a long time; the time in the excited state E_3 is very short, about 10^{-8} s, so the excited state is unstable. Some substances have metastable state E_2, which provides conditions for realizing the inversion of particle number. When electrons jump from E_1 to E_3, they will quickly transit to metastable state E_2 because excited state E_3 is unstable. When electrons jump between the E_1 and E_2 energy levels, the absorbed or radiated photon energy must satisfy the Bohr condition, $E_2 - E_1 = hf_{12}$, where Planck constant $h = 6.628 \times 10^{-34}$ J s and f_{12} denotes the photon frequency of absorption or radiation. If the number of atoms in the lower E_1 and the higher E_2 energy levels ($E_2 > E_1$) is N_1 and N_2, respectively. When the system is in thermal equilibrium, the following formula exists:

$$\frac{N_2}{N_1} = \exp\left(-\frac{E_2 - E_1}{kT}\right) \tag{12.1}$$

where Boltzmann constant $k = 1.381 \times 10^{-23}\ \mathrm{J\,K^{-1}}$ and T is temperature. Since $(E_2 - E_1) > 0$, $T > 0$, it is always $N_1 > N_2$. For absorption materials, $N_1 > N_2$, the stimulated absorption is greater than the stimulated radiation. For laser materials, N_2 will become larger than N_1 ($N_2 > N_1$) under certain pumping methods; that is, the stimulated radiation is greater than the stimulated absorption. Laser materials

Perovskite Light Emitting Diodes: Materials and Devices, First Edition. Hong Meng.

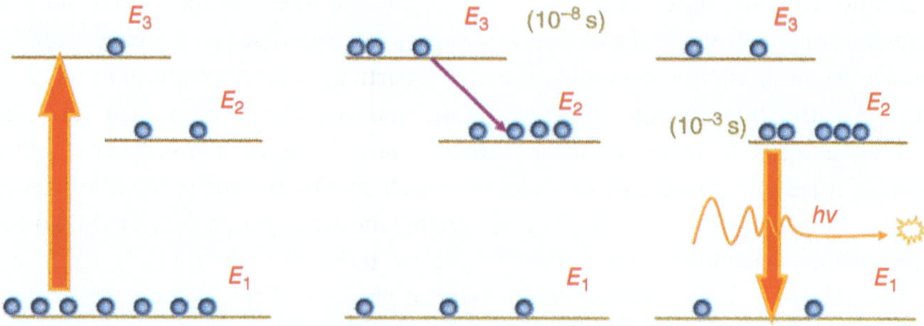

Figure 12.1 Energy level and electron transition of laser.

will produce amplified spontaneous emission (ASE) or lasing when stimulated radiation is greater than the stimulated absorption. The distribution of $N_2 > N_1$ is the opposite of the normal distribution ($N_1 > N_2$), so it is called the particle number inversion. The principle of semiconductor ASE/lasing can be described as follows: under one of certain excitation means (have 3 kinds: electric injection, electron beam pumping, and light pumping), particle number reversal distribution occurs between the conduction band and valence band, resulting in the recombination of a large number of electrons and holes and subsequent ASE. ASE will produce coherent light with the same frequency, phase position, and polarization and have same propagation direction as incident light. Thus, the ASE or lasing emission peak has a very narrow linewidth of about 0.1–10 nm.

The lasing emission is closely related to the carrier dynamics of materials. The carriers have two kinds of free electron–hole pairs and free excitons. For 3D bulk materials, free charge carrier recombination dominates the luminescence process [2]. The photogenerated charge carrier decay could be described with the following equation [3]:

$$\frac{d_n(t)}{d_t} = G - k_1 n - k_2 n^2 - k_3 n^3 \tag{12.2}$$

Here, $n(t)$ is the photogenerated charge carrier density, and k_1 is the extrinsic trap-assisted monomolecular individual carrier (electron or hole) and recombination rate constant. The k_2 is the bimolecular charge-carrier recombination rate constant, reflecting intrinsic free electron–hole recombination, which depends on both electron (n_e) and hole (n_h) densities. The k_3 is the three-body Auger recombination constant; the Auger recombination is a non-radiative recombination. When an electron and a hole recombine, the energy or momentum is transferred to another electron or another hole through collision, resulting in a three-particle recombination process, namely, Auger recombination. Under steady-state electrical injection, the radiative emission quantum yield ($\eta_{(n)}$) is given by [3, 4]:

$$\eta(n) = \frac{nk_2}{k_1 + nk_2 + n^2 k_3} \tag{12.3}$$

Therefore, the luminescence quantum yield (QY) should be strongly dependent on charge carrier density n. In free-carrier systems, at low carrier densities, the

QY should first increase with increasing n due to the gradual dominance of the second-order bimolecular recombination over the first-order charge carrier trapping. At high carrier densities, the three-particle Auger recombination dominates over the bimolecular recombination, and the QY will in turn decrease with increasing n. In general, the building of particle number inversion requires very high injected charge carrier density, which results in Auger recombination. Hence, k_1, k_2, and k_3 are crucial for analyzing the free charge carrier dynamics. Perovskites are excellent gain materials because they have a low lasing threshold of a few $\mu J\,cm^{-2}$, a low ASE carrier density of about $10^{18}\,cm^{-3}$, and a low Auger recombination constant (3×10^{-28} $cm^6\ s^{-1}$) [5, 6]. For 3D bulk organic–inorganic hybrid perovskites, the free charge carriers, and excitonic recombination coexist because they have stable excitons at room temperature and a high exciton binding energy (41 meV) [7].

For low-dimensional metal–halide perovskites (2D perovskites), 1D perovskite nanowires (NWs) , and 0D perovskite nanocrystals (NCs), the excitons dominate the luminescence process, which could be described with [2]:

$$\frac{dn(t)}{dt} = G - k_1 n - k_2 n^2 \tag{12.4}$$

Here, the first-order process k_1 includes exciton-trapping and exciton radiative recombination, and the second-order process k_2 represents biexciton Auger recombination. An exciton is an electron–hole pair, and this bound state system is formed by the Coulomb interaction between electrons and holes. Excitons have a short lifetime and will eventually disappear because of their unstable state. For 2D lead-halide perovskites, light emission originates from excitonic recombination in the thick multi-quantum wells at a much higher decay rate [2]. For 0D metal–halide perovskite quantum dots (QDs)/NCs, the ASE/lasing originated from both single excitons and biexcitons [8, 9]. The QDs/NCs have many biexcitons and multiexcitons due to the quantum confinement effect. By reducing the crystal size toward their exciton Bohr radius, the Auger rates will be enormously increased. The biexciton Auger recombination rate is volume-related and can be expressed as $V^2/8\tau_{XX}$, V is the QDs volume, and τ_{XX} is the biexciton lifetime. The typical biexciton Auger recombination rate of metal–halide perovskite QDs is about a few to hundreds of picoseconds, which linearly increases with the increase in QD volume and depends sensitively on the anion [10, 11]. The quantum confinement effect can improve the luminescence QY of low-dimensional materials depending on the carrier's high recombination rate. But low-dimensional materials have a high non-radiative Auger recombination rate that is detrimental to ASE or lasing emission under high-density carrier injection. Above the injected carrier density of $10^{16}\,cm^{-3}$, the luminescence QY of perovskite multi-quantum wells significantly decreases with increasing carrier densities [2]. The NWs also have the quantum confinement effect that indicates they have exciton recombination dominant luminescence and high Auger recombination [12, 13]. The Auger recombination rate can be reduced by reducing the exciton binding energy (E_b) of materials because the Auger recombination rate is proportional to the E_b [14]. Thus, the achievement of ASE or lasing in low-dimensional materials requires low carrier density injection or reduced exciton binding energy.

Photoluminescence (PL) can be divided into down-conversion luminescence and up-conversion luminescence. Down-conversion is the process of converting a high-energy short-wavelength photon into two or more low-energy long-wavelength photons and emitting them. Dexter first proposed the concept of down-conversion in 1957, which demonstrated that if two acceptors simultaneously receive half of the energy of the same donor, they can achieve quantum efficiency of over 100% [15]. The term down-conversion can also be called one-photon absorption (OPA) and emission, which is a quantum-cutting process caused by energy transfer, which involves the conversion of one photon into multiple photons [16]. The down-conversion luminescence refers to the phenomenon of PL that follows Stokes' law, where the energy of emitted photons is lower than that of absorbed photons. Ultraviolet (UV) light is generally used as an excitation light source for down-conversion/one-photon pumps because the energy of a UV photon is more than twice that of a visible photon. Therefore, one-photon pumped ASE or lasing emissions generally pumped are excited by UV light sources in the range of 200–400 nm. The average exciton density generated by OPA (n_1) [17],

$$n_1 = f_1 \times \frac{1 - T_1}{\hbar wl} \tag{12.5}$$

where f_1 is one-photon pumped power, $T_1 = e^{-\alpha l}$, α is the OPA coefficient at UV light wavelength, and l is the sample thickness. If f_1 is the lasing or ASE threshold, the one-photon pumped average threshold exciton density is n_1. Compared with linear one-photon/down-conversion absorption, multiple-photon/up-conversion absorption is nonlinear and has the advantages of larger penetration depth, high spatial resolution, and less damage to the target sample [18]. There are three types of luminescence mechanisms for the up-conversion process: excited state absorption, energy transfer process, and photon avalanche process. The typical multiple-photon absorption is two-photon absorption (TPA), which was first proposed by Maria Göppert Mayer in 1931 and was first proved in CaF_2:Eu^{2+} by ruby laser in 1961 [19, 20]. TPA is the process by which electrons can simultaneously absorb two photons, thereby transitioning from a low energy level to a high energy level. Infrared light is generally used as a two-photon pumped excitation light source because the energy of an infrared photon is lower than that of a visible photon. Therefore, two-photon pumped ASE, or lasing emissions generally pumped, are excited by an infrared light source in the range of 700–1000 nm. The average exciton density generated by TPA (n_2) [17]

$$n_2 = f_2 \times \frac{1 - T_2}{2\hbar wl} \tag{12.6}$$

where f_2 is two-photon pumped power, $T_2 = [\sqrt{\pi}\beta I_0 l]^{-1} \int_{-\infty}^{\infty} \log[1 + \beta I_0 l e^{-x^2}]dx$, β is the TPA coefficient at infrared light wavelength, and l is the sample thickness. If f_2 is the lasing or ASE threshold, the two-photon pumped average threshold exciton density is n_2. The difference between OPA and TPA is that TPA involves the simultaneous interaction of two photons, so it increases with the square of the incident light intensity, while OPA is linearly related to the incident light intensity [17]. Hence, TPA transitions can only be observed in strong laser beams, which produce very high instantaneous photon density.

12.2 Perovskite Laser for Different Morphologies

12.2.1 Laser of Perovskite Films

In 2014, the coherent light emission (ASE and lasing) properties of perovskite were first demonstrated by the spin-coated 65-nm-thick $MAPbI_3$ thin films [6]. These thin films exhibit tunable and durable ASE from 390–790 nm with a low threshold value of $\sim 12\,\mu J\,cm^{-2}$ (Figure 12.2a), which stem from their large absorption coefficients ($\alpha = 5.7 \times 10^4\ cm^{-1}$), ultralow bulk trap densities ($5 \times 10^{16}\ cm^{-3}$), low bimolecular recombination, and slow Auger recombination. The ASE is a fast recombination process within a short lifetime of <10 ps (Figure 12.2b). The ASE threshold carrier density for these films is calculated to be about $1.7 \times 10^{18}\ cm^{-3}$, which is significantly less than many high-temperature-grown inorganic semiconductors [17]. Following this discovery, the lasing performance of perovskite materials has attracted extensive research interest. Compared to the cavity-induced

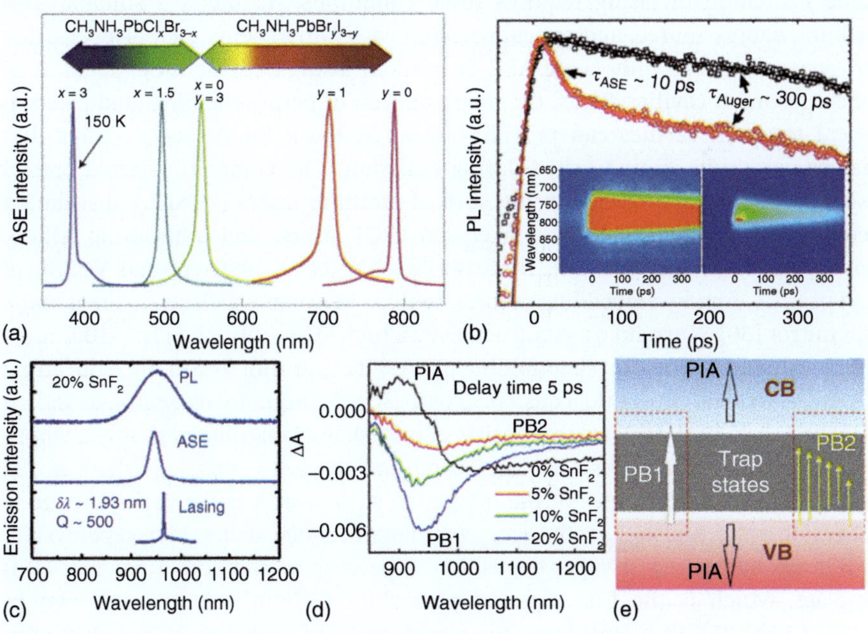

Figure 12.2 Figure 12.2 (a) The coherent light emission of $MAPbX_3$ (X = Cl, Br, I) perovskite; (b) the typical ASE lifetime of $MAPbX_3$ (X = Cl, Br, I) perovskite. Source: (a,b) Adapted from Xing et al. [6]. (c) The ASE and lasing emission of $CsSnI_3$ films; (d) the transient absorption spectroscopy of $CsSnI_3$ films with different SnF_2 addition ratios; (e) the reduction of trap state in SnF_2 treated CsSnI3 films. Source: Reproduced with permission [21], Copyright 2016, Wiley-VCH.

lasing emission, the PL and ASE behavior of cavity-free materials can provide a more explicit comparison for their suitability as gain media. For example, the bulk and surface trap densities of $MAPbI_3$ thin films (10^{16}–10^{17} cm^{-3}) are superior to those of solution-processed organic thin films (10^{19} cm^{-3}) [6, 22]. SnF_2 additive can effectively suppress the trap states in $CsSnI_3$ films that allow ultralow-threshold (≈6 μJ cm^{-2}) and large-gain (200 cm^{-1}) stimulated emission (Figure 12.2c) [21]. The free electron–hole bimolecular recombination dominates the emission from the SnF_2-treated $CsSnI_3$ films because of their quadratic behavior at higher pump fluence and small E_b (0.1–18 meV). Transient absorption (TA) spectroscopy shows that the photobleaching (PB) signature at around 950 nm of SnF_2-added $CsSnI_3$ films becomes progressively stronger than that of untreated $CsSnI_3$ films (Figure 12.2d). The 950 nm PB valley near the bandgap can be attributed to near-bandgap-stimulated emission of the carriers and state-filling of the band edge states. The nearly flat and broad PB band of untreated $CsSnI_3$ films represents state-filling of the trap states such as Sn vacancies. Thus, the great reduction of the amplitude ratio of PB2 to PB1 provides clear direct evidence for the reduction of the trap state in SnF_2-treated $CsSnI_3$ films (Figure 12.2e). The incorporation of ZnO nanoparticles in $CsPbBr_3$ films can reduce crystal grains, improve the reflection probability of light, and enable a compact and smooth surface, which leads to a high photoluminescence quantum yield (PLQY) and low ASE threshold [23].

The formation of lasing requires three conditions, including a suitable gain medium, energy source, and optical resonant cavity. In general, the polycrystalline perovskite films can only show ASE or random lasing because they are short of optical resonant cavities. Thus, the incorporation of perovskite films and external optical resonant cavities can provide optical feedback for perovskite films that enables perovskite films to show lasing emission. The common external optical resonant cavities are vertical cavity surface emitting lasers (VCSEL), distributed feedback (DFB) cavities, photonic crystal (PC) lasers, and whispering gallery mode (WGM) cavities [28, 29]. In 2014, Deschler et al. first reported VCSEL of $MAPbI_{3-x}Cl_x$ film by embedding perovskite films between a dielectric mirror and a gold mirror [30]. Nurmikko's group achieved a high Q (quality factor) of 1100, and a low-threshold of 7.6 μJ cm^{-2} monolithic perovskite thin-film VCSEL by embedding $MAPbI_3$ thin solid films into a custom GaN-based high-quality resonator, as shown in Figure 12.3a [24]. DFB cavity consists of grid-shaped distributed grooves, whose scattering behavior follows the λ_{Bragg} condition, $2n_{eff}\Lambda = m\lambda_{Bragg}$, where n_{eff} is the effective refractive index of the integrated structure, Λ is the periodicity, m is diffraction order, and λ_{Bragg} is the λ_{Bragg} wavelength. Saliba et al. achieved perovskite DFB lasers by coating or vapor deposition of perovskite film onto a polymer DFB template, which is the first reported perovskite DFB microlaser, as shown in Figure 12.3b [25]. In a DFB laser, the wavelengths of lasing peaks red shift with the increase of the λ_{Bragg} grating period Λ. The preparation of perovskite DFB laser can also be achieved by inkjet printing and chemical cavity engineering methods [31, 32]. Chen et al. demonstrated the perovskite laser within a 2D hexagonal lattice PC cavity by coating the pinhole-free and closely packed perovskite film on a repatterned quartz PC cavity, as shown in Figure 12.3c [26]. The preparation

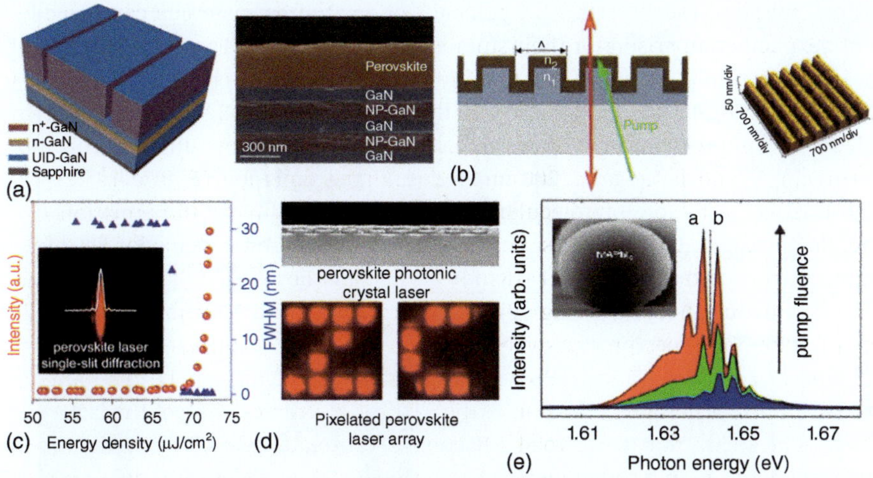

Figure 12.3 (a) The VCSEL cavity perovskite film laser, Reproduced with permission. Source: Chen et al. [24]/Reproduced with permission from John Wiley & Sons, Inc. (b) The perovskite film DFB lasers, Source: Reproduced with permission [25], Copyright 2016, Wiley-VCH. (c) Sharp threshold at input pulsed pump energy density of $68.5 \pm 3.0\,\mu J/cm^2$ is seen in emission, accompanied by collapse of the spectral line width from the spontaneous emission regime. Inset highlights single slit diffraction pattern imaged after propagating the PePhC output laser beam though a narrow rectangular aperture. (d) Cross-sectional SEM image of perovskite film on 2D PhC pattern.The scale bar is 400 nm. Different spatial laser beam patterns generated from the array using preprogrammed scanning protocol, here showing the letters "Z" and "C" [26], Copyright 2016, American Chemical Society.
(e) Spectral dependence of perovskite-coated microsphere emission on pump power at $T = 80$ K for three different pump fluences (100 $\mu J\,cm^{-2}$, blue; 175 $\mu J\,cm^{-2}$, green; and 300 $\mu J\,cm^{-2}$, red). Distinct laser modes are observed with a free spectral range of 700 GHz. Source: Sutherland et al. [27]/Reproduced from American Chemical Society.

of WGM lasers is achieved by incorporating perovskite films into a polygonal, spherical, circular, ring resonator cavity. Sutherland et al. reported perovskite WGM lasers by coating $MAPbI_3$ films onto a silica microsphere, as shown in Figure 12.3d [27]. Cegielski et al. achieved the lasing emission by integrating perovskite film with a stadium-shaped ring resonator [33]. Liu's group prepared a microdisk array by a focused ion beam etching method, and a single-mode laser was achieved on a 3 μm diameter disk with a low threshold of 1.6 $\mu J\,cm^{-2}$ [34]. The microlaser array mainly consists of many of the same perovskite film WGM lasers, which were prepared by template method, lithography, or laser direct writing [35–37].

12.2.2 Laser of Perovskite Nanowires

The perovskite nanowires (NWs) and nanorods can naturally form active F–P mode cavities because the smooth structural facets act as mirrors and the perovskite crystals themselves are the gain medium. In a F-P mode cavity, the mode spacing can be expressed as $\Delta\lambda = \lambda^2/2n_gL$, where λ is the lasing wavelength, n_g is the group index, and L is the cavity length. In general, the mode spacing is proportional to $1/L$. The various perovskite NW lasers with excellent performance have been widely

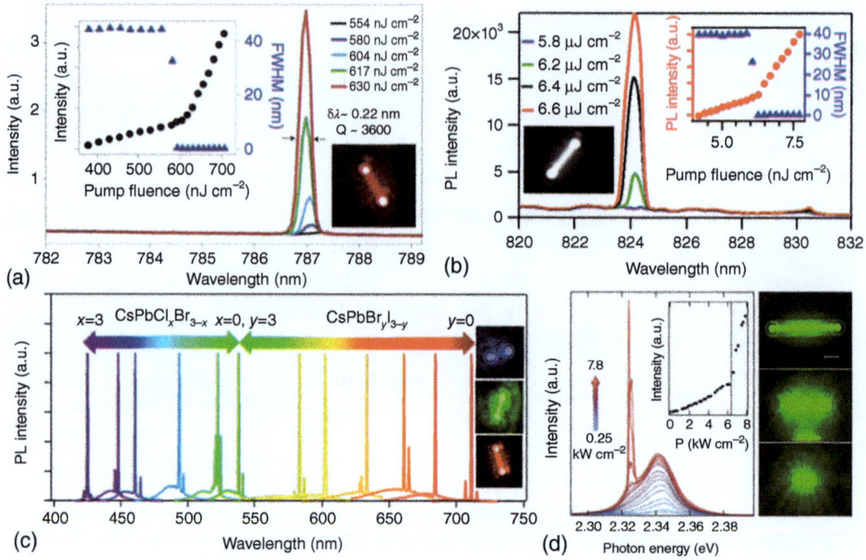

Figure 12.4 (a) The $MAPbX_3$ perovskite NW lasers with low thresholds and high quality factors, Source: Zhu et al et al. [38]/Reproduced from Springer Nature. (b) Lasing performance of $FAPbX_3$ perovskite NWs, Source: Reproduced with permission [39], Copyright 2016, American Chemical Society. (c) The all-inorganic $CsPbX_3$ NWs with low thresholds and tunable wavelength, Source: Fu et al. [40]/Reproduced with permission from American Chemical Society. (d) The CW lasing of $CsPbBr_3$ perovskite NWs at 77 K, Source: Evans et al. [41]/Reproduced with permission from John Wiley & Sons, Inc.

developed in recent years [38–48]. Zhu et al. reported the first demonstration of $MAPbX_3$ perovskite NW lasers with very low lasing thresholds (220–600 nJ cm^{-2}), high Qs ($Q \sim 3600$, evaluated by $Q = \lambda/\lambda_{\text{linewidth}}$), and broad tunability covering the near-infrared to visible wavelength region (Figure 12.4a) [38]. The smooth end facets, high QYs, high crystallization quality, and ultralow trap density are responsible for the superior lasing performance. Jin's group reported the lasing performance of $FAPbX_3$ (FA is $CH(NH_2)_2$) perovskite NWs with low thresholds of 6–30 nJ cm^{-2}, a tunable wavelength of about 560–820 nm and a high Q of about 1500–2300 (Figure 12.4b) [39]. $FAPbI_3$ shows better lasing photostability and thermal stability than $MAPbI_3$ due to the enhanced hydrogen bonding between FA^+ and the I^- of Pb-I octahedral. The same group obtained the more stable all-inorganic $CsPbX_3$ NWs with optically pumped low thresholds (3–10 nJ cm^{-2}) and tunable lasing (420–710 nm) that sustain at least 8 h or 7.2×10^9 laser shots under continuous illumination of 402 nm pulsed laser at room temperature (Figure 12.4c) [40]. All-inorganic $CsPbX_3$ perovskites exhibit better thermal stability than organic-inorganic hybrid perovskites because the all-inorganic structure supports a higher decomposition temperature than organic ones. Yan's group developed the vapor phase synthesis of composition-graded $CsPbBr_xI_{3-x}$ NWs with tunable dual-color lasing through the desynchronized deposition of metal halides and temperature-controlled anion exchange reaction [42]. Recently, Zhu's group achieved continuous wavelength (CW) lasing in $CsPbBr_3$ perovskite NWs at 77 K with a CW laser diode with 1 W output at 450 nm, which originates from polariton modes near the bottleneck region on the lower polariton branch (Figure 12.4d) [41]. Strong exciton–photon coupling and polariton lasing have been demonstrated in $CsPbBr_3$ and $MAPbBr_3$ NWs [43, 44]. The polariton condensate needs a large Rabi splitting with strong oscillator strength that is positively related to exciton binding energy and a sufficiently low dephasing rate with strong coupling and screening of Coulomb. The large vacuum Rabi splitting of $\Omega = 200$ meV of $CsPbBr_3$ NWs suggests strong light-matter interaction with expected polariton condensate. Thus, polariton lasing can emerge from the steady-state leakage of an exciton–polariton condensate below the threshold for population inversion because the small mode volume and enhanced cavity oscillator strength in the NW lead to a large Rabi splitting [49]. The CW lasing stems from the enhanced population inversion at low excitation densities of 6 kW cm^{-2}.

12.2.3 Laser of Perovskite Nanoplates and Microplates

The micro-nanospheres, nanocuboid, polygonal, and circular perovskite microplates/microdisks can naturally form active WGM cavities because the internal total reflection and perovskite crystals themselves are the gain medium. In a WGM cavity, photons propagate along with side planes by total internal reflection along the resonator-surrounding interface, which provides efficient optical feedback. And light can be trapped within a gain material with a higher refractive index than the environment, which enables WGM cavities to show high Q and small model volume. In WGM resonators, the light resonant condition is $2\pi nR = a\lambda$, where R is the resonator radius, a is an integer of wavelengths in a round trip of the cavity, n is the effective refractive index of the resonator, and λ is the light wavelength. The light is confined in WGM cavities since an effective refractive index and the thickness are orders of magnitude smaller than the in-plane dimensions. The various perovskite micro-nanoplates, micro-nanospheres, and nanocuboids have been demonstrated to exhibit excellent WGM lasing performance [50–59]. In 2014, Zhang et al. first demonstrated WGM lasers from $MAPbX_3$ nanoplates, which are synthesized by vapor deposition methods through the in situ conversion of triangular or hexagonal PbX_2 nanoplatelets into $MAPbX_3$ (Figure 12.5a) [50]. They achieve near-infrared lasing emission at ~780 nm with a low threshold of 37 μJ cm^{-2} and a high Q of 650. The same group subsequently developed all-inorganic perovskite $CsPbX_3$ WGM lasers with ultralow thresholds of 2–10 μJ cm^{-2}, tunable wavelength of 400–700 nm, and a high Q of ~4500 (Figure 12.5b) [51]. The high-crystallized-quality planar WGM cavities with low non-radiative recombination ensure adequate gain and efficient optical feedback for low-threshold in-plane lasing. The lasing originates from excitonic recombination because the threshold carrier density of 1.5×10^{17} cm^{-3} is smaller than the Mott density of $CsPbBr_3$ ($\approx 1.8 \times 10^{17}$ cm^{-3}). The high spectra coherence (Q or linewidth) can be attributed to room-temperature stable exciton with large E_b and high-crystallized quality WGM resonator. Compared with F-P microcavity unsupported total reflection in two end planes for the leakage of light,

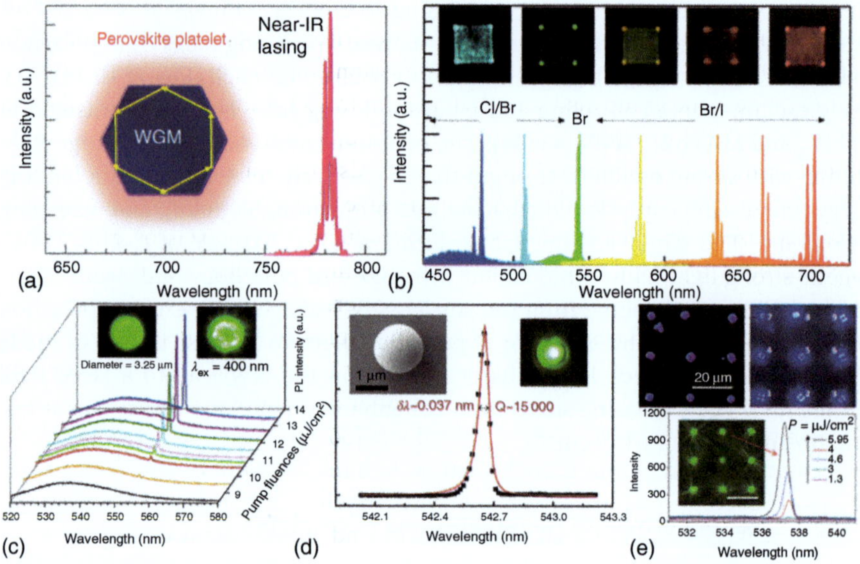

Figure 12.5 (a) WGM lasers of hexagonal $MAPbX_3$ nanoplates, Source: Zhang et al. [50]/Reproduced with permission from John Wiley & Sons, Inc. (b) All-inorganic perovskite $CsPbX_3$ WGM lasers with low thresholds, tunable wavelength, and high *Q*, Source: Zhang et al. [51]/Reproduced with permission from John Wiley & Sons, Inc. (c) Single-mode lasing from quasi-circular $FAPbBr_3$ microdisk with a low threshold and a high *Q*, Source: Li et al. [52]/Reproduced with permission from John Wiley & Sons, Inc. (d) Single-mode WGM laser of $CsPbBr_3$ microspheres with an ultrahigh *Q* of 15 000, Source: Tang et al. [53]/Reproduced with permission from John Wiley & Sons, Inc. (e) The WGM laser arrays consist of template-confined solution-growthed square $CsPbX_3$ microdisks. Source: He et al. [56]/Reproduced with permission from John Wiley & Sons, Inc.

the polygonal WGM microcavity has better light confinement ability for high *Q* value because WGM supports total reflection in all the side planes. Compared with other polygonal WGM microcavities, circular microcavities have better light confinement abilities due to the elimination of scattering at edges and corners. Li et al. demonstrated that the quasi-circular microdisk exhibits higher *Q* than square one due to the less boundary wave and pesudointegrable light leakage [52]. They achieved single-mode lasing from a quasi-circular $FAPbBr_3$ microdisk with a low threshold of $10.5\,\mu J\,cm^{-2}$, a relatively high *Q* of 3455, and a linewidth 0.16 nm (Figure 12.5c). The lasing behavior of these microdisks occurred within a few picoseconds with an avalanche of carrier recombination according to the observed ultrafast PL decay component above the lasing threshold. Zhang's group developed a $CsPbBr_3$ microspheres-based single-mode WGM laser with a very narrow linewidth of 0.037 nm, corresponding to an ultrahigh *Q* of 1.5×10^4 under two-photon (800 nm) excitation (Figure 12.5d) [53]. Compared with polygonal and circular WGM cavities, spherical cavities can provide a nearly ideal platform for internal total reflection and 3D confinement of the lasing field, leading to much lower optical and light scattering losses. Thus, the ultrahigh Q in the microsphere can be attributed to the ideal spherical shape, smooth surface, high gain ($\approx$450 cm^{-1}), and high spontaneous emission coupling efficiency β ($\approx$0.32) of $CsPbBr_3$. The same group also achieved one-photon (400 nm) pumped single-mode WGM lasers with low thresholds (0.42–1.65 μJ cm^{-2}), a high Q of ~6100, and a tunable lasing wavelength (425–715 nm) from $CsPbX_3$ spheres with tunable sizes of 0.2–10 μm [54]. It is generally hard to realize lasing in a subwavelength-scale cavity because shortening the optical path introduces large optical losses. However, the sufficiently high optical gain and optical confinement can overcome the cavity loss to support the lasing from $CsPbX_3$ nanospheres and nanocuboid cavities with subwavelength scale [53–55]. The electron–hole plasma mechanism is responsible for stimulated emission in the subwavelength nanocuboid $CsPbX_3$ laser [55]. Compared with single lasing, lasing arrays offer a scalable strategy for large-area light-emitting and optical gain applications [56–58]. Fu's group developed a template-confined solution-growth method to build WGM laser arrays (1 × 1 cm^2) of square $CsPbX_3$ microdisks with tunable wavelength (Figure 12.5e) [56]. The uniform square microdisks exhibit uniform and bright lasing with low lasing thresholds of 3–12 μJ cm^{-2}.

12.2.4 Laser of Perovskite Nanocrystals or Quantum Dots

Similar to perovskite films, the lasing in perovskite NCs is mainly achieved by random lasing or coupling perovskite NC films with external optical feedback cavities such as DFB, WGM, and vertical-cavity surface-emitting lasers [9, 60–64]. In 2015, Kovalenko's group demonstrated the ASE properties with a tunable emission wavelength (440–700 nm), low pump thresholds of 5 μJ cm^{-2}, and a high modal net gain of 450 cm^{-1} in $CsPbX_3$ NCs (Figure 12.6a) [9]. Then they achieved random lasing from perovskite NC films based on internal light scattering and WGM lasing from the external cavity by coating $CsPbX_3$ NCs on single silica microsphere resonators. Sun's group fabricated the VCSEL with low thresholds of 9–25 μJ cm^{-2} by spin-coating and inserting the $CsPbX_3$ NCs between two distributed λ_{Bragg} reflectors (Figure 12.6b) [61]. Tsukruk's group developed the high-resolution large-area arrays of $CsPbX_3$ NCs microdisk lasers and multicolor pixels by an orthogonal lithography process [62]. The $CsPbX_3$ NCs were filled into the microscale trenches of predesigned patterns fabricated from photoresist template-assistant standard photolithography. Tsukruk's group achieved room-temperature continuous wave lasing with a tunable color of 484–614 nm from situ-fabricated perovskite NCs by incorporating $MAPbX_3$-based NCs embedded in polyacrylonitrile films into a DFB cavity (Figure 12.6c,d) [63]. The in situ fabricated perovskite NCs embedded in polyacrylonitrile films exhibited high transmittance and PLQYs of ~90%, a very smooth surface with a roughness of 0.784 nm, a high thermal conductivity of ~1.2 W m^{-1} K^{-1}, and charge spatial separation effects in NCs and matrix. These merits enable easy realization of $MAPbX_3$ NCs-based green, blue, and red room-temperature CW lasers with low thresholds of 15, 58, and 24 W cm^{-2}, respectively.

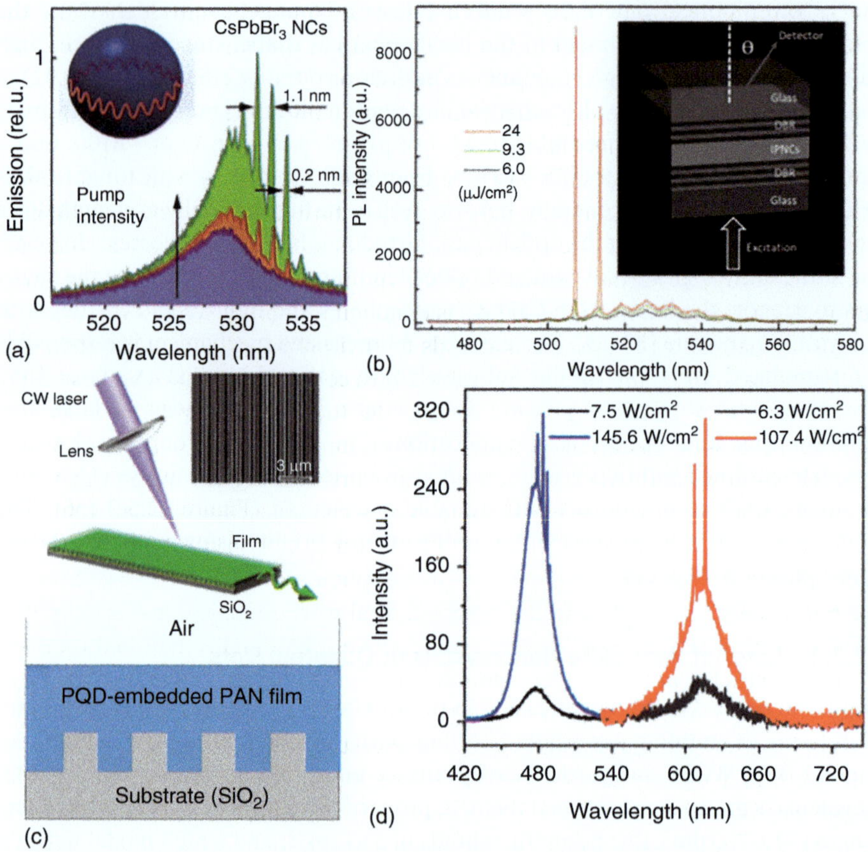

Figure 12.6 (a) $CsPbX_3$ NCs lasing emission with microsphere cavity, Source: Yakunin et al. [9]/Springer Nature/CC BY 4.0. (b) $CsPbX_3$ NCs lasing emission with VCSEL cavity, Source: Reproduced with permission [61], Copyright 2017, Wiley-VCH. (c) and (d) Tunable $MAPbX_3$ NCs lasing emission by embedding NCs polyacrylonitrile films into a DFB cavity, Source: Wang et al. [63]/Reproduced with permission from American Chemical Society.

12.3 Conclusion and Outlook

In conclusion, perovskite materials have demonstrated immense potential in the realm of laser technologies. Their outstanding optoelectronic properties, superior light-absorbing characteristics, impressive carrier mobility, and easily tunable bandgaps make them prime candidates for the next generation of laser materials and devices.

This chapter has explored various morphologies of perovskite lasers, each offering unique advantages. Perovskite films, with their simplicity in fabrication and control over thickness, have laid the groundwork for solid-state perovskite lasers. The exploration of perovskite NWs has provided insights into potential high-aspect-ratio laser structures, exhibiting enhanced stability, and the capability for integrated photonics applications. Perovskite nanoplates and microplates, with their 2D nature, have shown the feasibility of single-mode lasing, high optical gain, and lower lasing thresholds, presenting a significant step forward in the pursuit of miniaturized, low-power optical devices. Lastly, perovskite NCs or QDs, owing to their quantum confinement effect and size-dependent emission tunability, offer a promising avenue for tunable and low-threshold lasers.

While the field of perovskite lasers presents compelling prospects, it is not without challenges. Environmental stability, continuous operational reliability, long-term robustness, and electrically driven lasers are critical areas that warrant further exploration. Future research directions are manifold and offer exciting avenues for advancement. Top of the list is enhancing stability, which also involves addressing the resistance of these materials to environmental factors. Further, the environmental impact and toxicity associated with certain perovskite compositions, particularly those containing lead, necessitate the exploration of alternative, nontoxic perovskite materials. Greater insight into the lasing mechanisms in perovskites – how band structures, excitonic properties, and quantum confinement interplay with different material morphologies – can accelerate the development of even more efficient perovskite lasers. The integration of perovskite lasers with other advanced technologies, such as silicon photonics or flexible electronics, also holds the promise of fostering practical, real-world applications. Scaling the manufacturing techniques to ensure reproducible, high-quality perovskite lasers is another crucial milestone for commercialization. Moreover, exploring novel device designs to harness the unique properties of perovskite materials could pave the way for advanced functionalities. Lastly, investigating less-explored perovskite materials could unlock superior properties for lasing applications. While these future research endeavors present challenges, they also hold the potential to propel the field of perovskite lasers into new frontiers of innovation.

The journey to explore and perfect perovskite laser materials and devices is ongoing. However, the progress made so far offers a hopeful glimpse into a future where these materials could revolutionize the landscape of laser technologies, catalyzing advancements in various fields such as telecommunications, data storage, sensing technologies, and medical treatments. This exciting prospect serves as a potent motivation for the continued exploration of perovskite materials, forging ahead in the quest for next-generation laser technologies.

References

1 Schawlow, A.L. and Townes, C.H. (1958). Infrared and optical masers. *Physical Review* 112: 1940–1949.

2 Xing, G., Wu, B., Wu, X. et al. (2017). Transcending the slow bimolecular recombination in lead-halide perovskites for electroluminescence. *Nature Communications* 8: 14558.

3 Yamada, Y., Nakamura, T., Endo, M. et al. (2014). Photocarrier recombination dynamics in perovskite $CH_3NH_3PbI_3$ for solar cell applications. *Journal of the American Chemical Society* 136: 11610–11613.

4 Wehrenfennig, C., Eperon, G.E., Johnston, M.B. et al. (2013). High charge carrier mobilities and lifetimes in organolead trihalide perovskites. *Advanced Materials* 26: 1584–1589.

5 Wei, Q., Li, X., Liang, C. et al. (2019). Recent progress in metal halide perovskite micro- and nanolasers. *Advanced Optical Materials* 7: 1900080.

6 Xing, G., Mathews, N., Lim, S.S. et al. (2014). Low-temperature solution-processed wavelength-tunable perovskites for lasing. *Nature Materials* 13: 476–480.

7 Yamada, T., Aharen, T., and Kanemitsu, Y. (2018). Near-band-edge optical responses of $CH_3NH_3PbCl_3$ single crystals: photon recycling of excitonic luminescence. *Physical Review Letters* 120: 057404.

8 Tang, X., Hu, Z., Chen, W. et al. (2016). Room temperature single-photon emission and lasing for all-inorganic colloidal perovskite quantum dots. *Nano Energy* 28: 462–468.

9 Yakunin, S., Protesescu, L., Krieg, F. et al. (2015). Low-threshold amplified spontaneous emission and lasing from colloidal nanocrystals of caesium lead halide perovskites. *Nature Communications* 6: 8056.

10 Fengrui, H., Chunyang, Y., Huichao, Z. et al. (2016). Slow Auger recombination of charged excitons in nonblinking perovskite nanocrystals without spectral diffusion. *Nano Letters* 16: 6425–6430.

11 Li, Y., Luo, X., Ding, T. et al. (2020). Size- and halide-dependent auger recombination in lead halide perovskite nanocrystals. *Angewandte Chemie* 132: 14398–14401.

12 Wang, X., Shoaib, M., Wang, X. et al. (2018). High-quality in-plane aligned $CsPbX_3$ perovskite nanowire lasers with composition-dependent strong exciton-photon coupling. *ACS Nano* 12: 6170–6178.

13 Tang, Y., Yin, C., Jing, Q. et al. (2022). Quantized exciton motion and fine energy-level structure of a single perovskite nanowire. *Nano Letters* 22: 2907–2914.

14 Jiang, Y., Cui, M., Li, S. et al. (2021). Reducing the impact of Auger recombination in quasi-2D perovskite light-emitting diodes. *Nature Communications* 12: 336.

15 Dexter, D.L. (1957). Possibility of luminescent quantum yields greater than unity. *Physical Review* 108: 630–633.

16 Wegh, R.T., Donker, H., Oskam, K.D. et al. (1999). Visible quantum cutting in $LiGdF_4:Eu^{3+}$ through downconversion. *Science* 283: 663–666.

17 Li, X., Wei, Q., Wang, K. et al. (2019). Charge carrier dynamics and broad wavelength tunable amplified spontaneous emission in $Zn_xCd_{1-x}Se$ nanowires. *The Journal of Physical Chemistry Letters* 10: 7516–7522.

18 Duan, X., Huang, Y., Agarwal, R. et al. (2003). Single-nanowire electrically driven lasers. *Nature* 421: 241–245.

19 Göppert-Mayer, M. (1931). Über elementarakte mit zwei quantensprüngen. *Annalen der Physik* 401: 273–294.

20 Kaiser, W. and Garrett, C.G.B. (1961). Two-photon excitation in $CaF_2:Eu^{2+}$. *Physical Review Letters* 7: 229–231.

21 Xing, G., Kumar, M.H., Chong, W.K. et al. (2016). Solution-processed tin-based perovskite for near-infrared lasing. *Advanced Materials* 28: 8191–8196.

22 Willa, K., Häusermann, R., Mathis, T. et al. (2013). From organic single crystals to solution processed thin-films: charge transport and trapping with varying degree of order. *Journal of Applied Physics* 113: 133707.

23 Li, C., Zang, Z., Han, C. et al. (2017). Highly compact $CsPbBr_3$ perovskite thin films decorated by ZnO nanoparticles for enhanced random lasing. *Nano Energy* 40: 195–202.

24 Chen, S., Zhang, C., Lee, J. et al. (2017). High-q, low-threshold monolithic perovskite thin-film vertical-cavity lasers. *Advanced Materials* 29: 1604781.

25 Saliba, M., Wood, S.M., Patel, J.B. et al. (2015). Structured organic-inorganic perovskite toward a distributed feedback laser. *Advanced Materials* 28: 923–929.

26 Chen, S., Roh, K., Lee, J. et al. (2016). A photonic crystal laser from solution based organo-lead iodide perovskite thin films. *ACS Nano* 10: 3959–3967.

27 Sutherland, B.R., Hoogland, S., Adachi, M.M. et al. (2014). Conformal organohalide perovskites enable lasing on spherical resonators. *ACS Nano* 8: 10947–10952.

28 Wang, K., Wang, S., Xiao, S. et al. (2018). Recent advances in perovskite micro- and nanolasers. *Advanced Optical Materials* 6: 1800278.

29 Zhang, Q., Shang, Q., Su, R. et al. (2021). Halide perovskite semiconductor lasers: materials, cavity design, and low threshold. *Nano Letters* 21: 1903–1914.

30 Deschler, F., Price, M., Pathak, S. et al. (2014). High photoluminescence efficiency and optically pumped lasing in solution-processed mixed halide perovskite semiconductors. *The Journal of Physical Chemistry Letters* 5: 1421–1426.

31 Mathies, F., Brenner, P., Hernandez-Sosa, G. et al. (2018). Inkjet-printed perovskite distributed feedback lasers. *Optics Express* 26: 144.

32 Dong, Q., Fu, X., Seyitliyev, D. et al. (2022). Cavity engineering of perovskite distributed feedback lasers. *ACS Photonics* 9: 3124–3133.

33 Cegielski, P.J., Neutzner, S., Porschatis, C. et al. (2017). Integrated perovskite lasers on a silicon nitride waveguide platform by cost-effective high throughput fabrication. *Optics Express* 25: 13199.

34 Zhong, Y., Liao, K., Du, W. et al. (2020). Large-scale thin $CsPbBr_3$ single-crystal film grown on sapphire via chemical vapor deposition: toward laser array application. *ACS Nano* 14: 15605–15615.

35 Duan, Z., Wang, Y., Li, G. et al. (2017). Chip-scale fabrication of uniform lead halide perovskites microlaser array and photodetector array. *Laser & Photonics Reviews* 12: 1700234.

36 Chen, J., Wu, Y., Li, X. et al. (2017). Simple and fast patterning process by laser direct writing for perovskite quantum dots. *Advanced Materials Technologies* 2: 1700132.

37 Wu, J., Chen, J., Zhang, Y. et al. (2017). Pinhole-free hybrid perovskite film with arbitrarily-shaped micro-patterns for functional optoelectronic devices. *Nano Letters* 17: 3563–3569.

38 Zhu, H., Fu, Y., Meng, F. et al. (2015). Lead halide perovskite nanowire lasers with low lasing thresholds and high quality factors. *Nature Materials* 14: 636–642.

39 Fu, Y., Zhu, H., Schrader, A.W. et al. (2016). Nanowire lasers of formamidinium lead halide perovskites and their stabilized alloys with improved stability. *Nano Letters* 16: 1000–1008.

40 Fu, Y., Zhu, H., Stoumpos, C.C. et al. (2016). Broad wavelength tunable robust lasing from single-crystal nanowires of cesium lead halide perovskites ($CsPbX_3$, X = Cl, Br, I). *ACS Nano* 10: 7963–7972.

41 Evans, T.J.S., Schlaus, A., Fu, Y. et al. (2017). Continuous-wave lasing in cesium lead bromide perovskite nanowires. *Advanced Optical Materials* 6: 1700982.

42 Huang, L., Gao, Q., Sun, L.-D. et al. (2018). Composition-graded cesium lead halide perovskite nanowires with tunable dual-color lasing performance. *Advanced Materials* 30: 1800596.

43 Wenna, D., Shuai, Z., Jia, S. et al. (2017). Strong exciton-photon coupling and lasing behavior in all-inorganic $CsPbBr_3$ micro/nanowire Fabry-Pérot cavity. *ACS Photonics* 5: 2051–2059.

44 Zhang, S., Shang, Q., Du, W. et al. (2017). Strong exciton-photon coupling in hybrid inorganic-organic perovskite micro/nanowires. *Advanced Optical Materials* 6: 1701032.

45 Xing, J., Liu, X.F., Zhang, Q. et al. (2015). Vapor phase synthesis of organometal halide perovskite nanowires for tunable room-temperature nanolasers. *Nano Letters* 15: 4571–4577.

46 Schlaus, A.P., Spencer, M.S., Miyata, K. et al. (2019). How lasing happens in $CsPbBr_3$ perovskite nanowires. *Nature Communications* 10: 265.

47 Eaton, S.W., Lai, M., Gibson, N.A. et al. (2016). Lasing in robust cesium lead halide perovskite nanowires. *PNAS* 113: 1993–1998.

48 Liu, P., He, X., Ren, J. et al. (2017). Organic-inorganic hybrid perovskite nanowire laser arrays. *ACS Nano* 11: 5766–5773.

49 van Vugt, L.K., Piccione, B., Cho, C.-H. et al. (2011). One-dimensional polaritons with size-tunable and enhanced coupling strengths in semiconductor nanowires. *PNAS* 108: 10050–10055.

50 Zhang, Q., Ha, S.T., Liu, X. et al. (2014). Room-temperature near-infrared high-q perovskite whispering-gallery planar nanolasers. *Nano Letters* 14: 5995–6001.

51 Zhang, Q., Su, R., Liu, X. et al. (2016). High-quality whispering-gallery-mode lasing from cesium lead halide perovskite nanoplatelets. *Advanced Functional Materials* 26: 6238–6245.

52 Li, X., Wang, K., Chen, M. et al. (2020). Stable whispering gallery mode lasing from solution-processed formamidinium lead bromide perovskite microdisks. *Advanced Optical Materials* 8: 2000030.

53 Tang, B., Sun, L., Zheng, W. et al. (2018). Ultrahigh quality upconverted single-mode lasing in cesium lead bromide spherical microcavity. *Advanced Optical Materials* 6: 1800391.

54 Tang, B., Dong, H., Sun, L. et al. (2017). Single-mode lasers based on cesium lead halide perovskite submicron spheres. *ACS Nano* 11: 10681–10688.

55 Zhengzheng, L., Jie, Y., Juan, D. et al. (2018). Robust subwavelength single-mode perovskite nanocuboid laser. *ACS Nano* 12: 5923–5931.

56 He, X., Liu, P., Zhang, H. et al. (2017). Patterning multicolored microdisk laser arrays of cesium lead halide perovskite. *Advanced Materials* 29: 1604510.

57 Liu, X., Niu, L., Wu, C. et al. (2016). Periodic organic-inorganic halide perovskite microplatelet arrays on silicon substrates for room-temperature lasing. *Advanced Science* 3: 1600137.

58 Feng, J., Yan, X., Zhang, Y. et al. (2016). “Liquid knife” to fabricate patterning single-crystalline perovskite microplates toward high-performance laser arrays. *Advanced Materials* 28: 3732–3741.

59 Liao, Q., Hu, K., Zhang, H. et al. (2015). Perovskite microdisk microlasers self-assembled from solution. *Advanced Materials* 27: 3405–3410.

60 Chen, J., Du, W., Shi, J. et al. (2019). Perovskite quantum dot lasers. *InfoMat* 2: 170–183.

61 Wang, Y., Li, X., Nalla, V. et al. (2017). Solution-processed low threshold vertical cavity surface emitting lasers from all-inorganic perovskite nanocrystals. *Advanced Functional Materials* 27: 1605088.

62 Lin, C.H., Zeng, Q., Lafalce, E. et al. (2018). Large-area lasing and multicolor perovskite quantum dot patterns. *Advanced Optical Materials* 6: 1800474.

63 Wang, L., Meng, L., Chen, L. et al. (2019). Ultralow-threshold and color-tunable continuous-wave lasing at room-temperature from in situ fabricated perovskite quantum dots. *The Journal of Physical Chemistry Letters* 10: 3248–3253.

64 Li, X., Wang, Y., Sun, H. et al. (2017). Amino-mediated anchoring perovskite quantum dots for stable and low-threshold random lasing. *Advanced Materials* 29: 1701185.

Index

Perovskite Light Emitting Diodes: Materials and Devices, First Edition. Hong Meng.

l

m

n

o

p